Lecture Notes in Physics

Springer

Berlin
Heidelberg
New York
Barcelona
Hong Kong
London
Milan
Paris
Singapore
Tokyo

The Editorial Policy for Proceedings

The series Lecture Notes in Physics reports new developments in physical research and teaching – quickly, informally, and at a high level. The proceedings to be considered for publication in this series should be limited to only a few areas of research, and these should be closely related to each other. The contributions should be of a high standard and should avoid lengthy redraftings of papers already published or about to be published elsewhere. As a whole, the proceedings should aim for a balanced presentation of the theme of the conference including a description of the techniques used and enough motivation for a broad readership. It should not be assumed that the published proceedings must reflect the conference in its entirety. (A listing or abstracts of papers presented at the meeting but not included in the proceedings could be added as an appendix.)

When applying for publication in the series Lecture Notes in Physics the volume's editor(s) should submit sufficient material to enable the series editors and their referees to make a fairly accurate evaluation (e.g. a complete list of speakers and titles of papers to be presented and abstracts). If, based on this information, the proceedings are (tentatively) accepted, the volume's editor(s), whose name(s) will appear on the title pages, should select the papers suitable for publication and have them refereed (as for a journal) when appropriate. As a rule discussions will not be accepted. The series editors and Springer-Verlag will normally not interfere with the detailed editing except in fairly obvious cases or on technical matters.

Final acceptance is expressed by the series editor in charge, in consultation with Springer-Verlag only after receiving the complete manuscript. It might help to send a copy of the authors' manuscripts in advance to the editor in charge to discuss possible revisions with him. As a general rule, the series editor will confirm his tentative acceptance if the final manuscript corresponds to the original concept discussed, if the quality of the contribution meets the requirements of the series, and if the final size of the manuscript does not greatly exceed the number of pages originally agreed upon. The manuscript should be forwarded to Springer-Verlag shortly after the meeting. In cases of extreme delay (more than six months after the conference) the series editors will check once more the timeliness of the papers. Therefore, the volume's editor(s) should establish strict deadlines, or collect the articles during the conference and have them revised on the spot. If a delay is unavoidable, one should encourage the authors to update their contributions if appropriate. The editors of proceedings are strongly advised to inform contributors about these points at an early stage.

The final manuscript should contain a table of contents and an informative introduction accessible also to readers not particularly familiar with the topic of the conference. The contributions should be in English. The volume's editor(s) should check the contributions for the correct use of language. At Springer-Verlag only the prefaces will be checked by a copy-editor for language and style. Grave linguistic or technical shortcomings may lead to the rejection of contributions by the series editors. A conference report should not exceed a total of 500 pages. Keeping the size within this bound should be achieved by a stricter selection of articles and not by imposing an upper limit to the length of the individual papers. Editors receive jointly 30 complimentary copies of their book. They are entitled to purchase further copies of their book at a reduced rate. As a rule no reprints of individual contributions can be supplied. No royalty is paid on Lecture Notes in Physics volumes. Commitment to publish is made by letter of interest rather than by signing a formal contract. Springer-Verlag secures the copyright for each volume.

The Production Process

The books are hardbound, and the publisher will select quality paper appropriate to the needs of the author(s). Publication time is about ten weeks. More than twenty years of experience guarantee authors the best possible service. To reach the goal of rapid publication at a low price the technique of photographic reproduction from a camera-ready manuscript was chosen. This process shifts the main responsibility for the technical quality considerably from the publisher to the authors. We therefore urge all authors and editors of proceedings to observe very carefully the essentials for the preparation of camera-ready manuscripts, which we will supply on request. This applies especially to the quality of figures and halftones submitted for publication. In addition, it might be useful to look at some of the volumes already published. As a special service, we offer free of charge LATEX and TEX macro packages to format the text according to Springer-Verlag's quality requirements. We strongly recommend that you make use of this offer, since the result will be a book of considerably improved technical quality. To avoid mistakes and time-consuming correspondence during the production period the conference editors should request special instructions from the publisher well before the beginning of the conference. Manuscripts not meeting the technical standard of the series will have to be returned for improvement.

For further information please contact Springer-Verlag, Physics Editorial Department II, Tiergartenstrasse 17, D-69121 Heidelberg, Germany

Series homepage – http://www.springer.de/phys/books/lnpp

G. Grindhammer B. A. Kniehl G. Kramer (Eds.)

New Trends in HERA Physics 1999

Proceedings of the Ringberg Workshop
Held at Tegernsee, Germany,
30 May - 4 June 1999

Springer

Editors

G. Grindhammer
Max-Planck-Institut für Physik
Föhringer Ring 6
80805 München, Germany

B. A. Kniehl
G. Kramer
II. Institut für Theoretische Physik
Universität Hamburg
Luruper Chaussee 149
22761 Hamburg, Germany

Library of Congress Cataloging-in-Publication Data

New trends in HERA physics 1999 : proceedings of the Ringberg workshop held at Tegernsee, Germany, 30 May-4 June 1999 / G. Grindhammer, B.A. Kniehl, G. Kramer (eds.).
    p. cm. -- (Lecture notes in physics, ISSN 0075-8450 ; 546)
    Includes bibliographical references.

    1. Nuclear physics--Congresses. 2. Particles (Nuclear physics)--Congresses. I. Grindhammer, G. (Günter), 1946- II. Kniehl, B. A. (Bernd A.) III. Kramer, Gustav, 1932- IV. Series.

QC770 .N437 2000
539.7'2--dc21

00-026908

ISSN 0075-8450
ISBN 978-3-642-08647-2   e-ISBN 978-3-540-46522-5

Springer-Verlag is a company in the BertelsmannSpringer publishing group. © Springer-Verlag Berlin Heidelberg 2010
Printed in Germany

Cover design: *design & production*, Heidelberg
Printed on acid-free paper

Preface

The international workshop entitled *New Trends in HERA Physics 1999* took place between 30 May and 4 June 1999 at Ringberg Castle, which overlooks Lake Tegernsee in the foothills of the Bavarian Alps, one of the most picturesque locations to be found in the whole of Germany. The castle was built during the first half of this century by Duke Luitpold in Bavaria (Herzog Luitpold in Bayern), a member of the Wittelsbach family which ruled Bavaria over 800 years, and his friend Friedrich Attenhuber, an all-round artist, architect, and interior decorator. The castle is entirely their creation, from the massive Renaissance-inspired exterior right down to the fittings and furniture, which, in every detail, were designed by Attenhuber himself and executed by local craftsmen. Attenhuber also painted every single picture hanging in the castle. He found his models in the farmhouses around Lake Tegernsee. The castle embodies all trends of art and styles which dominated the first half of this century, combined with local Alpine originality and the individual creative power of its constructors. In accordance with the Duke's last will, the castle passed into the hands of the Max Planck Society after his death, in 1973. The castle was then transformed into a conference venue, where scientists can exchange their latest ideas and discuss problems with their colleagues from all over the world in beautiful surroundings and in a relaxed mountain atmosphere, high above the daily business activities.

This was the second event in a series of Ringberg workshops on HERA physics, which started two years ago with the workshop entitled *New Trends in HERA Physics*. In fact, at the end of that workshop, many participants expressed the opinion that this was a successful endeavour to bring theorists and experimentalists together in order to interpret the latest HERA data, and that it would be useful to organize a follow-up workshop in the same spirit.

On the occasion of the 1999 Ringberg workshop, thirty-eight experts of elementary-particle physics, both theorists and experimentalists, from twenty-two universities and research institutions in ten countries congregated to present their latest results on the various aspects of HERA physics. Specifically, there were twenty theorists and eighteen experimentalists, the latter representing the H1 and ZEUS collaborations at HERA and the collaborations at LEP and the Tevatron. The topics included: proton structure function; polarized ep scattering; final states in deep-inelastic scattering (DIS), with special emphasis on jet production at low x, power corrections in DIS, soft particle production, and instanton effects; photon structure function; photoproduction of jets and hadrons;

heavy-flavour and charmonium production; elastic and diffractive *ep* scattering; and new physics at HERA. We hope that the high-energy-physics community will benefit from these proceedings, in which the ongoing efforts in understanding the nature of the strong interactions, with particular emphasis on HERA physics, are documented.

We wish to thank all our friends and colleagues who have contributed their share to these proceedings. We are indebted to the workshop secretary, Mrs. Renate Saffert, for her assistance before, during, and after the workshop. We are grateful to Dr. Annette Holtkamp for technical assistance in the editorial work. The costs of catering and lodging at Ringberg Castle and the charges for the publication and dissemination of these proceedings were covered in equal parts by the Deutsches Elektronen-Synchrotron at Hamburg and the Max-Planck-Institut für Physik at Munich.

Hamburg,   *Günter Grindhammer*
November 1999   *Bernd A. Kniehl*
                                              Gustav Kramer

Contents

List of Participants

Johannes Blümlein
DESY Zeuthen Theory Group
Platanenallee 6
D-15738 Zeuthen
bluemlein@ifh.de

Nick Brook
Department of Physics and Astronomy
University of Glasgow
Glasgow, G12 8QQ
United Kingdom
n.brook@physics.gla.ac.uk

Wilfried Buchmüller
DESY
Notkestraße 85
D-22603 Hamburg
wilfried.buchmueller@desy.de

Gerd Buschhorn
Max-Planck-Institut für Physik
Föhringer Ring 6
D-80805 München
gwb@mppmu.mpg.de

Carsten Coldewey
DESY Zeuthen
Platanenallee 6
D-15738 Zeuthen
coldewey@mail.desy.de

Michael Düren
Universität Erlangen-Nürnberg
Erwin-Rommel-Str. 1
D-91058 Erlangen
michael.dueren@desy.de

Dirk Graudenz
Paul Scherrer Institut
CH-5232 Villigen PSI
Switzerland
dirk.graudenz@cern.ch

Timothy Greenshaw
Oliver Lodge Laboratory
Liverpool University
Liverpool L69 7ZE
England
green@hep.ph.liv.ac.uk

Günter Grindhammer
Max-Planck-Institut für Physik
Föhringer Ring 6
D-80805 München
guenter.grindhammer@desy.de

Brian Harris
Argonne National Laboratory ANL
9700 South Cass Avenue
Bldg. 362, HEP
Argonne, IL 60439
USA
harris@hep.anl.gov

Leif Joensson
Physics Department
Lund University
Box 118
S-22100 Lund
Sweden
leif@quark.lu.se

Christian Kiesling
Max-Planck-Institut für Physik
Föhringer Ring 6
D-80805 München
cmk@mppmu.mpg.de

Birger Koblitz
Max-Planck-Institut für Physik
Föhringer Ring 6
D-80805 München
koblitz@mail.desy.de

Michael Klasen
Argonne Nation Laboratoy
9700 South Cass Avenue
Argonne, IL 60439
USA
klasen@hep.anl.gov

Bernd Kniehl
II. Institut für Theoretische Physik
Universität Hamburg
Luruper Chaussee 149
D-22761 Hamburg
kniehl@mail.desy.de

Henri Kowalski
DESY
Notkestraße 85
D-22603 Hamburg
kowalski@mail.desy.de

Gustav Kramer
II. Institut für Theoretische Physik
Universität Hamburg
Luruper Chaussee 149
D-22761 Hamburg
kramer@mail.desy.de

Masahiro Kuze
Institute of Particle and Nuclear
Studies, KEK
Tanashi
188-8501 Tokyo
Japan
masahiro.kuze@desy.de

Peter Landshoff
DAMTP
Silver Street
Cambridge CB3 9EW
England
pvl@damtp.cambridge.ac.uk

Jungil Lee
II. Institut für Theoretische Physik
Universität Hamburg
Luruper Chaussee 149
D-22761 Hamburg
jungil@mail.desy.de

Eugene Levin
HEP Department
School of Physics
Tel Aviv University
Ramat Aviv, Tel Aviv
69978 Israel
leving@post.tau.ac.il

Alan Martin
University of Durham
South Road
Durham City, DH1 3LE
England
a.d.martin@durham.ac.uk

Kristal Mauritz
Iowa State University
Ames IA 50011
USA
kristal@d0phy1.fnal.gov

Joachim Meyer
DESY
Notkestraße 85
D-22603 Hamburg
jmeyer@mail.desy.de

Vincenzo Monaco
Instituto di Fisica
University of Torino
Via Pietro Giuria 1
I-10125 Torino

Italy
monaco@to.infn.it

Guido Nellen
Max-Planck-Institut für Physik
Föhringer Ring 6
D-80805 München
guido.nellen@desy.de

Richard Nisius
EP-Division
CERN
CH-1211 Geneva 23
Switzerland
richard.nisius@cern.ch

Wolfgang Ochs
Max-Planck-Institut für Physik
Föhringer Ring 6
D-80805 München
wwo@mppmu.mpg.de

John Outhwaite
Department of Physics
University of Durham
Durham, DH1 3LE
England
john.outhwaite@durham.ac.uk

Daniel Pitzl
DESY
Notkestraße 85
D-22603 Hamburg
pitzl@mail.desy.de

Björn Pötter
Max-Planck-Institut für Physik
Föhringer Ring 6
D-80805 München
poetter@mppmu.mpg.de

Alexander Proskuryakov
Moscow State University
Deparment of High Energy Physics
Vorobjovy Gory
119899 Moscow
Russia
alexander.proskuryakov@desy.de

Cong-Feng Qiao
II. Institut für Theoretische Physik
Universität Hamburg
Luruper Chaussee 149
D-22761 Hamburg
qiaocf@mail.desy.de

Andreas Ringwald
DESY
Notkestraße 85
D-22603 Hamburg
ringwald@mail.desy.de

Thomas Schörner
Max-Planck-Institut für Physik
Föhringer Ring 6
D-80805 München
schorner@mppmu.mpg.de

Laurel Sinclair
Department of Physics and Astronomy
University of Glasgow
Glasgow, G12 8QQ
United Kingdom
sinclair@desy.de

Hubert Spiesberger
Institut für Physik
WA ThEP
Universität Mainz
D-55099 Mainz
hspiesb@thep.physik.uni-mainz.de

Marco Stratmann
Department of Physics
University of Durham
Durham DH1 3LE
England
marco.stratmann@durham.ac.uk

Thomas Teubner
DESY
Notkestraße 85
D-22603 Hamburg
teubner@mail.desy.de

Valentin Zakharov
Max-Planck-Institut für Physik
Föhringer Ring 6
D-80805 München
xxz@mppmu.mpg.de

Fabian Zomer
IN2P3-CNRS and Université de
Paris-Sud
LAL, Bâtiment 200

F-91898 BP 34 Orsay Cedex
France
zomer@dice2.desy.de

Lennart Zwirner
II. Institut für Theoretische Physik
Universität Hamburg
Luruper Chaussee 149
D-22761 Hamburg
zwirner@mail.desy.de

Part I

Proton Structure

Inclusive Deep Inelastic Scattering at HERA and Related Phenomenology

Fabian Zomer

IN2P3-CNRS and Université de Paris-Sud, Laboratoire de l'Accélérateur Linéaire, Bâtiment 200, F-91898 BP 34 Orsay Cedex, France

Abstract. Recent measurements of inclusive deep inelastic scattering differential cross section in the range 1.5 GeV$^2 \leq Q^2 \leq$ 30000 GeV^2 and $5 \cdot 10^{-6} \leq x \leq 0.65$ are presented. Phenomenological analyses performed from these measurements are also described.

1 Introduction

In the Deep Inelastic Scattering (DIS) processes observed at HERA, a lepton $\ell = e^{\pm}$ of 27.5 GeV interacts with a proton P of 920 GeV yielding a lepton ℓ' and a set of hadrons X in the final state. Following the nature of ℓ' the interaction proceeds via a neutral ($\ell' = e^{\pm}$) current (NC) or a charged ($\ell' = \nu_e, \bar{\nu}_e$) current (CC). DIS events are collected in the H1 and ZEUS experiments [1] which are located at the two $e^{\pm}P$ interaction points of HERA.

The kinematics of the DIS inclusive processes, $\ell(k) + P(p) \to \ell'(k') + X$, is determined by two independent kinematic variables, besides the energy of the incoming lepton and proton. One usually chooses them among the four Lorentz invariants

$$Q^2 \equiv -q^2 = -(k-k')^2, \; x = \frac{Q^2}{2p \cdot q}, \; y = \frac{p \cdot q}{p \cdot k}, \; W^2 = (q+p)^2,$$

whereby at HERA one can neglect the lepton and proton masses so that the useful relation $Q^2 = xys$ holds. These kinematic variables are obtained experimentally by measuring the momentum and/or the hadronic energy, the direction of the scattered lepton and/or the hadronic energy flow.

In this report we shall restrict ourselves to the cross section measurements at HERA in the medium 1.5 GeV$^2 \leq Q^2 <$ 150 GeV^2 and high 150 GeV$^2 \leq Q^2 \leq$ 30000 GeV^2 domain of the DIS regime. During the past, a large number of precise measurements have been performed in the medium Q^2 region by fixed target experiments [2]. With HERA, three major improvements may be noticed:

- an extension of the Q^2 domain to very high Q^2 (10^4 GeV^2) but also to very small x ($\approx 10^{-6}$);
- an almost hermetic (4π) detection of the final state leading not only to the determination of the energy and angle of the scattered lepton but also of the produced hadrons;
- from the previous items it follows that the detection of both NC and CC is feasible in the same detector and during the same data taking period;

The somewhat arbitrary distinction between low and high Q^2 is related to different physics interests. In both regions perturbative Quantum-Chromo-Dynamics (pQCD) is expected to describe the HERA data [3]. The pQCD analysis of medium Q^2 data is part of a long tradition [4] from which the parton distributions of the nucleon and the strong coupling constant α_s have been extracted. On the top of that, very high Q^2 ($\approx M_Z^2$) NC and CC data open a field of research in electroweak physics up to now reserved, in DIS, to neutrino beam experiments.

The rest of this report is organized as follows. In section 2 the measurements of NC and CC differential cross sections are described. Section 3 is devoted to a phenomenological analysis of these measurements.

2   Measurement of NC and CC cross sections

Neutral current events, at medium and high Q^2, are basically identified by the presence of an electron (or a positron) in the final state. This is done by using tracking and calorimetric devices covering the range $7^\circ < \theta_e < 177^\circ$ and $E'_e > 4$ GeV (at HERA the forward direction $\theta_e = 0^\circ$ corresponds to the direction of the incoming proton).

The differential cross section measurement is done by counting the number of events within a kinematic interval in, say x and Q^2. Therefore one of the experimental problems is to achieve a good reconstruction of these kinematic variables from the detector information. Both, H1 and ZEUS, can use the outgoing lepton and hadronic final state information, namely the polar angles, the momenta and the deposited energies. It is then possible to define the kinematics of each event by using different (and independent) combinations of experimental information.

In ZEUS the double angle method is used

$$Q_{da}^2 = 4E_e^2 \frac{\sin \gamma_h (1 + \cos \theta_e)}{\sin \gamma_h + \sin \theta_e - \sin(\gamma_h + \theta_e)}, \quad y_{da} = \frac{\sin \theta_e (1 - \cos \gamma_h)}{\sin \gamma_h + \sin \theta_e - \sin(\gamma_h + \theta_e)}$$

$$x_{da} = \frac{E_e}{E_p} \frac{\sin \gamma_h + \sin \theta_e + \sin(\gamma_h + \theta_e)}{\sin \gamma_h + \sin \theta_e - \sin(\gamma_h + \theta_e)}.$$

The hadronic polar angle γ_h is defined by $\tan \gamma_h/2 = \sum_i (E_i - p_{z,i})/P_{t,h}$, where E_i and $p_{z,i}$ are the energy and longitudinal momentum of the final state hadron i and where $P_{t,h}$ is the total transverse momentum of the hadronic final state particles. In H1, the electron method is used

$$Q_e^2 = \frac{(E'_e)^2 \sin^2 \theta_e}{1 - y_e}, \quad y_e = 1 - \frac{E'_e}{E_e} \sin^2(\theta_e/2)$$

to determine Q^2 and x only at $y > 0.15$ since $dx/x = 1/y \, dE'_e/E'_e$ while at $y \leq 0.15$ the Σ method is used

$$Q_\Sigma^2 = \frac{(E'_e)^2 \sin^2 \theta_e}{1 - y_\Sigma}, \quad y_\Sigma = \frac{\sum_i (E_i - p_{z,i})}{\sum_i (E_i - p_{z,i}) + E'_e(1 - \cos \theta_e)}.$$

The reason for the differences between the methods used by H1 and ZEUS are related to the calorimeter performances: H1 possesses finely segmented electromagnetic calorimeters and ZEUS a very good hadronic calorimetry.

The redundancy in the determination of the kinematic variables is a crucial point and presents many advantages: minimization of the migration between the 'true' and the measured kinematic variable by choosing one particular method; cross calibration of the various calorimeter devices, and studies of photon radiation from the lepton line by comparing leptonic and hadronic information.

Once the collected events are gathered in x-Q^2 bins, besides the subtraction of photoproduction background, correction factors are applied for: the efficiency of the event selection; detector acceptance; wrong reconstruction of the kinematics due to detector effects, and the contribution of higher order electroweak processes. When possible, these correction factors are determined and/or cross checked from the data themselves. If this is not possible, then they are determined from a full simulation of the DIS and background processes including the detector response.

For the medium Q^2 data we shall describe the preliminary results of the high statistics 1997 data (more than 10000 events per x-Q^2 bin for H1). For high Q^2, the combined 1994-1997 (e^+ beam) data published in ref. [5] are presented.

At medium Q^2 and for the H1 measurements, the main systematic uncertainties are: the electron energy scale ($\approx 0.3\%$), the hadronic energy scale ($\approx 2-3\%$), the electron polar angle (≈ 0.3 mrad), the photoproduction background at high-y only ($\approx 3\%$ effect on the measurements) and the correction factors (see above) applied to the data (each one is of the order of 1-2%). The overall data normalization (including the luminosity measurement) uncertainty is 1.5 %. The systematic uncertainty is, in total, of the order of 3% and is larger than the statistical uncertainties which are at the level of 1 % for $Q^2 < 100$ GeV^2.

At high Q^2 the systematic uncertainties are essentially the same. In ZEUS the statistic and systematic uncertainties amount to 3-5% for the kinematic range 400 GeV$^2 < Q^2 < 30000$ GeV^2 considered in the analysis.

In charged current events, the outgoing neutrino escapes the detection. Such events are then characterized by missing transverse energy $p_{t,miss}$ Analysis in ZEUS (H1) have demanded $p_{t,miss} > 10$ GeV ($p_{t,miss} > 12$ GeV). For the reconstruction of the kinematic variables, one can only use information from the hadronic final state, i.e. the Jacquet-Blondel method, giving,

$$y_{JB} = \frac{\sum_i (E_i - p_{z,i})}{2E_e}, \quad Q^2_{JB} = \frac{p^2_{t,miss}}{1 - y_{JB}}.$$

The CC events statistics is still low, ≈ 900 events for $Q^2 > 400$ GeV^2 in ZEUS. The systematic uncertainty is dominated by the hadronic energy scale, which induces an effect of the order of 10%, except at very high Q^2 and very high x where the effect is above 20%. Other systematic sources related to the $p_{t,miss}$ cut, acceptance correction and photoproduction background subtraction (in the lowest Q^2 bins) lead to measurement uncertainties between 4% and 8%.

3 Phenomenological analysis of inclusive measurements at HERA

As mentioned in the introduction, we shall distinguish the phenomenological analysis of the medium Q^2 data from the high Q^2 data. As we are interested in the HERA data, it should be noted that we are considering the region of large $W^2 \gg 10$ GeV^2. Therefore, we will not be concerned by the non-perturbative effects and the higher twist effects appearing in this region so that the symbol pQCD, appearing below, refers to the leading twist of pQCD.

For all the mathematical details which cannot be given here we refer to ref. [6] and references therein.

3.1 Analysis of the medium Q^2 NC data

In the one boson exchange approximation, the NC differential cross section reads

$$\frac{d\sigma^{e^{\pm}p}}{dxdQ^2} = \frac{2\pi\alpha_{em}Y_+}{xQ^4}\sigma_r, \quad \sigma_r = F_2(x, Q^2) - \frac{y^2}{Y_+}F_L(x, Q^2) \mp \frac{Y_-}{Y_+}xF_3(x, Q^2), \quad (1)$$

where $Y_{\pm} = 1 \pm (1-y)^2$. The nucleon structure functions are modeled using the quark-parton model and pQCD. In the so called naive parton model one writes

$$F_2(x) = \sum_{i=1}^{n_f} A_i(Q^2)x[q_i(x) + \bar{q}_i(x)], \quad F_3(x) = \sum_{i=1}^{n_f} B_i(Q^2)[q_i(x) - \bar{q}_i(x)]$$

where q_i ($\bar{q}_i$)is the density function of the quark (anti-quark) of flavor i, n_f is the number of active flavors and $F_L = 0$. The functions A_i [4] depend on the electric charge e_i ($A_i = e_i^2$ for $Q^2 \ll M_Z^2$) and embody the effects of the Z exchange and $\gamma - Z$ interference in their Q^2 dependence. The same holds for the functions B_i [4] except that they vanish at $Q^2 \ll M_Z^2$.

Going beyond the simple parton model, higher order contributions in α_s are taken into account. However mass singularities appear in the initial state of DIS processes and cannot be regularised without resumming the whole perturbative series. This resummation is done in a restricted kinematic region where $\alpha_s \log Q^2$ is large [6]. This latter region is defined by $Q^2 \gg \Lambda^2 \approx 0.3^2$ GeV^2, and the pQCD calculations are safe for Q^2 above a few GeV^2. In this domain, the parton density functions (pdf) are given by the solution of the DGLAP equations [6]:

$$M_F\frac{\partial q_{i,NS}^{\pm}(x, M_F^2)}{\partial M_F} = P_{NS}^{\pm} \otimes q_{i,NS}^{\pm}(x, M_F^2)$$

$$M_F\frac{\partial}{\partial M_F}\begin{pmatrix}\Sigma(x, M_F^2) \\ g(x, M_F^2)\end{pmatrix} = \begin{pmatrix}P_{qq} & n_f P_{qg} \\ P_{gq} & P_{gg}\end{pmatrix} \otimes \begin{pmatrix}\Sigma(x, M_F^2) \\ g(x, M_F^2)\end{pmatrix} \quad (2)$$

with $A \otimes B \equiv \int_x^1 A(z)B(x/z)dz/z$ and where $\Sigma = \sum_{i=1}^{n_f}(q_i + \bar{q}_i)$ is the singlet quark density, $q_{i,NS}^- = q_i^v \equiv q_i - \bar{q}_i$ and $q_{i,NS}^+ = q_i + \bar{q}_i - \Sigma/n_f$ are the two non singlet densities and g is the gluon density. The splitting functions $P_{i,j} = \alpha_s(M_R^2)P_{i,j}^{(0)} + \alpha_s^2(M_R^2)P_{i,j}^{(1)}$ describe the branching of parton j from parton i, and they can be computed with pQCD up to the second order. In eq. (2) M_F is the factorization scale (below which the mass singularity is resummed) and M_R is the renormalisation scale (related to the ultra-violet singularity). As the two scales must be chosen somehow, a natural choice for M_F is $\sqrt{Q^2}$, i.e. the virtual mass of the probe. We shall, as usual, also set $M_R = M_F$ for convenience. It is worth mentioning that the DGLAP equations are universal, i.e. that they are independent of the specific hard process.

Eq. (2) embodies the mass singularity resummation and therefore it only describes the so called light parton, i.e. the parton of flavour i and mass m_i such that $m_i^2/Q^2 \ll 1$. In the medium Q^2 range one can take the gluon, the up, down and strange quarks as the light partons. For the heavy quarks (charm and beauty) one needs to specify a special scheme. We have chosen the fixed-flavor-scheme (FFS) [7] – suitable in the HERA medium Q^2 range – where beauty is neglected, and where the charm contribution is computed from the boson-gluon-fusion process $\gamma g \to c\bar{c}$ plus the α_s^2 corrections. In this scheme charm is produced 'outside' the hadron. The relation between the pdfs and the structure functions depends on the renormalisation scheme. In Next-to-Leading-Log-Approximation (NLLA) and in the $\overline{\text{MS}}$ scheme one obtains:

$$F_i(x, Q^2) = x \sum_{j=1}^{n_f} \left[\left(1 + \frac{\alpha_s(Q^2)}{2\pi} C_{j,q} \right) \otimes e_j^2 (q_j(x, Q^2) + \bar{q}_j(x, Q^2)) \right. $$
$$\left. + 2\frac{\alpha_s(Q^2)}{2\pi} C_{j,g} \otimes g \right] + F_i^{c\bar{c}}(x, Q^2)$$

for $n_f = 3$ and where $i = 1, 2$ (there is a similar expression for F_3 with $F_3^{c\bar{c}} = 0$); $C_{j,q}$ and $C_{j,g}$ are the coefficient functions depending on the hard process; $F_i^{c\bar{c}}$ is the charm contribution [8]. It suffices here to say that it depends on m_c^2 and on a renormalisation scale that we choose to be $\sqrt{m_c^2 + Q^2}$. Note that $F_L = F_2 - 2xF_1 \neq 0$ in the NNLA.

To solve the system of integro-differential equations (2), one must provide some initial conditions, i.e. some input functions of x at a given Q^2 for each pdf. Since these functions reflect some unknown non-perturbative mechanism, one must parameterize with the help of a set of parameters. As we shall see below, these parameters are determined by comparing the calculations to the experimental data. However, the inclusive DIS data alone cannot constrain all light flavours since the structure functions are linear combinations of the pdfs: introducing the singlet and a non-singlet $xu^+ = xu + x\bar{u} - x\Sigma/3$ densities in order to write $\sum_{i=1}^{n_f=3} e_i^2 x[q_i(x, Q^2) + \bar{q}_i(x, Q^2)] = 2/3xu^+ + 1/9x\Sigma$. There is one important property of the DGLAP kernels $P_{i,j}$: the average total momentum carried by the partons, $\int_0^1 (x\Sigma + xg)dx$, is independent of Q^2. This quantity is called the momentum sum rule and is usually fixed to 1.

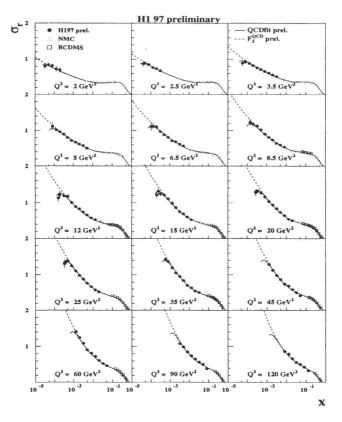

Fig. 1. H1 preliminary measurements of σ_r together with the result of a pQCD fit (see text).

So in principle one may be able to describe the inclusive HERA data by parameterizing the three functions xg, $x\Sigma$ and xu^+. But we found that such a description is not adequate for the following reasons:

- it leads systematically to a too large fraction of the total momentum carried by the gluons $\int_0^1 xg\,dx > 60\%$, in contradiction with the results of global fits including specific constrain on xg at high x [9];
- since the DGLAP equations involve some integrals of the pdfs from x to 1, one must also introduce some constraints at higher x, i.e. the fixed target hydrogen data from NMC and BCDMS [2];
- even when fixed target hydrogen data are included, one is unable to constrain the total momentum carried by the gluons. One must in addition include the fixed target deuterium data. In this case a second non-singlet density must be parameterized (essentially $u + \bar{u} - d - \bar{d}$), but now the valence counting rules, $\int_0^1 (u - \bar{u})dx = 2$ and $\int_0^1 (d - \bar{d})dx = 1$, can be applied, under certain assumptions, in order to constrain the momentum sum rule at high x.

H1 and ZEUS have used the latter option but with different assumptions. We will describe here the fits performed to the 1994 data and to the preliminary 1997 data. Both experiments include their own inclusive measurements and the NMC and BCDMS data.

In H1 two assumptions are made: $\bar{u} = \bar{d}$ and $\bar{s} = s = \bar{d}/2$. The first constraint is in contradiction with the global fit results [9] including the Drell-Yan data but we have found that it does not modify significantly the gluon density at below $x \approx 10^{-2}$. The second assumption comes from the results of the dimuon events of CCFR [10]. Finally xg, $x\bar{u}$, xu_v and xd_v are parameterized at a given value of Q^2 using the mathematical function $Ax^B(1-x)^C P(x)$ with $P(x) = 1 + Dx + E\sqrt{x}$. The momentum sum-rule and the quark counting rules are applied so that there are 16 free parameters in the H1 fit.

In ZEUS, the two valence quarks are taken from the MRS parameterization [9] and $\bar{s} = s = (\bar{d} + \bar{u})/2$ is also applied. xg, $x(\bar{u} - \bar{d})$ and $x(\bar{u} + \bar{d})$ are parameterized using the above mathematical functions.

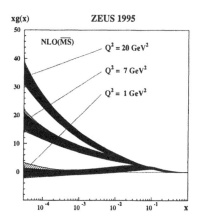

Fig. 2. xg extracted from the ZEUS fit to the 1994 data for three values of Q^2. Error bands contain the experimental error propagation and a theoretical error estimation (see text).

Concerning the data-theory comparison, from which the input pdfs have to be determined, both H1 and ZEUS use a χ^2 minimization procedure. The main steps of the fitting procedure are summarized below. For each iteration: 1) the pdf's are parameterized at a given value of Q^2 denoted Q_0^2, it is chosen to be 7 GeV^2 in the ZEUS fit and 4 GeV^2 in the H1 fit, 2) the DGLAP equations are solved numerically in x-space [11]. 3) the evolved pdf's are convoluted with the coefficient functions to obtain the structure functions. Assuming that all experimental uncertainties are normally distributed a χ^2 is computed. A crucial point of the analysis is the χ^2 expression which permits the use of the correlations introduced by some of the systematic uncertainties:

$$\chi^2 = \sum_{exp}\sum_{dat}\frac{[\mathcal{O}^{dat}_{exp} - \mathcal{O}^{fit} \times (1 - \nu_{exp}\sigma_{exp} - \sum_k \delta^{dat}_k(s^{exp}_k))]^2}{\sigma^2_{dat,stat} + \sigma^2_{dat,uncor}}$$
$$+ \sum_{exp}\nu^2_{exp} + \sum_{exp}\sum_k (s^{exp}_k)^2$$

where $\mathcal{O}$ stands for the observables (structure functions and differential cross

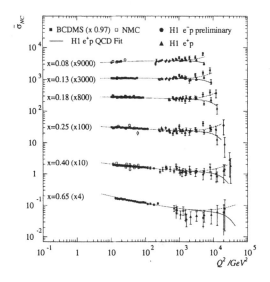

Fig. 3. High Q^2 H1 measurements of $\tilde{\sigma} \equiv \sigma_r$ (eq. 1) compared with pQCD fit results (see text).

sections). The first two sums run over the data (dat) of the various experiments (exp); σ_{exp} is the relative overall normalization uncertainty of the experiment exp; $\sigma_{dat,stat}$ and $\sigma_{dat,uncor}$ are the statistical error and the uncorrelated systematic error, respectively, corresponding to the data point dat; ν_{exp} is the number of standard deviations corresponding to the overall normalization of the experimental sample exp; $\delta^{dat}_k(s^{exp}_k)$ is the relative shift of the data point dat induced by a change by s^{exp}_k standard deviations of the kth correlated systematic uncertainty source of the experiment exp. It is estimated by

$$\delta^{dat}_k(s^{exp}_k) = \frac{\mathcal{O}^{dat}_{exp}(s^{exp}_k = +1) - \mathcal{O}^{dat}_{exp}(s^{exp}_k = -1)}{2\mathcal{O}^{dat}_{exp}}s^{exp}_k +$$
$$\left[\frac{\mathcal{O}^{dat}_{exp}(s^{exp}_k = +1) + \mathcal{O}^{dat}_{exp}(s^{exp}_k = -1)}{2\mathcal{O}^{dat}_{exp}} - 1\right](s^{exp}_k)^2 ,$$

where $\mathcal{O}_{exp}^{dat}(s_k^{exp} = \pm 1)$ is the experimental determination of $\mathcal{O}_{exp}^{dat}$ obtained varying by $\pm 1\sigma$ the kth source of uncertainty. Parameters ν_{exp} and s_k^{exp} can be determined by the χ^2 minimization or they can be fixed to zero during the minimization but released during the χ^2 error matrix calculation. In the first case one uses all the experimental information relying on the correctness of the estimate of the systematic uncertainties.

ZEUS CC 1994-97

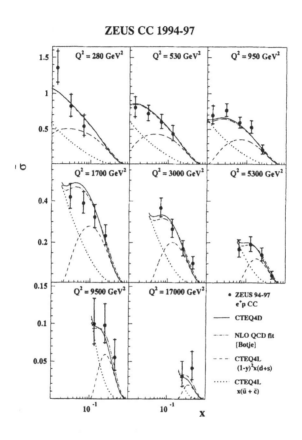

Fig. 4. ZEUS measurements of $\tilde{\sigma} \equiv \Phi_+$ together with various pQCD calculations (see text).

The result of the H1 fit is shown in fig. 1 together with the data. The agreement between data and pQCD is excellent. The gluon density obtained from the ZEUS fit (to the 1994 data) is shown in fig. 2. The error bands of the gluon density include the experimental error propagation as defined in ref. [12] and a theoretical uncertainty which includes the variation of: α_s, the charm mass, the pdf input parameterization form, the value of Q_0^2 and the factorization scale. With the 1997 data one can expect a reduction of the experimental uncertainty by a factor of two. The theoretical uncertainties will then dominate in the determination of xg, i.e. the third order splitting functions are needed.

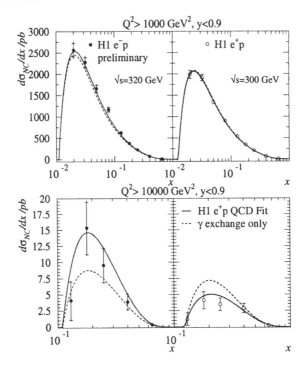

Fig. 5. H1 measurements of $d\sigma^{NC}/dx$ together with results from pQCD fits and different assumptions on electroweak contributions (see text).

3.2 Analysis of the high Q^2 NC and CC data

The fits applied to the high Q^2 data differ from the one described in the previous section by a different calculation of the contribution of the heavy quarks to the structure functions. As $m_c \approx 1.5$ GeV, one has $m_c/Q^2 \ll 1$ at high Q^2. The large term $\alpha_s^n \log^n(Q^2/m_c^2)$ – dominating the calculation of $F_2^{c\bar{c}}$ – must be resummed already at $Q^2 \approx 20$ GeV^2. The massless scheme is therefore used and only data with $Q^2 \geq 10$ GeV^2 are included in the fit. In the massless scheme, charm and beauty are considered as partonic constituents of the proton and their density functions are obtained by solving the DGLAP equations with the initial conditions $c(x, Q^2 \leq m_c^2) = 0$ and $b(x, Q^2 \leq m_b^2) = 0$. Such fits describe the HERA NC and CC (see fig. 3 and 4) data above $Q^2 = 10$ GeV^2.

In fig. 3 one can observe the different behavior of e^-p and e^+p cross sections at very high Q^2. This is related to the different sign of the contributions of F_3 to σ_r. Fig. 5 shows $d\sigma/dx$ together with the results of two pQCD fits including or not the Z exchange and $\gamma - Z$ interference. With the present data, sensitivity to electroweak effects in NC is for the first time observed at HERA.

Up to now we have only described the NC cross sections and related structure functions. For CC processes, in the one boson exchange approximation, one has

$$\frac{d\sigma_{CC}^{e^{\pm}p}}{dxdQ^2} = \frac{G_F^2}{2\pi x}\frac{M_W^4}{(M_W^2 + Q^2)^2}\Phi_{\pm}(x, Q^2),\qquad(3)$$

where G_F is the Fermi constant, and where the functions $\Phi_{\pm}$ depends on CC structure functions (see [4] for example). From eq. (3) one can first remark that the Q^2 slope of the CC differential cross section (see fig. 6) permits a

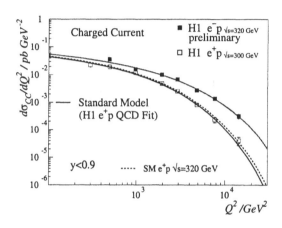

Fig. 6. H1 measurements of $d\sigma^{CC}/dQ^2$ together with the pQCD fit results.

determination of M_W, assuming (or not) the precisely measured value for G_F [13]. To extract M_W, H1 and ZEUS have used two different procedures. In H1, M_W is taken as an extra free parameter (G_F is fixed) of the pQCD fit and in ZEUS, the pdfs of CTEQ [9] are used in order to extract M_W and G_F (variations of the pdf choice is taken into account within the errors). The results are

$$\text{H1}: M_W = 80.9 \pm 3.3(stat.) \pm 1.7(syst.) \pm 3.7(theo)\,\text{GeV}$$
$$\text{ZEUS}: M_W = 80.4^{+4.9}_{-2.6}(stat.)^{+2.7}_{-2.0}(syst.)^{+3.3}_{-3.0}(pdf)\,\text{GeV}$$

and treating G_F as free, ZEUS obtain

$$M_W = 80.8^{+4.9}_{-4.5}(stat.)^{+5.0}_{-4.3}(syst.)^{+1.4}_{-1.3}(pdf)\,\text{GeV},$$
$$G_F = [1.171 \pm 0.034(stat.)^{+0.026}_{-0.032}(syst.)^{+0.016}_{-0.015}(pdf)] \times 5 \cdot 10^{-5}\,\text{GeV}^{-2}.$$

Let us point out that, concerning the H1 result, the theoretical uncertainty is dominated by the variation of the results when varying the ratio $\bar{d}/\bar{u}$ in the pQCD fit, and by the choice of the nuclear corrections applied to the deuterium target data entering the fit. These results, in good agreement with the world average values [13], show that the standard model gives a good description of both

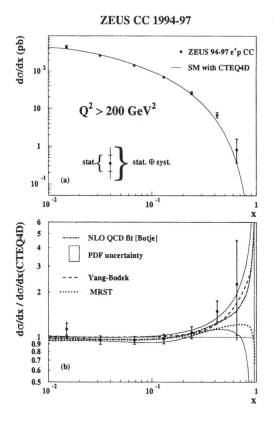

Fig. 7. ZEUS measurements of $d\sigma^{CC}/dx$ (for $e^{+}p$) together with various pQCD results (see text).

space-like (CC in DIS) and time-like (W production in $p\bar{p}$ and $e^{+}e^{-}$ collisions) processes.

In order to see the sensitivity of the CC cross section to the pdfs, we write $\Phi_{\pm}$ in LO

$$\Phi_{+} = x\bar{U} + (1-y)xD; \quad \Phi_{-} = xU + (1-y)x\bar{D}$$

with $U = u + c$ and $D = d + s$. From these expressions and from fig. 4 one can remark that: with positron (electron) beams one can determine d^{v} (u^{v}) at high x and small-y and $\bar{u} + \bar{c}$ ($\bar{d} + \bar{s}$) at small y. Let us mention that d_{v} and the sea quarks are basically determined in the global pQCD fits by μd and $\nu(\bar{\nu})F_{e}$ fixed target data, which require some nuclear corrections. Therefore, with the HERA $e^{\pm}p$ CC events one may have, with more statistics, a unique means to determine properly these quark densities.

In fig. 7, $d\sigma/dx$ is shown together with the error band determined by the ZEUS pQCD fit (without the CC and NC data described in the present article), and with the results of a recent analysis where an ansatz $d/u \neq 0$ as $x \rightarrow 1$ [14]

was introduced. Although the statistics is still low, one can notice from fig. 7 that this latter hypothesis is not required by the HERA data.

In fig. 8, the preliminary 1998 measurement of $d\sigma^{e^-p}/dx$ is shown. The error

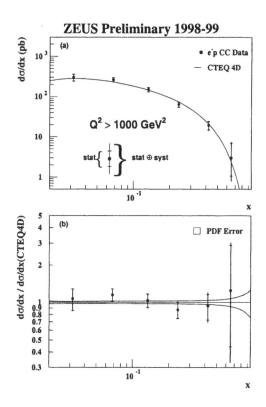

Fig. 8. ZEUS measurements of $d\sigma^{CC}/dx$ (for e^-p) together with the pQCD fit result.

band of the pQCD is much smaller than in fig. 7, therefore one can expect a better determination of electroweak parameters. The size of the error bands reflect that u^v is much better constrained than d^v in the pQCD fits.

3.3   Extraction of F_L

The longitudinal structure function is very hard to determine. It requires to combine data in a given x-Q^2 bin from different beam energies. However, from eq. (1), one observes that at high y the cross section receives a contribution both from F_2 and F_L. Therefore, taking F_2 from the result of a pQCD fit (see previous section) applied to the low y ($y < 0.35$) data one can determine F_L at high y by subtracting F_2, extrapolated to high y. The result of this operation is shown in fig. 9. To reach lower Q^2, where pQCD is not reliable, another method is used. Writing

$$\frac{\partial \sigma_r}{\partial \log y} = \frac{\partial F_2}{\partial \log y} - 2y^2 \frac{2-y}{Y_+^2} F_L - \frac{y^2}{Y_+} \frac{\partial F_L}{\partial \log y},$$

neglecting $\partial F_L / \partial \log y$, and assuming that $\partial F_2 / \partial \log y$ is a linear function of $\log y$, one obtains the results also shown in fig. (9). This determination is consistent with the LO calculation of pQCD.

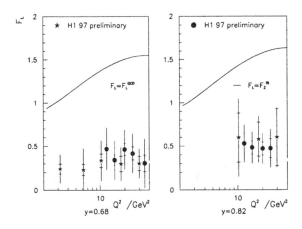

Fig. 9. H1 determination of F_L. The full points correspond to the subtraction method and the stars to the derivative method (see text).

4   Conclusion

Recent measurements of medium and high Q^2 differential cross sections at HERA have been presented, and a determination of F_L was also described. These new inclusive DIS data cover four orders of magnitude in Q^2 and five orders of magnitude in x.

In order to test Quantum Chromodynamics, fits based on the DGLAP equations have been performed successfully to the NC and CC data sets presented in this article. The extraction of the gluon density was described and the result of the analysis of the 1994 data was shown.

At medium Q^2 HERA has reached the limit where the systematic uncertainties dominate the statistical ones. With the 1997 data, the gluon density will be determined by the pQCD fit at the few percent level of accuracy, and a determination of both xg and α_s, is now foreseen.

Concerning the high Q^2 data, although the statistics of NC and CC events is still low, a sensitivity of the results to the effects of Z boson and $\gamma - Z$ interference in NC and to M_W in CC was observed. A determination of M_W from

$d\sigma^{CC}/dQ^2$ in space-like DIS was reported and a good agreement was found with the world average value from measurements in the time-like regions. Furthermore, comparing $d\sigma^{CC}/dxdQ^2$ with the pQCD calculation, we pointed out that such a measurement offers a unique possibility to pin down – independently of any nuclear effects – d^v and the different components of the proton sea.

Finally, comparing the measurements of $d\sigma^{e^+p}/dxdQ^2$ and $d\sigma^{e^-p}/dQ^2$ we observed, for the first time, the sensitivity of the NC to F_3.

With the high Q^2 NC and CC events a new field of research is touched here. It will be covered, with more precision, by the HERA-2000 upgrade with the help of an increase of luminosity and longitudinal polarization of the lepton beam.

References

1. H1 Coll., Nucl. Instr. Meth A336, **310** and **348** (1997); The ZEUS detector, status report DESY-1993.
2. BCDMS Coll., Phys. Lett. B223, **485** (1989); Phys. Lett. B237, **592** (1989); NMC Coll., Nucl. Phys. B483, **3** (1997).
3. H1 Coll. Nucl. Phys. B470, **3** (1996); ZEUS Coll. Eur. Phys. J. C7, **609** (1999), see also M. Botje, DESY 99-038 for the ZEUS pQCD fits.
4. See for a recent review: A.M. Cooper-Sarkar, R.C. Devenish and A. De Roeck, Int. J. Mod. Phys. A13, **3385** (1998).
5. H1 Coll., DESY 99-107; ZEUS Coll., DESY 99-56 and DESY 99-59.
6. W. Furmanski and R. Petronzio, Z. Phys. C1, **293** (1982).
7. M. Glück, E. Reya and M. Stratmann, Nucl. Phys. B422, **37** (1994).
8. E. Laenen et al., Nucl. Phys. B392, **162** and **229** (1993); Phys. Lett. B291, **325** (1992).
9. CTEQ4 Coll., Phys. Rev. D55, **1280** (1997); MRS Coll., Phys. Rev. D51, **4756** (1995).
10. CCFR Coll., Z. Phys. C65, **189** (1995).
11. C. Pascaud and F. Zomer, DESY 96-266; J. Blümlein et al., in Proc. of the Workshop on Future Physics at HERA, G. Ingelman, A. De Roeck and R. Klanner eds., **23** DESY (1996).
12. C. Pascaud and F. Zomer, LAL 95-05.
13. Particle Data Group, Eur. Phys. J. C3, **1** (1998).
14. U.K. Yang and A. Bodek, Phys. Rev. Lett. 82, **2467** (1999).
15. See for instance Proc. of the Workshop on Future Physics at HERA, G. Ingelman, A. De Roeck and R. Klanner eds., DESY (1996).

Measurements of F_2 and the total photon-proton cross sections at low x and low Q^2

Vincenzo Monaco

Università di Torino, INFN, Via Pietro Giuria 1, 10125 Torino, Italy

On behalf of the H1 and ZEUS Collaborations

Abstract. The HERA measurements of the proton structure function F_2 at low Q^2 and very low values of the Bjorken variable x are presented. These measurements cover the transition region between the photoproduction limit ($Q^2 = 0$) and the DIS regime at $Q^2 > 1$ GeV^2. Whereas the behaviour of F_2 at moderate and high Q^2 is described by DGLAP evolution, in the low Q^2 region, the F_2 behaviour can be described by phenomenological models inspired by Regge theory and the Vector Dominance Model.

1 Introduction

One of the early results from HERA was the discovery that the proton structure function F_2 is characterized by a steep rise with decreasing values of the Bjorken variable x [1,2]. This behaviour persists at low Q^2 values of the order of $Q^2 \sim 1.5$ GeV^2 [3,4] and can be interpreted as a purely perturbative phenomenon in terms of an increase of the gluon (xg) and sea quark ($x\bar{q}$) densities when smaller and smaller values of x are probed. Glück, Reya and Vogt (GRV) [5] predicted this behaviour well before the first HERA measurements; in their model the parton distributions are assumed to be non-singular (valence-like) at a very low Q^2 starting scale ($Q_0^2 \simeq 0.34$ GeV^2). The rise of xg and $x\bar{q}$ with decreasing x at higher Q^2 is generated dynamically via the DGLAP [6] evolution.

The proton structure function F_2 is related to the total virtual photon-proton cross section $\sigma_{tot}^{\gamma^* p}$ through the following relation, valid at low x and low Q^2

$$\sigma_{tot}^{\gamma^* p}(W, Q^2) \simeq \frac{4\pi^2 \alpha}{Q^2} F_2(x, Q^2) \,, \tag{1}$$

where $W^2 \simeq Q^2/x$ is the squared center of mass energy of the $\gamma^* p$ system.

From (1) it follows that a measurement of $F_2(x, Q^2)$ is equivalent to a measurement of the total cross section $\sigma_{tot}^{\gamma^* p}(W, Q^2)$, and that the rise of the structure function F_2 with decreasing x at moderate and high Q^2 and at low x corresponds to a rise of the total cross section with W at high values of W.

The W dependence of $\sigma_{tot}^{\gamma^* p}$ is shown in Fig. 1, where a partial compilation of measured cross sections is plotted as a function of W^2 for different Q^2 bins. The steep W dependence of the cross section at high and moderate Q^2 is well described by global next-to-leading order (NLO) perturbative QCD (pQCD) fits (the prediction of the MRS(R1) [7] parameterization is shown in Fig. 1). The photoproduction cross section at $Q^2 = 0$, shown in the same figure, exhibits a much softer W dependence.

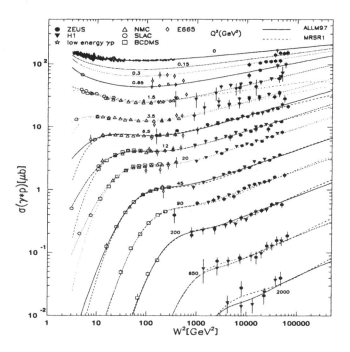

Fig. 1. Total $\gamma^* p$ cross section as a function of the $\gamma^* p$ center-of-mass energy in different bins of Q^2. The curves represent the ALLM97 [13] (*full line*) and the MRS(R1) [7] (*dotted line*) parameterizations. The points are data from H1, ZEUS and fixed target experiments

As shown by Donnachie and Landshoff (DL) [8] the W dependence of the photoproduction cross section is similar to the W dependence of the cross section of hadronic processes, described by a two-component Regge parameterization

$$\sigma_{tot}^{\gamma p}(W) = A_{IR}(W^2)^{\alpha_{IR} - 1} + A_{IP}(W^2)^{\alpha_{IP} - 1} , \qquad (2)$$

where IR and IP denote the Reggeon and Pomeron contributions, with intercepts $\alpha_{IR} \simeq 0.5$ and $\alpha_{IP} \simeq 1.08$ respectively. At high W, where the Reggeon term is negligible, the energy dependence of the total cross section can be parameterized as $\sigma_{tot}(W) \propto W^{2\lambda}$, with $\lambda = 1 - \alpha_{IP}$ small and close to 0 in the photoproduction limit. Conversely, the behaviour of the DIS data with $Q^2 > 1$ GeV^2 at high W corresponds to a much higher effective power λ (between 0.2 and 0.4).

A transition between the different W dependences of the cross section in the two regimes is expected. One of the reasons for expecting new phenomenology in this region comes from the observation that the total photoproduction cross section is finite; this, because of (1), implies that $F_2 \to 0$ as $Q^2 \to 0$, which entails strong scaling violations at low Q^2. Therefore, the description of the structure functions based on pQCD, which still works at a Q^2 scale of a few GeV^2, should fail at lower Q^2. It is interesting to see down to which low Q^2 value the NLO DGLAP formalism can be extended; at very low Q^2, however, pQCD predictions are not reliable at all and non-perturbative models have to be used.

At HERA, low Q^2 values correspond to very low values of x, where the average longitudinal distance travelled by the hadronic fluctuation of the photon in the proton rest frame $(l \sim 1/(2m_N x))$ is much longer than the target size. Then ep scattering can be viewed as the interaction of a color dipole $q\bar{q}$ with the proton, and the soft Regge-like behaviour of the photoproduction cross section is expected to hold also at finite but low values of Q^2. However, the Regge formalism cannot be pushed to too high Q^2 values without introducing hard contributions.

The question is at which Q^2 and x values the behaviour of the structure function changes from being described by perturbative QCD to being dominated by non-perturbative contributions. Different models and parameterizations exist to interpolate between the soft and hard behaviour of the cross section [9–14].

Experimental methods

The structure function F_2 and the cross section $\sigma_{tot}^{\gamma^* p}$ have been measured in the transition region using $e^+ p$ collisions from the 1995-1997 HERA runs.

The 4-momentum transfer squared, Q^2, is related to the energy E'_e and the angle ϑ'_e of the scattered positron by

$$Q^2 = 4E_e E'_e \cdot \sin^2 \frac{\vartheta'_e}{2} , \qquad (3)$$

where ϑ'_e is measured with respect to the direction of the positron beam and E_e is the positron beam energy. The limited angular acceptance of the main calorimeters of H1 and ZEUS fixes the lowest Q^2 to ~ 1.5 GeV^2. In order to measure F_2 at lower values of Q^2, the final state positron must be identified at very small scattering angles ϑ'_e or the beam energy E_e must be reduced.

"Shifted Vertex" runs have been provided by HERA in 1995; in these runs the nominal interaction vertex was shifted in the proton beam direction by +70 cm, thereby increasing the angular acceptance of the detectors to smaller positron scattering angles. Both the H1 and ZEUS experiments have measured F_2 using data collected in these runs, extending the coverage of the kinematic plane down to Q^2 values of ~ 0.35 GeV^2 (ZEUS) [15] and ~ 0.6 GeV^2 (H1) [16].

In 1995 ZEUS extended the kinematic acceptance to low angles by installing two small beam pipe calorimeter modules (BPC) on two sides of the beam pipe at ~ 3 m from the interaction point in the rear direction. Each module is a tungsten-scintillator sampling calorimeter, with an energy resolution of $\Delta E/E = 17\%/\sqrt{E}$. The BPC acceptance is $18 \leq \vartheta'_e \leq 32$ mrad. Data collected in 1995 with the BPC were used for the measurement of F_2 [17] in the kinematic range $0.11 \leq Q^2 \leq 0.65$ GeV^2 and $2 \times 10^{-6} \leq x \leq 6 \times 10^{-5}$.

Another method to extend the measurement of F_2 to lower Q^2 makes use of ep events with photons radiated collinear to the incident positron, resulting in an effective beam energy E_e reduced with respect to "non-radiative events". This method allows extension of the detector acceptance to low values of Q^2 in a complementary higher x range with respect to the BPC and shifted vertex measurements, but it suffers from large statistical and systematic errors.

The typical systematic error of the Shifted Vertex and BPC analyses is between 6 and 12 %, depending on the bin. The statistical errors are much lower.

2   Phenomenological analysis of the results

The results of the F_2 measurements based on the 1995 data are shown in Fig. 2 in Q^2 bins between $Q^2 = 0.11$ and 6.0 GeV^2. The results from the Shifted Vertex runs are labeled as H1 SVX95 [16], ZEUS SVX95 [15] those from BPC as ZEUS BPC95 [17]; some points from previous ZEUS analyses and from the E665 experiment [18] are also shown. The curves are the parameterizations from two fits of the ZEUS data based on non-perturbative models (ZEUSREGGE fit) and on NLO pQCD (ZEUSQCD fit). The data indicate that the rise of F_2 with decreasing x is prominent for values of $Q^2 > 1$ GeV^2, but becomes less steep for smaller Q^2 values.

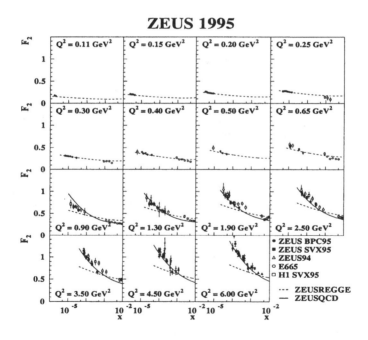

Fig. 2. $F_2(x, Q^2)$ at low Q^2 from "Shifted Vertex" runs (H1 SVX95 and ZEUS SVX95), ZEUS BPC and E665 data. The curves are described in the text

Figure 3 shows the cross section as a function of W^2 derived from the measured F_2 using (1). The two curves shown are the predictions of Donnachie and

Landshoff (DL) [8] and of GRV94 [5]. The DL model describes the W dependence of the data at very low values of Q^2, but above $Q^2 = 0.65$ GeV^2 it predicts a shallower rise of the cross section than the data exhibit. The increasing rise of $\sigma_{tot}^{\gamma^* p}$ with W^2 for Q^2 values above 1 GeV^2 is described by the GRV94 curves.

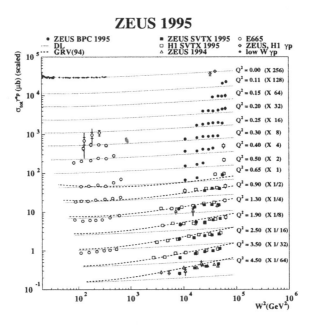

Fig. 3. The total $\gamma^* p$ cross section, $\sigma_{tot}^{\gamma^* p}$, as a function of W^2 for different Q^2 bins. Low Q^2 DIS measurements and photoproduction cross sections [20] are compared with the predictions of DL (*dotted line*) and GRV94 (*dashed line*)

In [15], a simplified Generalized Vector Dominance Model (GVDM) [19] approach was used to parameterize the Q^2 dependence of the cross section in the BPC range; in the fit described in [15] the longitudinal $\gamma^* p$ cross section is neglected, while the transverse cross section $\sigma_T(W^2, Q^2)$ is related to the corresponding cross section at $Q^2 = 0$, $\sigma_0(W^2)$, by

$$\sigma_T(W^2, Q^2) = \frac{M_0^2}{M_0^2 + Q^2} \sigma_0(W^2) , \qquad (4)$$

where only the contribution from continuum vector states from a cut-off-mass M_0^2 has been kept. A fit of the BPC F_2 points to (4) gives $M_0^2 = 0.53$ GeV^2.

The W dependence of $\sigma_0(W^2)$ is parameterized by Regge theory through (2). The Q^2 and W dependence of the cross section $\sigma_{tot}^{\gamma^* p}$ used in the ZEUSREGGE fit is obtained by combining (2) and (4)

$$\sigma_{tot}^{\gamma^* p}(W^2, Q^2) = \left(\frac{M_0^2}{M_0^2 + Q^2} \right) \left(A_{IR}(W^2)^{\alpha_{IR}-1} + A_{IP}(W^2)^{\alpha_{IP}-1} \right) . \qquad (5)$$

The ZEUS BPC data and low W photoproduction data from fixed target experiments [20] were fitted to (5), with A_{IR}, A_{IP} and α_{IP} left as free parameters. The intercept of the Pomeron is determined to be $\alpha_{IP} = 1.097 \pm 0.002$, compatible with the value of the DL fit. The fit result is shown in Fig. 2; the ZEUSREGGE fit gives a good description of the low Q^2 data in the BPC region $Q^2 \leq 0.65$ GeV^2, but at larger Q^2 the curves fall below the data.

The continuous curve in Fig. 2 (ZEUSQCD) is the result of a NLO QCD fit to the ZEUS data with $Q^2 > 1$ GeV^2; data from the NMC [21] and the BCDMS [22] experiments with $W^2 > 10$ GeV^2, not shown in Fig. 2, are also included to constrain the fit at high x. The details of the fit are described in [15]. The NLO DGLAP evolution equations are solved in the $\overline{MS}$ scheme. The input scale is chosen to be $Q_0^2 = 7$ GeV^2, and backward evolution is performed to fit the data with $Q^2 < Q_0^2$. The quality of the fit is good (the χ^2 is 1474 for 1120 data points and 11 free parameters) as can be seen in Fig. 2, where the resulting F_2 is shown together with the measured values.

Figure 4 shows the gluon (xg) and singlet ($x\Sigma$) distributions obtained from the fit as a function of x for Q^2 at 1, 7 and 20 GeV^2. At the smallest Q^2 the gluon distribution is almost flat and compatible with zero, while the quark singlet distribution is still rising at small x.

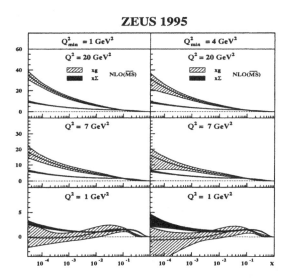

Fig. 4. The quark singlet momentum distribution $x\Sigma$ (*shaded*) and the gluon momentum distribution xg (*hatched*) as a function of x at $Q^2 = 1$, 7 and 20 GeV^2 from the ZEUSQCD fit. The left-hand plots show the results of the fit including data with $Q^2 > 1$ GeV^2; the results in the right-hand plots include data with $Q^2 > 4$ GeV^2. The error bands correspond to the quadratic sum of all the error sources

The quark singlet distribution is dominated at low x by the contribution from the $q\bar{q}$ sea distribution xS, and its evolution in Q^2 is coupled to that of the gluons. It seems rather unphysical that the gluon does not drive the sea evolution at low Q^2; this behaviour (also observed, for instance, in the result of the fit by Martin et al. (MRST) [23]) is incompatible with the hypothesis that the rapid rise in F_2 is driven by the rapid increase of the gluon density at small x induced by parton splitting. However, no breakdown of the technical validity of the NLO QCD fit has been found in the kinematic range explored; the experimental data are well described down to $Q^2 \sim 1$ GeV^2.

The low x behaviour of F_2 as a function of x and Q^2 can be quantified through the slopes $d\ln F_2/d\ln(1/x)$ at fixed Q^2 and $dF_2/d\ln Q^2$ at fixed x.

The slope $\lambda_{eff} = d\ln F_2/d\ln(1/x)$, shown in Fig. 5, is extracted by fitting the data at fixed Q^2 to the form $F_2 = Ax^{-\lambda_{eff}}$, equivalent to the parameterization $\sigma_{tot}^{\gamma^* p}(Q^2, W^2) = A' \cdot W^{2\lambda_{eff}}$ for the total cross section. The slope λ_{eff} at $Q^2 < 1$ GeV^2 is compatible with the Q^2-independent Regge prediction $\lambda = 1-\alpha_{IP} \simeq 0.1$, while at higher Q^2 it gradually increases with Q^2. This tendency at high Q^2 is not followed by the dominant Pomeron term of soft Regge phenomenology, but is qualitatively described by pQCD fits. As an example, the prediction of the GRV94 parameterization and the result of the ZEUSQCD fit for $Q^2 > 1$ GeV^2, obtained using the same x range of the data, are shown in Fig. 5.

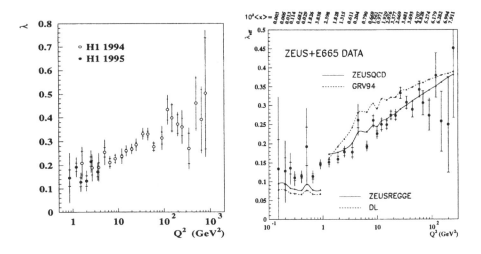

Fig. 5. The slope $\lambda_{eff} = d\ln F_2/d\ln(1/x)$ as a function of Q^2 as determined by a fit to H1 (**left**) and ZEUS (**right**) data

The logarithmic slope $dF_2/d\ln Q^2$ is derived from the data by fitting $F_2 = a + b\ln Q^2$ in bins of fixed x. The slope $dF_2/d\ln Q^2$ at low x is dominated by the convolution of the splitting function P_{qg} and the gluon density $(dF_2/d\ln Q^2 \propto \alpha_S P_{qg} \otimes xg)$, and can be considered as a quantitative evaluation of the scaling violations caused by gluon bremsstrahlung and quark pair creation.

The dependence of the logarithmic Q^2 slope on x, as determined from the fit of the ZEUS data, is shown in Fig. 6, where for each bin the mean value of Q^2 is

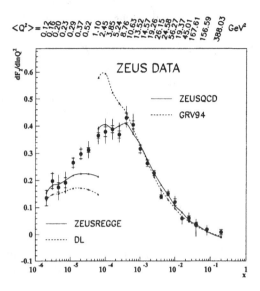

Fig. 6. The slope $dF_2/d\ln Q^2$ as a function of x as determined by the fit of ZEUS data

also shown. For values of x down to 3×10^{-4} the slope increases as x decreases. In this x range, where the mean values of Q^2 are high enough that pQCD fits are expected to be reliable, the ZEUSQCD fit and the GRV94 parameterization are in good agreement with the data. The trend of the slope in this region reflects the rapid increase of the gluon density with decreasing values of x obtained by the same ZEUSQCD fits.

At lower values of x and Q^2 the slope has the opposite behaviour, it decreases for decreasing x, with a "turn-over" at $x \sim 3 \times 10^{-4}$ and at a mean Q^2 value of a few GeV^2. It should be noticed that the exact position of the "turn-over" could depend on the correlations between the x and Q^2 values of the data points included in the fit. In the same Q^2 range the slope determined from fixed-target data continues to increase with decreasing values of x, but these data are at larger values of x.

The parameterization of DL and the ZEUSREGGE fit describe the x dependence of the slope only at very low values of x. As shown by the result of the ZEUSQCD fit, the standard NLO DGLAP equations are able to describe the behaviour of the slope down to $Q^2 \sim 1\ GeV^2$.

It is not surprising that the GRV94 parameterization, obtained by a fit not constrained by data at low Q^2, does not follow the turn-over of the slope between $Q^2 \sim 1$ and $Q^2 \sim 10\ GeV^2$; in this region the evolution of the sea and gluon distributions predicted by the DGLAP equations depends strongly on the shape of the parton distributions at the starting scale and on the value of the strong coupling constant α_s. The ZEUSQCD fit, tuned to the same data from which the slopes have been determined, follows the x dependence of $dF_2/d\ln Q^2$ down to lower values of Q^2. The flattening of the slope at $Q^2 \sim 1$ GeV^2 reflects the flat x-dependence of the gluon distributions determined from the ZEUSQCD fit at this Q^2 scale.

While in the pQCD regime the logarithmic Q^2 slope is sensitive to the gluon distribution, the x dependence of the structure function F_2 is related to the sea quark distribution. The fact that at $Q^2 \sim 1$ GeV^2 F_2 continues to rise with decreasing values of x is consistent with the x dependence of the singlet quark distribution obtained at these Q^2 values from the ZEUSQCD fit.

3   New experimental results

In order to reduce the background and the systematic error of the F_2 measurement in the transition region, ZEUS installed a tracker (BPT), consisting of two silicon microstrip detectors, in front of the Beam Pipe Calorimeter in 1997.

The BPT complements the information of the BPC for the reconstruction of the scattered positron and the measurement of the scattering angle. There are several advantages to using a tracker in front of the Beam Pipe Calorimeter:

- The tracker can be used to reduce the main source of background in the BPC analysis, due to neutral particles in the final state from photoproduction events misidentified as positrons. The reduction of the background allows extension of the kinematic coverage to higher values of y with respect to the 1995 analysis.
- With the BPT the scattering angle can be measured more precisely and independently of the knowledge of the interaction vertex; the F_2 measurement is thus less dependent on the efficiency of the vertex reconstruction and on the simulation of the hadronic final state in the Monte Carlo.
- The use of the BPT in the reconstruction of the track coordinates gives a better control of the fiducial volume cuts and of the position dependent corrections in the BPC energy reconstruction. The energy calibration, performed on 1997 data using kinematic peak events, determines the energy scale of the calorimeter to within $\sim 0.3\%$; the energy scale is uniform within $\pm 0.3\%$ over the entire active area of the detector.

The data taken with the BPT during six weeks at the end of the 1997 running period, corresponding to an integrated luminosity of 3.9 pb^{-1}, have been used to measure F_2. The kinematic region covered by this analysis is $0.045 < Q^2 < 0.65$ GeV^2 and $6 \times 10^{-7} < x < 1 \times 10^{-3}$, corresponding to $25 < W < 270$ GeV and $0.007 < y < 0.8$. The kinematic coverage has been extended to lower and higher y with respect to the 1995 BPC analysis. The lowest y bins overlap with the measurements of the E665 experiment.

For $y > 0.08$, the event kinematics are reconstructed with the electron method. At lower y, the y resolution is improved by combining positron and hadronic final state variables with the $e\Sigma$ method [26].

The simulation of ep events in this analysis makes use of two Monte Carlo generators: non-diffractive events are generated with DJANGO [24], while diffractive

– low mass and low multiplicity – events are generated with RAPGAP [25]. The two Monte Carlo samples are mixed in a proportion determined by fitting distributions of hadronic variables in data with a weighted sum of DJANGO and RAPGAP events.

The value of $R = F_L/(F_2 - F_L)$ is taken from the BKS model [27]. Due to the low Q^2 of this analysis, the effect of F_L is small, reaching at most 3% in few bins at high y and moderate Q^2.

The data are shown in Fig. 7, together with two fits (ALLM97 [13], DL98 [14]) which include the 1995 measurements, as well as with the original DL parameterization.

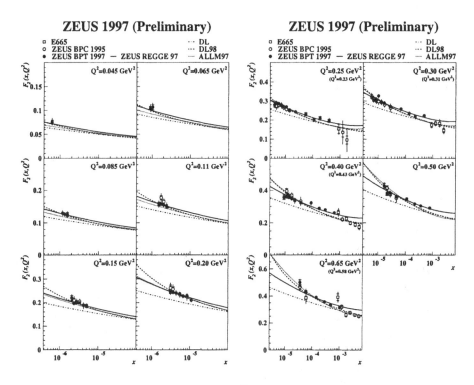

Fig. 7. F_2 as a function of x in bins of Q^2 from the BPT analysis compared with previous measurement from ZEUS (BPC 95) and E665 results

The average statistical error, including the statistical error of the Monte Carlo simulation, is 2.6%. The typical systematic error of 3.3% is similar to the statistical one and has been reduced by a factor 2 to 3 compared to the previous measurement.

The BPT F_2 data are converted to total $\gamma^* p$ cross sections using (1) and are shown in Fig. 8 as a function of Q^2 for different W bins. The measured cross sections for fixed W have been fitted to (4), with the values of the cross sections $\sigma_0(W^2)$ at $Q^2 = 0$ left as free parameters. The fit has a reasonable χ^2/dof of

1.3, taking into account statistical errors only; this indicates that the simplified GVDM model of (4) describes well the Q^2 dependence of the cross section.

The extrapolated cross sections σ_0 are shown in Fig. 9 as a function of the center of mass energy W; in the same figure the direct measurements from H1 [28] and ZEUS [29] and data from other experiments at low W [20] are also shown, and compared with the DL98 [14] and ALLM97 [13] parameterizations. The magnitude of the model dependence, estimated by extrapolating $\sigma_{tot}^{\gamma^*p}$ with the Q^2-dependence of DL, DL98 and ALLM97 parameterizations, is shown as a shaded band in Fig. 9.

A fit of the cross sections $\sigma_0(W)$ extrapolated from the BPT data, and of the low W measurements with the Regge-type form of (2) fixing $\alpha_{IR} = 0.5$, gives a value of the Pomeron intercept $\alpha_{IP} = 1.105 \pm 0.001(stat) \pm 0.007(sys)$.

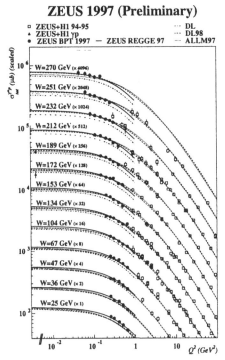

Fig. 8. $\sigma_{tot}^{\gamma^*p}$ versus Q^2 in bins of W from the BPT F_2 analysis and other measurements from ZEUS and H1

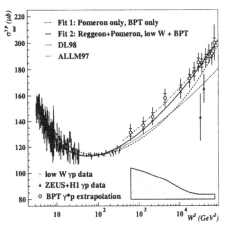

Fig. 9. Extrapolated and measured $\sigma_{tot}^{\gamma p}$ versus W^2. The curves are the ALLM97 and DL98 parameterizations and the result of the Regge fit described in the text

4   Conclusions

The HERA measurements of F_2 and $\sigma_{tot}^{\gamma^*p}$ at low Q^2 and x, in the transition region between DIS and the photoproduction limit, have been presented. The data with $Q^2 < 0.65$ GeV^2 can be described by a simple GVDM and Regge-inspired parameterization, which however fails at higher Q^2. The rapid fall of F_2 at small x for $Q^2 > 1$ GeV^2 is described by NLO QCD fits.

The slopes $d\ln F_2/d\ln(1/x)$ at fixed Q^2 and $dF_2/d\ln Q^2$ at fixed x have been detemined from the HERA data. In the regime where pQCD is reliable, the behaviour of the slopes is compatible with the results of a NLO QCD fit.

References

1. H1 Collab., I.Abt et al.: Nucl.Phys. B **407** (1993) 515;
   H1 Collab., T.Ahmed et al.: Nucl. Phys. B **439** (1995) 471
2. ZEUS Collab., M.Derrick et al.: Phys.Lett. B **316** (1993) 412; Z.Phys. C **65** (1995) 379; Z.Phys. C **72** (1996) 399
3. H1 Collab., S.Aid et al., Nucl.Phys. B **470** (1996) 4; H1 Collab., C.Adloff et al.: Nucl.Phys. B **497** (1997) 3
4. ZEUS Collab., M.Derrick et al.: Z.Phys. C **69** (1996) 607; Z.Phys. C **72** (1996) 399
5. M.Glück, E.Reya and A.Vogt: Z.Phys. C **48** (1990); Z.Phys. C **53** (1992) 127; Z.Phys. C **67** (1995) 433
6. V.N.Gribov and L.N.Lipatov: Sov.J.Nucl.Phys. **15** (1972) 438, 675; L.N.Lipatov: Sov.J.Nucl.Phys. **20** (1975) 95; Y.L.Dokshitzer: Sov.Phys.JETP **46** (1977) 641; G.Altarelli and G.Parisi: Nucl.Phys. B **126** (1977) 298
7. A.D.Martin,R.G.Roberts and W.J.Stirling: Phys.Lett B **387** (1996) 419
8. A.Donnachie, P.V.Landshoff: Phys.Lett. B **296** (1992) 227
9. A.Capella, A.Kaidalov,C.Merino, J.Tran Thanh Van: Phys.Lett. B **337** (1994) 358
10. B.Badelek and J.Kwiecinski: Phys.Lett. B **295** (1992) 263
11. K.Adel, F.Barreiro and F.J.Ynduráin: FTUAM 96-39, hep-ph/9610380
12. E.Gotsman, E.M.Levin and U.Maor: Phys.Lett B**425** (1998) 369
13. H.Abramowicz and A.Levy: DESY 97-251
14. A.Donnachie and P.V.Landshoff: Phys.Lett. B **437** (1998) 408
15. ZEUS Collab., J.Breitweg et al.: Eur.Phys.J. C **7** (1999) 609
16. H1 Collab., C.Adloff et al.: Nucl.Phys. B **497** (1997) 3
17. ZEUS Collab., J.Breitweg et al.: Phys.Lett B **407** (1997) 432
18. E665 Collab., M.R.Adams et al.: Nucl.Rev. D **54** (1996) 3006
19. J.J.Sakurai and D.Schildknecht: Phys.Lett. B **40** (1972) 121
20. D.O.Caldwell et al.: Phys.Rev.Lett. **40** (1978) 1222; S.I.Alekhin et al.: CERN-HERA 87-01 (1987)
21. NMC Collab., M.Arneodo et al.: Nucl.Phys. B **483** (1997) 3
22. BCDMS Collab., A.C.Benvenuti et al.: Phys.Lett. B **223** (1989) 485; Phys.Lett. B **237** (1990) 592
23. A.D.Martin et al.: Eur.Phys.J. C4 (1998) 463
24. K.Charchula, G.A.Schuler and H.Spiesberger: Comp.Phys.Commun. **81** (1994) 381; H.Spiesberger, DJANGO6 version 2.4 - A Monte Carlo Generator for Deep Inelastic Lepton Proton Scattering including QED and QCD Radiative effects (1996)
25. H.Jung: Comp.Phys.Comm. **86** (1995) 147
26. U.Bassler and G.Bernardi: DESY 97-131
27. B.Badelek et al.: Z.Phys. C**74** (1997) 297
28. H1 Collab.,A.Aid et al.: Z.Phys. C **69** (1995) 27
29. ZEUS Collab.,M.Derrick et al.: Z.Phys.C **63** (1994) 391

Diagonal and Skewed Parton Distributions

Alan Martin

University of Durham, Durham, DH1 3LE, UK

Abstract. We briefly review the global analyses that are used to determine parton distributions from data, and discuss recent topical issues that have arisen. We outline the properties of skewed parton distributions, and show that in the small x domain they are completely determined by the conventional distributions. The relevance to the description of vector meson production at HERA is emphasized.

1 Introduction

The cross section of hard scattering processes involving incoming protons (such as deep inelastic electron-proton scattering or the hadroproduction of jets of large transverse momentum) can be written as the sum of parton distributions multiplied by the cross sections of hard subprocesses calculated at the parton level using perturbative QCD. That is we can factor off the long distance (non-perturbative) effects into universal, process independent, parton distributions, $f_i(x, \mu^2)$ with $i = q, \bar{q}, g$, which represent the probability of finding parton i in the proton carrying a longitudinal fraction x of the proton's momentum. The parameter μ is a scale typical of the hard partonic subprocesses. Perturbative QCD determines the evolution of the parton distributions as a function of μ, provided they are known as some starting scale μ_0. Calculating the parton distributions from first principles at some low scale μ_0 is one of the most challenging problems of non-perturbative QCD. The most promising approach is lattice QCD, but much remains to be done. On the other hand, from a practical point of view, the parton distributions of the proton are determined with good precision from global analyses of deep inelastic and related hard scattering data. We will describe these analyses in Section 2 and focus on some of the recent issues that result.

Formally parton distributions can be expressed as matrix elements $\langle p|\hat{O}|p\rangle$ where $\hat{O}$ is a twist-2 quark or gluon operator, and p is the four momentum of the proton. Recently there has been much renewed interest in off-diagonal or skewed parton distributions which are given by matrix elements $\langle p'|\hat{O}|p\rangle$ in which the momentum p' of the outgoing proton is not the same as that of the incoming proton. For example, the *amplitudes* for processes such as deeply virtual Compton scattering ($\gamma^* p \rightarrow \gamma p$) or vector particle electro- or photo-production ($\gamma^* p \rightarrow Zp$ or $J/\psi p$) depend on off-diagonal distributions. Since $p \neq p'$ the parton returning to the proton has a different momentum to the one which is outgoing, and so we need two momentum variables to specify the skewed distributions. In Section 3 we will discuss the phenomenological developments involving skewed distributions (concentrating entirely on the small x domain) and emphasize their potential role in the global parton analyses.

2   Global parton analyses

As mentioned above, the parton densities $f_i(x, Q^2)$ are determined from a global analysis of a wide range of data for deep inelastic and related 'hard' scattering processes, which are listed in the first column of Table 1. The procedure is to parametrize the x dependence of the 'starting' distributions in physically motivated forms, such as

$$x f_i(x, Q_0^2) = A_i x^{-\alpha_i}(1 - x)^{\beta_i}(1 + \gamma_i x^{\frac{1}{2}} + \delta_i x) \tag{1}$$

at a low scale Q_0^2, but which is nevertheless sufficiently high to be in the perturbative region. Next-to-leading order DGLAP equations are then used to evolve up in Q^2 to determine $f_i(x, Q^2)$ at all the x, Q^2 at which data are to be fitted, and the parameters $A_i, \alpha_i, \ldots \delta_i$ varied until an optimum description of the data is achieved, subject to satisfying the momentum and flavour sum rules. The last column of Table 1 gives an indication of the parton properties that are constrained by particular data sets. Two independent groups have carried out global analyses: MRST and CTEQ. Their latest sets of partons are known as MRST [1], MRST99 [2] and CTEQ5 [3]. Semi-global analyses are also being undertaken by the experimental collaborations (see, for example, [4]). In this section we review topical issues concerning these global parton analyses. We adopt the conventional notation $u \equiv f_u, d \equiv f_d$ etc.

2.1   Light quark distributions: u, d, s

The u and d distributions are pinned down by the DIS data, with the slope of d/u being constrained for $x \lesssim 0.3$ by the CDF $W^\pm$ rapidity asymmetry data. In fact d/u is determined up to $x \sim 0.7$ by the NMC F_2^n/F_2^p data, *provided* that we assume that there are no deuterium binding corrections for $x > 0.3$. However, Yang and Bodek [5] have recently stressed a point made by Melnitchouk and Thomas [6] that the F_2^n data may have a nuclear binding correction which would lead to a larger d/u ratio at large x. This leads to considerable uncertainty in the d distribution at large x. Descriptions with $d/u \to 0$ or 0.2 as $x \to 1$ are equally possible. In principle measurements of charged current DIS $e^+ p \to \bar\nu X$ (and $e^- p \to \nu X$) at HERA, or $W^\pm$ asymmetry at large rapidity at the LHC, can resolve the issue [2].

The ratio $\bar d/\bar u$ of sea quark distributions is determined for $0.05 \lesssim x \lesssim 0.3$ by the E866 data for $pp, pn \to \mu^+\mu^- X$ [10]. Information is also coming from the HERMES data for $ep, en \to e\pi^\pm X$. It is found that $\bar d/\bar u$ rises from unity at $x = 0$ to about 1.5 at $x = 0.15$ and then falls sharply back through unity around $x \sim 0.3$.

Table 1. Processes studied in global analyses. The essential features of the partons that are probed by the particular data sets are indicated in the last column.

Process/ Experiment	Leading order subprocess	Parton behaviour probed
DIS ($\mu N \to \mu X$) $F_2^{\mu p}, F_2^{\mu d}, F_2^{\mu n}/F_2^{\mu p}$ (SLAC, BCDMS, NMC, E665)	$\gamma^* q \to q$	Four structure functions $\to$ $u + \bar{u}$ $d + \bar{d}$ $\bar{u} + \bar{d}$
DIS ($\nu N \to \mu X$) $F_2^{\nu N}, x F_3^{\nu N}$ (CCFR)	$W^* q \to q'$	s (assumed $= \bar{s}$), but only $\int x g(x, Q_0^2) dx \simeq 0.35$ and $\int (\bar{d} - \bar{u}) dx \simeq 0.1$
DIS (small x) F_2^{ep} (H1, ZEUS)	$\gamma^* (Z^*) q \to q$	λ $(x\bar{q} \sim x^{-\lambda_S}, \; xg \sim x^{-\lambda_g})$
DIS (F_L) NMC, HERA	$\gamma^* g \to q\bar{q}$	g
$\ell N \to c\bar{c} X$ F_2^c (EMC; H1, ZEUS)	$\gamma^* c \to c$	c $(x \gtrsim 0.01; \; x \lesssim 0.01)$
$\nu N \to \mu^+ \mu^- X$ (CCFR)	$W^* s \to c$ $\hookrightarrow \mu^+$	$s \approx \frac{1}{4}(\bar{u} + \bar{d})$
$pN \to \gamma X$ (WA70, UA6, E706, ...)	$qg \to \gamma q$	g at $x \simeq 2p_T^\gamma/\sqrt{s} \to$ $x \approx 0.2 - 0.6$
$pN \to \mu^+ \mu^- X$ (E605, E772)	$q\bar{q} \to \gamma^*$	$\bar{q} = ...(1 - x)^{\eta_S}$
$pp, pn \to \mu^+ \mu^- X$ (E866, NA51)	$u\bar{u}, d\bar{d} \to \gamma^*$ $u\bar{d}, d\bar{u} \to \gamma^*$	$\bar{u} - \bar{d}$ $(0.04 \lesssim x \lesssim 0.3)$
$ep, en \to e\pi X$ (HERMES)	$\gamma^* q \to q$ with $q = u, d, \bar{u}, \bar{d}$	$\bar{u} - \bar{d}$ $(0.04 \lesssim x \lesssim 0.2)$
$p\bar{p} \to W X (Z X)$ (UA1, UA2; CDF, D0)	$ud \to W$	u, d at $x \simeq M_W/\sqrt{s} \to$ $x \approx 0.13; \; 0.05$
$\to \ell^\pm$ asym (CDF)		slope of u/d at $x \approx 0.05 - 0.1$
$p\bar{p} \to t\bar{t} X$ (CDF, D0)	$q\bar{q}, gg \to t\bar{t}$	q, g at $x \gtrsim 2m_t/\sqrt{s} \simeq 0.2$
$p\bar{p} \to$ jet $+ X$ (CDF, D0)	$gg, qg, qq \to 2j$	q, g at $x \simeq 2E_T/\sqrt{s} \to$ $x \approx 0.05 - 0.5$

In principle $\nu N \to \mu^- X$ and $\bar{\nu} N \to \mu^+ X$ DIS data can determine s and $\bar{s}$ distributions separately. However so far the global analyses have used the studies of the CCFR dimuon production data, which yields

$$s + \bar{s} \simeq (\bar{u} + \bar{d})/2 \tag{2}$$

independent of x at $Q^2 = 1$ GeV^2, with typically a $\pm 10\%$ error [2].

2.2   Heavy quark distributions: c and b

The improved theoretical treatments have recently enabled these distributions to be put on a sound basis in the global analyses (see, for example, [7]). For a heavy quark distribution (say c) we should match, at $Q^2 = m_c^2$, the fixed flavour number scheme, describing $\gamma g \to c\bar{c}$ with $n_f = 3$ to the variable flavour number scheme with $n_f = 4$ flavours for $Q^2 > m_c^2$. MRST [1] use the Thorne-Roberts prescription [8] in which F_2^c and $\partial F_2^c/\partial \ln Q^2$ are required to be continuous at $Q^2 = m_c^2$, whereas CTEQ [3] use an alternative 'ACOT' prescription [9]. In this way the charm distribution is generated perturbatively with the mass of the charm quark m_c as the only free parameter. Not surprisingly, due to the $g \to c\bar{c}$ transition, the charm distribution mirrors the form of the gluon distribution. At some level we may expect an 'intrinsic' component of the charm distribution.

2.3   The gluon distribution g

In principle many processes are sensitive to the gluon distribution, but it is still difficult to determine for $x \gtrsim 0.2$ where it becomes increasingly small. In this region MRST [1] used the WA70 prompt photon data to determine the gluon. However this process suffers from scale dependence (reflecting the higher order corrections), effects of intrinsic k_T (needed to describe the E706 prompt photon data) and uncertainties due to fragmentation effects and isolation criteria. In an attempt to provide an acceptable spread of gluon distributions at large x, MRST presented three sets of partons ($g \uparrow$, the default MRST set, $g \downarrow$) corresponding, respectively, to an average intrinsic $k_T = 0, 0.4$ and 0.64 GeV for the WA70 data. The arrows indicate the relative sizes of the gluons in the large x region. The corresponding $\langle k_T \rangle$ needed to obtain a good description of the E706 prompt photon data is worryingly large at about 1 GeV.

Due to the uncertainties associated with prompt photon production, the recent CTEQ analysis [3] omits these data and instead determines the large x behaviour of the gluon using the single jet inclusive E_T distributions measured at the Tevatron. The above three MRST sets of partons 'predict' the shape of the jet E_T distributions remarkably well, but the predicted normalization is 7%, 13%, 17% below the data respectively with $g \uparrow$ being closest to CTEQ5.

Turning now to the small x domain, the gluon is well constrained by the observed behaviour of $\partial F_2/\partial \log Q^2$. To be precise the H1 and ZEUS data determine the gluon for $x \lesssim 0.01$, and the NMC data for $x \lesssim 0.1$. Inspection of F_2

as a function of $\ln Q^2$ at low fixed x [11,4] shows that the data flatten out at low Q^2, indicating a smaller value of

$$\partial F_2/\partial \ln Q^2 \simeq \alpha_S(Q^2) P_{gg} \otimes g. \qquad (3)$$

Also α_S increases with decreasing Q^2, and as a consequence $g(x)$ becomes valence-like for $Q^2 \simeq 1$ GeV^2, which is about the lowest value of Q^2 for a DGLAP description. In fact $Q^2 \simeq 2$ GeV^2, where the gluon becomes flat (that is $g(x) \to$ constant as $x \to 0$), is a more reasonable lower limit for DGLAP evolution.

2.4 W production at the Tevatron and the LHC

The total cross sections for W and Z hadroproduction are known to NNLO [12] and the input electroweak parameters are known to high accuracy. The main uncertainty in predicting the size of the cross sections comes from the parton distributions and, to a lesser extent, from the value of α_S. The MRST predictions are [2]

$$B(W \to \ell\nu)\sigma_W = \begin{cases} 2.45 \text{ nb } (\pm 3\%) \text{ at the Tevatron} \\ 20.3 \text{ nb } (\pm 5\%) \text{ at the LHC.} \end{cases} \qquad (4)$$

The errors arise from considering the spread of predictions arising from taking the full range of parton sets, including a ± 0.005 uncertainty in $\alpha_S(M_Z^2)$. In principle the predictions allow a check on the luminosity of the collider. Of course, there are uncertainties in the parton distributions due to the normalization errors on present data. To make an attempt to allow for this, the range of parton sets used to specify the errors in (4) includes sets obtained from global fits with the HERA deep inelastic data renormalized up by 2.5%, and then down by 2.5%.

2.5   Comparison of MRST and CTEQ partons

The predictions for σ_W serve another purpose. The long evolution length to $Q^2 = M_W^2$, together with the expected precision of the predictions, can expose small defects in the evolution codes used for the global analyses. An unexpectedly large difference was found between the cross sections predicted using MRST98 [1] and CTEQ5 [3] partons. The value for σ_W obtained from CTEQ5 was 10% greater than from MRST98 partons at the LHC energy. It means that the evolution codes should be carefully checked. Recall that neither CTEQ, nor MRS, agreed completely with the "standard HERA workshop code [13]" for a test evolution from $Q^2 = 4$ to 100 GeV^2. In the interval $10^{-4} \lesssim x \lesssim 10^{-1}$, the CTEQ sea was some 4% high, and the MRS sea was 1% low and the gluon about 1.5% low. MRST checked their code and found that it had to be corrected by a factor $C_F \to C_A$ in one of the terms appearing in the NLO P_{gg} splitting function. The effect is to make the gluon, and hence the sea, evolve a little more rapidly at small x. With the correction the MRST99 code [2] is found to agree precisely with the "standard" evolution code and, moreover, is consistent with exact momentum conservation as a function of Q^2, as it should be. However the discrepancy for σ_W remains: the CTEQ5 HQ prediction is about 6% (10%) higher at Tevatron (LHC) energies [2]. Differences between MRST and CTEQ partons can arise from the choice of the data sets to be included in the global fit, the Q^2, W^2 cuts used, the different renormalization of some data sets, the different treatment of heavy flavours, etc. However none of these differences appears sufficient to explain the discrepancy in the predictions for σ_W, which sample partons at $Q^2 = M_W^2$ with $x \simeq M_W/\sqrt{s}$.

2.6   Outstanding problems of global analyses

There is a need to determine realistic errors on the parton distributions extracted in a global analysis. This is far from easy. In particular, among other things, we need to quantify the uncertainties due to (i) heavy target and other nuclear corrections, (ii) the possible effects of higher twist contributions, (iii) the choice of the form of parametrization of the initial distributions, and (iv) the possible effects of k_T smearing on the analysis of exclusive processes.

A second, and related, problem is the need for a next-to-next-leading order (NNLO) analysis. The P_{gg} splitting function is not yet known to this order. Also, as data at smaller and smaller x are included, there is a need to resum $\log(1/x)$ effects, and to allow for parton shadowing corrections.

3   Skewed (or off-diagonal) parton distributions

Data of higher and higher precision are becoming available for processes which are described by off-diagonal (or skewed) parton distributions in the small x domain. A relevant example is diffractive vector meson production at HERA, $\gamma^* p \to V p$ with $V = \rho, J/\psi$ or Υ (see, for example, [14]). At high $\gamma^* p$ c.m. energy, W, the cross section is dominated by the two-gluon exchange amplitude

$$\frac{d\sigma}{dt}(\gamma^* p \to V p)\bigg|_{t=0} = \ldots [x_2 g(x_1, x_2, \mu^2)]^2 \tag{5}$$

where g is the off-diagonal ($x_1 \neq x_2$) gluon distribution with

$$x_1 = (Q^2 + M_{q\bar{q}}^2)/W^2,$$
$$x_2 = (M_{q\bar{q}}^2 - M_V^2)/W^2 \ll x_1, \tag{6}$$

see [15]. The process is shown schematically in Fig. 1. $M_{q\bar{q}}$ is the mass of the $q\bar{q}$ system produced by a photon of virtuality Q^2. The relevant scale is $\mu^2 = z(1-z)Q^2 + k_T^2 + m_q^2$ where $z, 1-z$ and $\pm k_T$ specify the momenta of the q and $\bar{q}$ which form the vector meson of mass M_V. The quadratic dependence of g in (5) shows that these data may offer a sensitive constraint on the gluon. Indeed we will see that the off-diagonal distributions are fixed by the conventional (diagonal) parton distributions of Section 2, so that the data can, in principle, be included in a global parton analysis.

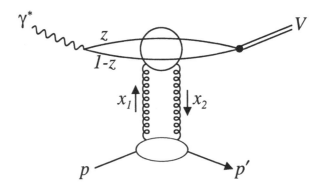

Fig. 1. Schematic diagram for diffractive vector meson production at HERA, $\gamma^* p \to Vp$. The longitudinal fractions x_1 and x_2 of the in-going and out-going proton momentum are given by (6). z and $1-z$ are the longitudinal fractions of the q and $\bar{q}$ momenta. There are four possible couplings of the two gluons to the q and $\bar{q}$, represented by the upper circle.

3.1   Ji's symmetrized distributions

It is convenient to use the skewed parton distributions $H(x,\xi) \equiv H(x,\xi,t,\mu^2)$, with support $-1 \leq x \leq 1$, introduced by Ji [16], with the minor difference that the gluon $H_g = x\bar{H}_g^{\mathrm{Ji}}$ [17]. The distributions depend on the momentum fractions

$$x_{1,2} = x \pm \xi \tag{7}$$

carried by the emitted and absorbed partons at each scale μ^2, and on the momentum transfer variable $t = (p-p')^2$. The variables do not change as we evolve the distributions up in the scale μ^2. In the limit $\xi \to 0$ they reduce to the conventional parton distributions

$$H_q(x,0) = \begin{cases} q(x) & \text{for } x > 0 \\ -\bar{q}(-x) & \text{for } x < 0 \end{cases}$$

$$H_g(x,0) = xg(x), \tag{8}$$

and satisfy DGLAP evolution. In the limit $\xi \to 1$ they obey ERBL evolution [18,19]. If we consider H_q at arbitrary values of ξ, then for $x > \xi$ and $x < -\xi$ we have DGLAP-like evolution for quarks and antiquarks respectively, while for $-\xi < x < \xi$ we have ERBL-like evolution for the emitted $q\bar{q}$ pair.

On account of the $x_1 \leftrightarrow x_2$ symmetry the distributions H_q and H_g are symmetric in ξ

$$H_i(x,\xi) = H_i(x,-\xi). \tag{9}$$

We also have symmetry relations in terms of the x variable

$$
\begin{aligned}
H_q^{NS}(x,\xi) &= H_q^{NS}(-x,\xi) \\
H_q^S(x,\xi) &= -H_q^S(-x,\xi) \\
H_g(x,\xi) &= H_g(-x,\xi)
\end{aligned}
\tag{10}
$$

where the superscripts S and NS denote singlet and non-singlet quarks respectively.

3.2 $H(x,\xi)$ in terms of conformal moments

In order to relate the skewed distributions to the conventional parton densities it is convenient to work in terms of the so-called conformal moments[1] of the skewed distributions

$$
O_N(\xi,\mu^2) = \int_{-1}^{1} dx \, R_N(x_1,x_2) H(x,\xi,\mu^2).
\tag{11}
$$

At leading order, these moments are not mixed by evolution. In fact

$$
O_N(\xi,\mu^2) = O_N(\xi,\mu_0^2) \left(\frac{\mu^2}{\mu_0^2}\right)^{\gamma_N},
\tag{12}
$$

where γ_N are the same anomalous dimensions as for diagonal partons. The R_N are known polynomials of degree N

$$
R_N = \sum_{k=0}^{N} \binom{N}{k} \binom{N+2p}{k+p} x_1^k x_2^{N-k}
\tag{13}
$$

with $p = 1, 2$ for quarks and gluons respectively. The O_N reduce to the usual moments in the limit $\xi \to 0$. For example for quarks

$$
O_N \to M_N = \int_0^1 x^N q(x) dx,
\tag{14}
$$

up to a normalizing factor $R_N(1,1)$.

The crucial step is to find the inverse relation to (11). That is to reconstruct $H(x,\xi)$ from a knowledge of its conformal moments. The result, due to Shuvaev [23], is

$$
H(x,\xi) = \int_{-1}^{1} dx' K(x,\xi;x') f(x')
\tag{15}
$$

---

[1] Conformal moments were introduced in [18,20] for $\xi = 1$, and in [21] for $\xi \neq 1$; see also [22].

where the kernel K is a known integral [23,24] and f is the Mellin transform

$$f(x') = \int \frac{dN}{2\pi i}(x')^{-N}O_N(\xi)/R_N(1,1). \tag{16}$$

The function f reduces to the conventional diagonal parton distribution for $\xi^2 \ll 1$. This follows since [16]

$$O_N(\xi) = \sum_{k=0}^{[(N+1)/2]} O_{Nk}\xi^{2k}$$

$$\simeq O_{N0} = O_N(0) = M_N R_N(1,1) \tag{17}$$

for small ξ^2. So the skewed distribution H is completely determined in terms of the diagonal distribution f via (15). This identification can be made at any scale, and so there is no need to solve the off-diagonal evolution equations to obtain the μ^2 dependence of $H(x,\xi,\mu^2)$. That is the normal DGLAP evolution of $f(x,\mu^2)$ is sufficient to generate the x,ξ and μ^2 dependence of $H(x,\xi,\mu^2)$ at small x and ξ.

3.3   A good small x, ξ approximation

We can simplify (15) further if we assume that the diagonal partons have the forms

$$xq(x) = N_q x^{-\lambda_q}, \quad xg(x) = N_g x^{-\lambda_g}, \tag{18}$$

which should be valid for very small x. Then the x' integration in (15) can be performed analytically, giving

$$H_i(x,\xi) = \xi^{-\lambda_i-p}F_i\left(\frac{x}{\xi}\right)$$

with $p = 1, 0$ for $i = q, g$ respectively, where the F_i are known in terms of N_i and λ_i. A full set of results can be found in [24] for the off-diagonal/diagonal ratios,

$$R_i(x,\xi) = H_i(x,\xi)/H_i(x+\xi,0), \tag{19}$$

for which the only free parameter is λ_i. In [24] the ratios $R_q^{NS,S}$ and R_g are plotted as functions of x/ξ for different values of λ_i. The scale dependence of the off-diagonal distributions, $H_i(x,\xi)$ of (3.3), and hence of the R_i, is hidden in the μ^2 dependence of the λ_i. Both λ_g and λ_q increase with increasing μ^2, see Fig. 13 of [1].

3.4   Application to $\gamma^* p \to V p$

The above results are exactly what we need to describe vector meson production at HERA at large $\gamma^* p$ c.m. energies W. From (6) we see that x_1 is small and $x_2 \approx$

0, that is $x \approx \xi$. The gluon is the relevant parton and the skewed distribution is larger than the conventional gluon by a factor

$$R_g(x = \xi) = \frac{H_g(\xi, \xi)}{H_g(2\xi, 0)} = \frac{2^{2\lambda_g + 3}}{\sqrt{\pi}} \frac{\Gamma\left(\lambda_g + \frac{5}{2}\right)}{\Gamma(\lambda_g + 4)}. \tag{20}$$

The cross section formula (5) may then be expressed in terms of the conventional gluon distribution g,

$$\left. \frac{d\sigma}{dt}(\gamma^* p \to V p) \right|_{t=0} = \ldots \left[R_g x_1 g(x_1, \mu^2) \right]^2, \tag{21}$$

where all the off-diagonal effects are contained in the known (enhancement) factor R_g^2. Of course to calculate the cross section properly we must use the unintegrated gluon distribution and integrate over the transverse momenta of the exchanged gluons and of the q and $\bar{q}$ forming the vector meson, see [14].

To obtain the scale dependence of R_g, we first obtain the μ^2 dependence of λ_g of (18) from the behaviour of the gluon found in the global parton analyses. For example, the MRST partons [1] have $\lambda_g = 0.205$ and 0.38 at $\mu^2 = 4$ and 100 GeV^2 respectively. The appropriate scale for the diffractive process $\gamma^*(Q^2)p \to V(q\bar{q})p$ is $\mu^2 \simeq m_q^2 + Q^2/4$. In this way, for diffractive J/ψ and Υ photoproduction at HERA we find from (20) that the off-diagonal enhancement, R_g^2, is $(1.15)^2$ and $(1.32)^2$ respectively. However, for Υ photoproduction, x is not sufficiently small (~ 0.01) and we have to improve the assumption made in (18). If we take $xg \sim x^{-\lambda_g}(1-x)^6$ and perform the x' integration in (15) numerically, then we find an enhancement of $(1.41)^2$ for Υ photoproduction [25]. Moreover note that the skewedness, and hence the enhancement, increase with Q^2. This is particularly noticeable in $\gamma^* p \to \rho p$, see [14].

3.5   Validity of the skewed → diagonal parton connection

Before we can be sure that the skewed distribution $H(x, \xi)$, at any scale, is fully determined at small x, ξ by knowledge of the diagonal parton distribution, at the same scale, we must consider the points below. First, we have checked that the analytic continuation of the conformal moments O_N in N is allowed [24]. A second consideration is that, from a formal point of view, we may add to the off-diagonal distribution any function which exists only in the ERBL-like region, $|x| < \xi$. In [24] we show such a contribution is negligible $O(\xi^2)$ at small ξ. So far our distributions allow the calculation of the imaginary part of the amplitude for the process. At small x and ξ it turns out that the real part may be calculated easily using a dispersion relation in the c.m. energy squared, W^2, and that the amplitude

$$A = i \mathrm{Im} A \frac{1 + e^{-i\pi\lambda}}{1 + \cos \pi \lambda}, \tag{22}$$

where $A \propto (W^2)^\lambda$. Finally we note that our result remains valid at NLO, since there is no conformal mixing for $\xi^2 \ll 1$.

3.6   Conclusions

We conclude that, at small x, ξ, the skewed distributions $H(x, \xi; \mu^2)$ are completely known in terms of conventional partons. Thus data for processes which are described by such distributions can, in principle, be included in a conventional global analysis to better constrain the low x behaviour of the partons. In particular, data for diffractive vector meson production, $\gamma^* p \to V p$, which depend on the *square* of the gluon distribution, have the potential to put a tight constraint on the gluon at small x. Before this can be done it is necessary to calculate the NLO contributions. Finally here we have been concerned only with the skewed distributions in the small x domain. Skewed distributions, in general, have become an area of intense activity with a large literature (see, for example, [16,26,27] and references therein).

Acknowledgements

I thank Bernd Kniehl, Günter Grindhammer and Gustav Kramer for their efficient organisation of the Ringberg Workshop. I also thank Krzysztof Golec-Biernat, Dick Roberts, Misha Ryskin, Andrei Shuvaev, James Stirling, Thomas Teubner and Robert Thorne for enjoyable collaborations on the subject of this review.

References

1. A.D. Martin, R.G. Roberts, W.J. Stirling and R.S. Thorne, Eur. Phys. J. **C4** (1998) 463.
2. A.D. Martin, R.G. Roberts, W.J. Stirling and R.S. Thorne, hep–ph/9907231.
3. CTEQ collaboration: H.L. Lai et al., hep–ph/9903282.
4. F. Zomer, these proceedings.
5. U.K. Yang and A. Bodek, Phys. Rev. Lett. **82** (1999) 2467.
6. W. Melnikchouk and A.W. Thomas, Phys. Lett. **B377** (1996) 11.
7. B. Harris, these proceedings.
8. R.S. Thorne and R.G. Roberts, Phys. Rev. **D57** (1998) 6781; Phys. Lett. **B421** (1998) 303.
9. M.A.G. Aivazis, J.C. Collins, F.I. Olness and W.-K. Tung, Phys. Rev. **D50** (1994) 3102.
10. E866 collaboration: E.A. Hawker et al., Phys. Rev. Lett. **80** (1998) 3715; D. Isenhower, Proc. of DIS99, Zeuthen.
11. V. Monaco, these proceedings.
12. W.L. van Neerven et al., Nucl. Phys. **B345** (1990) 331; **B359** (1991) 343; **B382** (1992) 11.
13. J. Blümlein et al., Proc. of 1995/6 HERA Physics Workshop, eds. G. Ingelman et al., Vol. 1, p.23, hep–ph/9609400.
14. T. Teubner, these proceedings.
15. A.D. Martin and M.G. Ryskin, Phys. Rev. **D57** (1998) 6692.
16. X. Ji, Phys. Rev. Lett. **78** (1997) 610; Phys. Rev. **D55** (1997) 7114; J.Phys. **G24** (1998) 1181.

17. K. Golec-Biernat and A.D. Martin, Phys. Rev. **D59** (1999) 014029.
18. A.V. Efremov and A.V. Radyushkin, Phys. Lett. **B94** (1980) 245.
19. G.P. Lepage and S.J. Brodsky, Phys. Rev. **D22** (1980) 2157.
20. M. Chase, Nucl.Phys. **B174** (1980) 109.
21. Th. Ohrndorf, Nucl.Phys. **B198** (1982) 26.
22. A.P. Bukhvostov, G.V. Frolov, L.N. Lipatov and E.A. Kuraev, Nucl.Phys. **B258** (1985) 601.
23. A. Shuvaev, hep-ph/9902318.
24. A. Shuvaev, K. Golec-Biernat, A.D. Martin and M.G. Ryskin, Phys. Rev. **D60** (1999) 014015.
25. A.D. Martin, M.G. Ryskin and T. Teubner, Phys.Lett. **B454** (1999) 339.
26. A.V. Radyushkin, Phys. Rev. **D56** (1997) 5524.
27. P.A.M. Guichon and M. Vanderhaeghen, Prog. Part. Nucl. Phys. **41** (1998) 125.

QCD Evolution of Structure Functions at Small x

Johannes Blümlein

DESY Zeuthen, Platanenallee 6, D–15735 ZEUTHEN, Germany

Abstract. The status of the resummation of small x contributions to the unpolarized and polarized deep inelastic structure functions is reviewed.

1 Introduction

The measurement of the nucleon structure functions in deep inelastic scattering provides important tests of the predictions of Quantum Chromodynamics (QCD) on the short–distance structure of nucleons. The experiments at the ep–collider HERA allowed to extend the kinematic region to very small values of $x \sim 10^{-4}$ at photon virtualities of $Q^2 \geq 10 \text{GeV}^2$ measuring the structure function $F_2(x, Q^2)$ at an accuracy of $O(1\%)$. These precise measurements allow dedicated tests of QCD. Also the polarized deep inelastic experiments approach smaller values of x with a higher accuracy. To obtain a description in this kinematic range potentially large contributions to the evolution kernels were studied during the last two decades and the resummations of 'leading term' contributions were performed. Here mainly two directions were followed.

In one approach [1] a non–linear resummation of fan–diagrams of single ladder cascades is performed in the double logarithmic approximation [2][1]. Corrections of this type may ultimately become important at very small values of x to restore unitarity. Numerical studies of this equation were performed in Refs. [4]. One important assumption in the solution of this equation was that the non-perturbative input distributions for the N–ladder terms are given by the Nth power of the single gluon distribution at some starting scale. This would imply a strong constraint on the hierarchy of higher twist distributions. Later it was found [5] that the approximation [1] has to be supplemented by further color correlations even in the double logarithmic approximation, which cannot be cast into a non–linear equation anymore. As implied by the operator product expansion, the contributions due to different twist renormalize independently. The corresponding input distributions are likely to be unrelated between the different twists. Still saturation effects of the structure functions at very small x may be caused due to higher twist contributions. However, the detailed dynamics is yet unknown.

---

[1] This approximation has to be considered as qualitative and leads often to an overestimate of the scaling violations, cf. [3].

Ladder equations also form the basis of other approaches. In a physical gauge the emission of gluons along a single ladder–cascade describes[2] in *leading order* (LO) the evolution of a parton density as predicted by the renormalization group equation if the emissions along the ladder are strongly ordered in the transverse momentum $k_{\perp,1} \ll ...k_{\perp,i} \ll k_{\perp,i+1}...$ If these emissions are evaluated in the approximation $x_1 \gg ...x_i \gg x_{i+1}...$ instead, using effective vertices, one obtains the BFKL–resummation [6] in LO. This particular aspect led sometimes to the impression that these two approximations were of competing nature. As we will show below this is, however, not the case as far as the description of the scaling violations of structure functions are concerned. One may study this process under a more general point of view and consider angularly ordered emissions covering both the above cases [7], which allows for interesting applications through Monte Carlo studies. Whereas this unified treatment is possible at LO, higher order corrections cannot be cast into this form in general. The renormalization group equation for the mass singularities, on the other hand, allows to perform consistent higher order calculations beyond these approximations accounting for the resummation of the small x contributions in the anomalous dimensions and coefficient functions.

In the second main approach these resummations are studied. Resummations were performed in leading order for the unpolarized singlet case [6,8], the non–singlet structure functions [9,10], and the polarized singlet distributions [11]. Applications were studied in the case of QED for the flavor non–singlet contributions to radiative corrections [12,13]. The quarkonic next–to–leading order (NLO) contributions in the unpolarized singlet case were calculated in [14]. Recently also the NLO resummed gluon anomalous dimension [15,16] in the DIS-Q_0 scheme [17] was obtained. If the evolution kernels are written in terms of a series in $\alpha_s/(N - N_s)$, where N_s denotes the position of the leading pole, the individual terms are large and require resummation.

One of the central questions for the understanding of the deep inelastic structure functions at small x is therefore to analyse the impact and rôle of these small x resummations and their potential corrections in even higher order. These terms have to be viewed in comparison with the known fixed order results used in the current analyses of the scaling violations of the twist–2 contributions to the structure functions.

In the present paper we review the status of the latter resummations and their impact on the scaling violations of deep inelastic structure functions.

2   The Evolution Equations

The twist-2 contributions to the structure functions in inclusive deep-inelastic scattering can be described in terms of the QCD-improved parton model. Their scaling violations are governed by renormalization group equations which can be formulated to all orders in the strong coupling constant. All small x resummations are based on perturbative QCD. As in the fixed order calculations one has

---

[2] The virtual contributions have to be added.

to factorize the collinear or mass singularities, which are absorbed into the non–
perturbative input distributions. The soft- and virtual singularities cancel order
by order according to the Bloch–Nordsiek theorem. A second renormalization
group equation describes the scale dependence of the strong coupling constant
$\alpha_s(\mu^2)$. The perturbative all-order small x resummations may turn out to yield
important contributions to the *scaling violation* of the deep inelastic structure
functions. Predictions on the shape of the parton densities at small x are, however, *beyond* a perturbative treatment, even in resummed form, since generally
low scales are involved and partonic approaches have to fail.

The small x resummations can be tested with respect to their prediction
on the scaling violations of deep inelastic structure functions as $F_2(x, Q^2)$ and
$F_L(x, Q^2)$. The evolution equations for the parton densities $f_i(x, \mu^2)$ are given
by

$$\frac{\partial}{\partial \log(\mu^2)} f_i(x, \mu^2) = P_i^j(x, a_s) \otimes f_j(x, \mu^2) . \tag{1}$$

Here $\otimes$ denotes the Mellin convolution. The splitting functions $P_{ij}(x, a_s)$ contain
besides the completely known LO and NLO contributions the LO and NLO small
x resummed terms to all orders in $a_s = \alpha_s/(4\pi)$,

$$P_{ij}(x, a_s) = a_s P_{ij}^{(0)}(x) + a_s^2 P_{ij}^{(1)}(x) + \sum_{k=2}^{\infty} a_s^{k+1} \widehat{P}_{ij, x \to 0}^{(k)}(x) + \sum_{k=2}^{\infty} a_s^{k+2} \widehat{\widehat{P}}_{ij, x \to 0}^{(k)}(x) . \tag{2}$$

Similarly, the coefficient functions take the form

$$c_i\left(x, \frac{Q^2}{\mu^2}\right) = \delta_{iq} \delta(1 - x) a_s c_i^{(0)}(x) + a_s^2 c_i^{(1)}(x)$$
$$+ \sum_{k=2}^{\infty} a_s^{k+1} \widehat{c}_{i, x \to 0}^{(k)}(x) + \sum_{k=2}^{\infty} a_s^{k+2} \widehat{\widehat{c}}_{i, x \to 0}^{(k)}(x) . \tag{3}$$

In this way the effect of the small x resummations is consistently included. As
these contributions do not a priori account for Fermion number and energy–
momentum conservation these conditions have to be imposed for the contributions beyond $O(a_s^2)$. The structure functions $F_A(x, Q^2)$ are finally obtained as

$$F_A(x, Q^2) = c_{q,A}\left(x, \frac{Q^2}{\mu^2}\right) \otimes f_q\left(x, \frac{\mu^2}{M^2}\right) + c_{g,A}\left(x, \frac{Q^2}{\mu^2}\right) \otimes f_g\left(x, \frac{\mu^2}{M^2}\right) . \tag{4}$$

The factorization scale dependence (μ^2) cancels order by order.

3 Small x Resummation of the Anomalous Dimensions

All resummations studied below are based on scale–invariant equations in leading
order. If one considers the renormalization group equation for an operator matrix

element E_k^n

$$\left[\mu \frac{\partial}{\partial \mu} + \beta \frac{\partial}{\partial g} + \gamma_m m \frac{\partial}{\partial m} + \gamma_{O_k} - n \gamma_\Phi \right] E_k^n = 0 \tag{5}$$

scale invariant solutions are obtained in the massless case ($m = 0$) and if the β-function is set to zero :

$$E_k^n(\mu^2) = E_k^n(\mu_0^2) \left(\frac{\mu^2}{\mu_0^2} \right)^{\frac{1}{2}\left(\gamma_{O_k} - n \gamma_\Phi \right)} . \tag{6}$$

Within this approach the coupling constant a_s is fixed. The scale invariant part of the anomalous dimension has the representation

$$\gamma_{O_k} - n \gamma_\Phi = \sum_{l=1}^{\infty} \gamma_O^{(l)} a_s^l \tag{7}$$

and exponentiates to all orders. The representation (6) applies also for higher order resummations under the above requirements. In this way one may derive in the different subsequent resummations the LO small x resummed anomalous dimensions. In higher than LO scale breaking effects emerge in QCD. Therefore a thorough treatment along these lines is no longer possible. Still one may try to identify those contributions of the anomalous dimension which are scale invariant applying a diagonalization as in (6).

4   Less Singular Terms

For most of the applications only the resummation of the leading singular terms is known. The contributions which are less singular by one or more powers in N may yield substantial contributions. This has been known for long [18,19], cf. also [20], and can easily be seen in the case of $F_L(x, Q^2)$ in $O(\alpha_s)$ as an example. If one disregards the second factor in the leading order coefficient function $c_g^{(0)} = C_g^{(0)} x^2(1 - x)$ in view of a small x approximation the value of F_L may be overestimated by a factor of four [18].

To get an estimate of the effect of the terms suppressed by one order or more orders in the Mellin moment N one may study some models. The possible size of these terms may be inferred expanding the LO and NLO anomalous dimensions and coefficient functions into series in $1/N$ comparing the expansion coefficients. Estimates of this kind were performed in [3,10,12,13,21–25]. Possible ansätze for the next order terms are

$$\begin{aligned} \Gamma(N, a_s) &\to \Gamma(N, a_s) - \Gamma(1, a_s) \\ \Gamma(N, a_s) &\to \Gamma(N, a_s)(1 - N) \\ \Gamma(N, a_s) &\to \Gamma(N, a_s)(1 - N)^2 \\ \Gamma(N, a_s) &\to \Gamma(N, a_s)(1 - 2N + N^3) . \end{aligned} \tag{8}$$

If one formally expands the LO and NLO anomalous dimensions in the above manner one finds [3], irrespectively of the factorization scheme, that at least four expansion terms are needed to represent the exact result on the 5% level, cf. figure 1. If one compares the respective NLO resummed coefficients with the LO resummed ones in the cases they were calculated even larger effects than indicated by the above estimate are found (see below).

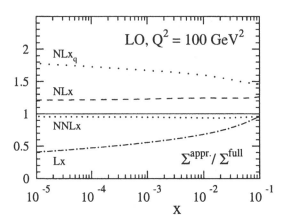

Fig. 1. Different approximation steps in the $1/(N-1)$ expansion of the complete LO unpolarized singlet distribution over four orders (Lx, NLx$_q$, NLx, NNLx), cf. Ref. [3].

5   Non–Singlet Structure Functions

The most singular contributions to the Mellin transforms of the structure–function evolution kernels $K^{\pm}(x, a)$ at all orders in a can be obtained from the positive and negative signature amplitudes $f_0^{\pm}(N, a)$ studied in [9] for QCD via

$$\mathcal{M}\left[K_{x\to 0}^{\pm}(a)\right](N) \equiv \int_0^1 dx\, x^{N-1} K_{x\to 0}^{\pm}(x, a) \equiv -\frac{1}{2}\, \Gamma_{x\to 0}^{\pm}(N, a) = \frac{1}{8\pi^2} f_0^{\pm}(N, a)\, . \tag{9}$$

These amplitudes are subject to the quadratic equations:

$$f_0^+(N, a) = 16\pi^2 a_0 \frac{a}{N} + \frac{1}{8\pi^2}\frac{1}{N}\left[f_0^+(N, a)\right]^2\, , \tag{10}$$

$$f_0^-(N, a) = 16\pi^2 a_0 \frac{a}{N} + 8b_0^- \frac{a}{N^2} f_V^+(N, a) + \frac{1}{8\pi^2}\frac{1}{N}\left[f_0^-(N, a)\right]^2\, . \tag{11}$$

Here $f_V^+(N, a)$ is obtained as the solution of the Riccati differential equation

$$f_V^+(N, a) = 16\pi^2 a_V \frac{a}{N} + 2b_V \frac{a}{N}\frac{d}{dN} f_V^+(N, a) + \frac{1}{8\pi^2}\frac{1}{N}\left[f_V^+(N, a)\right]^2\, . \tag{12}$$

The coefficients a_i and b_i in the above relations read for the case of QED, cf. sect. 7,

$$a_0 = 1, \quad b_0^- = 1, \quad a_V = 1, \quad b_V = 0, \tag{13}$$

and for QCD [9]

$$a_0 = C_F, \quad b_0^- = C_F, \quad a_V = -\frac{1}{2N_c}, \quad b_V = C_A, \tag{14}$$

with $C_F = 4/3$ and $C_A = N_c = 3$. In QED Eq. (12) further simplifies to an algebraic equation with the same coefficients as (10). The solutions of (10) and (11) were derived in [9] for the QCD case[3]. They are given by

$$\Gamma_{x \to 0}^+(N, a) = -N \left\{ 1 - \sqrt{1 - \frac{8aC_F}{N^2}} \right\} \tag{15}$$

$$\Gamma_{x \to 0}^-(N, a) = -N \left\{ 1 - \sqrt{1 - \frac{8aC_F}{N^2} \left[1 - \frac{8aN_c}{N} \frac{d}{dN} \ln \left(e^{z^2/4} D_{-1/[2N_c^2]}(z) \right) \right]} \right\}$$

where $z = N/\sqrt{2N_c a}$, and $D_p(z)$ denotes the function of the parabolic cylinder.

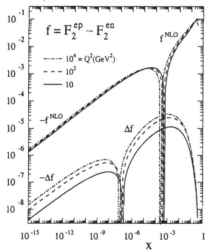

Fig. 2. The small-x Q^2-evolution of the unpolarized non–singlet structure function combination $F_2^{ep} - F_2^{en}$ in NLO and the absolute corrections to these results due to the resummed kernel. The initial distributions were chosen at $Q_0^2 = 4 \, \text{GeV}^2$, cf. Refs. [10,12].

Numerical results on the impact of the small x resummations were obtained in Refs. [10,12]. The new terms contribute at $O(a_s^3)$ and higher. If compared to the fixed order contributions in NLO the effect is of $O(1\%)$ or less, which is shown for the scaling violations of $F_2^{ep} - F_2^{en}$ in Fig. 2. Other examples as for xF_3

---

[3] Note a few misprints in Eq. (4.7) of ref. [9].

and the ±–evolutions for polarized non–singlet structure functions show a rather similar behaviour. Comparable results were obtained in Ref. [26]. The effect is expected to be rather small due to the typical shape of the input distributions in the non–singlet case. The size of the (small) correction does further vary significantly in dependence of the inclusion of less singular terms, cf. sect. 4, or if conservation laws are imposed. Large effects as anticipated in Refs. [27,28] are not confirmed.

6   Polarized Singlet Structure Functions

The LO small x evolution kernels in the case of the polarized singlet evolution were derived in [11]. The resummed splitting function is given by

$$P(x, a_s) \equiv \sum_{l=0}^{\infty} P_{x \to 0}^{(l)} a_s^{l+1} \log^{2l} x = \frac{1}{8\pi^2} \mathcal{M}^{-1}[F_0(N, a_s)](x). \tag{16}$$

The matrix valued function $F_0(N, a_s)$ is obtained as the solution of

$$F_0(N, a_s) = 16\pi^2 \frac{a_s}{N} M_0 - \frac{8a_s}{N^2} F_8(N, a_s) G_0 + \frac{1}{8\pi^2} \frac{1}{N} F_0^2(N, a_s) \tag{17}$$

with

$$F_8(N, a_s) = 16\pi^2 \frac{a_s}{N} M_8 + \frac{2a_s}{N} C_A \frac{d}{dN} F_8(N, a_s) + \frac{1}{8\pi^2} \frac{1}{N} F_8^2(N, a_s), \tag{18}$$

where

$$M_0 = \begin{pmatrix} C_F & -2T_R N_f \\ 2C_F & 4C_A \end{pmatrix}, \quad M_8 = \begin{pmatrix} C_F - C_A/2 & -T_R N_f \\ C_A & 2C_A \end{pmatrix}, \quad G_0 = \begin{pmatrix} C_F & 0 \\ 0 & C_A \end{pmatrix}. \tag{19}$$

Eq. (16) obeys [21]

$$P_{qg}^{(l)}/(T_R N_f) = -P_{gq}^{(l)}/C_F \tag{20}$$

to all orders, where $T_R = 1/2$ and N_f denotes the number of flavors. The leading contributions of the fixed order results in LO and NLO ($\overline{\text{MS}}$) are correctly described. In the supersymmetric limit $C_A = C_F = N_f = 1, T_R = 1/2$ the relations

$$P_{qq}^{(l)} + P_{gq}^{(l)} = P_{qg}^{(l)} + P_{gg}^{(l)} \tag{21}$$

are obeyed for all l and Eq. (16) can be given in a simple analytic form [21].

The impact of the resummation (16) on the evolution of the polarized singlet and gluon density and the structure function $g_1(x, Q^2)$ have been studied in Ref. [21]. As shown in Fig. 3 the corrections are much larger than the

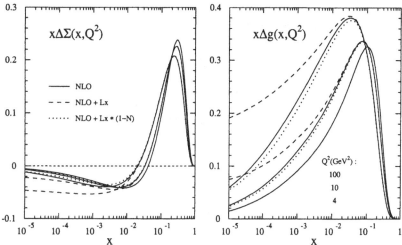

Fig. 3. The Q^2 evolution of the polarized quark singlet and gluon momentum distributions evolving from $Q_0^2 = 4\text{GeV}^2$, Ref. [21].

$O((a_s \ln^2 x)^l)$ corrections in the non–singlet case. Taking into account less singular terms of the type

$$P_{ij}^{(l>1)} \rightarrow P_{ij}^{(l>1)} \cdot (1 - N) , \tag{22}$$

as suggested by the analytic structure of the fixed order LO and NLO anomalous dimensions, this enhancement reduces, however, again to the value of the fixed order evolution in NLO (dotted line in Fig. 3).

7   QED Corrections

The non–singlet small x resummation was applied to resum the $O((\alpha \log^2 x)^l)$ terms in the QED corrections to deep inelastic scattering in Ref. [13]. These corrections are negative and amount to $O(10\%)$ in the high y range for $x = 10^{-4}...10^{-2}$, see Fig. 4.

They diminish the $O((\alpha \log(Q^2/m_e^2))^l)$ corrections which are very large in this domain. The first non–trivial contribution of $O(\alpha^2 \log^2 x)$ is in agreement with the result found in [29]. From the latter calculation also the next less singular term of $O(\alpha^2 \log(x))$ can be derived. Up to this term the evolution kernel reads $(a = \alpha/(4\pi))$

$$\mathcal{M}[P_{z \to 0}](N, a) = \frac{2a}{N} - 12\frac{a^2}{N^3}\left(1 - \frac{2}{9}N\right) + ... \tag{23}$$

If compared to the case of QCD this less–singular term is stronger suppressed.

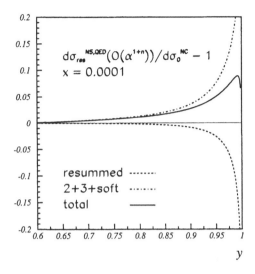

Fig. 4. 2nd and higher order QED initial state radiative corrections to deep inelastic ep scattering. Dashed line: small x resummed contribution, dash-dotted line: LO contributions up to $O(\alpha^3)$ and soft photon exponentiation, full line: resulting correction, Ref. [13].

8   Unpolarized Singlet Distributions

The LO resummation for the evolution kernel of the unpolarized singlet distributions was derived in [6]. Jaroszewicz [8] showed that the eigenvalue

$$(N-1) = \frac{\alpha_s N_c}{\pi}\chi_0(\gamma_L) \equiv \frac{\alpha_s N_c}{\pi}\left[2\psi(1) - \psi(\gamma_L) - \psi(1-\gamma_L)\right] \qquad (24)$$

represents the LO resummed gluon-gluon anomalous dimension $\gamma_L = \gamma_{gg}^{(0)}(N, a_s)$. The resummed LO gluon-quark anomalous dimension is given by $\gamma_{gq}^{(0)}(N, a_s) = (C_F/C_A)\gamma_L$ and the quarkonic terms do not contribute in $O((a_s/(N-1))^l$. Eq. (24) can be solved iteratively demanding $\gamma_L(N, a_s) \to \overline{\alpha}_s/(N-1)$ as $|N| \to \infty$ for $N \in \mathbf{C}$, which selects the physical branch of the resummed anomalous dimension,

$$\gamma_L \equiv \gamma_{gg,0}(N, \alpha_s) = \frac{\overline{\alpha}_s}{N-1}\left\{1 + 2\sum_{l=1}^{\infty}\zeta_{2l+1}\gamma_{gg,0}^{2l+1}(N, a_s)\right\}. \qquad (25)$$

Here we rewrite $\overline{\alpha}_s = N_c\alpha_s/\pi$. γ_L has the serial representation

$$\gamma_{gg,0}(N, \alpha_s) = \frac{\overline{\alpha}_s}{N-1} + 2\zeta_3\left(\frac{\overline{\alpha}_s}{N-1}\right)^4 + 2\zeta_5\left(\frac{\overline{\alpha}_s}{N-1}\right)^6 + 12\zeta_3^2\left(\frac{\overline{\alpha}_s}{N-1}\right)^7 + \cdots$$

$$(26)$$

Under the above conditions one may calculate $\gamma_L(N, a_s)$ in the whole complex plane. It is a bounded function of $\rho = (N-1)/\overline{\alpha}_s$, the singularities of which are branch points at [30,22]

$$\rho_1 = 4\log 2, \qquad \rho_{2,3} = -1.41048 \pm 1.97212\, i, \qquad (27)$$

cf. [30,3] for detailed representations. The LO BFKL anomalous dimension possesses **no poles**. Since the known NLO resummed anomalous dimensions are functions of $\gamma_L(N, a_s)$ which introduce no further singularities the contour integral around the singularities of the problem has to cover the three BFKL branch points, the singularities of the input distributions along the real axis to the left of 1, and the remaining singularities of the fixed order anomalous dimensions at the non–positive integers [30,3]. Note, that the resummed form of $\gamma_L(N, a_s)$ removes *all* the fixed–order pole singularities of Eq. (26) into branch cuts. Any finite correction to γ_L may thus lead to essential changes of the corresponding numerical results. Early numerical studies on the impact of the LO resummed anomalous dimensions were performed in [31]. More recent analyses have been performed in Refs. [3,22,23].

The next-to-leading order resummed anomalous dimensions are given by

$$\hat{\gamma}_{NL}(N, \alpha_s) = -2\left(\begin{matrix} \dfrac{C_F}{C_A}\left[\gamma_{qg}^{NL} - \dfrac{8}{3}a_s T_F\right]\gamma_{qg}^{NL} \\ \gamma_{gq}^{NL} \qquad\qquad \gamma_{gg}^{NL} \end{matrix}\right), \qquad (28)$$

with $T_F = T_R N_f$. The quarkonic contributions were calculated in Ref. [14], as well as the resummed coefficient functions $c_2(N, a_s)$ and $c_L(N, a_s)$. Recently γ_{gg}^{NL} was derived in [15,16] and γ_{gq}^{NL} is yet unknown[4]. In the DIS–scheme γ_{qg}^{NL} is found to be an analytic, scale–independent function of $\gamma_L(N, a_s)$ and reads

$$\gamma_{qg}^{NL,DIS}(N, \alpha_s) = T_F \frac{\alpha_s}{6\pi}\frac{2 + 3\gamma_L - 3\gamma_L^2}{3 - 2\gamma_L}\frac{[B(1-\gamma_l, 1+\gamma_L)]^3}{B(2+2\gamma_L, 2-2\gamma_L)}R(\gamma_L), \qquad (29)$$

where

$$R(\gamma) = \left[\frac{\Gamma(1-\gamma)\chi_0(\gamma)}{-\gamma\Gamma(1+\gamma)\chi_0'(\gamma)}\right]^{1/2}\exp\left[\gamma\psi(1) + \int_0^\gamma dz\frac{\psi'(1) - \psi'(1-z)}{\chi_0(z)}\right]. \qquad (30)$$

The NLO resummed gluon anomalous dimension γ_{gg}^{NL} was calculated in the Q_0–scheme[5]. One has to solve the Bethe–Salpeter equation

$$(N-1)G_N(q_1, q_2) = \delta^{D-2}(q_1 - q_2) + \int d^{D-2}q_3 K(q_1, q_2)G_N(q_3, q_2) \qquad (31)$$

with

$$K(q_1, q_2) = \delta^{D-2}(q_1 - q_2)2\omega(q_1) + K_{real}(q_1, q_2) + K_{virtual}(q_1, q_2). \qquad (32)$$

---

[4] As pointed out in Ref. [3] its quantitative influence is likely to be minor.
[5] For a transformation into the DIS–scheme cf. [3].

For $q_1^2 \gg q_2^2$ one diagonalizes as in the LO case using *formally* the same ansatz :

$$\int d^{D-2}dq_2 K(q_1,q_2)\left(q_2^2\right)^{\gamma-1} = \overline{\alpha}_s\left[\chi_0(\gamma) - \frac{\overline{\alpha}_s}{4}\delta(\gamma,q_1^2,\mu^2)\right]\left(q_1^2\right)^{\gamma-1} . \quad (33)$$

Here the scale–invariant LO eigenvalue $\overline{\alpha}_s\chi_0(\gamma)$ is supplemented by the NLO correction term $(\overline{\alpha}_s^2/4)\delta(\gamma,q_1^2,\mu^2)$,

$$
\begin{aligned}
\delta(\gamma,q_1^2,\mu^2) = &-\left(\frac{67}{9} - 2\zeta(2) - \frac{10}{27}N_f\right)\chi_0(\gamma) + 4\Phi(\gamma) - \frac{\pi^3}{\sin^2(\pi\gamma)}\\
&+\frac{\pi^2}{\sin^2(\pi\gamma)}\frac{\cos(\pi\gamma)}{1-2\gamma}\left[(22-\beta_0) + \frac{\gamma(1-\gamma)}{(1+2\gamma)(3-2\gamma)}\left(1+\frac{N_f}{3}\right)\right]\\
&+\frac{\beta_0}{3}\chi_0(\gamma)\log\left(\frac{q_1^2}{\mu^2}\right) + \left[\frac{\beta_0}{6} + \frac{d}{d\gamma}\right]\left[\chi_0^2(\gamma) + \chi_0'(\gamma)\right] - 6\zeta_3, \quad (34)
\end{aligned}
$$

with

$$\Phi(\gamma) = \int_0^1 \frac{dz}{1+z}\left[z^{\gamma-1} + z^\gamma\right]\left[\text{Li}(1) - \text{Li}(z)\right] . \quad (35)$$

Whereas the contributions in the first two lines of Eq. (34) do contain contributions to the anomalous dimension up to $O(a_s^2)$ the third line contributes only in three–loop order. The former terms are *scale–invariant* and are in agreement with the known fixed order results. Eq. (34) therefore makes a prediction on the small x contributions of the yet unknown gluon anomalous dimension in three–loop and higher order, which will be tested in the future. Besides the scale–dependent term $(\beta_0/3)\chi_0(\gamma)\log(Q^2/\mu^2)$ also the second addend depends on the choice of scales, since it is not invariant against the interchange of q_1^2 and q_2^2, cf. [16]. The third addend $6\zeta_3$, being numerically large, contains contributions of the gluonic contribution to the trajectory function $\omega(q_1^2)$. The result given in Ref. [32] was confirmed in a different calculation by Ref. [33]. A departing value was reported in [34].

Numerical results on the impact of the leading and next–to–leading anomalous dimensions and coefficient functions were provided in a series of detailed studies, see e.g. [22,3,25] and references therein. The matrix formalism for the solution of the all order evolution equations, extending a first approach in Ref. [35] to all orders, both for hadronic and photon structure functions, is described in Ref. [3] in detail. The quarkonic contributions lead to a strong enhancement of both $F_2(x,Q^2)$ and $F_L(x,Q^2)$ at small x during the evolution. However, already simple choices for the yet unknown less singular contributions diminish these effects sizably so that a final conclusion cannot be drawn at present. In the case of the resummed gluon anomalous dimension the NLO contributions are found to be extremely large and negative. The large rise due to the LO BFKL term is already canceled to the level of the fixed order contributions by the purely quarkonic contribution to γ_{gg}^{NL}, see Fig. 5. Adding also the gluonic contribution leads to negative values for the resummed splitting function already for

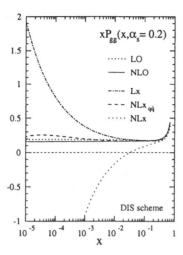

Fig. 5. Different contributions to the resummed splitting function $xP_{gg}(x, \alpha_s)$ in the DIS–scheme (overlayed), Ref. [25].

$\alpha_s = 0.2$ and $x \simeq 0.01$ which has to be regarded as unphysical. The LO and NLO resummed contributions to the gluon anomalous dimension seem to represent the first terms of a diverging series, which might be eventually resummed. This can, however, only be achieved reliably if several more less singular series are calculated *completely*, but not at the present stage.

9   An Exactly Soluble Model

The effect of potential subleading contributions to the LO anomalous dimension was estimated in the previous sections in the case of QCD. In ϕ^3 theory in $D = 6$ dimensions these terms can be determined in explicit form. ϕ_6^3 theory is rather similar to QCD (gluo–dynamics) due to the triple boson interaction and being an asymptotic free field theory. The leading order resummed anomalous dimension can be calculated for *all* values of x solving the Bethe–Salpeter equation [36]

$$T(p, q) = \frac{2^{2-D}}{\pi^{D/2}\Gamma((D-2)/2)} \frac{\lambda_D^2}{(p-q)^2} + \frac{\lambda_D^2}{(2\pi)^D} \int d^D k \frac{T(k, p)}{(q-k)^2 (k^2)^2} , \qquad (36)$$

with $q^2, p^2 < 0$, q the momentum transfer and p^2 a hadronic mass scale. For $D = 6$ the quantity $q.p\, T(p, q)$ is scale invariant and one may expand Eq. (36) into partial waves with

$$p.q\, T_N(p, q) = \left(\frac{q^2}{p^2}\right)^{-(N+1)/2} \left(\frac{q^2}{p^2}\right)^{-\gamma_L(N,a_s)/2} , \qquad (37)$$

where $a_s = \lambda_6^2/(4\pi)^3 = $ const. The anomalous dimension $\gamma_L(N, a_s)$ is given by

$$\gamma_L(N, a_s) = \sqrt{(N+2)^2 + 1 - 2\sqrt{(N+2)^2 + 4a_s}} - (N+1) . \qquad (38)$$

Note that $\gamma_L(N, a_s)$ possesses *no* poles but only branch cuts for $N \in \mathbf{C}$. The anomalous dimension γ_L covers all conformal contributions in leading order. If one expands this quantity it yields in first order in a_s the *complete* leading order anomalous dimension $\gamma_{SS}^{(0)}(N)$, up to an eventual term due to 4–momentum conservation which is easily imposed,

$$\gamma_{SS}^{(0)}(N) = -\frac{2}{(N+1)(N+2)} + \frac{1}{6} . \tag{39}$$

Furthermore, all the fixed–order leading poles at $N = -1$ are resummed in this representation. This has been verified by an explicit calculation up to 3–loop order [37]. The complete NLO fixed order anomalous dimension reads [37]

$$\gamma_{SS}^{(1)}(N) = -\frac{1}{6}\frac{22 + 111N + 211N^2 + 138N^3 + 28N^4}{(N+1)^3(N+2)^3} + \frac{5}{3}\frac{S_1(N)}{(N+1)(N+2)}$$

$$-\frac{1}{2}\left[1 + (-1)^N\right]\frac{2}{(N+1)^2(N+2)^2} + \frac{13}{216} . \tag{40}$$

One may now derive from Eq. (38) the small–x resummed anomalous dimension, covering the fixed–order leading pole contributions only

$$\gamma_L^{N \to -1}(N, a_s) = (N+1)\left[\sqrt{1 - \frac{4a_s}{(N+1)^2}} - 1\right] , \tag{41}$$

which again contains *no* poles for all $N \in \mathbf{C}$. This quantity corresponds to the LO BFKL anomalous dimension in QCD.

In deriving (41) one obtains as well the respective resummed subleading terms. As was shown in Ref. [37] the weight coefficient of these terms are of alternating sign with growing coefficients, which indicates already that the re-summation of the leading pole terms $(N = -1)$ does not yield the dominant contribution. This is expected, since neither $\gamma_L(N, a_s)$ nor $\gamma_L^{N \to -1}(N, a_s)$ have a pole singularity – as is also the case in the LO BFKL resummation, where a sim-ilar behaviour might be expected. Fig. 6 shows the behaviour of the respective splitting functions after the Mellin transform to x–space, normalized to the lead-ing order splitting function $P_0(x) = 2x(1-x)$. $P_L^{x \to 0}(x, a_s)$ is nowhere dominant and departs encreasingly from the complete solution $P_L(x, a_s)$ as $x \to 0$.

10   Conclusions

As in the case of the fixed order calculations the renormalization group equation, through which the factorization of the mass singularities is described, implies the evolution equations for the parton densities including the resummation of the small x terms. Due to the Mellin convolution between the respective evolution kernels and the extended input distributions the detailed knowledge of the ker-nels at medium x is as important. This is particularly the case for input densities

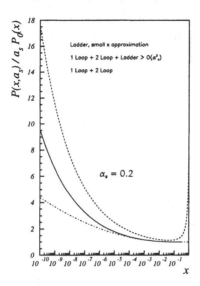

Fig. 6. Fixed-order and resummed splitting functions $P(x, a_s)$, normalized to $a_s P_{SS}^{(0)}(x)$, for $a_s = 0.2$. Dash-dotted line : $P = a_s P_{SS}^{(0)} + a_s^2 P_{SS}^{(1)}$; solid line : $P = P_L$; dashed line : $P = P_L^{x \to 0}$, Ref. [37].

with a large rise towards small x, as the gluon distribution. Less singular contributions to the evolution kernels turn out to have a sizable impact onto the scaling violations. In the example of ϕ_6^3 theory these contributions were calculated for the leading order resummation and turn out to be even more important than the leading pole terms $(N = -1)$. The reason for this behaviour is that the resummed anomalous dimension, as also the resummed $(N = -1)$-fixed-order pole contribution, possess *no poles* anymore. This is also the case for the leading order resummed BFKL anomalous dimension and the known resummed NLO contributions.

In a quantitative description of the scaling violation of structure functions the conservation laws as Fermion number conservation in the non–singlet case and energy–momentum conservation have to be obeyed. These integral relations imply strong relations between the small x and medium x contributions also for the resummed evolution kernels. A study of the known fixed–order results in leading and next–to–leading order shows furthermore that the evolution kernels, if approximated in a leading pole representation, require to take into account at least four orders which is likely to be the case for the small x resummed terms as well. The conformal part of the known terms of the small x resummations behaves stable but is not necessarily dominant.

An important future check of the small x resummed calculations is their prediction of the leading and next–to–leading order small x contributions to the 3–loop anomalous dimensions, which are yet unknown. The intimate interplay between small and medium x effects requires to continue consistent calculations of the anomalous dimensions and coefficient functions to even higher order and to compare these results with the scaling violations measured by experiment.

Acknowledgement. For discussions I would like to thank A. Vogt, V. Ravindran and W.L. van Neerven. This work was supported in part by EC contract FMRX–CT98–0194.

References

1. L.V. Gribov, E.M. Levin, and M.G. Ryskin: Nucl. Phys. **B188**, 555 (1983)
   A. Mueller and J. Qiu: Nucl. Phys. **B268**, 427 (1986)
2. A. De Rujula, S.L. Glashow, H.D. Politzer, S.B. Treiman, F. Wilczek, and A. Zee: Phys, Rev. **D10**, 1649 (1974)
   T. De Grand: Nucl. Phys. **B151**, 485 (1979)
   J.P. Ralston and D.W. McKay, in: *Physics Simulations at High Energies*, ed. by V. Barger (World Scientific, Singapore, 1987)
   J. Blümlein: Surv. High Energy Phys. **7**, 161 (1994)
   R.D. Ball and S. Forte: Phys. Lett. **B336**, 77 (1994)
3. J. Blümlein and A. Vogt: Phys. Rev. **D58**, 014020 (1998) (1997)
4. J. Bartels, J. Blümlein, and G. Schuler: Z. Phys. **C50**, 91 (1991)
   J. Collins and J. Kwiecinski: Nucl. Phys. **B335**, 89 (1990)
   M. Altmann, M. Glück, and E. Reya: Phys. Lett. **B285**, 359 (1992)
5. J. Bartels: Phys. Lett. **B298**, 204 (1993)
6. L.N. Lipatov: Sov. J. Nucl. Phys. **23**, 338 (1976)
   E.A. Kuraev, L.N. Lipatov, and V.S. Fadin: Sov. Phys. JETP **45**, 199 (1977)
   I.I. Balitskii and L.N. Lipatov: Sov. J. Nucl. Phys. **28**, 822 (1978)
   M. Ciafaloni: Nucl. Phys. **B296**, 49 (1988)
7. G. Marchesini, in: *QCD at 200 TeV*, ed. by L. Ciffarelli and Yu.L. Dokshitser, (Plenum Press, New York, 1992) pp. 183 and references therein
8. T. Jaroszewicz: Phys. Lett. **B116**, 291 (1982)
9. R. Kirschner and L.N. Lipatov: Nucl. Phys. **B213**, 122 (1983)
10. J. Blümlein and A. Vogt: Phys. Lett. **B370**, 149 (1996)
11. J. Bartels, B.I. Ermolaev, and M.G. Ryskin: Z. Phys. **C76**, 241 (1997)
12. J. Blümlein and A. Vogt: Acta Phys. Pol. **B27**, 1309 (1996)
13. J. Blümlein, S. Riemersma, and A. Vogt: Eur. Phys. J. **C1**, 255 (1998)
14. S. Catani and F. Hautmann: Nucl. Phys. **B427**, 475 (1994)
15. V.S. Fadin and L.N. Lipatov: Phys. Lett. **B429**, 127 (1998)
    V.S. Fadin: Preprint BUDKERINP-98/55; hep-ph/9807528
16. G. Camici and M.Ciafaloni: Nucl. Phys. **B496**, 305 (1997)
    G. Camici and M.Ciafaloni: Phys. Lett. **B430**, 349 (1998)
17. M. Ciafaloni: Phys. Lett. **B356**, 74 (1995)
18. J. Blümlein: J. Phys. **G19**, 1623 (1993)
19. W.L.van Neerven: Talk, DESY Theory Workshop Sept. 1993
20. B.M. McCoy and T.T. Wu: Phys. Lett. **B71**, 97 (1977)
21. J. Blümlein and A. Vogt: Phys. Lett. **B386**, 350 (1996)
22. R.K. Ellis, F. Hautmann, and B. Webber: Phys. Lett. **B348**, 582 (1995)
23. J. Blümlein, S. Riemersma, and A. Vogt: Nucl. Phys. **B** (Proc. Suppl.) **51C**, 30 (1996); Acta Phys. Pol. **B28**, 577 (1997)
24. J. Blümlein and A. Vogt: Phys. Rev. **D57**, R1 (1998) (1997)

25. J. Blümlein, V. Ravindran, W.L. van Neerven and A. Vogt: 'The Unpolarized Singlet Anomalous Dimension at Small x'. In: *DIS98, 6th International Workshop on Deep Inelastic Scattering and QCD, Brussels, Belgium, April, 1998*, ed. by Gh. Coremans and R. Rosen (World Scientific, Singapore, 1998), pp. 211–216, hep-ph/9806368

26. Y. Kiyo, J. Kodaira, and H. Tochimura: Z. Phys. **C74**, 631 (1997)

27. B.I. Ermolaev, S.I. Manyenkov, and M.G. Ryskin: Z. Phys. **C69**, 259 (1996)

28. J. Bartels, B.I. Ermolaev, and M.G. Ryskin: Z. Phys. **C70**, 273 (1996)

29. F. Berends, W.L. van Neerven, and G. Burgers: Nucl. Phys. **B297**, 429 (1988); E: **B304**, 921 (1988)

30. J. Blümlein, '$k_\perp$ dependent parton densities in the photon and proton.' In: *Proc. of the XXX Renc. de Moriond*, Les Arcs, France, March 1995, ed. J. Tran Than Van (Edition Frontieres, Paris, 1995), pp. 191–197, hep-ph/9506446

31. J. Kwiecinski: Z. Phys. **C29**, 561 (1985)

32. V.S. Fadin, R. Fiore, and M. Kotsky: Phys. Lett. **B387**, 593 (1996)

33. J. Blümlein, V. Ravindran, and W.L. van Neerven: Phys. Rev. **D58**, 091502 (1998)

34. I.A. Korchemskaya and G.P. Korchemsky: Phys. Lett. **B287**, 346 (1996)

35. R.K. Ellis, E.M. Levin, and Z. Kunszt: Nucl. Phys. **B420**, 517 (1994); E: **B433**, 498 (1995)

36. C. Lovelace: Phys. Lett. **B55**, 187 (1975); Nucl. Phys. **B95**, 12 (1975)

37. J. Blümlein and W.L. van Neerven: Phys. Lett. **B450**, 412 (1999)

Screening Effects on F_2 at Low x and Q^2

Eugene Levin

HEP Department
School of Physics and Astronomy
Tel Aviv University, Tel Aviv 69978, Israel
and
DESY, Theory Group
22603, Hamburg, Germany

Abstract. In this talk we discuss how deeply the region of high parton densities has been studied experimentally at HERA. We show that the measurements of deep inelastic structure functions at HERA confirm our theoretical expectation that at HERA we face a challenging problem of understanding a new system of partons: quarks and gluons at short distances with so large densities that we cannot treat this system perturbatively. We collect all experimental indications and manifestations of specific properties of high parton density QCD.

1   What Are Shadowing Corrections?

In the region of low x and low Q^2 we face two challenging problems which have to be resolved in QCD:

1. The matching of "hard" processes, that can be successfully described in perturbative QCD (pQCD), and "soft" processes, that should be described in non-perturbative QCD (npQCD), but actually, we have only a phenomenological approach for them;

2. The theoretical approach for the high parton density QCD (hdQCD) which we reach in the deep inelastic scattering at low x but at sufficiently high Q^2. In this kinematic region we expect that the typical distances will be small but the parton density will be so large that a new non perturbative approach shall be developed for understanding this system.

We are going to advocate the idea that these two problems are correlated and the system of partons always passes the stage of hdQCD before (at shorter distances) it goes to the black box, which we call non-perturbative QCD and which, practically, we describe in old fashion Reggeon phenomenology. In spite of the fact that there are many reasons to believe that such a phenomenology could be even correct, the distance between the Reggeon approach and QCD is so large we are loosing any taste of theory doing this phenomenology. In hdQCD we still have a small parameter (running QCD coupling α_S) and we can start to approach this problem using the developed methods of pQCD [1]. However, we should realize that the kernel of the hd QCD problems is a non-perturbative one, and therefore, approaching hdQCD theoretically we are preparing new training grounds for searching methods for npQCD.

First, let me recall that DIS experiment is nothing more than a microscope and we have two variables to describe its work. The first one is the resolution of the microscope, namely, $\Delta x \approx 1/Q$ where Q^2 is the virtuality of the photon. It means that our microscope can see all constituents inside a target with the size larger that Δx. The second variable is time of observation. It sounds strange that we have this new variable, which we do not use, working with a usual light microscope. However, we are dealing here with the relativistic system which can produce hadrons (partons). So, for everyday analogy, we should consider rather a box with flies which multiply and their number is, certainly, different in different moments of time. To estimate this time we can use the uncertainty principle $\Delta t \propto 1/\Delta E$ where ΔE is the change of energy, namely, $\Delta E = E_{initial} - E_{final}$, and for a system of quark and antiquark $\Delta E = q_0 - q_1 - q_2 = q_0 - q = \frac{(q_0-q)(q_0+q)m}{2q_0 m} = \frac{Q^2 m}{W^2} = mx$, where m is the mass of the target and q_0 and q is the energy and momentum of the virtual photon. Finally, $t = 1/mx$ with $x = \frac{Q^2}{W^2}$ where W is the energy of the photon - target interaction.

Therefore, the question, that we are asking in DIS at low x, is what happens with constituents of rather small size after a long time. It is clear that the number of these constituents should increase since in QCD each parton can decay in two partons with the probability $P_i = \frac{N_c \alpha_S}{\pi} \frac{dE_i}{E_i} \frac{d^2 k_{i,t}}{k_{i,t}^2}$ where E_i and k_i are energy and momentum of an emitted parton i.

This growth we can describe by introducing the so called structure function $(xG(x,Q^2))$ or the number of partons that can be resolved with the microscope with definite Q^2 and x. Indeed,

$$\frac{\partial^2 xG(x,Q^2)}{\partial \ln(1/x)\, \partial \ln Q^2} = \frac{N_c \alpha_S}{\pi} xG(x,Q^2) . \tag{1}$$

This equation is the DGLAP [2] evolution equation in the region of low x. It has an obvious solution $xG(x,Q^2) \propto exp\left(2\sqrt{\frac{N_c \alpha_S}{\pi} \ln(1/x) \ln(Q^2/Q_0^2)}\right)$. Therefore, we expect the increase of the parton densities at $x \to 0$.

In Fig. 1 we picture the parton distributions in the transverse plane. At $x \approx 1$ there are several partons of a small size. The distance between partons is much larger than their size and we can neglect interactions between them.

However, at $x \to 0$ the number of partons becomes so large that they are populated densely in the area of a target. In this case, you cannot neglect the interactions between them which was omitted in the evolution equations (see Eq. (1)). Therefore, at low x we have a more complex problem of taking into account both emission and rescatterings of partons. Since the most important in QCD is the three parton interaction, the processes of rescattering is actually a process of annihilation in which one parton is created out of two partons (gluons).

Therefore, at low x we have two processes

1. The emission induced by the QCD vertex $G + G \to G$ with the probability which is proportional to $\alpha_S \, \rho$ where ρ is the parton density in the transverse plane , namely

$$\rho = \frac{xG(x, Q^2)}{\pi R^2} \, , \tag{2}$$

where πR^2 is the target area;

2. The annihilation induced by the same vertex $G + G \to G$ with the probability which is proportional to $\alpha_S \sigma_0 \, \rho^2$, where α_S is probability of the processes $G + G \to G$, σ_0 is the cross section of two parton interaction and $\sigma_0 \propto \frac{\alpha_S}{Q^2}$. $\sigma_0 \, \rho$ gives the probability for two partons to meet and to interact, while $\alpha_S \sigma_0 \, \rho^2$ gives the probability of the annihilation process.

Finally, the change of parton density is equal to [1] [3]

$$\frac{\partial^2 \rho(x, Q^2)}{\partial \ln(1/x) \, \partial \ln Q^2} = \frac{N_c \alpha_S}{\pi} \rho(x, Q^2) - \frac{\alpha_S^2 \gamma}{Q^2} \rho^2(x, Q^2) \tag{3}$$

or in terms of the gluon structure function

$$\frac{\partial^2 xG(x, Q^2)}{\partial \ln(1/x) \, \partial \ln Q^2} = \frac{N_c \alpha_S}{\pi} xG(x, Q^2) - \frac{\alpha_S^2 \gamma}{\pi R^2 Q^2} \left(xG(x, Q^2) \right)^2 \, , \tag{4}$$

where γ has been calculated in pQCD [3].

Therefore, Eq.(4) is a natural generalization of the DGLAP evolution equations. The question arises, why we call such a natural equation for a balance

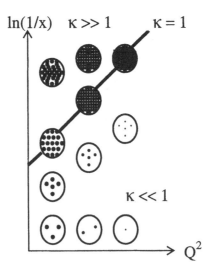

Fig. 1. Parton distribution in the transverse plane

of partons due to two competing processes shadowing and/or screening corrections (SC). To understand this let us consider the interaction of the fast hadron with the virtual photon at rest (Bjorken frame). In the parton model, only the slowest ("wee") partons interact with the photon. If the number of the "wee" partons N is not large, the cross section is equal to $\sigma_0 N$. However, if we have two "wee" partons with the same energies and momenta, we overestimate the value of the total cross section using the above formula. Indeed, the total cross section counts only the number of interactions and, therefore, in the situation when one parton is situated just behind another we do not need to count the interaction of the second parton if we have taken into account the interaction of the first one. It means that the cross section is equal to

$$\sigma_{tot} = \sigma_0 N \left\{ 1 - \frac{\sigma_0}{\pi R^2} \right\} , \tag{5}$$

where R is the hadron radius. One can see that we reproduce Eq.(4) by taking into account that there is a probability for a parton not to interact being in a shadow of the second parton.

2   What Have We Learned about Sreening Corrections?

During the past two decades high parton density QCD has been under the close investigation of many theorists [1] [3][4][5] and we summarize here the result of their activity.

- The parameter which controls the strength of SC has been found and it is equal to

$$\kappa = \frac{3\pi^2\alpha_s}{2Q^2} \times \frac{xG(x, Q^2)}{\pi R^2} = \sigma_0 \times \rho(x, Q^2) . \tag{6}$$

  The meaning of this parameter is very simple. It gives the probability of interaction for two partons in the parton cascade or, better to say, a packing factor for partons in the parton cascade.

- We know the correct degrees of freedom at high energies: colour dipoles [6]. By definition, the correct degrees of freedom is a set of quantum numbers which mark the wave function that is diagonal with respect to the interaction matrix. Therefore, we know that the size and the energy of the colour dipole are not changed by the high energy QCD interaction.

- A new scale $Q_0^2(x)$ for hdQCD has been traced in the pQCD approach which is a solution to the equation

$$\kappa = \frac{3\pi^2\alpha_s}{2Q_0^2(x)} \times \frac{xG(x, Q_0^2(x))}{\pi R^2} = 1 . \tag{7}$$

  This new scale leads to the effective Lagrangian approach which gives us a general non-perturbative method to study hdCD.

- We know that the GLR equation (see Eqs.(3) and (4)) describes the evolution of the dipole density in the full kinematic region [7]. We understand that the Mueller-Glauber approach for colour dipole rescattering gives the initial condition to the GLR equation.

- The new, non-perturbative approach, based on the effective Lagrangian [5], have been developed for hdQCD which gives rise to the hope that hdQCD can be treated theoretically from first principles.

- We are very close to understanding of the parton density saturation [1].

In general, we think that the theoretical approach to hdQCD in a good shape now.

3   The HERA Puzzle: Where Are the Sreening Corrections?

The wide spread opinion is that HERA experimental data for $Q^2 \geq 1\,GeV^2$ can be described quite well using only the DGLAP evolution equations, without any other ingredients such as shadowing corrections, higher twist contributions and so on (see, for example, the reviews [8]). On the other hand, the most important HERA discovery is the fact that the density of gluons (gluon structure function) becomes large in the HERA kinematic region [8][9]. The gluon densities extracted from HERA data are so large that the parameter κ (see Eq.(6)) exceeds unity in a substantial part of the HERA kinematic region (see Fig.2a). Another way to see this is to plot the solution to Eq.(7) (see Fig.2b). It means that in a large kinematic region $\kappa \geq 1$ (to the left from the line $\kappa = 1$ in Fig.2b), we expect that the SC should be large and important for a description of the experimental data. At first sight such expectations are in clear contradiction with the experimental data. Certainly, this fact gave rise to the suspicion or even mistrust that our theoretical approach to SC is not consistent. However, the revision and re-analysis of the SC , as has been discussed in the previous section, have been completed with the result, that κ is responsible for the value of SC.

Therefore, we face a puzzling question: *where are the SC?*. Actually, this question includes, at least, two questions: (i) why SC are not needed to describe the HERA data on F_2, and (ii) where are the experimental manifestation of strong SC. The answers for these two questions you will find in the next three sections but, in short, they are: SC give a weak change for F_2 in the HERA kinematic region, but they are strong for the gluon structure function . We hope to convince you that there are at least two indications in the HERA data supporting a large value of SC to the gluon density:

1. x_P - behaviour of the cross section for diffractive dissociation (σ^{DD}) in DIS;

2. Q^2 - behaviour of F_2 -slope ($\frac{\partial F_2(x,Q^2)}{\partial \ln Q^2}$).

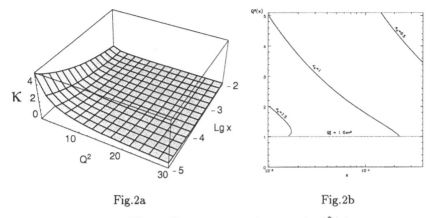

Fig.2a   Fig.2b

Fig. 2. Parameter κ and new scale $Q_0^2(x)$.

4   SC for F_2

It is well known, that the γ^* - hadron interaction goes in two stages: (i) the transition from virtual photon to colour dipole and (ii) the interaction of the colour dipole with the target. To illustrate how SC work, we consider the Glauber - Mueller formula which describes the rescatterings of the colour dipole with the target[10]:

$$F_2(x_B, Q^2) = \frac{N_c}{6\pi^3} \sum_1^{N_f} Z_f^2 \int_{Q_0^2}^{Q^2} dQ'^2 \int db_t^2 \{ 1 - e^{-\frac{1}{2}\frac{4}{9}\kappa(x_B, Q'^2) S(b_t)} \}$$

$$+ F_2(x_B, Q_0^2), \tag{8}$$

where $S(b_t) = e^{-\frac{b_t^2}{R^2}}$ is the target profile function in the impact parameter representation and $\frac{4}{9}\kappa(x_B, Q'^2) = \sigma(x_B, r_\perp^2 = 4/Q^2)$ is the cross section of the dipole scattering in pQCD.

One can see that Eq.(8) leads to

$$F_2(x_B, Q^2) \longrightarrow \frac{N_c}{6\pi^3} \sum_1^{N_f} Z_f^2 \, Q^2 R^2 \, . \tag{9}$$

However, we are sure that the kinematic region of HERA is far away from the asymptotic one. The practical calculations depend on three ingredients: the value of R^2, the value of the initial virtuality Q_0^2 and the initial F_2 at Q_0^2. We fix them as follows: $R^2 = 10 \, GeV^{-2}$ which corresponds to "soft" high energy phenomenology [11], $Q_0^2 = 1 \, GeV^2$ and $F_2^{input}(x_B, Q_0^2) = F_2^{GRV'94}(x_B, Q_0^2 = 1 \, GeV^2)$. Therefore, the result of the calculation should be read as "SC for colour dipoles with the size smaller than $r_\perp^2 \leq 4/GeV^2$ are equal to ..."

From Fig.3 one can see that the SC are rather small for F_2 but they are strong and essential for the gluon structure function. It means that we have to

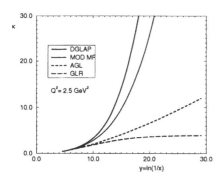

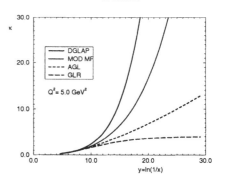

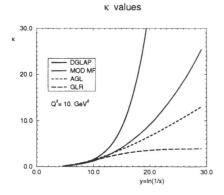

Results:

- SC are large for $xG(x, Q^2)$ but their values do not depend on the way how we take SC into account;

- SC are rather small for F_2 ;

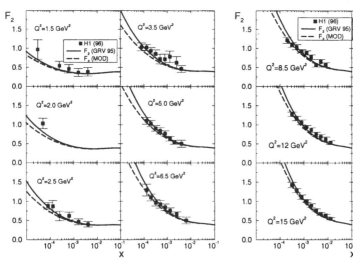

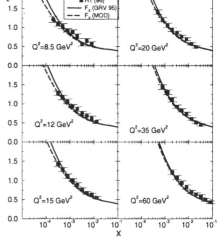

Fig. 3. SC for $xG(x, Q^2)$ and F_2 in the HERA kinematic region .

look for physical observables which will be more sensitive to the value of the gluon structure function than F_2.

5 x_P- Dependence of σ^{DD}

One of such observables is the cross section of the diffractive dissociation and, especially, the energy dependence of this cross section.

Data: Both H1 and ZEUS collaborations [8] found that

$$\sigma^{DD} \propto \frac{1}{x^{2\Delta_P}} , \tag{10}$$

where $\Delta_P = \alpha_P(0) - 1$ and the values of $\alpha_P(0)$ are:

- **H1 [12]** : $\alpha_P(0) = 1.2003 \pm 0.020(\text{stat.}) \pm 0.013(\text{sys})$;
- **ZEUS [13]**: $\alpha_P(0) = 1.1270 \pm 0.009(\text{stat.}) \pm 0.012(\text{sys})$.

It is clear that the Pomeron intercept ($\alpha_P(0)$) for diffractive processes in DIS is higher than the intercept of the "soft" Pomeron [11].

Why is it surprising and interesting? To answer this question we have to recall that the cross sections for diffractive production of a quark-antiquark pair have the following form in pQCD [14] [15]:

$$x_P \frac{d\sigma_{DD}^T(\gamma^* \to q + \bar{q})}{dx_P dt} \propto \int_{Q_0^2}^{\frac{M^2}{4}} \frac{dk_\perp^2}{k_\perp^2} \times \frac{\left(\alpha_S \, x_P \, G(x_P, \frac{k_\perp^2}{1-\beta}) \right)^2}{k_\perp^2} ; \tag{11}$$

$$x_P \frac{d\sigma_{DD}^L(\gamma^* \to q + \bar{q})}{dx_P dt} \propto \int_{Q_0^2}^{\frac{M^2}{4}} \frac{dk_\perp^2}{Q^2} \times \frac{\left(\alpha_S \, x_P \, G(x_P, \frac{k_\perp^2}{1-\beta}) \right)^2}{k_\perp^2} . \tag{12}$$

From Eqs.(11) and (12) you can see that the $k_\perp$ integration looks quite differently for a transverse or a longitudinal polarized photon: the latter one has a typical log integral over $k_\perp$ while the former has the integral which normally converges at small values of $k_\perp$. We have the same property for the production of a more complex system than $q\bar{q}$, for example $q\bar{q}G$ [15]. Therefore, we expect that the diffractive production should come from long distances where the "soft" Pomeron contributes. However, the experiment says a different thing, namely, that this production has a considerable contamination from short distances. How is it possible? As far as we know, there is the only one explanation: SC are so strong that $xG(x, k_\perp^2) \propto k_\perp^2 R^2$ (see Eq.(9)) Substituting this asymptotic limit in Eq.(11) one can see that the integral becomes convergent and it sits at the upper limit of integration which is equal to $k_\perp^2 = Q_0^2(x)$.

Finally, we have

$$x_P \frac{d\sigma_{DD}}{dx_P dt} \longrightarrow \left(x_P \, G(x_P, Q_0^2(x_P)) \right)^2 \times \frac{1}{Q_0^2(x_P)} \tag{13}$$

The calculation for $\frac{\partial x G(x, Q^2)}{\partial \ln(1/x)}$ is given in Fig.4 for the HERA kinematic region using the Glauber-Mueller formula [10] for SC. Taking into account that $Q_0^2(x)$ in Fig.2b can be fitted as $Q_0^2(x) = 1 \, GeV^2 \left(\frac{x}{x_0}\right)^{-\lambda}$ with $\lambda = 0.54$ and $x_0 = 10^{-2}$ we see from Eq.(13) and Fig.4 that we are able to reproduce the experimental value of $\alpha_P(0)$ and conclude that the typical $k_\perp^2$ which are dominant in the integral is not small ($k_\perp^2 \approx 1 - 2 \, GeV^2$ [15]). For the Golec-Biernat Wüsthoff approach, which we will discuss later, $\lambda = 0.288$ and the value of typical $k_\perp^2$ turns out to be higher.

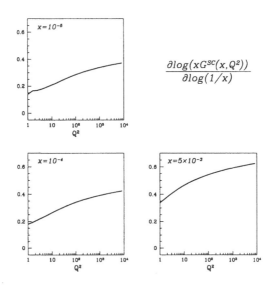

Fig. 4. Effective Pomeron intercept for the gluon structure function calculated using the Glauber-Mueller formula for SC

6   The Q^2- Dependence of the F_2 - Slope.

Data: The experimental data [13] for the F_2 - slope $dF_2(x, Q^2)/d\ln Q^2$ are shown in Fig.5a (Caldwell plot). These data give rise to a hope that the matching between "hard" (short distance) and "soft" (long distance) processes occurs at sufficiently large Q^2 since the F_2-slope starts to deviate from the DGLAP predictions around $Q^2 \approx 5 - 8 \, GeV^2$.

F_2-slope and SC: Our principle idea, as we have mentioned in the beginning of the talk, is that matching between "hard" and "soft" processes is provided by the hdQCD phase in the parton cascade or, in other words, due to strong SC. The asymptotic behaviour of $F_2 \propto Q^2 R^2$ for $Q^2 \leq Q_0^2(x)$ leads to

$dF_2(x, Q^2)/d\ln Q^2 \propto Q^2 R^2$ at $Q^2 \leq Q_0^2(x)$ (see Eq.(9)) and this behaviour supports our point of view [16][17].

However, we have two problems to solve before making any conclusion: (i) the experimental data are taken at different points (x, Q^2) and therefore could be interpreted as the change of x-behaviour rather than Q^2 one; and (ii) the value of the F_2-slope is quite different from the value of F_2 while for the asymptotic solution it should be the same. Therefore, we have to calculate the F_2 - slope to understand them. The result of the calculation using the Glauber-Mueller formula [16] is presented in Fig.5b. One can see that (i) the experimental data show rather the Q^2 - behaviour than the x-dependence, which is not qualitatively influenced by SC; and (ii) SC are able to describe both the value and the Q^2-behaviour of the experimental data. Fig.5b shows also that the ALLM'97 parameterization [18], which can be viewed as the phenomenological description of the experimental data, has the same features as our calculation confirming the fact that the data show the Q^2 - dependence but not the x-behaviour of the F_2-slope.

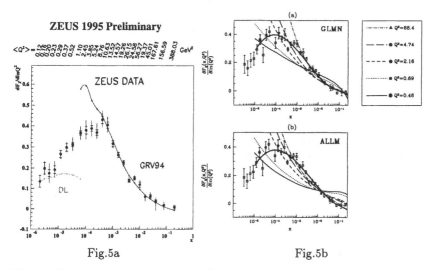

Fig.5a   Fig.5b

Fig. 5. F_2-slope: experimental data (Caldwell plot) (Fig.5a) and calculations using the Glauber-Mueller formula (Fig.5b) .

7   Golec-Biernat Wüsthoff Approach

Golec-Biernat and Wüsthoff [19] suggested a phenomenological approach which takes into account the key idea of hdQCD, namely, the new scale of hardness in the parton cascade. They use for $\gamma^* p$ cross section the following formula [19]

$$\sigma_{tot}(\gamma^* p) = \int d^2 r_\perp \int_0^1 dz \, |\Psi(Q^2; r_\perp, z)|^2 \, \sigma_{tot}(r_\perp^2, x) ; \qquad (14)$$

$$\sigma(x, r_\perp) = \sigma_0 \left\{ 1 - e^{-\frac{r_\perp^2}{R^2(x)}} \right\} ; \qquad (15)$$

$$R^2(x) = 1/Q_0^2(x) \quad with \quad Q_0^2(x) = Q_0^2 \left(\frac{x}{x_0} \right)^{-\lambda} . \qquad (16)$$

Extracting the parameters of their model from fitting of the experimental data, namely, $\sigma_0 = 23.03$ mb, $\lambda = 0.288$, $Q_0^2 = 1 GeV^2$ and $x_0 = 3.04 \, 10^{-4}$, they described quite well all the data on total and diffractive cross sections in DIS (see Fig.6).

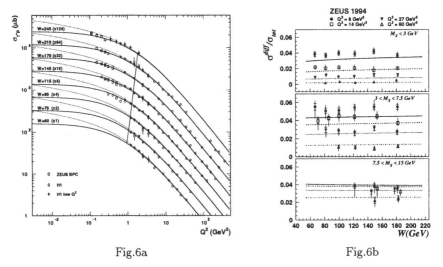

Fig.6a   Fig.6b

Fig. 6. $\sigma_{tot}(\gamma^* p)$ and the ratio σ^{DD}/σ_{tot} for DIS in the Golec-Biernat Wüsthoff model. Vertical line in Fig.6a is the $Q_0^2(x)$ given by Eq.(16).

8   Why Have We Only Indications?

The answer is: because we have or can have an alternative explanation of each separate fact. For example, we can describe the F_2 -slope behaviour changing the initial x-distribution for the DGLAP evolution equations [20]. Our difficulties in an interpretation of the experimental data is seen in Fig.6a where the new scale $Q_0^2(x)$ is plotted. One can see that $Q_0^2(x)$ is almost constant in the HERA kinematic region. It means that we can put the initial condition for the evolution equation at $Q_0^2 = < Q_0^2(x) >$ where $< Q_0^2(x) >$ is the average new scale in the HERA region. Therefore, SC can be absorbed to a large extent in the initial condition and the question, that can and should be asked, is how well motivated these conditions are. For example, I do not think that the initial gluon distribution in the MRST parameterization [20], needed to describe the F_2 - slope data, can be considered as a natural one.

9 Summary

We hope we convinced you that (i) hdQCD is in a good theoretical shape; (ii) the hdQCD region has been reached at HERA; (iii) HERA data do not contradict the strong SC effects; (iv) there are at least two indications on SC effects in the HERA data: the Q^2 behaviour of the F_2 slope and the x_P behaviour of diffractive cross section in DIS; and (v) the HERA data and the hdQCD theory gave an impetus for a very successful phenomenology for matching "hard" and "soft" physics.

We would like to finish this talk with a rather long citation: "Small x Physics is still in its infancy. Its relation to heavy ion physics, mathematical physics and soft hadron physics along with a rich variety of possible signatures makes it central for QCD studies over the next decade" (A.H. Mueller, B. Müller, G. Rebbi and W.H. Smith "Report of the DPF Long Range Planning WG on QCD"). Hopefully, we will learn more on low x physics at the next Ringberg Workshop.

References

1. L. V. Gribov, E. M. Levin and M. G. Ryskin: Phys.Rep. **100**, 1 (1983).
2. V.N. Gribov and L.N. Lipatov: Sov. J. Nucl. Phys. **15**,438 (1972); L.N. Lipatov: Yad. Fiz. **20**, 181 (1974; G. Altarelli and G. Parisi: Nucl. Phys. B **126**, 298 (1977); Yu.L. Dokshitser: Sov. Phys. JETP **46**, 641 (1977).
3. A.H. Mueller and J. Qiu: Nucl. Phys. B **268**, 427 (1986).
4. E. Laenen and E. Levin: Ann. Rev. Nucl. Part. **44**, 199 (1994) and references therein; A.H. Mueller: Nucl. Phys. B **437**, 107 (1995); G. Salam: Nucl. Phys. B **461**, 512 (1996); A.L. Ayala, M.B. Gay Ducati and E.M. Levin: Nucl. Phys. B **493**, 305 (1997) (1997); **510**, 355 (1998).
5. L. McLerran and R. Venugopalan: Phys. Rev. D **49**,2233,3352 (1994); **50**, 2225 (1994); **53**,458 (1996); J. Jalilian-Marian, A. Kovner, A. Leonidov and H. Weigert: Phys. Rev. D **59**,014014, 034007 (1999); J.Jalilian-Marian, A. Kovner, L. McLerran and H. Weigert: Phys. Rev. D **55**, 5414 (1997); A. Kovner, L. McLerran and H. Weigert: Phys. Rev. D **52**, 3809,6231 (1995); Yu. Kovchegov: Phys. Rev. D **54**,5463 (1996); **55**,5445 (1997); Yu. V. Kovchegov and A.H. Mueller: Nucl. Phys. B **529**, 451 (1998); Yu. V. Kovchegov, A.H. Mueller and S. Wallon: Nucl. Phys. B **507**, 367 (1997).
6. A.H. Mueller: Nucl. Phys. B **425**.471 (1994).
7. Yuri V. Kovchegov: *Small x F_2 Structure Function of a Nucleus Including Multiple Pomeron Exchange*, NUC-MN-99/1-T, hep-ph/9901281.

8. A.M. Cooper-Sarkar, R.C.E. Devenish and A. De Roeck: Int.J.Mod.Phys. A **13**, 3385 (1998); H.Abramowicz and A. Caldwell: *HERA Collider Physics* DESY-98-192, hep-ex/9903037, Rev. Mod. Phys. (in press).

9. ZEUS Collaboration, J. Breitweg et al.: Eur. Phys. J. C **7**,609 (1999).

10. A. H. Mueller: Nucl. Phys. B **335**, 115 (1990).

11. A. Donnachie and P.V. Landshoff: Nucl.Phys. B **244**, 322 (1984; **267**, 690 (1986); Phys. Lett. B **296**, 227 (1992; Z. Phys. C **61**, 139 (1994); E. Gotsman,E. Levin and U. Maor: Phys. Lett. B **452**, 287 (1999),**304**,199 (1992) ; Phys. Rev. D **49**, 4321 (1994); Z. Phys. C **57**,672 (1993).

12. H1 Collaboration, T. Ahmed et al.: Phys. Lett. B **348**, 681 (1995); C. Adloff et al.: Z. Phys. C **76**, 613 (1997).

13. ZEUS Collaboration, M. Derrick et al.: Z. Phys. C **68**, 569 (1995); J. Breitweg et al.: Eur. Phys. J. C **6**, 43 (1999).

14. J. Bartels, H. Lotter and M. Wüsthoff: Phys. Lett. B **379**, 239 (1996) and references therein.

15. E. Gotsman,E. Levin and U. Maor: Nucl. Phys. B **493**, 354 (1997).

16. E. Gotsman,E. Levin and U. Maor: Phys. Lett. B **425**, 369 (1998); E. Gotsman, E. Levin, U. Maor and E. Naftali: Nucl. Phys. B **539**, 535 (1999).

17. A.H. Mueller: *Small x and Diffraction Scattering*, DIS'98, eds. Ch. Coremans and R. Roosen, WS,1998, *Parton Saturation at Small x and in Large Nuclei*, CU-TP-937-99, hep-ph/9904404.

18. H. Abramowicz, E. Levin, A. Levy and U. Maor: Phys. Lett. B **269**, 465 (1991); H. Abramowicz and A. Levy: *ALLM'97*, DESY 97 -251, hep-ph/9712415.

19. K. Golec-Biernat and M. Wüsthoff: Phys. Rev. D **59**, 014017 (1999); *Saturation in Diffractive Deep Inelastic Scattering*, DTP-99-20,hep-ph/9903358.

20. A.D. Martin, R.G. Roberts, W.J. Stirling and R.S. Thorne: *Parton Distributions and the LHC: W and Z Production*, DTP-99-64, hep-ph/9907231.

Experimental Results and Perspectives of Polarized Lepton-Nucleon Scattering

Michael Düren

Universität Erlangen-Nürnberg, Erwin-Rommel-Str. 1,
D-91058 Erlangen, Germany

Abstract. The spin structure of the nucleon and its spin puzzle is introduced in this paper, followed by a summary of the recent results on polarized lepton nucleon scattering. The spin structure functions of the proton and neutron have been measured and their first moments have been determined. The recent experiments confirm that the Ellis-Jaffe sum rule is violated, whereas the Bjorken sum rule is valid within the precision of the measurement. Results are reported about the flavor decomposition of the quark polarization, a first direct measurement of a positive gluon polarization, the observation of a double-spin asymmetry in diffractive ρ^0 production, the polarization of Λ hyperons, the observation of transverse single-spin asymmetries and the measurement of the Gerasimov-Drell-Hearn sum rule. Prospects of future fixed target and collider facilities are discussed.

1 Introduction

The fact that the nucleon has a large anomalous magnetic moment proves that it is not a fundamental spin-1/2 Dirac particle. The approximate agreement of the measured magnetic moments of the members of the baryon multiplets with predictions of the $SU(3)_f \times SU(2)$ symmetric quark-parton model is an important confirmation of the quark model approach. Additional predictions about the spin structure of the nucleon were the Bjorken [1] and Ellis-Jaffe [2] sum rules, which however were not experimentally accessible in the first two decades of the quark model. In 1987 the EMC collaboration published a measurement of the spin structure function $g_1^p(x)$ of the proton and of its first moment, the Ellis-Jaffe sum [3]. The observed violation questioned our understanding of the spin structure of the nucleon in terms of the quark-parton model. It caused intense discussion in the community and demanded experimental and theoretical clarification.

Since then a series of new spin experiments has been performed at SLAC, CERN and DESY as summarized in [4]. They became feasible by improved experimental techniques to polarize beams and targets. At CERN and SLAC the improved figure of merit, which is the product of luminosity and the squared values of the polarization of beam and target, allowed for a precision measurement of the inclusive spin structure functions. The HERMES experiment [5] uses a completely novel method, a polarized internal storage cell target in a storage ring with longitudinally polarized electrons. The HERMES storage cell is a 40 cm long elliptical tube around the stored electron beam. Polarized atoms are continuously injected into the storage cell and leave the cell only after an average of a few hundred wall bounces. The main advantage of the storage cell technique is the ability to use pure, highly polarized atomic species (H, D, ^{3}He) in contrast to solid state targets where only a small fraction of the atomic species is polarizable.

The HERMES experiment is dedicated to semi-inclusive measurements, i.e. measurements which detect final state hadrons in coincidence with the scattered lepton. By tagging certain final state hadrons with different flavor content, the spin contributions Δu, Δd, Δs of the up, down and strange quark flavors can be disentangled.

The inclusive as well as the semi-inclusive results show that only a fraction of the nucleon spin is due to the spin contribution $\Delta \Sigma = \Delta u + \Delta d + \Delta s$ of the quarks. The rest is due to the contribution ΔG of the gluon spin and due to angular momentum contributions L_q and L_G of quarks and gluons moving with high speed in the nucleon. Current experiments show first evidence of a non-zero contribution by the gluon spin, however further experiments are needed to disentangle and understand all the contributions of the nucleon's spin $s_z^{\mathcal{N}}$, which are summarized in the helicity sum rule

$$s_z^{\mathcal{N}} = \frac{1}{2} = \frac{1}{2} \left(\Delta u + \Delta d + \Delta s \right) + L_q + \Delta G + L_G. \tag{1}$$

To improve our understanding of the spin structure of the nucleon, in future not only the collinear spin contributions have to be measured, but more emphasis has to be given to investigate transverse spin components, twist-3 contributions, spin-dependent off-forward parton distributions and spin effects in fragmentation.

2   Inclusive Deep Inelastic Scattering

2.1   Spin Structure Functions

The spin dependent part of the inclusive deep inelastic scattering cross section $e + \mathcal{N} \rightarrow e' + X$ is characterized by two spin structure functions $g_1(x, Q^2)$ and $g_2(x, Q^2)$. In the quark-parton model the Bjorken variable x is interpreted as the momentum fraction carried by the struck quark and $-Q^2$ is the squared four-momentum of the exchanged virtual photon and is related to the resolution of the scattering process. The first spin structure function $g_1(x, Q^2)$ is interpreted as

$$g_1(x, Q^2) = \frac{1}{2} \sum_f e_f^2 \Delta q_f(x, Q^2) \tag{2}$$

with e_f being the charge for a quark with flavor f in units of the elementary charge and $\Delta q_f(x, Q^2) = q_f^{\uparrow\uparrow}(x, Q^2) - q_f^{\uparrow\downarrow}(x, Q^2)$ is the polarized quark distribution function. $q_f^{\uparrow\uparrow(\uparrow\downarrow)}(x, Q^2)$ is the distribution function of quarks with spin orientation parallel (anti-parallel) to the spin of the nucleon. The second spin structure function $g_2(x, Q^2)$ is zero in the simple quark model. In general, $g_2(x, Q^2)$ contains the Wandzura-Wilczek contribution $g_2^{WW}(x, Q^2) = -g_1(x, Q^2) + \int_x^1 dx' \, g_1(x', Q^2)/x'$, a pure twist-3 contribution, and minor contributions from transverse parton polarizations [6].

The spin structure function g_1 has been measured since many years, and an impressive, precise data set has been collected by the experiments at SLAC, by SMC and by HERMES as shown in Fig. 1 [4]. At all three sites the spin structure function $g_1(x, Q^2)$ was measured for the proton and the neutron. The neutron spin asymmetry is obtained either from the subtraction of deuteron and proton data, or directly from a ^{3}He target. In the ^{3}He nucleus the spins of the protons are anti-parallel and do not contribute to the measured asymmetry (except for a small contribution which can be corrected). All experimental results are consistent and agree with the Q^2 evolution as predicted by QCD.

The measurement of g_2 is more difficult. Experiments at SMC and SLAC have shown that g_2 is consistent with g_2^{WW}, but data are also almost consistent with zero. The experiment E155x at SLAC ran in spring 1999 with the aim to provide more precise data on g_2 and to gain access to the interesting twist-3 component of g_2 that remains after subtraction of the Wandzura-Wilczek contribution g_2^{WW}.

2.2 Sum Rules

Using $SU(3)_f$ symmetry arguments, the first moment Γ_1 of the spin structure function $g_1(x)$ can be related to the axial couplings F and D known from baryon decays:

$$\Gamma_1^{p(n)} = +(-)\frac{F+D}{12} + \frac{3F-D}{36} + \frac{\Delta\Sigma}{9} + \text{QCD corrections terms} \qquad (3)$$

In addition, the total spin carried by quark spins can be expressed as

$$\Delta\Sigma = 3F - D + 3\Delta s. \qquad (4)$$

The Ellis-Jaffe sum rule [2] is obtained from Eq. (3) by the additional (reasonable but possibly false) assumption that the spin carried by strange quarks is negligible: $\Delta s = 0$. From this assumption follows that the total spin carried by quark spins is $\Delta\Sigma = 3F - D = 58\%$ (modulo QCD corrections at finite Q^2). Fig. 2 [7] demonstrates that the Ellis-Jaffe sum rule is violated with the consequence, that either the strange quark contribution to the proton spin is significant, or that the $SU(3)_f$ symmetry which was used to derive Eq. (3) is broken. The Bjorken sum rule [1], which is the non-singlet combination of the first moments of proton and neutron does not require these two assumptions and is experimentally verified at the 10% level (see Fig. 2). The precision of the sum rule tests is limited by theoretical uncertainties in the extrapolation of the spin structure functions at low x. Experimentally, the low x range will be accessible only by a future high energy spin experiment at a new facility as the proposed polarized HERA collider [8].

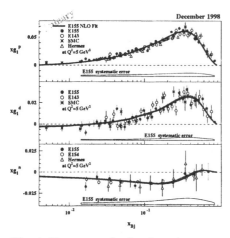

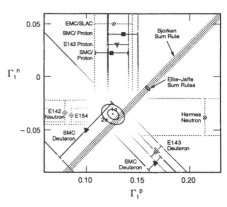

Fig. 1. Recent results for the spin structure functions $xg_1(x)$ at $Q^2 = 5$ GeV^2 for the proton, deuteron and neutron. The continuous lines are results of a QCD fit at NLO.

Fig. 2. The measured values of the first moments Γ_1^p and Γ_1^n of the spin structure functions g_1^p and g_1^n of the proton and neutron are in good agreement with the Bjorken sum rule, but deviate from the prediction of the Ellis-Jaffe sum rule.

A further sum rule, the Gerasimov-Drell-Hearn (GDH) sum rule, relates the polarization dependent part of the total photoproduction cross section to the anomalous magnetic moment of the nucleon. This important relation has been tested by a precision experiment at the tagged polarized photon beam of the microtron MAMI in Mainz, Germany. The energy range accessible by the experiment is 200-800 MeV. A preliminary analysis of a small subset of the data reports a number of 230 ± 20 µb, which accounts for most of the GDH prediction of $\int_{m_\pi}^{\infty} (\sigma_{3/2} - \sigma_{1/2})/\nu \, d\nu = 204$ µb [9]. The remaining difference might be due to contributions at higher energies, which are planned to be measured at ELSA/Bonn and at Jefferson Lab.

The GDH sum rule can be generalized for electroproduction. Data have been recently published by the HERMES collaboration [10], and a new measurement in the important resonance region was performed in spring 1999 as experiment E94-010 at Jefferson Lab [11].

3   Semi-inclusive Deep Inelastic Scattering

3.1   Flavor Decomposition

The observed violation of the Ellis-Jaffe sum rule made clear that the spin structure of the nucleon is not understood. More detailed experimental information is needed to disentangle the various possible contributions of quarks and gluons to the spin of the nucleon. Semi-inclusive data can be used to measure the sea polarization directly and to test SU(3)$_f$ symmetry by comparing the first moments of the flavor distributions to the SU(3)$_f$ predictions. In addition, semi-inclusive polarized DIS experiments can determine the separate spin contributions Δq_f of quark and antiquark flavors f to the total spin of the nucleon not only as a total integral but as a function of the Bjorken scaling variable x.

Hadron production in DIS is described by the absorption of a virtual photon by a point-like quark and the fragmentation into a hadronic final state. The two processes can be characterized by two functions: the quark distribution function $q_f(x, Q^2)$, and the fragmentation function $D_f^h(z, Q^2)$. The semi-inclusive DIS cross section $\sigma^h(x, Q^2, z)$ to produce a hadron of type h with energy fraction $z = E_h/\nu$ is then given by

$$\sigma^h(x, Q^2, z) \propto \sum_f e_f^2 q_f(x, Q^2) D_f^h(z, Q^2). \tag{5}$$

It is assumed that the fragmentation process is spin independent, i.e. that the probability to produce a hadron of type h from a quark of flavor f is independent of the relative spin orientations of quark and nucleon. The spin asymmetry A_1^h in the semi-inclusive cross section for production of a hadron of type h by a polarized virtual photon is given by

$$A_1^h(x, Q^2, z) = \frac{\sum_f e_f^2 \Delta q_f(x, Q^2) D_f^h(z, Q^2)}{\sum_f e_f^2 q_f(x, Q^2) D_f^h(z, Q^2)} \cdot \frac{1 + R(x, Q^2)}{1 + \gamma^2}. \tag{6}$$

The term $R = \sigma_L/\sigma_T$ is the ratio of the longitudinal to transverse photon absorption cross section and appears in this formula to correct for the longitudinal component. It is assumed that R is flavor and target independent and that the contribution from the second spin structure function $g_2(x, Q^2)$ can be neglected. The term $\gamma = \sqrt{Q^2}/\nu$ is a kinematic factor which enters from the $g_2 = 0$ assumption. Eq. (6) can be used to extract the quark polarizations $\Delta q_f(x)/q_f(x)$ from a set of measured asymmetries on the proton and neutron for positively and negatively charged hadrons.

Results on the decomposition of the proton spin into contributions from the valence spin distributions Δu_v and Δd_v and from the sea Δq_s have been previously reported by SMC [12]. New, more precise data from HERMES have been presented recently [13]. Fig. 3 shows the polarization $\Delta q/q$ of quarks in the proton, separated into flavors. The *up* flavor has a positive polarization which reaches about 40% at large x, whereas the *down* flavor has a polarization opposite to the proton spin, in excess of 20%. In the sea region at small x the up and down polarizations do not vanish completely. The sea polarization itself is compatible with zero as shown in the lower panel. The extraction of the sea was done under the assumption that the polarization of the sea quarks is independent of their flavor.

The first and second moments of the spin distributions have been determined by HERMES. In the measured region, the results of HERMES and SMC agree within the quoted errors. A simple Regge-type extrapolation has been applied at low x to obtain the total integrals as quoted in Table 1. The HERMES results for the first and second moment of Δu_v show a significant discrepancy with

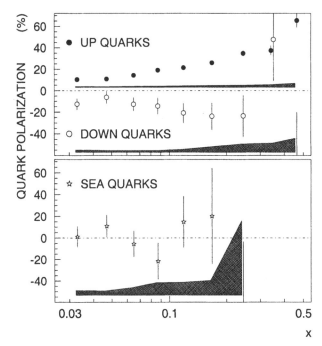

Fig. 3. The polarization of quarks in the proton has been measured by HERMES as a function of x, separately for the flavors up and down and for the sea. The error bars shown are the statistical and the bands the systematic uncertainties.

a prediction from quenched lattice QCD in Ref. [14]. The result for $\Delta u + \Delta \bar{u}$ is inconsistent with the result from the inclusive data based on $SU(3)_f$ flavor symmetry as in Ref. [15]. The inconsistency of the up flavor has its counterpart in the difference which is observed in the sea results. The inclusive analysis obtains a large negative strange sea compared to the zero sea in the semi-inclusive analysis. However, the uncertainties of the sea measurements are so large that the result is not conclusive.

Table 1. The integrals of various spin distributions as measured by HERMES for $Q^2 = 2.5$ GeV^2. Note that the entry for $\Delta s + \Delta \bar{s}$ does not represent a direct measurement of the strange sea but relies on the assumption that the sea polarization is flavor independent. An uncertainty of the Regge-type extrapolation at low x is not included in the quoted error.

	total integral
$\Delta u + \Delta \bar{u}$	$0.56 \pm 0.02 \pm 0.03$
$\Delta d + \Delta \bar{d}$	$-0.25 \pm 0.06 \pm 0.05$
$\Delta s + \Delta \bar{s}$	$-0.02 \pm 0.03 \pm 0.04$
Δq_0	$0.28 \pm 0.04 \pm 0.09$
Δq_3	$0.83 \pm 0.07 \pm 0.06$
Δq_8	$0.32 \pm 0.09 \pm 0.10$
Δu_v	$0.57 \pm 0.05 \pm 0.08$
Δd_v	$-0.21 \pm 0.10 \pm 0.13$
$x \Delta u_v$	$0.12 \pm 0.01 \pm 0.01$
$x \Delta d_v$	$-0.02 \pm 0.02 \pm 0.02$

Possible explanations of these differences are that either SU(3)$_f$ is violated, which would modify the inclusive result, or that the assumption about the flavor independence of the sea is wrong, which would modify the semi-inclusive result. To test the applicability of SU(3)$_f$ and SU(2)$_f$ flavor and isospin symmetry, the semi-inclusive results for the octet combination Δq_8 and the triplet combination Δq_3 of the polarized quark distributions have been compared to the predictions $\Delta q_8 = 3F - D$ and $\Delta q_3 = F + D$ (Bjorken sum rule). Both predictions agree with the HERMES results when the appropriate QCD corrections are taken into account. For a decisive conclusion about the origin of the discrepancy, the precision has to be further improved and the sea assumption has to be tested explicitly.

A significant improvement of the precision of the $\Delta d(x) + \Delta \bar{d}$ determination is expected in the near future from HERMES using the 1999 deuterium data set. The recently installed RICH will allow a direct measurement of $\Delta s(x)$ using kaon identification.

3.2 Transverse Asymmetries

The next step in polarized DIS, beyond the understanding of the collinear part of the quark and gluon polarization in the nucleon, is the understanding of the transverse polarization components. Single-spin asymmetries in polarized hadronic reactions are interpreted as effects of *time-reversal-odd* distribution functions (Sivers mechanism) or *time-reversal-odd* fragmentation functions (Collins mechanism) [16].

SMC presented recently the first measurement of semi-inclusive DIS hadron production on a transversely polarized target [17]. Leading hadron production has been analyzed in terms of the Collins angle and indeed a non-zero asymmetry $A_N = 11\% \pm 6\%$ has been found for positive hadrons, whereas the negative hadrons yield $-2\% \pm 6\%$.

A significant result has been reported by HERMES on a related quantity [18]. HERMES measured the asymmetry of hadron production on a longitudinally polarized target. Even in this case an asymmetry is expected in the azimuthal angle between the plane which contains the produced pion and the virtual photon and the plane which contains the scattered lepton and the virtual photon. Fig. 4 shows this single-spin asymmetry as a function of the azimuthal angle for positive and negative pions. A sinusoidal fit yields an asymmetry of $A_N = 2.0\% \pm 0.4\%$ for the positive and $A_N = -0.1\% \pm 0.5\%$ for the negative pions.

4   Diffractive Asymmetries

Results on double-spin asymmetries in diffractive ρ^0-production have been reported by SMC [19] and HERMES [20]. Naively, no spin asymmetry is expected in the approach where diffraction is described by the exchange of a pomeron with vacuum quantum numbers. In this frame one should expect that diffractive ρ^0 production does not know about the spin of the target nucleon. The SMC

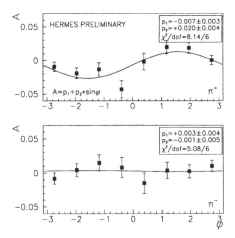

Fig. 4. Azimuthal dependence of the single-spin asymmetry in the cross section for π^+ (top) and π^- (bottom). The error bars are statistical uncertainties. The curves are sinusoidal fits to the data.

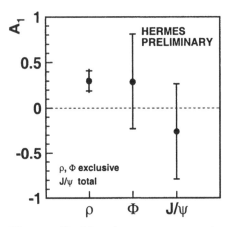

Fig. 5. Double-spin asymmetry in diffractive ρ, ϕ and J/ψ production from HERMES.

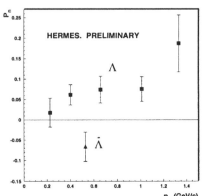

Fig. 6. The squares (triangles) indicate the transverse polarization of Λ ($\bar{\Lambda}$) in unpolarized photoproduction as function of the transverse momentum.

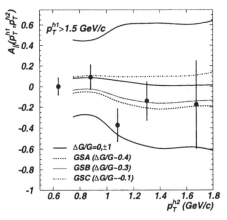

Fig. 7. Spin asymmetry in the production of hadron pairs with high p_T and opposite charge. One hadron with $p_T > 1.5$ GeV is required. The spin asymmetry is plotted as a function of the p_T of the second hadron and compared to Monte-Carlo predictions using various assumptions for the gluon polarization $\Delta G/G$.

result is in agreement with this prediction. No significant asymmetry has been observed.

HERMES reported a significant, unexpected positive asymmetry of $A_1^\rho = 0.30 \pm 0.11 \pm 0.05$ in exclusive ρ^0 production as shown in Fig. 5. The asymmetry in the production of other vector mesons, ϕ and J/ψ, was also measured, but with much less precision and is compatible with zero (see Fig. 5). The difference of the two results is possibly due to different reaction mechanisms at the different W^2

ranges of the two experiments: $W^2 \approx 225$ GeV^2 for SMC, and $W^2 \approx 25$ GeV^2 for HERMES.

4.1 Λ Polarization

Due to the parity violation of the weak decay of Λ hyperons, the angular distribution of the decay products can be used to extract the spin orientation of the Λ hyperon before its decay. This unique feature was used at HERMES to extract two interesting quantities [21].

The first one is the measurement of the polarization transfer in DIS scattering of longitudinally polarized electrons off unpolarized targets. A Λ polarization of $P_\Lambda = 0.03 \pm 0.06 \pm 0.03$ was reported, a number which however is not precise enough to distinguish between different predictions. The naive quark model which assumes 100% polarization of s-quarks in Λ hyperons predicts $P_\Lambda = 0.018$, whereas a SU(3)$_f$ symmetric model from Jaffe predicts $P_\Lambda = -0.057$, based on the measured results for the Ellis-Jaffe sum.

A much more precise result was reported concerning the transverse polarization of Λ hyperons in quasi-photoproduction off an unpolarized target:

$$\gamma^{(*)} p \to \Lambda X. \tag{7}$$

The polarization was measured with reference to the plane perpendicular to the Λ production plane. Fig. 6 shows the polarization as a function of the transverse momentum of the Λ and $\bar{\Lambda}$. A large positive polarization is observed for Λ hyperons, with the tendency to increase with their transverse momentum. The $\bar{\Lambda}$ antihyperons show a negative polarization. There is no straight-forward explanation of the observed asymmetries in QCD; however, similar polarizations have been found in hadronic collisions.

4.2   Gluon Polarization

The "most wanted" component of the nucleon spin is the polarization of gluons, as they are probably responsible for the spin deficit of the quarks. As the virtual photon does not couple directly to gluons, a measurement of the gluon polarization was up to now only very indirectly possible by using the QCD evolution equations, which relate the Q^2-dependence of the quark distributions to the gluon distribution. A QCD NLO analysis of recent data yields a gluon contribution to the proton spin of $\Delta G \approx 1.8 \pm 1.0$ [7].

For the first time a more direct measurement of the gluon polarization has been presented [22]. By selecting events with two hadrons with opposite charge and with large transverse momentum, HERMES was able to accumulate a sample of events which is enriched by photon-gluon fusion events. By requiring a large transverse momentum of 1.5 (1) GeV/c for the first (second) hadron, the subprocess, where the gluon splits into two quarks, has a hard scale and can be treated pertubatively. HERMES estimates from Monte-Carlo studies that the average squared transverse momentum of the quarks is 2.1 (GeV/c)^2. As long as the fragmentation process is spin independent, the spin asymmetry in the production of the quark-antiquark pair is the same as the spin asymmetry of the observed final state. The measured asymmetry is however affected by background processes. The unique signature of the HERMES result is the negative sign of the asymmetry. All background processes have a positive asymmetry, as long as they are dominated by the positive polarization of the up-quarks in the proton. The observed negative asymmetry can be explained by a significant positive gluon polarization. The change of sign comes from the negative analyzing power of the photon-gluon fusion diagram. Using a specific background Monte Carlo, HERMES obtains a value of the gluon polarization of $\Delta G/G = 0.41 \pm 0.18 \pm 0.03$ at $\langle x_G \rangle = 0.17$. The quantitative result depends however critically on the detailed understanding of the background processes. Fig. 7 shows the asymmetry together with Monte-Carlo predictions using various assumptions for the gluon polarization $\Delta G/G$.

5   Future Facilities

The future of polarized DIS will consist of both, fixed target [23] and collider experiments [8,24].

5.1   Fixed target

At SLAC the fixed target inclusive era will end with the precise measurement of $g_2^{p,d}$ at E155x. At lower energies, MAMI at Mainz, ELSA at Bonn and CEBAF at Jefferson Lab will continue to do spin physics. The main future players at higher energies will be HERMES at DESY and COMPASS at CERN. Both experiments will concentrate on semi-inclusive data.

Their main aims are the measurement of the gluon polarization, the flavor decomposition of polarized quark distributions, polarized vector meson production, polarized fragmentation functions, transversity, and, in the case of COM-PASS, also the angular momentum of quarks and gluons and off-forward parton distributions.

HERMES has upgraded its particle identification recently to achieve pion, kaon and proton separation in the full kinematic region, using a RICH detector. An improved muon acceptance and identification will allow for a better J/Ψ detection. A wheel of silicon detectors just behind the target cell will be installed, which improves the acceptance especially for Λ decay products. A recoil detector system is dedicated to low energy, large angle target fragments and spectator nucleons.

COMPASS will start in 2000 in the experimental area where the SMC experiment has been, however with an improved beam, improved target, high luminosity and, compared to the SMC experiment, with a much better and larger hadron acceptance and particle identification. In the final stage COMPASS will have two spectrometer magnets, two RICH detectors, two hadron and two electromagnetic calorimeters. Compared to HERMES, the high beam energy of COMPASS of

100-200 GeV enables measurements at smaller x and larger Q^2 and W^2. The large W^2 allows for charm production well above threshold and for the generation of hadrons with large transverse momentum. The production of open charm allows for a direct measurement of the gluon polarization.

A future fixed target machine is ELFE, a possible new European electron machine in the 15-30 GeV range, which is discussed in connection with the TESLA project at DESY and also at CERN as a machine which could re-use the cavities from LEP. Two further experiments, which both aimed for the measurement of the gluon polarization via charm production in photoproduction, were proposed some years ago, and are still discussed: E-156 at SLAC and APOLLON at ELFE.

5.2   Collider

Two colliders will govern the high energy part of spin physics in the future: the polarized proton collider RHIC at BNL [24], and possibly the HERA collider at DESY which may have polarized protons in future [8].

In both machines, the acceleration and storage of polarized protons is a major challenge to machine physicists. Several Siberian snakes will be needed which compensate depolarizing resonances of the beam polarization.

RHIC will start its physics program in 2000. Main points on the program are the measurement of the antiquark polarization and the gluon polarization. The antiquark polarization can be extracted using Drell-Yan production via W^+ and W^-. As W-production depends on flavor and on helicity, the experiment can extract Δu, $\Delta \bar{u}$, Δd and $\Delta \bar{d}$ separately. The gluon polarization is approached by the production of prompt photons, π^0's, jets and heavy quarks (charm).

The main aims of a polarized HERA collider are the measurement of the spin structure functions at low x, which will improve the precision of the verification of the fundamental Bjorken sum rule, and the polarization of the gluon. Fig. 8 shows the anticipated precision of this measurement. As at RHIC, the gluon polarization can be extracted from the production of heavy flavors and from jet production. Fig. 9 [8] summarizes the anticipated precision of the measurement of the gluon polarization at RICH, HERA, and COMPASS. Also included is the HERA-N option, which is a proposal to scatter HERA protons off a fixed target.

References

1. J.D. Bjorken, Phys. Rev. **148** (1966) 1467; Phys. Rev. D **1** (1970) 1376.
2. J. Ellis and R.L. Jaffe, Phys. Rev. D **9** (1974) 1444; Phys. Rev. D **10** (1974) 1669.
3. EMC, J. Ashman *et al.*, Phys. Lett. B **206** (1988) 364; Nucl. Phys. B **328** (1989) 1.
4. R. Windmolders, Proc. of the 7th Int. Workshop on Deep Inelastic Scattering and QCD (DIS99), Zeuthen, Germany, April 19-23, 1999, to appear in Nucl. Phys. B (Proc. Suppl.).
5. HERMES, K. Ackerstaff *et al.*, Nucl. Instr. Meth. A **417** (1998) 230.
6. W. Wandzura and F. Wilczek, Phys. Lett. B **172** (1977) 195.
7. E. Hughes in [4].

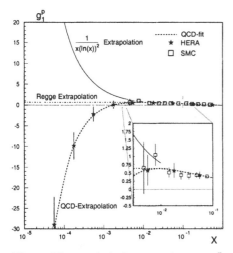

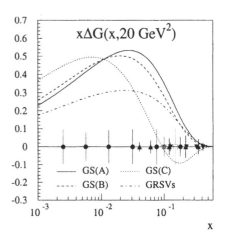

Fig. 8. The statistical uncertainty on g_1^p from possible measurements at HERA with 500 pb^{-1} is shown along with different theoretical predictions at low x. The center of the figure is expanded to show the low-x SMC points.

Fig. 9. Comparison of the statistical accuracy of $x\Delta G$ of the proposed future experiments as a function of x along with predictions from various theoretical models.

8. V. Hughes and D. von Harrach in [4].
9. A. Thomas in [4].
10. HERMES, K. Ackerstaff *et al.*, Phys. Lett. B **444** (1998) 531.
11. Z.-E. Meziani in [4].
12. SMC, D. Adeva *et al.*, Phys. Lett. B **420** (1998) 180.
13. HERMES, K. Ackerstaff et al., subm. to Phys. Lett.; hep-ex/9906035.
14. M. Göckeler *et al.*, Phys. Lett. B **414**, (1997) 340.
15. J. Ellis and M. Karliner, Phys. Lett. B **341** (1995) 397;
    Invited Lectures at the International School of Nucleon Spin Structure, Erice, August 1995, hep-ph/9601280.
16. J. Collins, Nucl. Phys. B **396** (1993) 161.
17. S. Bravar in [4].
18. H. Avakian in [4].
19. A. Tripet in [4].
20. F. Meißner in [4].
21. S. Belostotski in [4].
22. M. Amarian in [4].
23. E. Kabuß in [4].
24. Y. Goto and N. Saito in [4].

Polarized Parton Densities and Processes

Marco Stratmann

Department of Physics, University of Durham, Durham, DH1 3LE, England

Abstract. The main goals of 'spin physics' are recalled, and some theoretical and phenomenological aspects of longitudinally polarized deep inelastic scattering and other hard processes are reviewed. The spin dependent parton densities of protons and photons and polarized fragmentation functions are introduced, and the relevant theoretical framework in next-to-leading order QCD is briefly summarized. Technical complications typical for spin dependent calculations beyond the leading order of QCD, like a consistent γ_5 prescription, are sketched, and some recent results for jet and heavy quark production are discussed. Special emphasis is put on conceivable measurements at a future polarized upgrade of the HERA collider which is currently under consideration.

1   Introduction

One of the most fundamental properties of elementary particles is their spin. However, the vast majority of past and present experiments at high energy e^+e^-, ep, and pp colliders are performed with unpolarized beams thus neither exploiting the advantages of polarization, which were demonstrated, e.g., by the SLD experiment at SLAC, nor revealing any information on the spin dependence of fundamental interactions. Unlike lepton beams it is an extremely challenging task to maintain the polarization of protons throughout the acceleration to high energies, which explains the lack of polarized ep or pp collider experiments in the past. To circumvent this problem, a series of fixed target experiments with longitudinally polarized lepton beams scattered off, e.g., proton targets have been performed at comparatively low energies over the past few years [1].

Aiming at polarized deep inelastic scattering (DIS) these experiments have been used to extract first information about the spin dependent parton densities

$$\Delta f^H(x, Q^2) \equiv f_+^{H+}(x, Q^2) - f_-^{H+}(x, Q^2) \ , \tag{1}$$

where f_+^{H+} (f_-^{H+}) denotes the density of a parton f with helicity '+' ('−') in a hadron H with helicity '+'. It is important to notice that the Δf^H contain information *different* from that included in the more familiar unpolarized distributions f^H [defined by taking the sum on the r.h.s. of (1)], and their measurement is indispensable for a *complete* understanding of the partonic structure of hadrons. However, due to the lack of any experimental information apart from DIS and the limited kinematical coverage in x and Q^2 of the available measurements [1], our knowledge of the Δf is still rather rudimentary compared to the abundance of results on f.

Much experimental progress and, hopefully, exciting new results have to be expected in the next couple of years. Most importantly measurements of, for instance, jet, prompt photon, and W-boson production rates at the recently completed first polarized pp collider RHIC will vastly reduce our ignorance of the Δf. Ongoing efforts in the fixed target sector by HERMES [2] and (soon) by COMPASS [3] to study, in particular, semi-inclusive DIS and charm production, respectively, will contribute to a more complete picture of polarized parton densities as well. Here we will mainly focus on the prospects of a conceivable future polarized upgrade of the HERA ep collider [4], which is currently under scrutiny, and highlight on some important measurements uniquely possible at an ep collider.

Having pinned down the polarized parton densities (1) one can study one of the most fundamental aspects of polarization: the question of how the spin S_z of non-pointlike objects like nucleons is composed of the spin of their constituents, the quarks and gluons, and their orbital angular momentum $L_z^{q,g}$. The total contribution of quarks and gluons to S_z is determined by the first moments of (1), $\Delta f(Q^2) \equiv \int_0^1 \Delta f(x, Q^2)dx$, and S_z can be written as

$$S_z = \frac{1}{2} = \frac{1}{2}\Delta\Sigma(Q^2) + \Delta g(Q^2) + L_z^q(Q^2) + L_z^g(Q^2) \ , \tag{2}$$

where $\Delta\Sigma \equiv \sum_q(\Delta q + \Delta\bar{q})$ and Q denotes the 'resolution scale' at which the nucleon is probed. The so far unmeasured angular momentum contribution $L_z^{q,g}$ has attracted considerable theoretical interest recently, and it was suggested [5] that deeply virtual Compton scattering $\gamma^*(Q^2)p \to \gamma p'$ in the limit of vanishing momentum transfer $t = (p - p')^2$ may provide first direct information on $L_z^{q,g}$, however this subject is beyond the scope of this talk.

The definition of polarized parton densities (1) also holds true for the hadronic content of *photons*, Δf^γ, and can be easily extended to the time-like case, i.e., spin dependent fragmentation functions, ΔD_f, as well. Both densities have been measured in the unpolarized case, and their Q^2 evolution provides an important test of perturbative QCD. Needless to stress again that a measurement of Δf^γ and ΔD_f is required for a complete understanding of space- and time-like distributions. So far Δf^γ is completely unmeasured, and almost nothing is known experimentally about spin dependent fragmentation. It is argued below that a polarized HERA would be also an ideal place to learn more about these densities.

Our contribution is organized as follows: First we review the spin dependent proton structure and shall give an example of a recent QCD analysis of polarized DIS data [6]. Then the framework is extended to the case of Δf^γ and ΔD_f, and theoretical models for these densities are introduced. Next we turn to polarized processes and briefly sketch the basic technical framework and complications due to the appearance of γ_5. Finally we discuss the main results of two recently finished NLO calculations: jet [7,8] and heavy flavor production [9]. It should be noted that we have to omit several interesting topics such as L_z, transverse polarization and transversity distributions, single spin processes, etc. Some recent results and references can be found, e.g., in [10].

2   Polarized Proton Structure and DIS

Longitudinally polarized DIS can be described by introducing a structure function g_1, in analogy to F_2 and F_L in the helicity-averaged case. The NLO expression for g_1 reads (suppressing the obvious x and Q^2 dependence)

$$g_1 = \frac{1}{2} \sum_{q=u,d,s} e_q^2 \left[(\Delta q + \Delta \bar{q}) \otimes \left(1 + \frac{\alpha_s}{2\pi} \Delta C_q\right) + \frac{\alpha_s}{2\pi} \Delta g \otimes \Delta C_g \right], \qquad (3)$$

where $\Delta C_{q,g}$ are the spin dependent Wilson coefficients, and the symbol $\otimes$ denotes the usual convolution in x space. From (3) it is obvious that the available inclusive DIS data [1] can reveal only information on $\Delta q + \Delta \bar{q}$, but neither on Δq and $\Delta \bar{q}$ nor on Δg, which enters (3) only as an $\mathcal{O}(\alpha_s)$ correction. Thus all QCD analyses [11,12,6] have to impose certain *assumptions* about the flavor decomposition in order to be able to estimate other hard processes for upcoming experiments like RHIC. Alternatively one can stick, of course, to a comprehensive analysis of quantities accessible in polarized DIS [13,14].

The Δf obey the standard DGLAP Q^2 evolution equations – with all unpolarized quantities such as splitting functions replaced by their spin dependent counterparts (given in [15,16]) – which are readily solved analytically in Mellin n moment space. A subtlety arises in NLO in the non-singlet (NS) sector [17]. The independent NS combinations $q_- = q - \bar{q}$ and $q_+ \sim q - q'$ evolve in the unpolarized and the polarized case with the same but interchanged kernels, i.e., $P_\pm = \Delta P_\mp$. This simply reflects the fact that in the unpolarized case the first moment of q_-, the number of valence quarks, is conserved with Q^2, whereas in the polarized case $\Delta q_+(Q^2)$ refers to a conserved NS axial vector current. The Δf are constrained by the unpolarized densities via the positivity condition

$$\left| \Delta f(x, Q^2) \right| \leq f(x, Q^2), \qquad (4)$$

which is exploited in most of the QCD analyses. Of course, the bound (4) is strictly valid only in LO and is subject to NLO corrections [18] because the Δf become unphysical, scheme dependent objects in NLO. However the corrections are not very pronounced, in particular at large x [18], the only region where (4) imposes some restrictions in practice and hence (4) can be used also in NLO.

Figure 1 shows the result of a recent NLO QCD analysis [6] of all presently available data [1]. The fit is performed directly to the measured spin asymmetry

$$A_1(x, Q^2) \simeq \frac{g_1(x, Q^2)}{F_2(x, Q^2)/[2x(1 + R(x, Q^2))]}, \qquad (5)$$

where $R = F_L/2xF_1$, rather than to the extracted structure function g_1 itself. Eq. (5) is related to the polarized-to-unpolarized cross section ratio $\Delta\sigma/\sigma$, and experimental uncertainties like the absolute normalization conveniently drop out.

As mentioned above, each QCD fit has to rely on several assumptions. The shown GRSV analysis [6] is characterized by the choice of a low starting scale for the evolution, $Q_0 \approx 0.6\,\text{GeV}$, the $\overline{\text{MS}}$ scheme, and a simple but flexible ansatz

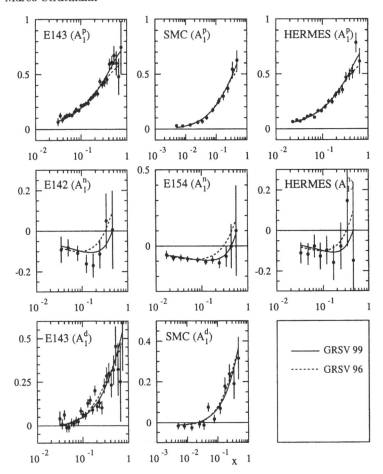

Fig. 1. Comparison of an updated NLO QCD analysis [6] in the GRSV framework [11] with available data sets [1] (the E155 data are not shown, but included in the fit). Also shown are the original GRSV results [11] based on older and fewer data sets.

for the polarized densities $\Delta f(x, Q_0^2) = N_f x^{\alpha_f} (1 - x)^{\beta_f} f(x, Q_0^2)$, *assuming* that $\Delta \bar{q} = \Delta \bar{u} = \Delta \bar{d}$ and $\Delta s = \Delta \bar{s} = \lambda \Delta \bar{q}$. For the unpolarized reference distributions f the updated GRV densities [19] have been used, which also fixes the choice of Q_0 (and $\alpha_s(M_z^2) = 0.114$). The remaining free parameters are determined by the fit after exploiting constraints for the first moments of the NS combinations Δq_+ (F and D values) and by *choosing* $\lambda = 1$, i.e., a $SU(3)_f$ symmetric sea.

The individual parton densities Δf resulting from the fit in Fig. 1 are shown in Fig. 2. To demonstrate that, in particular, the gluon density is hardly constrained at all by present data, two other fits based on additional *ad hoc* constraints on Δg are shown in Fig. 2. The '$\Delta g = 0$' fit starts from a vanishing gluon input, and the 'static Δg' is chosen in such a way that its first moment becomes roughly independent of Q^2. Both gluons give also excellent fits to the available

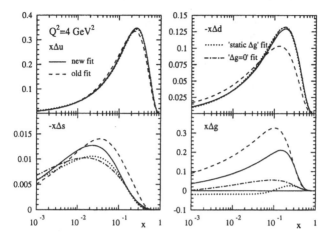

Fig. 2. The polarized NLO $\overline{\text{MS}}$ densities at $Q^2 = 4\,\text{GeV}^2$ as obtained in the new [6] and old [11] GRSV analyses. Also shown are the distributions obtained in two other fits employing additional constraints on Δg (see text).

data and do not affect the results for u and d. In fact one can obtain fits without changing χ^2 by more than one unit for an even wider range of gluon inputs. This uncertainty in Δg is compatible with the findings of other recent analyses such as [14]. In addition, similarly agreeable fits are obtained, e.g., for the choice $\lambda = 1/2$ as well as by using an independent x shape for Δs, reflecting the above mentioned uncertainty in the flavor separation. The range of results for the Δf obtained by the various QCD analyses [6,11–14] gives a rough measure of the theoretical uncertainties due to different assumptions used for the fits.

It is interesting to observe that for the 'best fit' gluon in the GRSV framework [6,11] the spin of the nucleon (2) is dominantly carried by quarks and gluons at the low bound-state like input scale Q_0, and only during the Q^2 evolution a large negative $L_z^g(Q^2)$ is being built up in order to compensate for the strong rise of $\Delta g(Q^2)$, see Fig. 5 in [20]. However, no definite conclusions can be reached yet because for the 'static Δg' the situation is completely different, and S_z is entirely of angular momentum origin for *all* values of Q^2, contrary to what is intuitively expected. In addition, direct measurements of $L_z^{q,g}$ are completely missing.

Inevitably the large uncertainty in Δg implies that the small x behaviour of g_1 is completely uncertain and not reliably predictable as is illustrated in Fig. 3. This translates also into a sizeable theoretical error for the $x \to 0$ extrapolation when calculating first moments of g_1, which play an important role in spin physics since they are related to predictions such as the Bjorken sum rule [21]. The situation is similar to our ignorance of the small x behaviour of F_2 in the pre-HERA era and can be resolved only experimentally. Needless to say that a polarized variant of HERA would be of ultimate help here. In addition, the high Q^2 region would be accessible for the first time at HERA. Here electroweak effects become increasingly important and new structure functions, which probe different combinations of parton densities, enter.

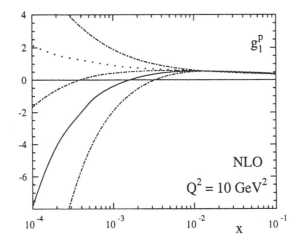

Fig. 3. Predictions for the small x behaviour of g_1 by extrapolating from the measured region $x > 0.01$ to smaller x values for different assumptions about Δg. The solid line is the result obtained using the 'best fit' Δg of [6] as shown in Fig. 2.

3   Polarized Photon Structure and Fragmentation

The complete NLO QCD framework for the Q^2 evolution of Δf^γ and the calculation of the polarized photon structure function g_1^γ, which would be accessible in $e\gamma$ DIS at a future polarized linear collider [22], was recently provided in [23]. Unlike the proton densities the Δf^γ obey an *inhomogeneous* evolution equation schematically given by

$$\frac{d\Delta q_i^\gamma}{d\ln Q^2} = \Delta k_i + (\Delta P_i \otimes \Delta q_i^\gamma) \;, \tag{6}$$

where Δq_i^γ stands for the flavor NS quark combinations or the singlet (S) vector $\Delta q_S^\gamma \equiv \begin{pmatrix} \Delta \Sigma^\gamma \\ \Delta g^\gamma \end{pmatrix}$, and Δk_i denotes the photon-to-parton splitting functions. Again, solutions of (6), which can be decomposed into a 'pointlike' (inhomogeneous) and a 'hadronic' (homogeneous) part, $\Delta q_i^\gamma = \Delta q_{i,PL}^\gamma + \Delta q_{i,had}^\gamma$, can be given analytically for n moments (cf. [24]). It should be noted that perturbative instabilities for g_1^γ in the $\overline{\text{MS}}$ scheme due to the $x \to 1$ behaviour of the photonic coefficient function ΔC_γ [23] can be avoided, as in the unpolarized case [24], by absorbing ΔC_γ into the definition of the quark densities [23] (DIS$_\gamma$ scheme).

At present the Δf^γ are unmeasured, and one has to fully rely on theoretical models. The only guidance is provided by the positivity constraint analogous to Eq. (4). The 'current conservation' (CC) condition [25], which demands a vanishing first moment of g_1^γ and is automatically fulfilled for the pointlike part [23], is not very useful without any data since it can be implemented at x values smaller than those one is interested in, say, at $x < 0.005$. To obtain a realistic estimate for the theoretical uncertainties in Δf^γ coming from the unknown

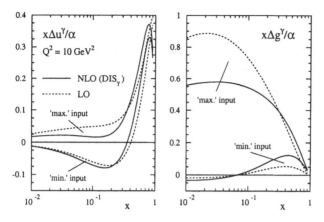

Fig. 4. $x\Delta u^{\gamma}/\alpha$ and $x\Delta g^{\gamma}/\alpha$ evolved to $Q^2 = 10\,\text{GeV}^2$ in LO and NLO (DIS$_{\gamma}$) using the two extreme models explained in the text.

hadronic input, one can consider two very different models [26,23] by either saturating the positivity bound (4) at $Q_0 \simeq 0.6\,\text{GeV}$ ('maximal scenario') with the phenomenologically successful unpolarized GRV photon densities [27] or by using a vanishing input ('minimal scenario'). The resulting Δf^{γ} for both scenarios are shown in Fig. 4 and will be applied below to estimate the prospects of measuring Δf^{γ} in photoproduction processes at a polarized HERA in the future.

Studies of *spin transfer reactions* could provide further invaluable insight into the field of spin physics. A non-vanishing twist-2 spin transfer asymmetry requires the measurement of the polarization of one outgoing particle, in addition to having a polarized beam or target, and is sensitive to spin dependent *fragmentation*. Λ baryons are particularly suited for such studies due to the self-analyzing properties of their dominant weak decay, which were successfully exploited at LEP [28] to reconstruct the Λ spin. In [29] a first attempt was made to extract the spin dependent Λ fragmentation functions, ΔD_f^{Λ}, by analyzing these data [28], which, however, turned out to be insufficient. Rather different, physically conceivable scenarios appear to describe the data equally well, and for the 'unfavoured' sea quark and gluon fragmentation functions one has to fully rely on mere assumptions. Clearly, further measurements are required to test the models proposed in [29], and, again, HERA can play an important role here.

The time-like (TL) ΔD_f^{Λ} are defined in a similar way as their space-like (SL) counterparts in Eq.(1) via

$$\Delta D_f^{\Lambda}(z, Q^2) \equiv D_{f_+}^{\Lambda+}(z, Q^2) - D_{f_+}^{\Lambda-}(z, Q^2) \ , \tag{7}$$

where, e.g., $D_{f_+}^{\Lambda+}(z, Q^2)$ is the probability for finding a Λ baryon with positive helicity in a parton f with positive helicity at a mass scale Q, carrying a fraction z of the parent parton's momentum. The Q^2 evolution of (7) is similar to the SL case, and it should be recalled only that the off-diagonal entries in the singlet

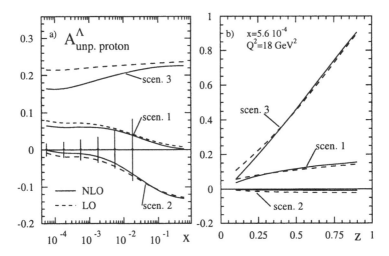

Fig. 5. The semi-inclusive DIS asymmetry A^A for unpolarized protons and polarized A's and leptons for the three distinct scenarios of ΔD_f^A of [29]. In **a)** the expected statistical errors for such a measurement at HERA are shown, assuming a luminosity of $500 \, \text{pb}^{-1}$, a lepton beam polarization of 70%, and a A detection efficiency of 0.1.

evolution matrices $\Delta \hat{P}^{(SL,TL)}$ interchange their role when going from the SL to the TL case, see, e.g., [30,31].

As a manifestation of the so-called Gribov-Lipatov relation [32] the SL and TL splitting functions are equal in LO. Furthermore they are related by analytic continuation (ACR) of the SL splitting functions (Drell-Levy-Yan relation [33]), which can be schematically expressed as ($z < 1$)

$$\Delta P_{ij}^{(TL)}(z) = z\mathcal{AC}\left[\Delta P_{ji}^{(SL)}(x = \frac{1}{z})\right], \tag{8}$$

where the operation $\mathcal{AC}$ analytically continues any function to $x \to 1/z > 1$ and correctly adjusts the color factor and the sign [31]. The breakdown of the ACR beyond the LO in the $\overline{\text{MS}}$ scheme can be understood in terms of a corresponding breakdown for the $n = 4 - 2\varepsilon$ dimensional LO splitting functions and can be easily accounted for by a simple factorization scheme transformation [31]. Alternatively, the ACR breaking can be calculated, of course, graph-by-graph [31] in the light-cone gauge method [34], which is of course much more cumbersome.

LO and NLO predictions for the semi-inclusive spin asymmetry A^A for the production of polarized A's in DIS of *unpolarized protons* off polarized leptons [29] is shown in Fig. 5 for three different conceivable models of the ΔD_f^A mentioned above (see [29] for details). Such types of spin measurements, which would help to pin down the ΔD_f^A more precisely, can be performed at HERA immediately after the spin rotators in front of H1 and ZEUS have been installed even *without* having a polarized proton beam. Similar studies can be done in the photoproduction case where an integrated luminosity of only about $100 \, \text{pb}^{-1}$ would be sufficient [35]. Helicity transfer reactions can be also examined in pp collisions at RHIC [36].

4 Polarized Processes

4.1 Some General Remarks, γ_5, and All That

To calculate longitudinally polarized cross sections one has to project onto the two independent helicity configurations of the incoming polarized partons (for simplicity we ignore here helicity transfer processes where the formalism applies in a similar way). This is achieved by using the standard relations (see, e.g., [37])

$$\epsilon_\mu(k, \lambda)\,\epsilon_\nu^*(k, \lambda) = \frac{1}{2}\left[-g_{\mu\nu} + i\lambda\epsilon_{\mu\nu\rho\sigma}\frac{k^\rho p^\sigma}{k \cdot p}\right] \tag{9}$$

for incoming bosons with momentum k and helicity λ, and where p denotes the momentum of the other incoming particle, and

$$u(k, h)\bar{u}(k, h) = \frac{1}{2}\,\not{k}(1 - h\gamma_5) \tag{10}$$

for incoming massless quarks with momentum k and helicity h. Using (9) and (10) one can calculate the cross sections for unpolarized *and* polarized beams *simultaneously* by taking the sum or the difference of the two helicity dependent squared matrix elements

$$\text{unpolarized}: \quad \overline{|M|}^2 = \frac{1}{2}\left[|M|^2\,(++) + |M|^2\,(+-)\right] \tag{11}$$

$$\text{polarized}: \quad \Delta|M|^2 = \frac{1}{2}\left[|M|^2\,(++) - |M|^2\,(+-)\right] \tag{12}$$

where $|M|^2\,(h_1, h_2)$ denotes the squared matrix element for any of the contributing subprocesses for definite helicities h_1 and h_2 of the incoming particles. The possibility to recover well-known unpolarized results 'for free' is usually regarded as a first important check on the correctness of the spin dependent results.

As usual the presence of IR, UV, and collinear singularities demands some consistent method to make them manifest. For this purpose one usually works in the well-established framework of n dimensional regularization (DREG), which immediately leads to complications in the polarized case since both γ_5 and the totally antisymmetric tensor $\epsilon_{\mu\nu\rho\sigma}$ in (9) and (10) are genuine *four* dimensional and have no straightforward continuation to $n \neq 4$ dimensions[1]. Since the use of a naive anticommuting γ_5 in n dimensions is known to lead to algebraic

---

[1] Sometimes a variant of DREG, dimensional *reduction* [38] (DRED), is preferred. Here the Dirac algebra is performed in *four* rather than n dimensions. However, extra counterterms have to be introduced to match the UV sectors of DREG and DRED [39,40]. Once this is done DREG and DRED are simply related by a factorization scheme transformation [40,41,16].

inconsistencies [42], one usually chooses to work in the HVBM scheme [43], which was shown to be internally consistent in n dimensions, and its peculiarities will be briefly reviewed below. Alternatively one can stick to an anticommuting γ_5 by abandoning the cyclicity of trace [44]. In this scheme a 'reading point' has to be defined from where *all* Dirac traces of a given process have to be started which can be a quite cumbersome procedure. Another prescription was suggested to handle traces with one γ_5 [45] by utilizing $\gamma_\mu \gamma_5 = i/(3!)\epsilon_{\mu\nu\rho\sigma}\gamma^\nu\gamma^\rho\gamma^\sigma$ and contracting the resulting Levi-Civita tensors in n dimensions. This avoids $(n-4)$ dimensional scalar products which show up in the HVBM scheme but results in more complicated trace calculations. Needless to say that in the end all consistent prescriptions should give the same result when used appropriately.

In the HVBM scheme [43] the four dimensional definition for γ_5 is maintained, and the ϵ-tensor is regarded as a genuinely four dimensional object. In this way the n dimensional space is splitted up into a four and a $(n-4)$ dimensional subspace, and $(n-4)$ dimensional scalar products ('*hat momenta*') can show up in $|M|^2 (h_1, h_2)$ apart from their usual n dimensional counterparts (i.e., Mandelstam variables). For single inclusive jet or heavy quark production, e.g., one can choose a convenient frame where all non-vanishing $(n-4)$ dimensional scalar products can be expressed by a single hat momenta combination $\hat{p}^2$. These terms deserve special attention when performing the $2 \to 3$ phase space integrations since the $(n-4)$ dimensional subspace cannot be integrated out trivially as in any unpolarized calculation. However, the modified phase space can be conveniently written as $\mathrm{dPS}_3 = \mathrm{dPS}_{3,\mathrm{unp}} \times \mathcal{I}(\hat{p}^2)$ such that it reduces to the well-known 'unpolarized' phase space formula $\mathrm{dPS}_{3,\mathrm{unp}}$ for the vast majority of terms in the matrix element which do not depend on $\hat{p}^2$; see [46,9] for details.

The remaining calculation is then standard and proceeds in the same way as for any unpolarized cross section with one further crucial exception concerning the factorization of mass singularities. It was observed [16] that the LO polarized splitting function in $n = 4 - 2\varepsilon$ dimensions in the HVBM prescription, $\Delta P_{qq}^{(0),n}$, is no longer equal to its unpolarized counterpart, i.e., it violates helicity conservation, $\Delta P_{qq}^{(0),n}(x) - P_{qq}^{(0),n}(x) = 4C_F\varepsilon(1-x)$. This unwanted property has to be accounted for by an additional factorization scheme transformation whenever a pole $\sim \Delta P_{qq}^{(0)}1/\varepsilon$ has to be subtracted [16]. When talking about the $\overline{\mathrm{MS}}$ scheme in the polarized case in connection with the HVBM prescription, it is always understood that this additional transformation is already done.

4.2   Some Recent Results: Jets, Heavy Quarks

Let us finally focus on some recent phenomenological results. The complete NLO QCD corrections for jet production in polarized pp [7] and ep [8] collisions have become available recently in form of MC codes which allow to study all relevant differential jet distributions. The photoproduction of jets at a polarized HERA is known to be an excellent tool to extract first information on the photonic densities Δf^γ by experimentally enriching that part of the cross section that stems from 'resolved' photons [47]. In case of single inclusive jet production this

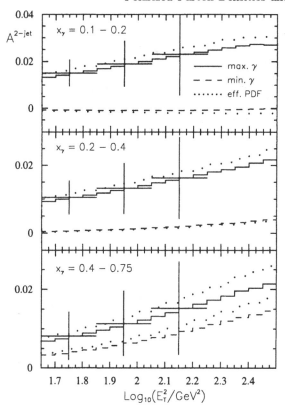

Fig. 6. Predictions for $A^{2-\text{jet}}$ for different bins in x_γ using the two scenarios for Δf^γ as described in the text and the LO GRSV distributions [11] for Δf^p. Also shown are the results using the effective parton density approximation and the expected statistical errors assuming a luminosity of $200\,\text{pb}^{-1}$ and 70% beam polarizations.

can be achieved by looking into the direction of positive jet rapidities (proton direction), and this feature was shown to be maintained also at NLO [8]. In addition, an improved dependence of the cross section on the factorization and renormalization scales, μ_f and μ_r, respectively, was found, and the LO jet spin asymmetries in [47] receive only moderate NLO corrections [8].

Similar studies of di-jet production have the advantage that the kinematics of the underlying hard subprocess can be fully reconstructed and the momentum fraction x_γ of the photon can be determined on an experimental basis. In this way it becomes possible to experimentally suppress the 'direct' photon contribution by introducing some suitable cut $x_\gamma \leq 0.75$ [48], or by scanning different bins in x_γ. Very encouraging results were found in [49], and it was shown that the LO QCD parton level calculations nicely agree with 'real' jet production processes including initial and final state QCD radiation as well as non-perturbative effects such as hadronization, as modeled using the spin dependent **SPHINX** MC [50].

Figure 6 shows the experimentally relevant di-jet spin asymmetry $A^{2-\text{jet}} \equiv d\Delta\sigma/d\sigma$ in LO for three different bins in x_γ [51], using similar cuts as in a cor-

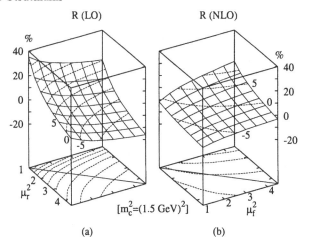

Fig. 7. $R = [\Delta\sigma^c_{\gamma p}(\mu^2_r, \mu^2_f) - \Delta\sigma^c_{\gamma p}(\mu^2_r = \mu^2_f = 2.5m^2_c)]/\Delta\sigma^c_{\gamma p}(\mu^2_r = \mu^2_f = 2.5m^2_c)$ in LO (a) and NLO (b) in percent for $\sqrt{S} = 10$ GeV. μ_f and μ_r are in units of the charm quark mass $m_c = 1.5$ GeV. The contour lines are in steps of 5% and for convenience a line corresponding to the usual choice $\mu_f = \mu_r$ is shown at the base of the plots.

responding unpolarized measurement [52]. $A^{2-\text{jet}}$ in NLO QCD has not been studied yet, but again only moderate corrections should be expected. Since it would be a very involved task to unfold the Δf^γ from such a measurement of $A^{2-\text{jet}}$ due to the wealth of contributing subprocesses and combinations of parton densities, the result of the so-called 'effective parton density approximation' is also shown in Fig. 6. This handy but still accurate approximation which is based on an old idea [53], allows to straightforwardly extract a specific 'effective' combination of the Δf^γ given by [51] $\Delta f^\gamma_{\text{eff}} \equiv \sum_q[\Delta q^\gamma + \Delta\bar{q}^\gamma] + \frac{11}{4}\Delta g^\gamma$, once the corresponding combination Δf^p_{eff} for proton densities is precisely known from, e.g., jet production in DIS at a polarized HERA [4] or from RHIC.

Finally, the calculation of the NLO QCD corrections to the polarized photoproduction of heavy quarks has been finished recently as well [9], and NLO results for the charm contribution g^{charm}_1 to the DIS structure function g_1 [54] and for the hadroproduction of heavy quarks [55] will become available very soon. Heavy flavor production is dominated by gluon initiated fusion processes and hence highly sensitive to the so far poorly known Δg. Unfortunately at HERA neither g^{charm}_1 nor the photoproduction of charm give sizeable enough contributions to be of any use in determining Δg. In the case of photoproduction of charm the prospects are much better for the upcoming fixed target experiment COMPASS at CERN [3]. The NLO corrections in this case appear to be sizeable but well under control, and, most importantly, the theoretical uncertainties due variations of the scale μ_f and μ_r are greatly reduced when going to the NLO of QCD [9] as is illustrated in Fig. 7.

Certainly the next couple of years will produce many new experimental results in the field of spin physics. In particular first data from the RHIC pp collider, but also results from HERMES and COMPASS, will considerably improve our knowledge of the spin structure of nucleons. But only a future polarized ep and a linear e^+e^- collider can ultimately resolve issues like the small x behaviour of g_1, the structure of polarized photons, and spin dependent fragmentation.

Acknowledgements

It is a pleasure to thank the organizers for inviting me to this interesting meeting at such an inspiring location.

References

1. E142 Collab., P.L. Anthony et al.: Phys. Rev. **D54**, 6620 (1996); E154 Collab., K. Abe et al.: Phys. Rev. Lett. **79**, 26 (1997); HERMES Collab., K. Ackerstaff et al.: Phys. Lett. **B404**, 383 (1997); E143 Collab., K. Abe et al.: Phys. Rev. **D58**, 112003 (1998); HERMES Collab., A. Airapetian et al.: Phys. Lett. **B442**, 484 (1998); SM Collab., B. Adeva et al.: Phys. Rev. **D58**, 112001 (1998). E155 Collab., P.L. Anthony et al.: SLAC-PUB-8041, 1999 [Phys. Rev. Lett.]

2. See, e.g., M. Düren: these proceedings

3. COMPASS Collab., G. Baum et al.: CERN/SPSLC-96-14, CERN-SPSLC-96-30

4. A. De Roeck and T. Gehrmann (eds.): *Physics with polarized protons at HERA*, DESY-PROCEEDINGS-1998-01

5. X. Ji: Phys. Rev. Lett. **78**, 610 (1997)

6. M. Stratmann: *hep-ph/9907465*, to appear in the proc. of the workshop *DIS '99*, Zeuthen, Germany, 1999 [Nucl. Phys. **B** Proc. Suppl.]

7. D. de Florian, S. Frixione, A. Signer, and W. Vogelsang: Nucl. Phys. **B539**, 455 (1999)

8. D. de Florian and S. Frixione: Phys. Lett. **B457**, 236 (1999)

9. I. Bojak and M. Stratmann: Phys. Lett. **B433**, 411 (1998); Nucl. Phys. **B540**, 345 (1999)

10. W. Vogelsang: *hep-ph/9906289*, to appear in the proc. of the workshop *DIS '99*, Zeuthen, Germany, 1999 [Nucl. Phys. **B** Proc. Suppl.]

11. M. Glück, E. Reya, M. Stratmann, and W. Vogelsang: Phys. Rev. **D53**, 4775 (1996)

12. T. Gehrmann and W.J. Stirling: Phys Rev. **D53**, 6100 (1996); D. de Florian, O.A. Sampayo, and R. Sassot: Phys. Rev. **D57**, 5803 (1998); E. Leader, A.V. Siderov, and D.B. Stamenov: Phys. Rev. **D58**, 114028 (1998)

13. G. Altarelli, R. Ball, S. Forte, and G. Ridolfi: Nucl. Phys. **B496**, 337 (1997); Acta Phys. Polon. **B29**, 1145 (1998)

14. SM Collab., B. Adeva et al.: Phys. Rev. **D58**, 112002 (1998)

15. R. Mertig and W.L. van Neerven: Z. Phys. **C70**, 637 (1996)

16. W. Vogelsang: Phys. Rev. **D54**, 2023 (1996); Nucl. Phys. **B475**, 47 (1996)

17. M. Stratmann, W. Vogelsang, and A. Weber: Phys. Rev. **D53**, 138 (1996)

18. G. Altarelli, S. Forte, and G. Ridolfi: Nucl. Phys. **B534**, 277 (1998)

19. M. Glück, E. Reya, and A. Vogt: Eur. Phys. J. **C5**, 461 (1998).

20. M. Stratmann: in proc. of the *2nd Topical Workshop on Deep Inelastic Scattering off Polarized Targets*, Zeuthen, 1997, J. Blümlein and W.-D. Nowak (eds.), p. 94.

21. J.D. Bjorken: Phys. Rev. **148**, 1467 (1966); Phys. Rev. **D1**, 1376 (1970)

22. M. Stratmann: *hep-ph/9907467*, to appear in the proc. of the workshop *Photon '99*, Freiburg, Germany, 1999 [Nucl. Phys. **B** Proc. Suppl.]

23. M. Stratmann and W. Vogelsang: Phys. Lett. **B386**, 370 (1996)

24. M. Glück, E. Reya, and A. Vogt: Phys. Rev. **D45**, 3986 (1992)

25. S.D. Bass: Int. J. Mod. Phys. **A7**, 6039 (1992); S. Narison, G.M. Shore, and G. Veneziano: Nucl. Phys. **B391**, 69 (1993); S.D. Bass, S.J. Brodsky, and I. Schmidt: Phys. Lett. **B437**, 417 (1998)

26. M. Glück and W. Vogelsang: Z. Phys. **C55**, 353 (1992); ibid. **C57**, 309 (1993); M. Glück, M. Stratmann, and W. Vogelsang: Phys. Lett. **B337**, 373 (1994)

27. M. Glück, E. Reya, and A. Vogt: Phys. Rev. **D46**, 1973 (1992)

28. ALEPH Collab., D. Buskulic et al.: Phys. Lett. **B374**, 319 (1996); paper submitted to the *XVIII International Symposium on Lepton Photon Interactions*, 1997, Hamburg, Germany, paper no. **LP279**; DELPHI Collab.: DELPHI 95-86 PHYS 521 [paper submitted to the *EPS-HEP '95* conference, Brussels, 1995]; OPAL Collab., K. Ackerstaff et al.: Eur. Phys. J. **C2**, 49 (1998)

29. D. de Florian, M. Stratmann, and W. Vogelsang: Phys. Rev. **D57**, 5811 (1998)

30. W. Furmanski and R. Petronzio: Phys. Lett. **97B**, 437 (1980)

31. M. Stratmann and W. Vogelsang: Nucl. Phys. **B496**, 41 (1997)

32. V.N. Gribov and L.N. Lipatov: Sov. J. Nucl. Phys. **15**, 438 (1972); ibid. **15**, 675 (1972)

33. S.D. Drell, D.J. Levy, and T.M. Yang: Phys. Rev. **187**, 2159 (1969); Phys. Rev. **D1**, 1617 (1970)

34. R.K. Ellis, H. Georgi, M. Machacek, H.D. Politzer, and G.G. Ross: Phys. Lett. **B78**, 281 (1978); Nucl. Phys. **B152**, 285 (1979)

35. D. de Florian, M. Stratmann, and W. Vogelsang: in *Physics with polarized protons at HERA*, A. De Roeck and T. Gehrmann (eds.), DESY-PROCEEDINGS-1998-01, p. 140

36. D. de Florian, M. Stratmann, and W. Vogelsang: Phys. Rev. Lett. **81**, 530 (1998)

37. N.S. Craigie, K. Hidaka, M. Jacob, and F.M. Renard: Phys. Rep. **99**, 69 (1983)

38. W. Siegel: Phys. Lett. **B84**, 193 (1979)

39. G.A. Schuler, S. Sakakibara, and J.G. Körner: Phys. Lett. **B194**, 125 (1987)

40. J.G. Körner and M.M. Tung: Z. Phys. **C64**, 255 (1994)

41. Z. Kunszt, A. Signer, and Z.Trocsanyi: Nucl. Phys. **B411**, 397 (1994); B. Kamal: Phys. Rev. **D53**, 1142 (1996)

42. M. Chanowitz, M. Furman, and I. Hinchliffe: Nucl. Phys. **B159**, 225 (1979)

43. G. 't Hooft and M. Veltman: Nucl. Phys. **B44**, 189 (1972); P. Breitenlohner and D. Maison: Comm. Math. Phys. **52**, 11 (1977)

44. J.G. Körner, D. Kreimer, and K. Schilcher: Z. Phys. **C54**, 503 (1992)

45. S.A. Larin and J.A.M. Vermaseren: Phys. Lett. **B259**, 345 (1991); S.A. Larin: Phys. Lett. **B303**, 113 (1993)

46. L.E. Gordon and W. Vogelsang: Phys. Rev. **D48**, 3136 (1993)

47. M. Stratmann and W. Vogelsang: Z. Phys. **C74**, 641 (1997); in proc. of the 1995/96 workshop on *Future Physics at HERA*, DESY, Hamburg, G. Ingelman et al. (eds.), p. 815

48. J.R. Forshaw and R.G. Roberts: Phys. Lett. **B319**, 539 (1993)

49. J.M. Butterworth, N. Goodman, M. Stratmann, and W. Vogelsang: in *Physics with polarized protons at HERA*, A. De Roeck and T. Gehrmann (eds.), DESY-PROCEEDINGS-1998-01, p. 120

50. S. Güllenstern et al.: *hep-ph/9612278*; O. Martin, M. Maul, and A. Schäfer: in *Physics with polarized protons at HERA*, A. De Roeck and T. Gehrmann (eds.), DESY-PROCEEDINGS-1998-01, p. 236

51. M. Stratmann and W. Vogelsang, *hep-ph/9907470*, to appear in the proc. of the workshop *Polarized Protons at High Energies - Accelerator Challenges and Physics Opportunities*, DESY, Hamburg, Germany, 1999
52. H1 Collab., C. Adloff et al.: Eur. Phys. J. **C1**, 97 (1998)
53. B.L. Combridge and C.J. Maxwell: Nucl. Phys. **B239**, 429 (1984)
54. J. Smith: to appear in the proc. of the workshop *DIS '99*, Zeuthen, Germany, 1999 [Nucl. Phys. **B** Proc. Suppl.]
55. I. Bojak and M. Stratmann: in preparation

Part II

Final States in Deep-Inelastic Scattering

Forward Jet Production at HERA

Björn Pötter

Max-Planck-Institut für Physik, Föhringer Ring 6, 80805 München, Germany

Abstract. We discuss forward jet production data recently published by the H1 and ZEUS collaborations at HERA. We review how several Monte-Carlo models compare to the data. QCD calculations based on the BFKL formalism and on fixed NLO perturbation theory with and without resolved virtual photons are described.

1 Introduction

It is by now firmly established that the proton structure function $F_2(x, Q^2)$ shows a steep rise in the small x-Bjorken region, i.e., below $x = 10^{-3}$ [1,2]. This rise is compatible with DGLAP evolution [3], where F_2 is fitted at a fixed input scale Q_0^2 and then evolved up to Q^2 by summing up $\ln Q^2$ terms. In the small x region an alternative way to describe the evolution might be to sum up $\ln(1/x)$ terms, as it is done in the BFKL approach [4]. The BFKL equation actually directly predicts the behaviour of F_2 as a function of x. An attempt to unify these two approaches is the CCFM evolution equation [5], which reproduces both, the DGLAP and the BFKL behaviour, in their respective regimes of validity.

The existing data on F_2 do not allow to unambigously determine whether the BFKL mechanism is needed in the x-range covered by HERA. Therefore, alternatively the cross section for forward jet production in deep inelastic scattering (DIS) has been proposed as a particularly sensitive means to investigate the parton dynamics at small x [6]. The proposal is based on the observation that the DGLAP and BFKL equations predict different ordering of the transverse momenta $k_{T,i}$ along the parton cascade developing in DIS jet production, see Fig. 1. While the DGLAP equation predicts a strong $k_{T,i}$ ordering and only a weak ordering in the longitudinal momentum fractions x_i, i.e.,

$$Q^2 = k_{T,n}^2 \gg \ldots \gg k_{T,1}^2 \qquad \text{and} \qquad x = x_n < \ldots < x_1 \qquad (1)$$

the BFKL approach predicts a strong ordering in the x_i, but no ordering in $k_{T,i}$. The idea is now to observe jets at very small x, with $x \ll x_1$ in the forward region with $Q^2 \simeq k_{T,1}^2$. In this way the DGLAP evolution is suppressed, whereas the BFKL evolution is left active.

The forward jet cross section has been measured recently at HERA [7,8] and in the following I will review the various attempts that have been made to describe this data.

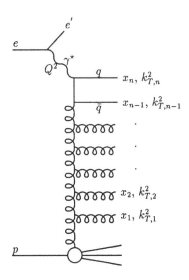

Fig. 1. Evolution of the transverse momenta $k_{T,i}$ and the longitudinal momentum fractions x_i along the ladder contributing to DIS jet production.

2   Data and Monte-Carlo models

The H1 and ZEUS collaborations have measured forward jet cross sections at small x for rather similar kinematical conditions [7,8]. The jet selection criteria and kinematical cuts are summarized in Tab. 1.

In Fig. 2 and 3 the collected data from ZEUS and H1 in the forward region is shown together with predictions from various Monte-Carlo models. The data from both groups is on the hadron level. For the ZEUS data, the Monte-Carlo models ARIADNE [9], LEPTO [10], HERWIG [11] and LDC [12] have been used for predictions. ARIADNE includes one of the main features present in the BFKL approach, which is the absence of the strong k_T ordering. LDC is based on the CCFM approach and finally LEPTO and HERWIG are based on conventional leading log DGLAP evolution. Except for HERWIG, the same

Table 1. Forward jet selection criteria by H1 and ZEUS

H1 cuts	ZEUS cuts
$E'_e > 11$ GeV	$E'_e > 10$ GeV
$y_e > 0.1$	$y_e > 0.1$
$E_{T,jet} > 3.5$ (5) GeV	$E_{T,jet} > 5$ GeV
$1.7 < \eta_{jet} < 2.8$	$\eta_{jet} < 2.6$
$0.5 < E^2_{T,jet}/Q^2 < 2$	$0.5 < E^2_{T,jet}/Q^2 < 2$
$x_{jet} > 0.035$	$x_{jet} > 0.036$

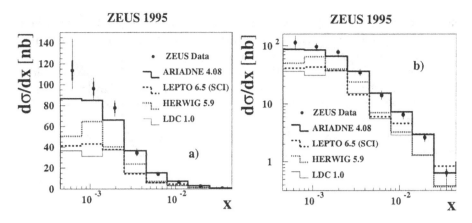

Fig. 2. Forward jet data of ZEUS [7] as a function of Bjorken-x for $E_{T_{jet}} > 5$ GeV. (a) linear scale, (b) logarithmic scale. The shaded band gives the error due to the uncertainty of the jet energy scale. The data is compared to various Monte-Carlo models.

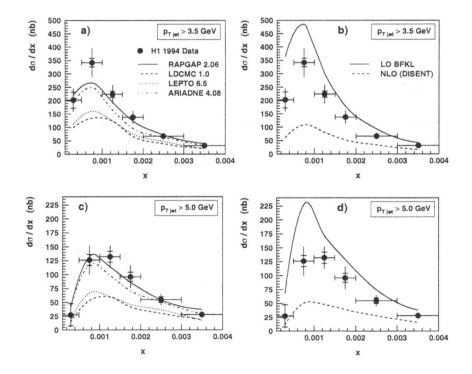

Fig. 3. Forward jet data of H1 [8] as a function of Bjorken-x for two $E_{T_{jet}}$ cuts of 3.5 GeV and 5.0 GeV. (a) and (c) contain Monte-Carlo model predictions, whereas (b) and (d) show the results of a LO BFKL (full) and a fixed NLO (dashed) calculation.

Monte-Carlo models are shown in the H1 plots. Instead of HERWIG, the RAP-GAP model [13] is shown, which is also based on DGLAP evolution but contains an additional resolved virtual photon component.

As is clear from Fig. 2, ARIADNE describes the forward jet cross section reasonably well, apart from the smallest x-bin where it gives slightly too small cross sections. The other three used models predict cross sections which lie significantly below the data. Similar results can be extracted from Fig. 3 (a) and (c). LDC and LEPTO lie below the data by a factor of 2, whereas ARIADNE gives a reasonable good description. As an interesting result, also RAPGAP gives a good description of the H1 data.

3   BFKL approach

Of course, attempts have been made to calculate the forward jet cross section directly within the BFKL formalism. BFKL calculations in LO by Kwieciński et al., Bartels et al. and Tang [14] based on the BFKL approach overshoot the older forward jet data [15]. This can also be seen in Fig. 3 (b) and (d), where the LO BFKL calculation on the parton level [16] (full line) is compared to the recent H1 data [8]. These older calculations suffer, however, from several deficiencies. They are asymptotic and do not contain the correct kinematic constraints of the produced jets. Furthermore they do not allow the implementation of a jet algorithm as used in the experimental analysis. Also NLO $\ln(1/x)$ terms in the BFKL kernel [17] predict large negative corrections which are expected to reduce the forward cross section as well.

Recently the BFKL calculations have been improved by taking into account higher order consistency conditions as a way of including sub-leading corrections to the BFKL equation [18]. The consistency constraint (CC) requires that the virtuality of the emitted gluons along the chain should arise predominantly from the transverse components of momentum. By including this CC the authors in [18] claim to recover a dominant part of higher order effects. Furthermore, the CC is said to subsume energy-momentum conservation over a wide range of the allowed phase space, which is another source of sub-leading contributions.

Including the CC conditions, good agreement between the predictions and the forward jet data is found, for both the H1 and the ZEUS data, as shown in Fig. 4. The predictions depend on the choice of scale and on an additional infrared cut-off parameter k_0. However, the k_0 dependence is much less than the uncertainty due to the choice of scales. Further details about the BFKL calculations from [18] can be found in these proceedings [19], also describing the calculation for forward π^0 production.

4   Fixed NLO QCD calculations

Calculations in fixed order perturbation theory have already been performed for the older H1 forward jet measurement [15] by Mirkes and Zeppenfeld [20] using

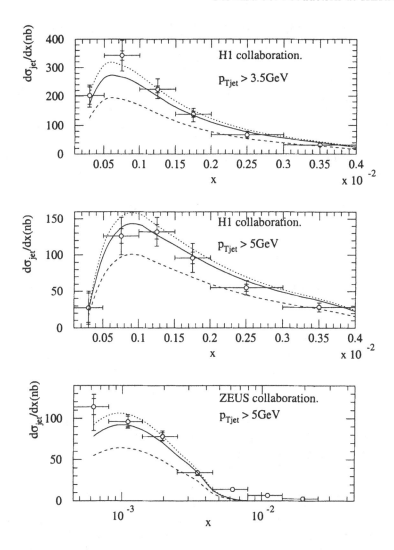

Fig. 4. Forward jet data of H1 [8] and ZEUS [7] compared to prediction based on the BFKL formalism including sub-leading corrections [18]. The three curves correspond to three choices of scales and infrared cut-off.

their fixed order program MEPJET [21]. The calculations where done in next-to-leading order (NLO) accuracy, i.e., taking the matrix elements up to $O(\alpha_s^2)$ into account. It was found that the NLO calculations are a factor of 2 to 4 below the data. Similar predictions have been made with help of the DISENT program [22] for the more recent H1 measurements [8], shown as the dashed lines in Fig. 3 (b) and (d), which confirm the earlier findings. Since the forward jet data can be succesfully described by the RAPGAP model which includes resolved virtual photons, it is fair to ask whether also fixed order calculations including a resolved virtual photon component will be able to describe the data.

4.1   Low Q^2 jet production in NLO

NLO calculations for jet production with slightly off-shell direct and resolved virtual photons have become available recently [23], extending calculations done in the photoproduction regime [24]. The NLO calculations [23] are performed with the phase space slicing method. As is well known the higher order (in α_s) contributions to the direct and resolved cross sections have infrared and collinear singularities. For the real corrections singular and non-singular regions of phase space are separated by a technical cut-off parameter y_s. Both, real and virtual corrections, are regularized by going to d dimensions. The NLO corrections to the direct process become singular in the limit $Q^2 \to 0$ in the initial state on the real photon side. For $Q^2 = 0$ these photon initial state singularities are usually also evaluated with the dimensional regularization method. Then the singular contributions appear as poles in $\epsilon = (4 - d)/2$ multiplied with the splitting function $P_{q\gamma}$ and have the form $-\frac{1}{\epsilon}P_{q\gamma}$ multiplied with the LO matrix elements for quark-parton scattering. These singular contributions are absorbed into PDF's $f_{a/\gamma}(x)$ of the real photon. For $Q^2 \neq 0$ the corresponding contributions are replaced by

$$-\frac{1}{\epsilon}P_{q\gamma} \to -\ln(s/Q^2)P_{q\gamma} \tag{2}$$

where $\sqrt{s}$ is the c.m. energy of the photon-parton subprocess. These terms are finite as long as $Q^2 \neq 0$ and can be evaluated with $d = 4$ dimensions, but become large for small Q^2, which suggests to absorb them as terms proportional to $\ln(M_\gamma^2/Q^2)$ in the PDF of the virtual photon. Parametrizations of the virtual photon have been provided by several groups [25]. By this absorption the PDF of the virtual photon becomes dependent on M_γ, which is the factorization scale of the virtual photon, in analogy to the real photon case. Of course, this absorption of large terms is necessary only for $Q^2 \ll M_\gamma^2$. In all other cases the direct cross section can be calculated without the subtraction and the additional resolved contribution. M_γ^2 will be of the order of E_T^2. But also when $Q^2 \simeq M_\gamma^2$, one can perform this subtraction. Then the subtracted term will be added again in the resolved contribution, so that the sum of the two cross sections remains unchanged. In this way also the dependence of the cross section on M_γ^2 must cancel, as long as the resolved contribution is calculated in LO only.

   In the general formula for the deep-inelastic scattering cross section, one has two contributions, the transverse $(d\sigma_{\gamma b}^U)$ and the longitudinal part $(d\sigma_{\gamma b}^L)$. Since only the transverse part has the initial-state collinear singularity the subtraction in [23] has been performed only in the matrix element which contributes to $d\sigma_{\gamma b}^U$. Therefore the longitudinal PDF's $f_{a/\gamma}^L$ are not needed. It is also well known that $d\sigma_{\gamma b}^L$ vanishes for $Q^2 \to 0$. The calculation of the resolved cross section including NLO corrections proceeds as for real photoproduction at $Q^2 = 0$, except that the cross section is calculated also for final state variables in the virtual photon-proton center-of-mass system.

The NLO calculations in the low Q^2 region are implemented in the fixed order program JETVIP [26]. Various measurements at HERA in which the jet analysis has been done with JETVIP point to the presence of a resolved virtual photon component up to moderate virtualities of $Q^2 \simeq 5 \text{ GeV}^2$ [27].

4.2   Comparison to forward jet data

Recently, we have performed a NLO calculation including the virtual resolved photon for the forward jet region [28] with the help of JETVIP. The results for the ZEUS kinematical conditions are shown in Fig. 5 a,b. In Fig. 5 a we plotted the full $\mathcal{O}(\alpha_s^2)$ inclusive two-jet cross section (DIS) as a function of x for three different scales $\mu^2 = \mu_R^2 = 3M^2 + Q^2, M^2 + Q^2$ and $M^2/3 + Q^2$ with a fixed $M^2 = 50 \text{ GeV}^2$ related to the mean E_T^2 of the forward jet and compared them with the measured points from ZEUS [7]. The choice $\mu_F^2 > Q^2$ is mandatory if we want to include a resolved contribution. Similar to the results obtained with MEPJET and DISENT, the NLO direct cross section is by a factor 2 to 4 too small compared to the data. The variation inside the assumed range of scales is small, so that also with a reasonable change of scales we can not get agreement with the data. In Fig. 5 b we show the corresponding forward jet cross sections with the NLO resolved contribution included, labeled DIR$_S$+RES, again for the three different scales μ as in Fig. 5 a. Now we find good agreement with the ZEUS data. The scale dependence is not so large that we must fear our results not to be trustworthy.

In Fig. 5 c,d we show the results compared to the H1 data [8] obtained with $E_T > 3.5 \text{ GeV}$ in the HERA system. In the plot on the left the data are compared with the pure NLO direct prediction, which turns out to be too small by a similar factor as observed in the comparison with the ZEUS data. In Fig. 5 d the forward jet cross section is plotted with the NLO resolved contribution included in the way described above. We find good agreement with the H1 data inside the scale variation window $M^2/3 + Q^2 < \mu^2 < 3M^2 + Q^2$. We have also compared the predictions with the data from the larger E_T cut, namely $E_T > 5.0 \text{ GeV}$, and found similar good agreement [28].

As described in [28] the NLO resolved contribution supplies higher order terms in two ways, first through the NLO corrections in the hard scattering cross section and second in the leading logarithmic approximation by evolving the PDF's of the virtual photon to the chosen factorization scale. This way the logarithms in E_T^2/Q^2 are summed, which, however, in the considered kinematical region is not an important effect numerically. Therefore, the enhancement of the NLO direct cross section through inclusion of resolved processes in NLO is mainly due to the convolution of the point-like term in the photon PDF with the NLO resolved matrix elements, which gives an approximation to the NNLO direct cross section without resolved contributions. One can therefore speculate that the forward jet cross section could be described within a fixed NNLO calculation, using only direct photons.

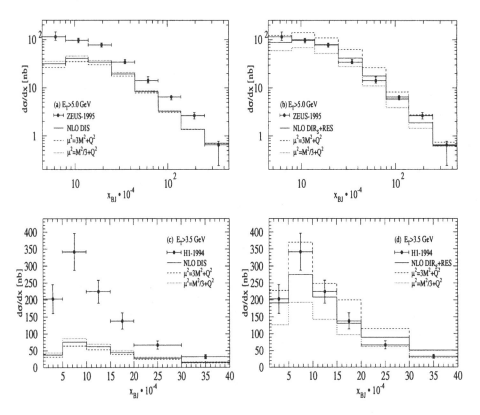

Fig. 5. Dijet cross section in the forward region compared to HERA data: (a) and (b) ZEUS; (c) and (d) H1. (a) NLO DIS, $E_T > 5$ GeV; (b) NLO DIR$_S$+RES, $E_T > 5$ GeV; (c) NLO DIS, $E_T > 3.5$ GeV; (d) NLO DIR$_S$+RES, $E_T > 3.5$ GeV.

5   Conclusions

We conclude that two alternative models exist for describing the forward jet data. The BFKL calculation including CC describes the data well. This is supportet by the ARIADNE model, where strong k_T ordering is absent and which also describes the data. It is however not clear, why the LDC model does not decribe the data, although BFKL dynamics should be included in the respective regime of validity.

Similarly, the NLO theory with a resolved virtual photon contribution as an approximation of the NNLO DIS cross section, which is presently not available, gives a good description of the forward jet data. Presumably, the RAPGAP model produces these higher order effects through parton shower contributions in the resolved cross section.

Acknowledgments. I thank the organizers for the kind invitation and the pleasant workshop athmosphere. The results of section 4 where obtained in collaboration with G. Kramer.

References

1. H1 Collaboration (I Abt et al.) Nucl. Phys. **B 407** (1993) 515; H1 Collaboration (T Ahmed et al.) Nucl. Phys. **B 439** (1996) 471
2. Zeus Collaboration (M Derrik et al.) Phys. Lett. **B 316** (1993) 412; Z. Phys. **C 65** (1995) 379; Z. Phys. **C 69** (1996) 607
3. V N Gribov, L N Lipatov, Sov. J. Nucl. Phys. **15** (1972) 438, 675; Y L Dokshitzer, Sov. Phys. JETP **46** (1977) 641; G Altarelli, G Parisi, Nucl. Phys. **B126** (1977) 298
4. E A Kuraev, L N Lipatov, Y S Fadin, Sov. Phys. JETP **45** (1977) 199; Ya Ya Balitzky, L N Lipatov, Sov. J. Nucl. Phys. **28** (1978) 288
5. M Ciafaloni, Nucl. Phys. **B 296** (1988) 49; S Catani, F Fiorani, G Marchesini, Phys. Lett. **B 234** (1990) 339, Nucl. Phys. **B 336** (1990) 18; G Marchesini, Nucl. Phys. **B 445** (1995) 49
6. A Mueller, Nucl. Phys. B (Proc. Suppl.) **18C** (1990) 125; J. Phys. **G17** (1991) 1443
7. ZEUS Collaboration (J Breitweg et al.) Eur. Phys. J. **C 6** (1999) 239
8. H1 Collaboration (C Adloff et al.) Nucl. Phys. **B 538** (1999) 3
9. L Lönnblad, Comp. Phys. Comm. **71** (1992) 15
10. G Ingelman, A Edin, and J Rathsman, Comp. Phys. Comm. **101** (1997) 108
11. G Marchesini et al., Comp. Phys. Comm. **82** (1992) 445
12. H Kharraziha, L Lönnblad, JHEP **9803** (1998) 6
13. H Jung, Comp. Phys. Comm. **86** (1995) 147;
    H Jung, L Jönsson, H Küster, Eur. Phys. J. **C 9** (1999) 383
14. J Kwieciński, A D Martin, J P Sutton, Phys. Rev. **D 46** (1992) 921;
    J Bartels, A De Roeck, M Loewe, Z. Phys. **C 54** (1992) 635;
    W K Tang, Phys. Lett. **B 278** (1992) 363
15. H1 Collaboration (S Aid et al.) Phys. Lett. **B 356** (1995) 118
16. J Bartels, V Del Duca, A De Roeck, D Graudenz, M Wüsthoff, Phys. Lett. **B 384** (1996) 300
17. V S Fadin, L N Lipatov, Phys. Lett. **B 429** (1998) 127
18. J Kwieciński, A D Martin, J J Outhwaite, Eur. Phys. J. **C 9** (1999) 611
19. J J Outhwaite, these proceedings
20. E Mirkes, D Zeppenfeld, Phys. Rev. Lett. **78** (1997) 428
21. E Mirkes, D Zeppenfeld, Phys. Lett. **B 380** (1996) 205
22. S Catani, M H Seymour, Phys. Lett. **B 378** (1996) 287; Nucl. Phys. **B 485** (1997) 291
23. M Klasen, G Kramer, B Pötter, Eur. Phys. J. **C1** (1998) 261;
    G Kramer, B Pötter, Eur. Phys. J. **C5** (1998) 665;
    B Pötter, Eur. Phys. J. Direct **C5** (1999) 1, hep-ph/9707319
24. M Klasen, these proceedings, hep-ph/9907366
25. M Glück, E Reya, M Stratmann, Phys. Rev. **D 51** (1995) 3220; G A Schuler, T Sjöstrand, Z. Phys. **C 68** (1995) 607; Phys. Lett. **B 376** (1996) 193; M Glück, E Reya, I Schienbein, Phys. Rev. **D 60** (1999) 054019
26. B Pötter, Comp. Phys. Comm. **119** (1999) 45

27. S Maxfield, B Pötter, L Sinclair, J. Phys. **G25** (1999) 1465;
    G Kramer, B Pötter, Lund-Workshop, Sweden 1998, ed. G Jarlskog and
    T Sjöstrand, p. 29, hep-ph/9810450; B Pötter, DIS98 Workshop, Belgium 1998,
    ed. Gh Coremans and R Roosen, p. 574, hep-ph/9804373
28. G Kramer, B Pötter, Phys. Lett. **B453** (1999) 295

Forward Jet Production
as a Probe of Low x Dynamics

John Outhwaite

Department of Physics, University of Durham, Durham, DH1 3LE, UK.

Abstract. The HERA facility at Hamburg has unveiled a wealth of physics spanning a broad kinematic domain. The question of how best to describe data obtained at low x remains open, the main problem being the difficulty of cleanly disentangling perturbative and non-perturbative effects. A recent set of measurements, looking at final states with high transverse momentum produced close to the proton remnant offers fresh insight into low x dynamics. The reasons why these configurations are especially interesting are motivated, and complementary approaches to describing the physics involved are outlined. We illustrate this by comparing predictions based on small x (QCD) dynamics with recent data for deep inelastic events containing forward jets or forward π^0 mesons. We quantify the effect of imposing the (higher order) consistency condition on the BFKL equation and study uncertainties inherent in the QCD predictions.

1 Introduction

The ability of perturbative QCD to describe a wide variety of HERA phenomena over a broad span of (x, Q^2) kinematic space has been well demonstrated. There is general consensus that the mechanisms which underpin the description of physics within the large Q^2 region, are based on the DGLAP evolution equations which effectively resum the $\alpha_s \log Q^2$ contributions. The small x limit of this approach resums the $\alpha_s \log Q^2 \log(1/x)$ contributions and is known as the Double Leading Log Approximation (DLLA).

In recent years, the focus of experimental and theoretical efforts have shifted to the region of low x and with medium Q^2, which goes beyond DLLA in that we must resum $\alpha_s \log(1/x)$ terms for all Q^2, and not just take the $\log Q^2$ limit. As HERA began to probe at ever decreasing values of Bjorken x $(1/x \sim s)$, it became apparent that at high energies, a perturbative expansion containing terms of the form $\alpha_s \log(1/x)$ at *all* Q^2, and not simply the $\log Q^2$ limit, is necessary. That is, for very low x, the small size of the strong coupling α_s is not sufficient to tame large $\log(1/x)$ terms (unaccompanied by $\log Q^2$). These would need to be resummed to all orders. The failure of the standard DGLAP approach to DIS phenomena, wherein large logarithms of $(\alpha_s \log Q^2)^2$ and also $\alpha_s (\alpha_s \log Q^2)^2$ are systematically resummed to all orders, would be a signal for the onset of a distinct low x dynamics. Is there experimental evidence for such effects? This question is not so easy to answer as it first appears.

An early idea, first mooted by Mueller [1], in which one scatters inelastically from a parton, and selects kinematic configurations that can isolate clearly the low x effects, was enthusiastically taken up by the experimental community at HERA [2,3]. Data of sufficient quality have now been obtained and it is now an interesting phenomenological challenge to see what the measurements imply about the underlying small x dynamics. This is the subject of the current review.

Resumming the large $\log(1/x)$ terms to leading order is performed within the Balitskij, Fadin, Kuraev, Lipatov (BFKL) formalism, first outlined in the late 1970's [4,5]. Since then, leading log BFKL based phenomenology was shown to be insufficient, since the next-to-leading corrections were calculated [6–8] and found to be large. A phenomenological modification to the leading BFKL formalism [9,10] has provided an economical, physically based, description of a variety of phenomena within the low x QCD framework. An alternative formalism [11,12], based on the concept of the parton content of the resolved photon, has shown similar success, over a similarly broad range of phenomena.

Unequivocably describing the physics of low x region, within the bounds of perturbative QCD, would certainly be a worthwhile achievement.

2   BFKL $\log(1/x)$ resummation

The BFKL prescription for the resummation of small x contributions has been performed for both leading log (LL) [4,5], terms of the form $\alpha_s^n \log^n(1/x)$, and next-to-leading log (NLL) [6–8], terms of the form $\alpha_s^{n+1} \log^n(1/x)$.

All $\log(1/x)$ contributions have to be considered, and so we must retain the full transverse momentum dependence of the gluon distribution, and not simply the $\log Q^2$ dependence arising from the strong ordering in k_T. The ordering in the transverse momentum is relaxed, and an unintegrated gluon distribution, ϕ_i, is introduced, related to the usual gluon distribution, g_i, of a photon of virtuality Q^2 and polarisation $i = T, L$, probed at a scale $\bar{Q}^2$ via:

$$ xg_i(x, Q^2, \bar{Q}^2) = \int^{\bar{Q}^2} \frac{dk_T^2}{k_T^2} \phi_i(x, k_T^2, Q^2) $$

x is the longitudinal momentum fraction of the virtual photon carried by the gluon, and k_T denotes its transverse momentum.

We begin with a brief overview of the BFKL technique. It is assumed that the dominant process in the high energy limit is gluon exchange in the t-channel. Higher order QCD corrections are collected together to form an 'effective' diagram. This comprises an effective gluon vertex, which subsumes real gluon emissions, whilst the t-channel gluon reggeizes with the virtual corrections. Then, using this 'rung', one can recursively build up a ladder diagram with reggeized t-channel gluonic propagators emitting regular gluons from effective vertices. The BFKL equation represents the sum of all such ladder diagrams.

$$\phi_i(x, k_T^2, Q^2) = \phi_i^{(0)}(x, k_T^2, Q^2)+$$

$$k_T^2 \int_x^1 \frac{dz}{z} \int_{k_0^2}^\infty \frac{dk_T'^2}{k_T'^2} \bar{\alpha}_s \left(\frac{\phi_i(x/z, k_T'^2, Q^2) - \phi_i(x/z, k_T^2, Q^2)}{\mid k_T^2 - k_T'^2 \mid} + \frac{\phi_i(x/z, k_T^2, Q^2)}{(4k_T'^4 + k_T^4)^{\frac{1}{2}}} \right)$$

$$(1)$$

where $i = T$ or L, the $\phi_i^{(0)}$ represent the driving terms of the equation, $\bar{\alpha}_s = 3\alpha_s/\pi$ and x is the fraction of the virtual photon's longitudinal momentum carried by the gluon. The term proportional to $\phi_i(x/z, k_T'^2, Q^2)$ corresponds to the real gluon emissions and the terms proportional to $\phi_i(x/z, k_T^2, Q^2)$ describe the virtual corrections generating reggeization of the gluons. We see that real and virtual contributions to the integrand cancel in the potentially singular limit of $k_T'^2 \longrightarrow k_T^2$. Notice the equation is invariant under a scaling of the transverse momentum, which allows an analytic solution in the case of fixed α_s to be calculated using Mellin transforms.

Taking α_s fixed, it is possible to calculate the characteristic exponent of the singular power behaviour in x of the unintegrated gluon distribution:

$$\phi \sim x^{-\lambda}$$

with exponents at leading and next-to-leading log, in the $\overline{\text{MS}}$ scheme, given by

$$\lambda_{LL} = \bar{\alpha}_s 4 \log 2$$

$$\lambda_{NLL} = \lambda_{LL} \left(1 - 3.4\bar{\alpha}_s - 0.15 \frac{N_f \alpha_s}{\pi} \right) \qquad (2)$$

[A brief aside: The BFKL singular behaviour of g_i as $x \longrightarrow 0$ cannot be the complete picture of low x physics. On general conceptual grounds, the Froissart bound forbids the growth of the cross-section faster than $\log^2(s)$. We can motivate this physically by recognising that as we go to higher energies, the density of gluons within the proton will increase to the point that they can no longer be treated as free partons. The gluons will recombine, and the cross-section will be suppressed. At current HERA energies there seems to be no evidence yet for the presence of these 'shadowing' or 'unitarity' corrections [13].]

It is clear that the NLL corrections are substantial and thus that LL based phenomenology is strictly unreliable. It is possible to impose higher order, physically motivated restrictions in the governing evolution equation *to all orders*. One important example is that the LL BFKL equation does not respect energy-momentum conservation, the violation being produced by subleading terms. In fact Orr and Stirling [14] are developing a BFKL Monte Carlo which will allow the effect of energy-momentum conservation to be quantified.

There exists a further physical constraint that is more stringent than energy-momentum conservation, and implemented in the BFKL equation rather simply.

The *consistency constraint* imposes the physical condition that the virtuality of exchanged gluons along a BFKL chain should arise predominantly from the transverse component of momentum [9].

$$k_T'^2 < \frac{k_T^2}{z}.$$

This ensures the validity of the small x approximation, by restricting the maximum transverse momentum of an emitted gluon. The BFKL equation was derived assuming this constraint, and in the region of phase space beyond that delimited by the constraint the BFKL equation is inappropriate. Note that, in particular, the gluon propagators which appear in the BFKL equation contain only the transverse momenta squared. It has been shown [9] that including this condition implicitly subsumes the neglected energy-momentum conservation over the greater part of the small x kinematic space.

It can be neatly encapsulated, and incorporated into the LL BFKL equation, through the inclusion of a Θ-function on the kernel component governing real gluon emission:

$$\Theta(k_T^2/k_T'^2 - z)$$

It transpires that limiting the LL BFKL formalism in this way induces an effect at all orders which if truncated to NLL gives the dominant part of the NLL expression of (2). One might conjecture that these represent the dominant higher order effects, and that the modified BFKL evolution resums these. Is it possible to obtain a good description of the HERA data by implementing this type of formalism?

3   Experimental signatures of BFKL-type dynamics

A definitive experimental identification of the low x physics 'smoking gun' has proven somewhat elusive. There exist several variations on the original 'gold plated' measurements proposed by Mueller[1], that in principle might exhibit the characteristic signs of a rapid rise in gluon distribution at low x, but uniquely separating the $\log(1/x)$ from non-perturbative effects is certainly non-trivial.

3.1   Inclusive quantities

The first dramatic evidence for the effects of small x resummations came from HERA measurements of the proton structure function $F_2(x, Q^2)$, which exhibit a strong rise with decreasing x. In [15] a quantitative analysis of the F_2 data within a united modified BFKL and DGLAP framework was presented.

The steep increase was initially lauded as a clear demonstration of the BFKL small x resummation effects, but it soon became apparent that it was not possible to ascribe this feature unambiguously to BFKL-type physics. It proved possible to adequately describe the unexpected features in the data as a consequence of DLLA resummations, in which terms of the form $\alpha_s^n \log^n(1/x) \log^n(Q^2)$ are resummed to all orders. There exist sufficient provisos in the theoretical descriptions of F_2 to compromise its discriminatory powers:

- DGLAP evolution, with a sufficiently 'tuned' non-perturbative input distribution can mimic the inverse power behaviour predicted by the BFKL approach.
- The disorder in k_T corresponds to a random walk in transverse momentum along the gluon chain. The kinematic bounds imposed by the F_2 measurement allow for this transverse momentum to significantly diffuse into the infra-red region, blurring clear interpretation.
- The parton distribution functions necessary to an analysis are not sufficiently well determined at all the required kinematical configurations for a quantitative analysis.

A more stringent process is required in order to draw firm conclusions.

3.2   Exclusive quantities

It is clear that inclusive measurements (F_2 etc...) are unable to disentangle cleanly low x parton dynamics from non-perturbative effects. In attempting to do so, model dependences are necessarily introduced. One should try and utilise more exclusive measurements. More specifically, final states restricted in kinematic (x, Q^2) space by the requirement that they are produced close to the proton remnant, prove to be of particular interest.

Mueller proposed that DIS events containing an identified forward jet would prove an excellent discriminator for low x dynamics. The detailed formalism was given in [16]. These particular measurements were made by the H1 and ZEUS collaborations at HERA [2,3]. Recent analyses can be found in [16,17]. Subsequently, it was realised that this process could be extended, and in some sense refined, by studying energetic forward π^0's produced close to the proton direction [17], and also that the forward dijet rate [18] would yield another possible window on the low x physics.

These processes share the same basic properties that make them a useful probe of low x dynamics:

- Experimentally, one requires that $Q^2 \sim k_T^2$. This ensures that DGLAP type evolution is suppressed, and that features in the data arise solely from small x dynamics.
- By restricting the study to configurations where the final state is produced close to the beam pipe, one insists that the jet carries a large fraction of the longitudinal momentum of the parent proton. This ensures the evolution parameter x/x_j is small, necessary for the validity of the BFKL approach, and also that x_j is sufficiently large for the parton distribution functions to be determined at regions in the phase space where they are well determined from global analyses.

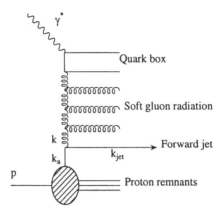

Fig. 1. Generic BFKL formalism for the Mueller process. Virtual photon-gluon fusion realised through the quark box, and soft gluon radiation between that and the struck parton produced close to the proton remnants.

- In principle, the large k_T^2 of the process suppresses the diffusion into the infra-red region due to the random walk in transverse momentum along the gluon chain.

In summary, Mueller's proposal was to perform deep inelastic scattering on parton constituent of the proton, and so to neatly sidestep non-perturbative hadronization problems.

In addition one can examine the subset of events that produce π^0's in the forward direction [2,19]. These measurements possess additional qualities that complement the basic Mueller process:

- The π^0 measurement samples more energetic forward jets than could be measured explicitly in jets. By allowing the struck parton fragment collinearly into a pion with longitudinal momentum fraction $x_\pi < x_j$ and transverse momentum $p_{\pi T} < k_{jT}$ we sample a greater amount of the relevant phase space.
- Experiments take measurements at the *hadron* level, theoreticians make calculations at the *parton* level, so there is a certain amount of ambiguity involved in comparing the predictions with the data... In studying final state particles, hadronisation uncertainties are swept into parameterisations of fragmentation functions, and we can be certain of comparing like quantities.
- Experimentally, it is easier to unambiguously identify a π^0 rather than a jet. The weak model dependence from the algorithms used to specify exactly what constitutes a 'jet' is also eliminated by looking at a specific final state.
- The conceptual 'gains' outlined above must be tempered with a practical 'loss'. By requiring a single energetic fragment of the jet, we reduce the cross-section and so the observed event rate.

Finally, we can obtain an additional measure of the low x region by looking at the subset of forward jet events that contain *two* jets in the forward region.

4   The BFKL approach to forward jet production

Unlike the previous analyses [16,17], ref. [10] uses the modified BFKL equation to evolve the gluon distribution from the photon end down to the struck parton. The virtual photon is connected to the gluon chain through a quark box, which is calculable in perturbative QCD. This comprises the driving term for the unintegrated gluon distribution, $\phi^{(0)}$ in the BFKL equation. The modified BFKL equation is solved numerically for the unintegrated gluon distribution of the virtual photon, with the quark box as the inhomogeneous driving terms, and allowing α_s to run. Longitudinal momentum fractions are related at the jet vertex through the *strong ordering* condition

$$k_{aT}^2 \ll k_{jT}^2, \quad x_j^P \gg x$$

and utilising the jet on-shell condition $k_j^2 = 0$ The singular behaviour of the unintegrated gluon distribution in the BFKL approach then feeds through into observables via k_T factorisation

$$\frac{\partial^2 F_i}{\partial x_j^P \partial k_{jT}^2} = \bar{\alpha}_s k_{jT}^4 \left(\sum_a f_a(x_j^P, k_{jT}^2) \right) \Phi_i \left(\frac{x}{x_j^P}, k_{jT}^2, Q^2 \right) \tag{3}$$

where $i = 1, 2$ and $\sum_a f_a$ is the sum over parton distribution functions in the effective structure function approximation. A recent leading order paramaterisation of the parton distributions is used [20].

These differential structure functions are related to differential cross-sections in the usual way

$$\frac{\partial \sigma_j}{\partial x \partial Q^2 \partial x_j^P \partial k_{jT}^2} = \frac{4\pi\alpha^2}{xQ^4} \left((1-y) \frac{\partial^2 F_2}{\partial x_j^P \partial k_{jT}^2} + \frac{1}{2} y^2 \frac{\partial^2 F_T}{\partial x_j^P \partial k_{jT}^2} \right)$$

The experimental cuts applied by the H1 [2] and ZEUS [3] collaborations in order to restrict the sampling to purely forward jet events are imposed on the numerical evaluation of the cross-sections.

The analgous cross-sections for π^0 production are constructed by convoluting the DIS + forward jet differential cross-sections with the π^0 fragmentation function parameterisations. These are parameterisations of the non-perturbative hadronization of the parton into a final state π^0. We use the LO $\pi^\pm$ fragmentaion functions of Binnewies et al. [21], and the π^0 functions are then taken to be $\frac{1}{2}(\pi^+ + \pi^-)$ distributions. The π^0 is assumed to carry a fraction $z = x_\pi/x_j^P$ of the jet's longitudinal momentum in a direction collinear to the parent parton. A sample of the results are shown in Figs. 2 and 3. Plots are shown for different choices of the scales (in the driving term $\phi^{(0)}$ of (1), and in α_s of (3)) and of the infrared cut-off k_0^2 in (1), taken respectively to be:

$$
\begin{array}{lll}
\text{(i)} & (k_T^2 + m_q^2)/4, \ k_T^2/4, \ k_0^2 = 0.5\text{GeV}^2 & \text{(upper dashed)} \\
\text{(ii)} & (k_T^2 + m_q^2)/4, \ k_T^2/4, \ k_0^2 = 1\text{GeV}^2 & \text{(continuous)} \\
\text{(iii)} & (k_T^2 + m_q^2), \quad k_T^2, \quad k_0^2 = 0.5\text{GeV}^2 & \text{(lower dashed)}.
\end{array} \tag{4}
$$

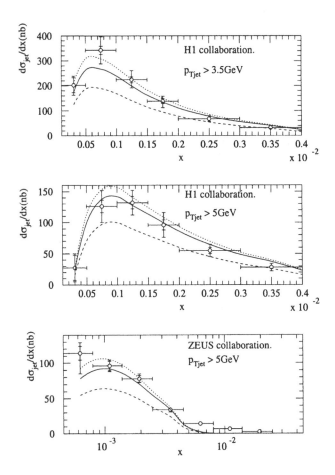

Fig. 2. The DIS + forward jet differtential cross-section versus Bjorken-x as measured at the hadron level by the H1 [2] and ZEUS [5] collaborations. The curves are predictions, at the parton level, based on the BFKL formalism including sub-leading corrections, corresponding to the three choices of the scales and infrared cut-off given in (4).

We note that the calculations describe the forward jet, forward π^0 and forward dijet measurements well, for physically reasonably scales and cut-off. We see that they exhibit a weak residual dependence on a transverse momentum cut-off parameter and a stronger dependence on changes in the choice of scale.

In Fig. 4 we contrast the predictions of the modified BFKL formalism with that of the leading-log($1/x$) approximation. The data prefer the inclusion of the resummation to all orders of subleading terms through the consistency constraint, especially since converting the data down to the parton level is expected to lead to a decrease in cross-section of some 15-20%.

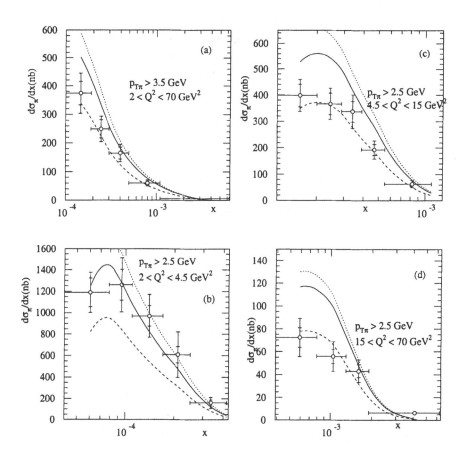

Fig. 3. The π^0 differential cross-section versus Bjorken-x obtained from H1 1996 data [19]. The curves are predictions, based on the BFKL formalism including sub-leading corrections, corresponding to the three choices of the scales and infrared cut-off given in (4).

5 The resolved and direct photon description of forward jet production

The resolved and direct photon approach splits the process into two distinct mechanisms according to the size of the characteristic scale μ^2 of the subprocess involved. When $\mu^2 < Q^2$, the inverse size of the photon, a standard perturbative approach, with point-like photon interactions is taken. However, when $\mu^2 > Q^2$, the partonic structure of the virtual photon is said to have been resolved. This is then parameterised by a set of parton density functions analgous to the familiar proton parton density functions.

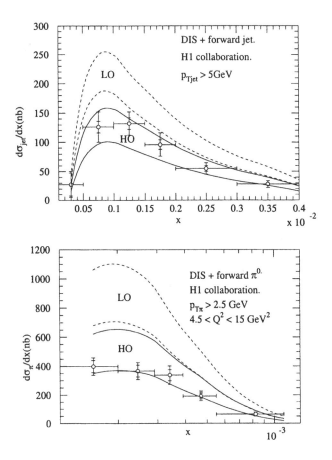

Fig. 4. The differential cross-section for (a) DIS + forward jet, (b) DIS + forward π^0, events. The continuous and dashed curves correspond to the inclusion and omission of the consistency constraint. The lower set of curves corresponds to the scales and infrared cut-off of (i) in (4), and the upper set of curves to (iii). The jet data are at hadron level; converting to the parton level is expected to lower the jet data points by 15-20%.

Two groups [11,12] have implemented the concept of a resolved photon contribution to describe forward jets in DIS. Both have produced Monte Carlo event generators (RAPGAP and JetViP) that describe the surplus of events over the naive (direct, point-like photon) DIS expectation.

RAPGAP [11] uses purely leading order matrix elements, and allows parton showering between both the hard subprocess and the resolved photon, and the subprocess and proton. It is thought that these simulate higher order processes, as the leading order matrix elements alone prove insufficient to describe the data.

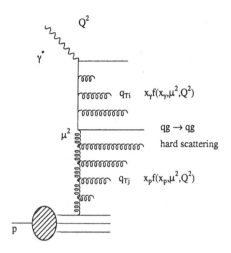

Fig. 5. Deep inelastic scattering and a forward jet within the LO resolved + direct photon formalism of [11], with DGLAP evolution from both the proton and virtual photon ends towards a $qg \longrightarrow qg$ subprocess.

The total cross-section is then calculated as the sum of resolved and direct photon contributions. Whilst we have transverse momentum ordering within each of the DGLAP chains, this formalism allows for an overall disordering in k_T^2 between the virtual photon and proton. This k_T^2 disorder appears to be a 'necessary' condition in producing the rise in the cross-sections at low x. It features in many of the models that best describe the DIS + forward jet data (c.f. BFKL phenomenology, this approach and the color dipole model (CDM) implemented in ARIADNE)

JetViP [12] treats the resolved and direct contributions consistently using NLO matrix elements, taking care to count only once those parts from the NLO direct contribution which are implicitly included from the resolved photon. It was found that the significant part of the NLO corrections arose from the NLO contribution from the resolved photon. This piece corresponds to an extra rung (and so includes one 'disorder') on the gluon ladder from the photon remnant and is essentially calculable within the framework of perturbative QCD.

Both formalisms including a resolved photon contribution into the DIS + forward jet process find good agreement between HERA forward jet data and calculation.

6   Conclusions

One cannot obtain a sufficient increase with decreasing x of the forward jet cross-section from fixed order QCD calculations. NLO predictions [22,23] show a large deficit compared with the data. It is clear that the modified BFKL, and the resolved photon (both NLO and LO + parton showers.) theoretical approaches to processes designed to clarify low x physics provide a good description of HERA measurements. Resumming subleading effects to *all* orders in BFKL is clearly necessary given the large NLL corrections. Unfortunately, it is not known how to perform such a resummation exactly, indeed it is not clear if such a process is a feasible goal.

The consistency constraint provides a route by which we can circumvent the need for a full resummation, and instead implement the rather simpler all-orders resummation of the *dominant* subleading corrections. It is then possible to make phenomenological predictions for forward jets and π^0's.

Alternatives to this approach exist, based on supplementing the 'traditional' direct photon contribution with a situation where the photon itself exhibits partonic structure and is said to be resolved. The different formalisms, although at first glance appear quite distinct, share some of their conceptual ideas, and it is an informative exercise to contrast them. One might think of the modified BFKL approach, with gluon evolution extending *downwards* from the virtual photon-gluon fusion, as representing in some sense the unintegrated gluon content of the virtual photon.

For the resolved photon approach, the parton showering included at LO level is thought to approximate the explicit higher order effects in the full NLO calculation. In both cases significant enhancement of the cross-section arises from additional gluon emission from virtual photon corrections.

The LO resolved photon formalism can be thought of as mimicking, in a limited way, the complete k_T disordering of the BFKL approach, to the extent that the two approaches are indistinguishable at HERA.

Finally it is worth emphasising here the chief conceptual distinctions. The resolved photon analysis unilaterally divides the underlying mechanism according to the relative sizes of the scales (Q^2 and k_T^2) involved, whereas the modified BFKL approach treats in a unified way all of the possible kinematical configurations. In the BFKL approach, the Q^2 dependence of the parton 'content' of the virtual photon is dynamically generated through a perturbative calculation of the quark box, rather than paramaterised through the virtual photon parton distribution functions.

References

1. A. Mueller, *Nucl. Phys.* **B** (*Proc. Suppl*) **18C** (1990) 125; *J. Phys.* **G17**, 1443 (1991).
2. H1 Collaboration, C. Adloff et al., *Nucl. Phys.* **B538** (1999) 3.
3. ZEUS Collaboration, *Eur. Phys. J.* **C6** (1999) 239.
4. E.A. Kuraev, L.N. Lipatov and V.S. Fadin, *Sh. Eksp. Teor. Fiz.* **72** (1977) 373, (*Sov. Phys. JETP* **45** (1977) 199); Ya. Ya. Balitskij and L.N. Lipatov, *Yad. Fiz.* **28** (1978) 1597, (*Sov. J. Nucl. Phys.* **28** (1978) 822); J.B. Bronzan and R.L. Sugar, *Phys. Rev.* **D17** (1978) 585; T. Jaroszewicz, *Acta. Phys. Polon.* **B11** (1980) 965.
5. L.N. Lipatov, in "Perturbative QCD", edited by A.H. Mueller (World Scientific, Singapore 1989), p.441.
6. V.S. Fadin, M.I. Kotskii and R. Fiore, *Phys. Lett.* **B359** (1995) 181; V.S. Fadin, M.I. Kotskii, L.N. Lipatov, hep-ph/9704267; V.S. Fadin, R. Fiore, A. Flachi and

M.I. Kotskii, *Phys. Lett.* **B422** (1998) 287; V.S. Fadin and L.N. Lipatov, hep-ph/9802290; *Phys. Lett.* **B429** (1998) 127; V.S. Fadin, hep-ph/9807527, hep-ph/9807528; M. Ciafaloni and G. Camici, *Phys. Lett.* **B386** (1996) 341; **B412** (1997) 396; **B417** (1998) 390 (E); *Phys. Lett.* **B430** (1998) 349; M. Ciafaloni, hep-ph/9709390.

7. D.A. Ross, *Phys. Lett.* **B431** (1998) 161.

8. G.P. Salam, *JHEP* **9807** (1998) 019.

9. J. Kwieciński, A.D. Martin and P.J. Sutton, *Z. Phys.* **C71**, 585 (1996).

10. J. Kwieciński, A.D. Martin and J.J. Outhwaite, *Eur. Phys. J.* **C9** (1999) 611.

11. H. Jung, L. Jönsson and H. Küster, hep-ph/9811368; *Eur. Phys. J.* **C9** (1999) 383.

12. G. Kramer and B. Pötter, *Phys. Lett.* **B453** (1999) 295.

13. L.V. Gribov, E.M. Levin and M.G. Ryskin, *Phys. Rep.* **100** (1980) 1.

14. L. H. Orr and W. J. Stirling, Contribution to ICHEP 98, Vancouver, July 1998, hep-ph/9811423.

15. J. Kwieciński, A.D. Martin and A. Stasto, *Phys. Rev.* **D56** (1997) 3991.

16. J. Bartels, A. De Roeck and M. Loewe, *Z. Phys.* **C54**, 635 (1992); J. Kwieciński, A.D. Martin and P.J. Sutton, *Phys. Rev.* **D46** (1992) 921; W.K. Tang, *Phys. Lett.* **B278** (1992) 363.

17. J. Kwieciński, S.C. Lang and A.D. Martin, *Eur. Phys. J.* **C6** (1999) 671.

18. J. Kwieciński, C.A.M. Lewis and A.D. Martin, *Phys. Rev.* **D57** (1998) 496.

19. E. Elsen, talk given at the Physics Research Committee at DESY, Hamburg, January 13th 1999; See also Thorsten Wengler, Ph.D. Thesis, University of Heidelberg, 27th January 1999.

20. A.D. Martin, R.G. Roberts, W.J. Stirling and R.S. Thorne, *Phys. Lett.* **B443** (1998) 301.

21. J. Binnewies, B.A. Kniehl and G. Kramer, *Phys. Rev.* **D52** (1995) 4947.

22. J. Bartels, V. Del Duca, A. De Roeck, D. Graudenz and M. Wüsthoff, *Phys. Lett.* **B384** (1996) 300.

23. E. Mirkes and D. Zeppenfeld, *Phys. Rev. Lett.* **78** 428 (1997).

Forward Jet and Hadron Production at Low x and Q^2

Timothy Greenshaw

Oliver Lodge Laboratory, Liverpool University, Liverpool L69 7ZE, ENGLAND

Abstract. Recent measurements of forward jet and single forward hadron production at HERA are described. An introduction is given to some of the consequences of various QCD approximation schemes used in the calculation of forward jet and particle production cross sections.

1 Introduction

1.1 HERA measurements and QCD

The H1 and ZEUS experiments at the HERA electron-proton (ep) collider have made impressive progress since HERA delivered first luminosity in 1992. Measurements of many aspects of ep interactions have been made with high precision over a large kinematic range. Here, we concentrate on measurements of jet and particle production in the forward region[1] and the description of these measurements provided by QCD. We shall work in the deep-inelastic scattering (DIS) regime, the regime in which both the negative squared four-momentum transferred to the proton from the electron $Q^2 = -q^2 \gg m_p^2$ and the final state hadronic mass squared $W^2 \gg m_p^2$, where m_p is the mass of the proton. Figure 1 illustrates the definition of some of these variables. For the measurements discussed here $Q^2 \ll m_Z^2$, where m_Z is the Z^0 mass, so the interactions can be considered to be mediated by virtual photons. In addition $W^2 \gg Q^2$, so the Bjorken scaling variable $x = Q^2/(Q^2 + W^2) \ll 1$. Recall that, in the frame in which the proton has infinite momentum, x is the fraction of the proton's momentum carried by the quark struck in the deep-inelastic interaction.

The HERA data are, on the whole, described extremely well by QCD. For example, the dijet cross section measurements presented in [1] are beautifully reproduced by NLO QCD calculations.

1.2 First forward jet measurements

If we now move to consider explicitly the forward region, the above no longer holds; NLO QCD [4] fails to describe the first HERA forward jet cross section measurements [2], as shown in figure 2. What is happening here?

---

[1] The HERA co-ordinate system has its origin at the nominal interaction point, its z axis in the proton direction and its y axis vertical; the x axis completes the right-handed orthogonal co-ordinate system. Polar angles are denoted by θ and azimuthal angles by ϕ. The term "forward" refers to the region for which $\theta < 90°$.

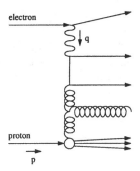

Fig. 1. A generic Feynman diagram for deep-inelastic ep scattering, illustrating the definition of some of the commonly used variables.

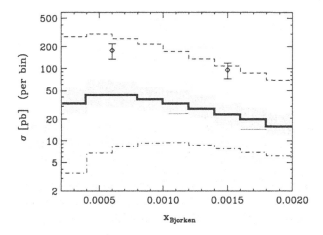

Fig. 2. The forward jet cross section in the kinematic regon described the text. The error bars illustrate the quadratic sum of the statistical and the systematic errors. The curves show NLO QCD calculations with an error band corresponding to a scale variation of two orders of magnitude, LO QCD calculations and the results of a BFKL calculation.

The key to understanding the problem lies in the definition of the kinematic region in which the forward jet measurement was made; it was required that the squared transverse momentum of the jet be similar to Q^2, $0.5 < p_{T\text{jet}}^2/Q^2 < 4$, and that the jets have a large longitudinal momentum in the proton direction. This latter condition arises from the requirements that the jets be found in the forward region, $6° < \theta_{\text{jet}} < 20°$, and that the ratio $x_{\text{jet}} = E_{\text{jet}}/E_p > 0.025$, Here, E_{jet} and E_p are the jet and proton energies, respectively. Hence, the parton emitted from the proton which caused the production of the forward jet must have carried a large proportion of the proton's momentum, and yet the quark struck by the virtual photon carried only a very small proportion of the proton's momentum, as the Bjorken x range considered was $2 \times 10^{-4} < x < 2 \times 10^{-3}$. The diagram in figure 1 may well be misleading. The phase space for the parton emerging from the proton to emit further partons is particularly large in this case, so the situation could be more like that illustrated in figure 3. Calculating

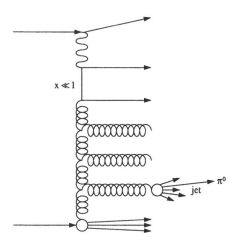

Fig. 3. A Feynman diagram illustrating low x DIS with the production of a forward jet carrying a large proportion of the proton's momentum, containing an energetic π^0.

to order α_s^2 allows us a maximum of 3 partons, as shown in figure 3, and such calculations are unlikely to be able to describe situations in which larger numbers of partons play a significant role. Is it possible for us to include the effects of these? Here, we use the Uncertainty Principle to try and construct an approximate picture of how we might expect these partons to behave, before going on compare the results of QCD calculations, made using various approximation schemes, to the HERA data. The calculations are discussed in more detail in other contributions to these proceedings [3].

Consider a fast parton, 1, which emits a further parton, 2, as illustrated in figure 4. The initially massless system has now acquired a mass given by:

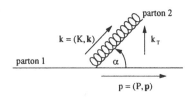

Fig. 4. A parton 1 emitting a parton 2, with four-momenta as illustrated.

$$M^2 = (p+k)^2 = 2(PK - PK\cos\alpha) \approx PK\alpha^2.$$

The magnitude of the resulting energy change, which determines the length of time the virtual system can exist before 2 is re-absorbed by 1, is $\Delta E = M$ in the rest frame of the system. Including the time dilation factor $\gamma \approx P/M$ gives the time for which this state can exist in the laboratory:

$$\Delta t = \frac{1}{\Delta E} = \frac{1}{M}\frac{P}{M}$$
$$= \frac{P}{PK\alpha^2} = \frac{1}{K\alpha^2}.$$

Using the fact that $\alpha \approx k_T/K$ we have $\Delta t \approx K/k_T^2$. Such emissions and re-absorptions will take place continuously in a proton prior to a deep-inelastic interaction. The interaction will "free" all the partons that exist at the time it takes place. We see that there will be some which can have existed for a long time before the interaction, those with large $\Delta t \approx K/k_T^2 \approx k_L/k_T^2$, and others that can only have existed for a very short time before the interaction, those with small $\Delta t \approx k/k_T^2$; it will appear that there is shower of time ordered partons $1, 2, 3 \ldots$ such that $K_1/k_{T1}^2 > K_2/k_{T2}^2 > K_3/k_{T3}^2 \ldots$

1.3   The DGLAP, BFKL and CCFM equations

The above argument suggests that we can think of the initial state partons as forming a cascade in which either the k_T increases strongly as we move towards the photon-quark vertex, with the longitudinal momentum relatively constant, or the longitudinal momentum decreases and the k_T remains approximately constant. A further alternative is to consider there to be an ordering of the emission angles[2], as illustrated in figure 5. These three possibilities correspond

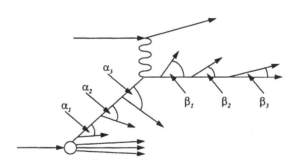

Fig. 5. Angular ordering in initial and final state parton cascades.

roughly to three approximation schemes within QCD which allow the inclusion

---

[2] For completeness sake, note that the emissions in the final state occur in order of decreasing angle.

of the effects of certain terms in the perturbative expansion of DIS processes to all orders. The case in which the cascade is considered as being composed of partons strongly ordered in k_T, $k_{T1}^2 \ll k_{T2} \ll k_{T3}^2 \ldots$, corresponds to using the Dokshitzer, Gribov, Lipatov, Altarelli and Parisi (DGLAP) equations [5], which sum to all orders, n, the terms $\sim \log^n Q^2$. How this is done is discussed in [6]. The case in which the cascade is considered to be composed of partons with decreasing longitudinal momenta, $x_1 \gg x_2 \gg x_3 \ldots$, corresponds to using the Balitski, Fadin, Kadeev and Lipatov equations [7], which are discussed in [8]. These sum to all orders the terms $\sim \log^n(1/x)$. Thinking of the partons as being ordered in emission angle, $\alpha_1 < \alpha_2 < \alpha_3 \ldots$, corresponds to using the Ciafaloni, Catani, Fiorani and Marchesini (CCFM) equations [9].

The success of the DGLAP equations is well known. For example, they give an excellent description of HERA $F_2(x, Q^2)$ measurements despite the range of more than four orders of magnitude in both Q^2 and x spanned by the HERA data, as is discussed in [10]. Can they describe the forward jet measurements? Looking at the kinematic region chosen, we must conclude that it is unlikely. The DGLAP equations describe data in which the parton cascade can be thought of as being ordered in k_T, the forward jet measurement, in which it is required that $k_{Tjet}^2 \sim Q^2$ precludes the development of such cascades as the k_T at both ends of the cascade is similar. Figures 7, 8a) and 8c) confirm our suspicions.

Let us turn instead to the BFKL equation. This should give accurate results where the cascade can be thought of as consisting of partons with strongly decreasing longitudinal momenta. This is exactly the situation here. The requirement that x_{jet} be large and x small ensures that the longitudinal momentum at the start of the cascade is large and that at the end small. Indeed, this is precisely the reason why these conditions were chosen for the measurement [11], the hope being that a kinematic region could be found in which there was clear evidence for the applicability of the BFKL equation. The evidence from the early forward jet measurements, as shown in figure 2, is that the BFKL solution [12] does a good job.

1.4   Monte Carlo Programs

Monte Carlo (MC) programs provide an alternative means of comparing QCD calculations with measurements. The programs of interest here are the following:

- LEPTO [13] matches LO QCD matrix elements with DGLAP inspired parton showers; the showers display k_T ordering.
- RAPGAP [14], originally developed to simulate diffractive DIS events, is very similar to LEPTO; the facility to simulate the effects of allowing the virtual photon to develop hadronic structure is of interest here, as is discussed in the following.
- HERWIG [15] produces DGLAP-like showers using angular ordering. The showers are corrected to incorporate the correct LO QCD results for the photon-parton interaction.

- ARIADNE [16] is an implementation of the Colour Dipole Model [17] (CDM), which treats QCD radiation as arising from the colour dipole generated in the deep-inelastic scattering process; the resulting parton cascades do not display k_T ordering and in this sense the results of the CDM are similar to those of the BFKL equation.
- The Linked Dipole Chain [18] (LDC) model relies on a re-formulation of the CCFM equations and should produce DGLAP-like and BFKL-like showers in the appropriate high Q^2 and low x limits, or, more accurately in the situations in which there is a large k_T or x "lever arm", respectively. It has been shown that, at the level of the leading log (LL) terms $\sim \log^n(1/x)$, the predictions of the CCFM equations for the hadronic final state in DIS at small x are identical to those of the BFKL equation [19].

One advantage of using MCs to compare theoretical predictions with data is that the MCs incorporate hadronisation models. This allows direct comparison of the predictions with data; parton level jet cross sections are likely to differ by 15 to 20% from the results expected following hadronisation. All the above models, barring HERWIG, use the JETSET [20] implementation of colour string fragmentation to perform hadronisation, following their various parton cascade schemes. The HERWIG model forms low mass colourless "pre-clusters" following its cascade procedure. These are then allowed to decay into hadrons.

2   New forward jet measurements and BFKL calculations

The H1 and ZEUS Collaborations have both produced new forward jet measurements since those discussed above. Both groups used a "cone" algorithm to identify the jets in the laboratory frame of reference. These algorithms maximise the transverse energy within a cone of radius $\sqrt{\Delta\eta^2 + \Delta\phi^2} = 1$, where $\Delta\eta$ is the difference in pseudorapidity, $\eta = -\ln(\tan(\theta/2))$, between the cone's axis and the calorimeter energy deposits under consideration, and $\Delta\phi$ is the difference in azimuthal angle. The algorithm used by ZEUS is described in [21]. The ZEUS measurement requires in addition that $x'_{\text{jet}} = p_{z\text{jet}}/E_p > 0.036$, $0.5 < E_{T\text{jet}}^2/Q^2 < 2$, $\eta_{\text{jet}} < 2.6$ (*i.e.* $\theta_{\text{jet}} > 8.5°$) and that $E_{T\text{jet}} > 5\,\text{GeV}$. Further, to explicitly remove events in which the forward jet is caused by the quark scattered in the deep-inelastic interaction, ZEUS require that the longitudinal momentum of the jet in the Breit frame of reference[3] $p_{z\text{jet}}^{\text{Breit}} > 0$. The Bjorken x region studied is $4.5 \times 10^{-4} < x < 4.5 \times 10^{-2}$. Despite the experimental difficulties involved in making this measurement, particularly the problems of measuring energy deposits close to the outgoing proton remnant and the forward beamhole in the calorimeter, the distribution of transverse energy about the forward jet axis following the above selection does reveal jet structure, as is illustrated in figure 6.

---

[3] The Breit frame is the frame of reference in which the four-momenta of the virtual photon and the proton are given by $(E, p_T, p_L) = (0, 0, -Q)$ and $(Q/2x, 0, Q/2x)$, respectively.

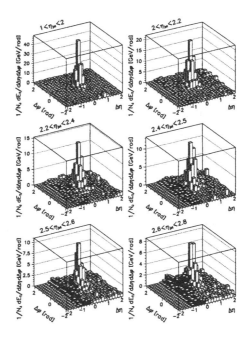

Fig. 6. The transverse energy flow around the forward jet axis averaged over all forward jets for various η_{jet} regions; grey indicates the E_T attributed to the jet, black the E_T around the calorimeter beam hole and white other E_T deposits.

Figure 7 shows the forward jet cross section [22] obtained by the ZEUS Collaboration[4]. Also shown in figure 7 are the predictions of several MC programs. Neither the DGLAP based MCs, LEPTO and HERWIG, nor the LDC MC are able to describe the data. The "pseudo-BFKL" MC prediction, that of ARIADNE, is the best of those shown.

The new H1 forward jet measurements [23] are shown in figure 8. The cone algorithm used here is described in [24]. The requirements made on the forward jets were that $x_{jet} > 0.035$, $0.5 < p_{Tjet}^2/Q^2 < 2$ and $7° < \theta_{jet} < 20°$. Results are shown in two regions of p_{Tjet}, namely $p_{Tjet} > 3.5\,\mathrm{GeV}$ and $p_{Tjet} > 5\,\mathrm{GeV}$. As in the above case, all MC predictions fall below the data, including those of the LDC model, with ARIADNE providing the best description of the data. NLO QCD calculations are again seen to fall significantly below the data, with a LO BFKL calculation tending to be too high. The discrepancy between the latter calculation and the LDC MC is intriguing given the formal equivalence of the CCFM and BFKL equations at the LL level. Are there significant sub-leading terms? Have we still to reach "low" x?

---

[4] In this and subsequent figures in which experimental data are shown, the inner error bars correspond to the statistical errors, the outer error bars to the quadratic sum of the statistical and systematic errors.

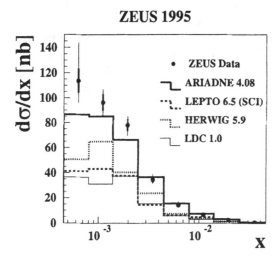

Fig. 7. The forward jet cross section as a function of x in the kinematic region described in the text with the additional requiremnts that the scattered electron energy $> 10\,\mathrm{GeV}$ and that $y > 0.1$. The curves show the predictions of various MC programs.

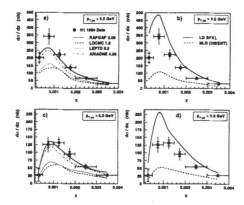

Fig. 8. The forward jet cross section as a function of x for $p_{T\mathrm{jet}} > 3.5\,\mathrm{GeV}$ and $p_{T\mathrm{jet}} > 5\,\mathrm{GeV}$; the curves in a) and c) are MC predictions, in b) and d) the full lines are analytic LO BFKL calculations without inclusion of the effects of the jet algorihm, the dashed lines are NLO QCD calculations made using DISENT [25] and include the effects of the jet algorithm at the parton level.

As mentioned above, and described in more detail in [26], the BFKL results used to this point rely on a summation of the LL terms. Recently, a calculation of the effects of the next-to-leading log (NLL) terms, $\sim \log^{n-1}(1/x)$, has become available [27] which has showed that they can be as large as the effects of the LL terms. The resulting debate on whether and how, despite this, BFKL-based phenomenology may be possible has yet to be resolved [28–30][5]. Here, we consider modified LL BFKL calculations, incorporating a "consistency constraint" which effectively includes part of the NLL terms [32]. The forward jet cross section determined using this approach is shown in figure 9, for various choices of scale, with the forward jet measurements of the ZEUS and H1 Collaborations. The

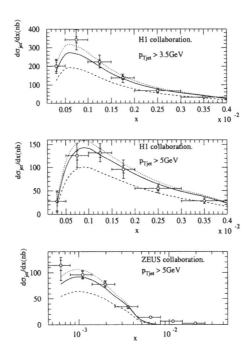

Fig. 9. The forward jet cross section measurements of ZEUS and H1 as a function of x compared with BFKL calculations including a "consistency constraint", for various choices of scale.

modified LL BFKL calculation is significantly better than the unadulterated LL BFKL result shown in figure 8.

The H1 forward dijet rates, measured in the kinematic range given above with the requirement that both jets satisfy $p_{Tjet} > 3.5\,\mathrm{GeV}$, is $\sigma_{2jet} = 6.0 \pm 0.8\,(\mathrm{stat.}) \pm 3.2\,(\mathrm{syst.})\,\mathrm{pb}$, which compares favourably with the prediction from the modified BFKL calculations of $\sigma_{2jet} = 2.7 \ldots 5.2\,\mathrm{pb}$, where the range results from varying the choice of scale.

---

[5] For discussion of the problems associated with the running of α_S in the BFKL formalism see, for example, [31]

3   Forward particle production

A problem that arises when trying to compare forward jet measurements with
BFKL based calculations is the uncertainty of 15 to 20% introduced by the effect
of hadronisation. This can be circumvented by comparing predictions for single
forward π^0 production with measurement. Here, measured π fragmentation functions are folded with the above modified BFKL calculations [32], as illustrated
in figure 3. The success of this approach is apparent in figure 10, in which the

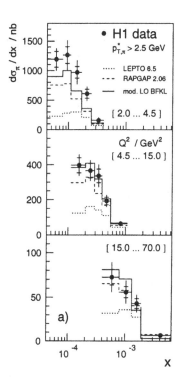

Fig. 10. The H1 forward π^0 cross section as a function of x compared with BFKL
calculations including a "consistency constraint", and with two DGLAP MCs, one
which allows for the effects of virtual photon structure (RAPGAP) and one which does
not (LEPTO).

calculations are compared with the H1 forward π^0 measurements [33].

4   Forward jets and virtual photon structure

Intriguingly, in figure 10 it is not only the BFKL calculation that provides a
reasonable description of the data, the results of the RAPGAP MC are also

respectable. RAPGAP is a DGLAP based MC, but allows for the inclusion of the effects of virtual photon structure. Such effects may become significant in the regime in which the k_T^2 at the photon end of the parton cascade exceeds the photon's virtuality Q^2. Recall that the transverse dimensions of the photon are $\sim 1/\sqrt{Q^2}$ and that the transverse wavelength of the partons $\lambda_T \sim 1/k_T$. If $\lambda_T < 1/\sqrt{Q^2}$ the parton is able to resolve the hadronic structure of the virtual photon which arises through the $\gamma \to q\bar{q}$ splitting. Indeed, such events have been used to determine the effective parton density in the virtual photon [34]. Is this relevant to forward particle and jet production? The answer is a resounding perhaps, as can be seen by the following argument. The requirement that $0.5 < p_{T\mathrm{jet}}^2/Q^2 < 2$ allows events in which $k_T > Q^2$ into the sample studied, both explicitly at the high end of the $p_{T\mathrm{jet}}^2/Q^2$ range and implicitly in that the correlation between $p_{T\mathrm{jet}}$ and k_T is far from 100%. This implies we are in a regime in which virtual photon structure may be significant. In addition, the fact that the parton entering the hard scattering process from the photon side carries only a fraction $x_\gamma = p_{\mathrm{part}}/p_\gamma$ of the momentum p_γ of the photon ensures that the resulting jets will tend to be more forward than in the situation in which $x_\gamma = 1$, as is illustrated in figure 11. In terms of parton cascades, if we choose to explain

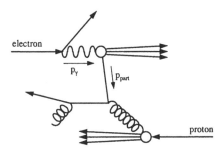

Fig. 11. A DIS event in the HERA laboratory frame in which the structure of the virtual photon is resolved leading to the production of forward jets and particles, *i.e.* jets and particles adjacent to the proton remnant.

the development of both the virtual photon structure "within" the photon and the structure within the proton using the DGLAP prescription, we will have cascades of partons with increasing k_T developing from both the photon and proton ends of the interaction, as illustrated in figure 12. Note that this picture bears some resemblance to the parton cascade produced by the ARIADNE MC, in which there is no k_T ordering. This is a possible explanation for the success of ARIADNE in describing the forward jet and forward particle measurements.

The MC predictions including the effects of virtual photon structure not only describe forward particle production but also provide a reasonable description of the forward jet results, as is shown in figure 13. In this figure comparison is made

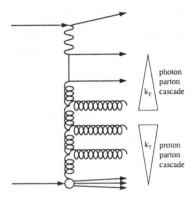

Fig. 12. An illustration of a DIS event in which the structure of both the virtual photon and the proton develops via a parton cascade described by the DGLAP prescription, with k_T ordering as indicated by the increasing width of the relevant triangles.

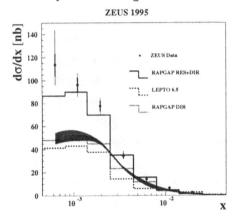

Fig. 13. The ZEUS forward jet measurements as a function of x shown with predictions from MCs with, RAPGAP RES+DIR, and without, LEPTO and RAPGAP DIR, the effects of virtual photon structure.

firstly with LEPTO and RAPGAP allowing only direct photons to contribute, that is those in which the structure of the photon is not resolved. These models are very similar, as is borne out by the results shown, and are not able to describe the data. Allowing the inclusion of resolved photon processes, those in which the partonic structure of the virtual photon is resolved, significantly improves the quality of the description.

This approach, allowing the inclusion of virtual photon structure, can also be taken in NLO QCD calculations, as is discussed in [35,36]. The results of such calculations are shown in figure 14 together with the H1 forward jet measurements. Again, the agreement is found to be good.

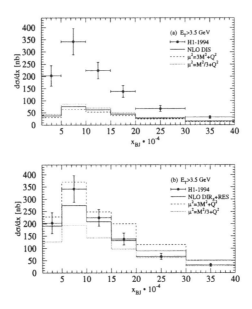

Fig. 14. The H1 forward jet measurements as a function of x shown with the results of NLO QCD calculations (a) without and (b) with the effects of virtual photon structure for various choices of scale.

5   Summary

HERA has provided new impetus to both theoretical and experimental investigations of low x QCD and recent progress has been impressive. We are now able to describe reasonably the measured rates of forward jets [23,22] using either NLO QCD calculations incorporating virtual photon structure [35], modified BFKL calculations [32] or MCs based on LO QCD allowing either directly, as in the case of RAPGAP [14], or indirectly, as in the case of ARIADNE [16], for virtual photon structure. The relative success of ARIADNE may alternatively be considered to be associated with the BFKL-like nature of the parton showers it produces. It is likely that NNLO QCD calculations will provide as good a description of data as the above without having to invoke virtual photon structure [36]. Forward π^0 production is also reasonably described within the modified BFKL framework. New low x MC programs are being developed, such as the LDC MC [18], which is based on a reformulation of the CCFM [9] equations and as such should reproduce BFKL predictions for the hadronic final state in DIS at small x [19]. Many problems remain to be solved, however. The recent calculation of the NLL corrections to the BFKL equation [27] raises difficulties associated with both the running of α_S and the unexpectedly large size of the corrections. The apparent disagreement between LDC MC predictions and LL BFKL calculations is a puzzle. Experimentally, measurements in the forward region remain a challenge. Improving the precision of current measurements and devising new ways of testing low x QCD will tax the ingenuity of physicists for many years to come.

Acknowledgments

The author would like to thank the organisers of Ringberg 99 for the invitation to spend a week in delightful surroundings in such stimulating company. Thanks also to Dr. S.J. Maxfield for his careful reading of this manuscript.

References

1. L. Jönsson, these proceedings.
2. S. Aid et al. [H1 Collaboration], Phys. Lett. **B356** (1995) 118.
3. See the contributions of to these proceedings of A. Martin, J. Blümlein, E. Levin, B. Pötter, J. Outhwaite, M. Klasen and D. Graudenz.
4. E. Mirkes and D. Zeppenfeld, Phys. Rev. Lett. **78** (1997) 428.
5. Yu.L. Dokshitzer, Sov. Phys. JETP **46** (1977) 641;
   V.N. Gribov and L.N. Lipatov, Sov. J. Nucl. Phys. **15** (1972) 438 and ibid. 675;
   G. Altarelli and G. Parisi, Nucl. Phys. **B126** (1977) 298.
6. F. Halzen and A. Martin, "Quarks and Leptons", John Wiley and Sons 1984.
7. E.A. Kuraev, L.N. Lipatov and V.S. Fadin, Sov. Phys. JETP **45** (1977) 199;
   Ya.Ya. Balitsky and L.N. Lipatov, Sov. J. Nucl. Phys. **28** (1978) 822;
   L.N. Lipatov, Sov. Phys. JETP **63** (1986) 904.
8. J. Forshaw and D. Ross, "QCD and the Pomeron", Cambridge Univ. Press 1997.
9. M. Ciafaloni, Nucl. Phys. **B296** (1987) 249;
   S. Catani, F. Fiorani and G. Marchesini, Phys. Lett. **B234** (1990) 339 and Nucl. Phys. **B336** (1990) 18.
10. F. Zomer, these proceedings.
11. A.H. Mueller, Nucl. Phys. **B** (Proc. Suppl.) **18C** (1991) 125.
12. J. Bartels, V. Del Duca, A. De Roeck, D. Graudenz and M. Wusthoff, Phys. Lett. **B384** (1996) 300.
13. G. Ingelman, Proc. of the HERA workshop, Ed.s W. Buchmüller and G. Ingelman, Hamburg 1992, Vol. 3, p. 1366.
14. H. Jung, Comp. Phys. Comm. **86** (1995) 147.
15. G. Marchesini, B.R. Webber, G. Abbiendi, I.G. Knowles, M.H. Seymour and L. Stanco, Comp. Phys. Comm. **67** (1992) 465.
16. L. Lönnblad, Comp. Phys. Comm. **71** (1992) 15.
17. B. Andersson, G. Gustafson and L. Lonnblad, Nucl. Phys. **B339** (1990) 393.
18. H. Kharraziha and L. Lönnblad, JHEP **03** (1998) 006.
19. G.P. Salam, JHEP **03** (1999) 009.
20. T. Sjöstrand, Comp. Phys. Comm. **82** (1994) 74.
21. J. Huth et al., Proc. of the 1990 DPF Summer Study on High Energy Physics, Snowmass, Colorado, Ed. E.L. Berger (World Scientific, Singapore, 1992) p. 134;
   F. Abe et al. [CDF Collaboration], Phys. Rev. **D45** (1992) 1448.
22. J. Breitweg et al. [ZEUS Collaboration], Eur. Phys. J. **C6** (1999) 239.
23. C. Adloff et al. [H1 Collaboration], Nucl. Phys. **B538** (1999) 3.
24. J.G. Contreras Nuno, unpublished Ph.D. thesis, Dortmund University, 1997.
25. S. Catani and M.H. Seymour, Nucl. Phys. **B485** (1997) 291.

26. J. Outhwaite, these proceedings.
27. V.S. Fadin and L.N. Lipatov, Phys. Lett. **B429** (1998) 127.
28. D.A. Ross, Phys. Lett. **B431** (1998) 161.
29. G.P. Salam, JHEP **07** (1998) 019.
30. J.R. Forshaw, D.A. Ross and A.S. Vera, Phys. Lett. **B455** (1999) 273.
31. N. Armesto, J. Bartels and M.A. Braun, Phys. Lett. **B442** (1998) 459.
32. J. Kwiecinski, A.D. Martin and J.J. Outhwaite, Eur. Phys. J. **C9** (1999) 611.
33. C. Adloff *et al.* [H1 Collaboration], Phys. Lett. **B462** (1999) 440.
34. C. Adloff *et al.* [H1 Collaboration], hep-ex/9812024.
35. G. Kramer and B. Pötter, Phys. Lett. **B453** (1999) 295;
36. B. Pötter, these proceedings.

Measurements of Jet Production
in Deep Inelastic Scattering

Leif Jönsson

Phys. Dept., Lund University, Box 118, 221 00 Lund, Sweden

Abstract. Some recent results based on jet production at the HERA ep-collider are presented. Inclusive jet cross sections have proven to be in good agreement with predictions from QCD Monte Carlo models up to very high momentum transfer, whereas di-jet cross sections are less well described. Recent next-to-leading order calculations turn out to reproduce the di-jet data very well which enables a direct determination in next-to-leading order of the gluon density in the proton and an improved measurement of the strong coupling constant.

1 Introduction

Considerable progress in our understanding of how matter is constructed has been made over the past decades mainly thanks to results from deep inelastic scattering (DIS) experiments. The basic idea behind these experiments dates as far back as to Rutherford's famous experiment from 1911 in which it was proven that the atom contains a massive nucleus. In modern experiments leptons are used to probe the structure of matter, the advantage being that the leptons, as far as we know, are pointlike particles and that their interaction with matter can be described by a well founded theory, the electroweak theory. The basic DIS process (zeroth order in QCD) is a purely electroweak interaction between a highly virtual photon emitted by the incoming lepton and a parton in the proton (Fig. 1a).

The HERA storage ring, which collides 27.5 GeV positrons (electrons) with 820 GeV protons, has provided a great increase in resolution by extending the accessible kinematic range by several orders of magnitude in the transverse momentum squared, Q^2, transferred by the virtual photon and in the Bjorken scaling variables, x and y.

The scattered electron from a DIS process carries information about the parton density in the proton as does also the particle jet from the scattered quark, which is usually called the current jet. Thus an inclusive measurement of either the scattered electron or the current jet will determine the proton structure function, F_2. Results from HERA have revealed a steep rise of the structure function as the fractional parton momentum decreases. This can be explained by a model where the emitted gluons split up into quark-antiquark pairs and thus create a large number of sea-quarks. The more partons sharing the proton momentum the less fraction each of them will carry. Since the gluons are driving the strong rise via the sea-quarks, the gluon density can be determined from the proton structure function assuming that the QCD splitting functions describing the parton emission are correct.

A more direct determination of the gluon density would be to measure the rate of boson-gluon fusion (BGF) processes, which is of first order in the strong coupling constant, α_S. The BGF process is an interaction between the virtual photon and a gluon which has fragmented into a quark-antiquark pair (Fig. 1c), and consequently the final state will contain two hard jets. Another first order α_S process is QCD-Compton scattering, where a gluon is emitted from the scattered quark (Fig. 1b). Being of first order in α_S, both processes are thus sensitive to the strong coupling constant and can be used to determine its value. Even though the di-jet cross section is proportional to α_S, leading order (LO) calculations alone suffer from large scale dependence. Thus higher order corrections are necessary to reduce the dependence of the cross section on the renormalization (μ_r) and factorization (μ_f) scales. It turns out that already NLO corrections leads to much smaller dependences on μ_r and μ_f which make reliable cross section calculations possible. In recent years several Monte Carlo programs, like MEP-JET [1], DISENT [2] and DISASTER++ [3], have been developed, providing cross section calculations on the parton level and allowing for the application of various kinematic cuts and choices of jet algorithms. In order to cancel infrared divergencies which appear in the analytic calculations due to soft or collinear gluon emissions two methods have been used. The MEPJET program uses the phase space slicing method [4] which requires the matrix elements for the real emission process to be approximated in the regions close to the singularities. These regions are defined by the introduction of non-physical cut-off parameters. The DISENT program is based on the subtraction method [5] which uses the exact matrix element expression in the whole phase space region. In this method the singularities due to soft and collinear gluon emission will explicitly cancel against the poles in the virtual diagrams.

Since the cross section for di-jet production at HERA energies is in the range 100 pb to few nb, enough statistics will be available for various precision tests of QCD.

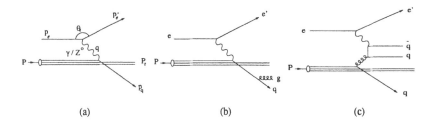

Fig. 1. Generic Feynman diagrams for zeroth and first order α_S processes.

The kinematics of a deep inelastic scattering event can be determined from two independent Lorentz variables, which can be any of the two Bjorken scaling variables x_{Bj} or y, the square of the momentum, Q^2, transferred by the exchanged boson, and the invariant mass squared of the hadronic system, W^2. These variables are defined in terms of the 4-momenta of the incoming proton, P, the incoming and scattered electron, p_e and p'_e, and of the virtual boson, q.

$$Q^2 \equiv -q^2 = -(p_e - p'_e)^2 \tag{1}$$

$$y \equiv \frac{P \cdot q}{P \cdot p_e} \tag{2}$$

$$x_{Bj} \equiv \frac{-q^2}{2P \cdot q} = \frac{Q^2}{ys} \tag{3}$$

$$W^2 \equiv (q + P)^2 = Q^2 \left(\frac{1 - x_{Bj}}{x_{Bj}} \right) + m_p, \tag{4}$$

where s is the centre-of-mass energy squared and m_p is the mass of the proton.

Other variables which are useful in describing the kinematic properties of di-jet events are the Lorentz invariant partonic scaling variables x_p and z_p.

$$x_p = \frac{Q^2}{2p \cdot q} = \frac{Q^2}{2\xi P \cdot q} = \frac{x_{Bj}}{\xi} \approx \frac{Q^2}{Q^2 + \hat{s}} \tag{5}$$

$$z_p = \frac{P \cdot p_{jet}}{P \cdot q} \approx \frac{E_{jet}(1 - cos\theta_{jet})}{\sum_i E_i(1 - cos\theta_i)}, \tag{6}$$

where ξ is the fraction of the proton momentum carried by the interacting parton and $\hat{s}$ is the invariant mass squared of the hard subsystem. E_{jet} and θ_{jet} are the energy and polar angle of the jet, whereas E_i and θ_i are the energy and polar angle of the individual hadrons in the hard subsystem. The variable x in Fig. 2 represents the fraction of the longitudinal proton momentum taken by the parton. For LO processes x is equivalent to ξ but for higher orders it becomes a non-measurable quantity which can only be defined within a specific theoretical frame. M_{jj}^2 is the invariant mass squared of the two jets produced in the hard scattering and is thus equivalent to $\hat{s}$.

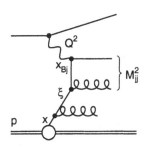

Fig. 2. Generic Feynman diagram illustrating the meaning of some kinematic variables.

The first order parton level process must be corrected for higher order perturbative QCD effects causing the emission of additional partons on a short space-time scale before hadronisation. For example the LEPTO [6] and HERWIG [7] event generators calculate processes up to the first order in α_S by using exact QCD matrix elements convulated with the parton densities in the proton. Additional parton emission is generated using the DGLAP [8] equations in a so called parton shower evolution. In the ARIADNE [9] Monte Carlo program gluons are emitted according to the colour dipole model [10] from independently radiating colour dipoles which are stretched between the partons.

Due to the bremsstrahlung nature of this radiation, most partons will be soft or collinear with the emitting parton and will thus not produce separately observable jets, but rather cause softening of the hadron momenta and broadening of the jet. The reconstruction of a jet must therefore account correctly for both the parton emission process and the hadronisation process.

There are different approaches to describe the fragmentation of partons into hadrons. The most naive picture is to assume that the different partons fragment independently (*independent fragmentation* [11]). This model is, however, connected with several problems like non-conservation of the total flavour, energy and momentum, and it is also not Lorentz invariant. In the *string fragmentation* [12] picture a colour flux tube is streched between a quark and an antiquark. If the transverse dimension of the tube is assumed to be uniform along its length, this automatically leads to confinement with a linearly rising potential. In case a gluon is emitted the colour string will be streched from the quark via the gluon to the antiquark and, if several gluons are emitted, the string picture becomes correspondingly more complicated. As the partons move apart the energy stored in the string will increase and eventually it will break up creating new quark antiquark pairs. The Lund string fragmentation is for example used in the LEPTO and ARIADNE programs. The concept of *cluster fragmentation* [13] contains three components. The generation of clusters by parton shower evolution, the fragmentation of large clusters into smaller ones, and the decay of clusters into hadrons. The clusters are assumed to be the basic units from which hadrons are produced. The decay is assumed to be isotropic in the rest frame of the cluster which gives a compact description with few parameters. This fragmentation model is used by the HERWIG generator.

The intuitive picture of a jet as a collimated flow of particles implies that a minimum energy must be available for the jet formation. In particular, the particle momenta along the initial parton direction must be much larger than the transverse momentum fluctuations induced in the hadronisation process. As the parton momentum increases, the jets tend to be more collimated which simplifies their identification. Nevertheless, the definition of a jet is not unambigous.

A number of jet-finding algorithms have been developed to provide ways of reconstructing jets but since the definition of a jet is not unique, the algorithms all include a parameter which specifies the desired resolution. The input to a jet algorithm are 4-vectors representing particles and/or energy clusters in the detector and as output the 4-momenta of the jets are given. If the resolution depends on the kinematic region in which the algorithm is applied, a suitable scaling variable has to be found to account for this kinematic dependence.

There are two general types of jet algorithms. The *cone algorithms* [14] are based on the definition of a cone with its tip in the interaction point and with a predefined opening angle, normally defined through $R = \sqrt{\eta^2 + \phi^2}$, where η is the pseudorapidity and ϕ the azimuthal angle. The cone is swept over the solid angle covered by the detector and a jet is defined by maximizing the total energy inside the cone. The *cluster algorithms* calculate the distances between pairs of particles or energy clusters using some distance measure which is specific to each algorithm. In an iterative process pairs may be merged until the distance measure exceeds a predefined cut-off value, specifying the resolution of the algorithm. The jet algorithms which will be mentioned here are two versions of the k_T algorithm and the modified Durham algorithm, which are all cluster algorithms. The factorizable k_T algorithm [15] uses two distance measures to determine whether a particle should be merged with another particle or with the proton remnant which is represented by an infinite momentum particle. The distance measures are defined as $k_{T,ij}^2 = 2min(E_i^2, E_j^2)(1 - cos\theta_{ij})$ and $k_{T,ir}^2 = 2E_i^2(1 - cos\theta_{ir})$, where i and j represent particles and r the proton remnant. The longitudinally invariant k_T algorithm [16] uses the same procedure but with cone like distance measures; $k_{T,ij}^2 = 2min(E_{T,i}^2, E_{T,j}^2)(\Delta\eta_{ij}^2 + \Delta\phi_{ij}^2)$ and $k_{T,ir}^2 = 2E_{T,i}^2$ The modified Durham algorithm is a modification of the JADE algorithm [17] with one distance measure, $k_{T,ij}^2 = 2min(E_i^2, E_j^2)(1 - cos\theta_{ij})$ and the cut-off parameter scaling with W^2.

2   Inclusive jet cross sections

The production of multi-jet events at high Q^2 can be used to test the predictions by perturbative QCD. The ZEUS collaboration has used a data sample corresponding to 42.5 pb^{-1} to study inclusive jet production. The jets were reconstructed in the laboratory system using the longitudinally invariant k_T algorithm on clusters in the calorimeter. Reconstructed jets were required to have a transverse momentum greater 14 GeV and only jets within the pseudorapidity range $-1 < \eta_{LAB} < 2$ were accepted, since this is the region where the calorimeter offers good acceptance. As an example of the comparisons between data and QCD models the differential jet cross section $d\sigma/dQ^2$ is shown in Fig 3a and b. It is seen that ARIADNE describes the data very well up to the highest Q^2 bin where a slight deviation can be observed. This is, however, probably due to the fact that the version of ARIADNE which was used doesn't give a proper description of the jet cross sections at high Q^2. It should also be kept in mind that since the jet reconstruction has been performed in the laboratory frame the tranverse energies of the jets are strongly influenced by event kinematics. Furthermore the event sample is dominated by zeroth order α_S events and therefore the sensitivity to QCD effects is suppressed.

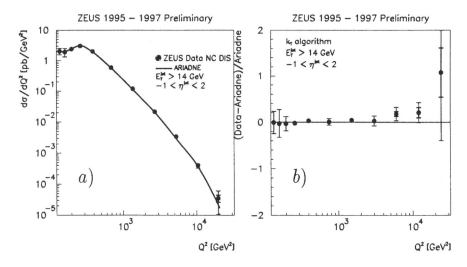

Fig. 3. The differential inclusive jet cross section $d\sigma/dQ^2$ compared to the predictions by ARIADNE.

3   A direct determination of the gluon density in NLO

A direct determination of the gluon density is not only motivated by its own interest, but a comparison with indirect measurements constitutes an important test of QCD. The lack of cross section calculations to next-to-leading order restricted an earlier analysis to a LO determination [18]. Since then several NLO programs, to which various jet algorithms and also experimental cuts can be applied, have become available. Although the QCD-Compton process has the same event topology as the BGF process and thus constitutes a background, the BGF process dominates in the kinematic region investigated.

A recent H1 analysis was based on an integrated luminosity of 36 pb^{-1} collected in the years 1994-97. It was performed in a kinematic region which covers a wide range in the transverse momentum transfer squared, $10 < Q^2 < 5000$ GeV^2, and was restricted to $0.2 < y < 0.6$ in order to get a good determination of y from the measurement of the scattered electron and to suppress backgrounds from photoproduction processes. The jet reconstruction was performed in the Breit frame using the longitudinally invariant k_T algorithm, since this turns out to give the smallest hadronisation corrections, as determined from Monte Carlo models. The jets were required to have a transverse energy sum, $E_{T,jet1} + E_{T,jet2}$, greater than 17 GeV in the Breit frame and simultaneously the transverse energy of each jet had to be greater than 5 GeV. With the additional cut $-1 < \eta_{jet} < 2.5$ in the laboratory system, it was ensured that the jets were well inside the hadronic calorimeter. The data were corrected for detector effects and initial and final state electroweak radiation by performing detailed simulations using DJANGO [19], which provides an interface between the ARIADNE and LEPTO events generators, repectively and HERACLES [20].

Since hadronisation corrections are small, the data have been compared directly to a NLO calculation performed with the DISENT program, using the CTEQ5M [21] parton density parametrization and its corresponding $\alpha_S(M_Z^2)$ value of 0.116. The factorization scale was chosen to be $\mu_f^2 = 200$ GeV^2, which is approximately equal to the average transverse energy squared of the jets in the Breit frame. Q^2 was taken as the renormalization scale squared, μ_r^2. Fig. 4a shows the corrected differential cross section as a function of the jet scattering angle, defined as $\eta' = \frac{\eta_{1,Breit} - \eta_{2,Breit}}{2}$, for three regions in $E_{T,Breit}$ of the jets. The agreement between the data and the NLO prediction is good and it can be observed that the jet production at larger values of pseudorapidity, η', is suppressed as E_T increases. In Fig. 4b the differential cross section is plotted against the fractional parton momentum, ξ, in four different Q^2 bins for comparisons of the data with both the NLO calculation and QCD model predictions. Again the NLO prediction reproduces the data, whereas at low Q^2 the LEPTO predictions fall below the data and at high Q^2 ARIADNE underestimates the cross section. HERWIG with a one loop α_S calculation, similar to LEPTO and ARIADNE, is unable to describe the data everywhere. It should also be noted that the data tend to overshoot the NLO prediction in the lowest ξ bin, which may be an indication that the gluon density has been underestimated in the calculation.

The jet cross section can be written as a convolution of perturbative coefficients with the parton densities in the proton:

$$\sigma_{jet} = \sum_n \alpha_S^n(\mu_r^2) \int_0^1 dx [c_{g,n}(\mu_r^2, \mu_f^2)g(x, \mu_f^2) + \sum_f c_{f,n}(\mu_r^2, \mu_f^2)q(x, \mu_f^2)], \quad (7)$$

where the cross section is calculated to all orders in α_S and for all quark flavours f. The perturbative coefficients are denoted $c_{g,n}$ for the gluon and $c_{f,n}$ for the quarks and the parton density functions are written $g(x, \mu_f^2)$ and $q(x, \mu_f^2)$ for the gluon and the quarks, respectively. The formula illustrates the correlation between the strong coupling constant and the parton densities, which prevents an unambigous determination of the gluon density from a deep inelastic measurement alone. In order to circumvent this problem the $\alpha_S(M_Z^2)$ value of 0.119 ± 0.005 from measurements at LEP [22] was used.

The perturbative coefficients are not all independent but can be reduced to three independent coefficients by linear combinations of the single flavour coefficients.

$$c_G = c_g,$$
$$c_\Delta = 3(c_u - c_d),$$
$$c_\Sigma = 1/3(4c_d - c_u). \quad (8)$$

This results in three independent parton densities:

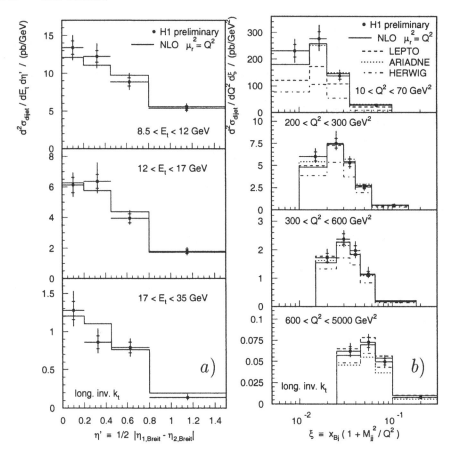

Fig. 4. The doubly differential di-jet cross section a) $d\sigma/dE_t d\eta'$ and b) $d\sigma/dQ^2 d\xi$ compared to the predictions by a NLO calculation and QCD models.

$$\Delta(x, \mu_f) = \sum_f e_f^2 (q_f(x, \mu_f) + \bar{q}(x, \mu_f)),$$

$$G(x, \mu_f) = g(x, \mu_f),$$

$$\Sigma(x, \mu_f) = \sum_f (q_f(x, \mu_f) + \bar{q}(x, \mu_f)). \tag{9}$$

Table 1 summarizes which parton densities contribute to which order in α_S for the inclusive, σ^{incl}, and the di-jet, σ^{dijet}, cross sections, respectively.

With the value of $\alpha_S(M_Z^2)$ fixed to 0.119 ± 0.005 a simultaneous fit of the gluon and quark densities in the proton was performed using data both on inclusive DIS and on di-jet production. The Q^2 region for the inclusive data was restricted to $200 < Q^2 < 650$ GeV^2, which is approximately equal to the jet transverse energy squared in the di-jet data. This also motivates a fixed value of 200 GeV^2 for the

Table 1. LO and NLO contributions of the various parton densities to the inclusive and di-jet cross sections, respectively.

	LO	NLO
σ^{incl}	$\Delta(x)$	$\Delta(x), G(x)$
σ^{dijet}	$G(x), \Delta(x)$	$G(x), \Delta(x), \Sigma(x)$

factorization scale, μ_f^2. In this kinematic region the inclusive data constrain the quark density strongly, whereas the di-jet data is mainly sensitive to the gluon density. The di-jet events were required to have $Q^2 > 200$ GeV^2 in order to avoid large NLO corrections and hadronisation effects. For the di-jet data fits were made to the double differential cross sections $d\sigma/dQ^2d\xi$ and $d\sigma/dQ^2dx_{Bj}$, and for the inclusive data to $d\sigma/dQ^2dx_{Bj}$, since they are the most sensitive ones to the x-dependence of the parton densities.

In Fig. 5 the result of the gluon density fit is compared to global fits using parton density parametrizations according to CTEQ4M, MRST and GRV94HO, and to a QCD analysis of the F_2 structure function as measured by the H1 collaboration. The direct measurement indicates a somewhat steeper rise with decreasing x than the global fits although the results are compatible within the error band. On the other hand, the direct measurement is in good agreement with the H1 QCD analysis of F_2. The errors are dominated by the uncertainty in the energy scales of the electromagnetic and hadronic calorimeters, and in the determination of the luminosity.

4   A study of differential di-jet rates

In an H1 analysis differential di-jet rates have been used for comparisons with various QCD models and to a NLO prediction. The differential di-jet rate is defined as the number of di-jet events which is obtained with a predefined cut-off value for the resolution parameter, y_{cut}, in the jet algorithm, divided by the total number of DIS events. In an iterative process the jet candidates with the smallest invariant masses are combined until exactly two hard jets remain. The maximum value of the resolution parameter at which two hard jets are resolved is denoted y_2.

The data used for the analysis by H1 were collected in the years 1995-97 and correspond to an integrated luminosity of 35 pb^{-1}. The events were restricted to the kinematic range $Q^2 > 150$ GeV^2 and $0.1 > y > 0.7$ which results in a data sample of 60000 events. The statistics allow a subdivision into three different ranges in Q^2.

Two jet algorithms, the factorizable k_T-algorithm and the modified Durham algorithm, have been used in the study. The k_T-algorithm was applied in the Breit frame with a cut-off value of $y_2 > 0.8$, which approximately corresponds to

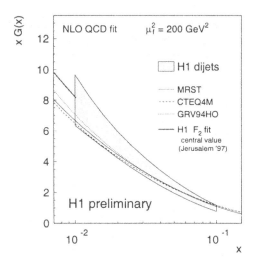

Fig. 5. The gluon density in the proton. The result of the direct determination is compared to global fits with various parton density parametrizations and to a QCD analysis of the proton structure function F_2.

a mean transverse jet energy of 10 GeV in the Breit frame. The modified Durham algorithm was used in the laboratory system. The resolution parameter was given by $y_2 = k_T^2/W^2$, and the cut-off value was chosen to be $y_2 = 0.005$, which again corresponds to a mean transverse jet energy of 10 GeV in the Breit frame. For both algorithms the jets were restricted to an angular region $10^o < \theta_{LAB} < 140^o$ to ensure good detector acceptance. In addition, jets reconstructed below 7° by the Durham algorithm were rejected.

The di-jet distributions presented have been corrected for detector effects and the influence of QED radiation. The major systematic error comes from the uncertainties in the energy scales of the electromagnetic and hadronic calorimeters.

Data have been compared to the LEPTO and ARIADNE models and to a NLO calculation performed with the DISENT program using Q^2 as renormalization scale. The hadronisation effects have been estimated from simulations with the QCD models and were found to be between 10% and 20%.

Fig. 6a shows that the y_2 distributions are in good agreement with both Monte Carlo models and the NLO prediction in all three Q^2 bins. The spectra get harder as Q^2 increases. The z_p distributions, however, are not so well described by the models which tend to overshoot the data as seen in Fig. 6b. Again the NLO calculation reproduces the data. In Fig. 7 the distribution of the polar angle for the most forward going jet is plotted and compared to the NLO prediction where also the BGF fraction is indicated. At high Q^2, where the BGF fraction is of the order of 30%, the agreement is good, whereas in the low Q^2 region, which contains 70% BGF, the NLO calculation seems to fall below the data point in the lowest θ bin. This might indicate an underestimation of the gluon density in the NLO calculation.

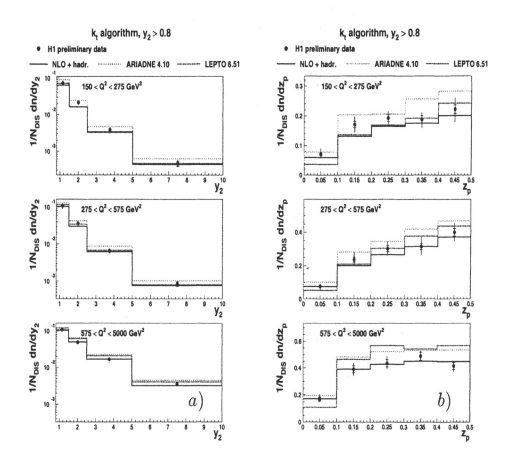

Fig. 6. The differential di-jet rates as a function of a) y_2 and b) z_p compared to a NLO calculation and QCD models.

5   Measurement of the strong coupling constant from di-jet rates

The ZEUS collaboration has used data collected in the years 1996-97 to investigate di-jet rates in deep inelastic scattering with the aim of determining the strong coupling constant, α_S. The data correspond to an integrated luminosity of 38 pb^{-1}. In order to avoid experimental and theoretical difficulties as far as possible the analysis was performed in the high Q^2 region, $470 < Q^2 < 20000$ GeV^2. The event kinematics was extracted using the double angle method [23]

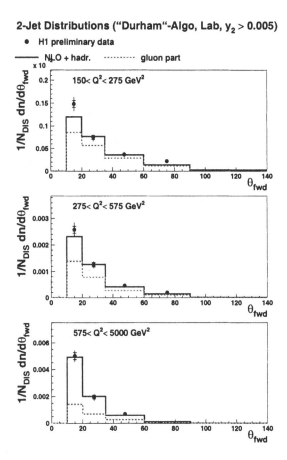

Fig. 7. The distribution in polar angle for the most forward going jet compared to a NLO calculation. The contribution from the BGF process is indicated.

which is the method least sensitive to errors in the absolute energy scale of the calorimeter. It was also proven to be the most accurate one in the Q^2 region investigated.

The jet reconstruction was performed in the Breit frame using the longitudinally invariant k_T-algorithm. Energy clusters in the calorimeter were used as jet seeds, and the jet reconstruction procedure followed the Snowmass scheme. Events with exactly two jets were selected, where both jets were required to have energies above 5 GeV and their energy sum to be above 17 GeV. This cut scenario avoids problems with divergencies, related to soft and collinear gluon emissions, in fixed order calculations. In order to ensure that the jets were well inside the acceptance of the detector they were restricted to the rapidity range $-1 < \eta_{LAB} < 2$.

Monte Carlo simulations have been used to correct the data for detector inefficiences and to take effects from QED radiation and hadronisation into account. Further, contributions from Z^o exchange had to be included since the Q^2 range extends as far up as 20000 GeV^2. Comparisons of the data have been made with two versions of the DJANGO program based on the LEPTO and ARIADNE Monte Carlo generators. Also the HERWIG program was selected for comparisons since it uses a different fragmentation algorithm to describe the hadronisation.

For calulations of NLO cross sections the DISENT program was used and it was checked to be consistent with the results of the DISASTER++ program to within a few percent.

Various control plots were produced in order to see how well the data were reproduced by Monte Carlo models and the NLO prediction. Fig. 8 shows the differential di-jet cross section as a function of the the transverse energies, E_t, and rapidities, η, of the two jets in the event. The uncertainty due to the hadronic energy scale of the calorimeter is the dominant systematic error. The error bars contain contributions from all other systematic errors and the statistical errors added in quadrature. The NLO calculations have been performed by DISENT in the $\overline{MS}$ renormalization scheme using the parton density parametrization according to CTEQ4M [24] and choosing $\mu_R^2 = \mu_F^2 = Q^2$ as renormalization and factorization scales. A closer look at the renormalization scale dependence of the NLO calculation motivates the lower Q^2 cut of 470 GeV^2 above which the dependence is very small. Also the dependence on the parton density functions is small in this region. An excellent agreement between data and the NLO calculation is observed in all plots, which is a necessary condition for a reliable extraction of the strong coupling constant.

In each Q^2-bin the ratio of di-jet events, $R_{2+1} = \sigma_{2+1}/\sigma_{tot}$, was calculated with the DISENT program. Five different values of the strong coupling constant, ranging from 0.110 to 0.122, were used in the CTEQ4M parametrization of the parton density functions. The MRST parametrization was also used for a consistency check. The di-jet rates obtained for the five α_S values were fitted with a straight line in each Q^2-bin to allow for a continuous variation of α_S with the di-jet rate. The α_S value was then extracted through a χ^2 fit of the NLO prediction to the measured di-jet ratio according to:

$$\chi^2 = \sum_{i=1}^{5} \frac{(R_{2+1,meas}^i - R_{2+1,theor}^i)^2}{\sigma_{2+1}^{2,i}} \tag{10}$$

The value obtained is:

$$\alpha_S(M_Z^2) = 0.120 \pm 0.003 (stat) \tag{11}$$

A summary of the systematic errors are given in Fig. 9. As has been mentioned previously, and which is clearly illustrated in the figure, the by far dom-

ZEUS 96-97 PRELIMINARY

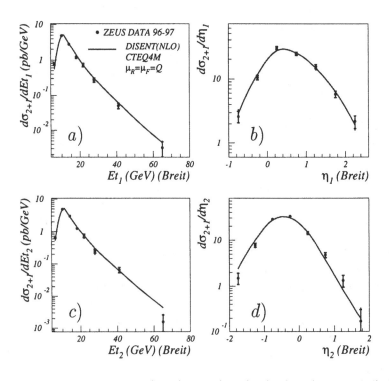

Fig. 8. The di-jet cross sections a) $d\sigma/dE_{t1}$, b) $d\sigma/d\eta_1$), c) $d\sigma/dE_{t2}$ and d) $d\sigma/d\eta_2$ compared to the predictions by the NLO calculation.

inant systematic error comes from the uncertainty in the jet energy scale. Another contribution to the experimental systematic error is due to the model dependence. The uncertainty due to a variation of the renormalization scale between $1/2Q^2$ and $2Q^2$, and the dependence on the parton density functions and the NLO program used, define the theoretical systematic error. The final result including all errors is:

$$\alpha_S(M_Z^2) = 0.120 \pm 0.003(stat) \pm_{0.006}^{0.005}(exp) \pm_{0.003}^{0.002}(th) \qquad (12)$$

5.1   Summary

Jet production in DIS provides an interesting testing ground of perturbative QCD, which in many respects offers higher sensitivity and contains more information about the proton structure than the inclusive measurements of the proton structure function. Over the past few years quite some progress has been made in the description of jet production in deep inelastic scattering. Several

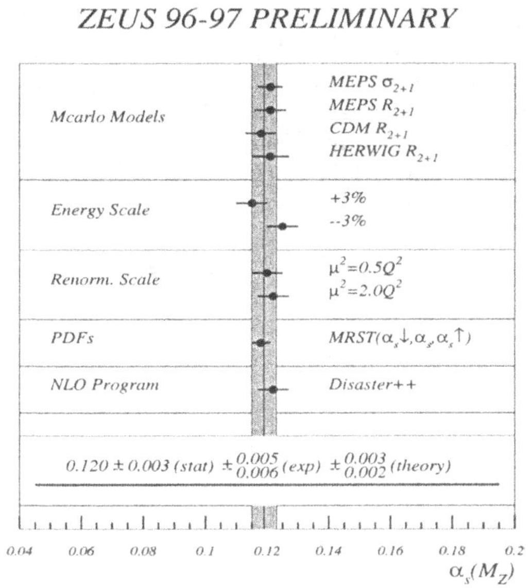

ZEUS 96-97 PRELIMINARY

Fig. 9. Contributions to the systematic errors on the measured α_S value.

Monte Carlo programs for calculations of next-to-leading order α_S jet cross sections have been developed, which allow different jet algorithms to be applied and experimental cuts to be defined. Although the new programs offer quantitative agreement at large Q^2, deviations were observed in the low Q^2 region. Only recently an explanation of these data could be achieved both by a Monte Carlo model [25] and by a NLO calculation [26] by, in addition to interactions with a pointlike photon, also including contributions where the interaction takes place with the parton content of the photon.

5.2   Acknowledgement

This presentation was given on behalf of the H1 and ZEUS collaborations. I would like to thank S. Schlenstedt, L. Sinclair, E. Tasso, M. Weber, N. Tobien and M. Wobisch for providing me with the relevant material.

References

1. E. Mirkes and D. Zeppenfeld, Phys. Lett. B380(1996)205.
2. S. Catani and M. Seymour, Nucl. Phys. B485(1997)291.
3. D. Graudenz, hep-ph/9710244.
4. K. Fabricius, G. Kramer, G. Schierholz and I. Schmitt, Z. Phys. C11(1981)315, G. Kramer and B. Lampe, Fortschr. Phys. 37(1989)161, W.T. Giele and E.W.N. Glover, Phys. Rev. D46(1992)1980, W.T. Giele, E.W.N. Glover and D.A. Kosower, Nucl. Phys. B403(1993)663.

154   Leif Jönsson

5. R.K. Ellis, D.A. Ross and A.E. Terrano, Nucl.Phys. B178(1981)421.
6. G. Ingelman, in Proc. Workshop on Physics at HERA, Hamburg, October 1991, eds. W. Buchmüller and G. Ingelman, vol. 3(1992)1366. G. Ingelman, A. Edin and J. Rathsman, Comput. Phys. Comm. 101(1997)108.
7. G. Marchesini et al., Comput. Phys. Comm. 67(1997)465.
8. Yu.L. Dokshitzer, Sov. Phys. JETP 46(1977)641, V.N. Gribov and L.N. Lipatov, Sov. J. Nucl. Phys.15(1972)438 and 675, G. Altarelli and G. Parisi, Nucl. Phys. 126(1977)297.
9. L. Lönnblad, Comput. Phys. Comm. 71(1992)15, L. Lönnblad, Z. Phys. C65(1995)285.
10. G. Gustafson, Phys. Lett. B175(1986)453, G. Gustafson and U. Petterson, Nucl. Phys. B306(1988)746, B. Andersson, G. Gustafson, L. Lönnblad and U. Petterson, Z. Phys. C43(1989)625.
11. R.D. Field and R.P. Feynman, Nucl. Phys. B136(1978)1, P. Hoyer et al., Nucl. Phys. B161(1979)349, A. Ali et al., Nucl. Phys. B168(1980)409, A. Ali et al., Phys. Lett. 93B(1980)155.
12. B. Andersson, G. Gustafson and C. Peterson, Z. Phys. C1(1979)105.
13. B.R. Webber, Nucl. Phys.B238(1984)492, G. Marchesini and B.R. Webber, Nucl. Phys. B238(1984)1.
14. CDF Collab., F. Abe et al., Phys Rev. Lett. 70(1993)713, M.H. Seymour, Z. Phys. C62(1994)127.
15. S. Catani, Yu.L. Dokshitzer and B.R. Webber, Phys. Lett. B285(1992)291.
16. S.D. Ellis and D.E. Soper, Phys. Rev. D48(1993)3160, S. Catani, Yu. L. Dokshitzer, M.H. Seymour and B.R. Webber, Nucl. Phys. B406(1993)187.
17. JADE Collab., W. Bartel et al., Z. Phys. C61(1986)23.
18. H1 Collab., S. Aid et al., Nucl. Phys. B449(1995)3.
19. K. Charchula, G. Schuler and H. Spiesberger, Comput. Phys. Comm. 81(1994)381.
20. A. Kwiatkowski, H. Spiesberger and H.-J. Möhring, Comp. Phys. Comm. 69(1992)155.
21. H.L. Lai et al., hep-ph/9903282
22. S. Catani, Int.Symposium on Lepton Photon Interactions, Hamburg, Germany (1997) and hep-ph/9712442.
23. S. Bentvelsen, J. Engelen and P. Kooijman, Proc. of the Workshop 'Physics at HERA', vol.1, eds. W. Buchmüller and G. Ingelman, DESY(1991)23.
24. H.L.Lai et al., Phys. Rev. D55(1997)1280.
25. H. Jung, L. Jönsson and H. Küster, DESY 98-051 and hep-ph/9805396, H. Jung, L. Jönsson and H. Küster, hep-ph/9811368 and Proc. of the Workshop 'Photon interactions and the photon structure', eds. G. Jarlskog and T. Sjöstrand, Lund(1998)229, H. Jung, L. Jönsson and H. Küster, Eur. Phys. J. C9(1999)383.
26. G. Kramer and B. Pötter, Eur. Phys. J. C1(1998)261, G. Kramer and B. Pötter, Eur. Phys. J. C5(1998)665, G. Kramer and B. Pötter, Phys. Lett. B453(1999)295.

Jet Production in DIS

Dirk Graudenz

Paul Scherrer Institut, 5232 Villigen PSI, Switzerland

Abstract. The status of next-to-leading-order corrections to jet cross sections in deeply inelastic scattering is reviewed. I give a brief introduction into the techniques which are used to calculate jet cross sections, and make a few remarks on a recent program comparison. A third part gives a few results of recent measurements of the strong coupling constant $\alpha_s(\mu^2)$ and the gluon density $f_{g/P}(\xi, \mu_f^2)$.

1 Introduction

The increased integrated luminosity at HERA permits the study of hadronic final states of deeply-inelastic processes in great detail [1,2]. Earlier analyses of this kind of process suffered mainly from the poorly understood emission of particles in the forward region. The situation is improved now, because a larger event sample allows for harder cuts on the photon virtuality Q and the transverse momentum E_T of the jets, the latter cut excluding jets in the forward region.

Besides testing QCD, the measurement of (2+1)-jet final states has two main applications: the measurement of the strong coupling constant $\alpha_s(\mu^2)$ and the determination of the proton's gluon density $f_{g/P}(\xi, \mu_f^2)$. Figure 1 shows the basic two leading-order processes which contribute to (2+1)-jet final states. Diagram (a) is the so-called QCD-Compton process, and (b) is called boson–gluon fusion[1]. The corresponding cross sections are of $\mathcal{O}(\alpha_s)$, and thus the strong coupling constant can be measured, preferably via the (2+1)-jet rate $R_{2+1} = \sigma_{2+1}/\sigma_{\text{tot}}$, if the parton densities are assumed to be known. Conversely, the cross section of process (b) is a convolution of a hard scattering cross section with the gluon density $f_{g/P}(\xi, \mu_f^2)$. By subtraction of the well-known quark-initiated contribution and a suitable unfolding procedure, the gluon density can be extracted.

Because of the scale dependence of the strong coupling constant α_s and the parton densities $f_{i/P}$, perturbative calculations have to be done beyond the leading order to be physically meaningful. Virtual corrections suffer from two problems. Large loop momenta introduce ultraviolet singularities, which are removed by renormalisation. Small loop momenta lead to infrared singularities. These cancel, in principle, against divergences from the real corrections, but they lead to certain technical problems in the calculations.

The outline of this paper is as follows: in Section 2 I describe the cancellation mechanism of infrared divergences in jet cross sections, and make a few remarks

---

[1] There are, of course, two additional diagrams: the gluon in (a) attached to the outgoing quark, and the diagram with quark and antiquark exchanged in (b).

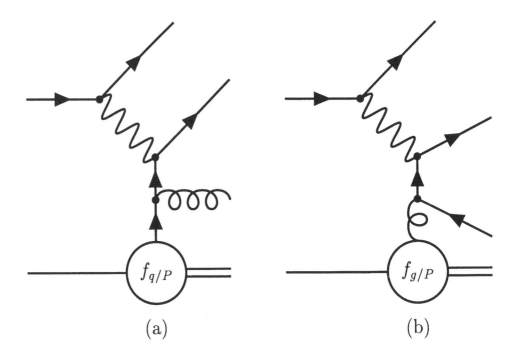

Fig. 1. The leading-order processes for (2+1)-jet production: (a) the QCD-Compton diagram, (b) the boson–gluon fusion diagram.

concerning the techniques used to achieve this. In Section 3, the available calculation for jet cross sections in DIS are discussed and compared. Recent results of data analyses of jets at HERA and a general discussion of scale choices are given in Section 4.

2   QCD at Work

QCD is the microscopic theory of the strong interactions. Perturbative QCD, the evaluation of the theory at large scales for small values of the strong coupling constant, is very successful in describing high-energy scattering processes. Except for very inclusive observables in e^+e^- annihilation, however, perturbation theory is not sufficient to arrive at phenomenological predictions, because non-perturbative input is required. Factorization theorems of QCD [3] guarantee that the non-perturbative part of the description is process-independent. In DIS, for example, parton densities $f_{i/P}(\xi, \mu_f^2)$ are required in order to describe the distribution of partons in the proton.

The standard description of processes in QCD is illustrated in Fig. 2. A hard scale, in DIS usually the photon virtuality $Q^2 \gg \Lambda_{\text{QCD}}^2$, allows for a perturbative calculation in fixed order of perturbation theory (I). In the case

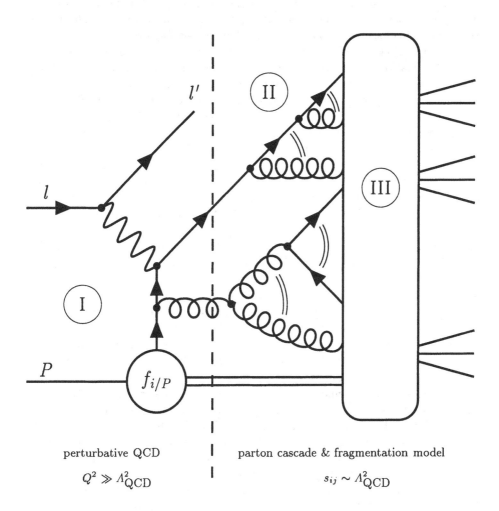

Fig. 2. The standard description of processes in QCD phenomenology consists of three stages. Processes are modelled by a hard scattering process (I), followed by a parton cascade (QCD in the collinear / leading logarithmic approximation, II) and a phenomenological fragmentation model (III). The quantity s_{ij} is the invariant mass of a typical pair of partons after the splitting process.

of generators, a parton cascade (II) is appended, which simulates QCD in the collinear limit of parton branching processes. Finally, in order to obtain final-state hadrons instead of partons, a phenomenological fragmentation model (III), such as the Lund string model [4], is used. Steps (II) and (III) involve certain choices of parameters, for example cut-off parameters for the parton cascade. In contrast, for step (I) there is, except for the parton density parametrization $f_{i/P}$, essentially just one parameter determining the process: the strong coupling constant. In QCD generators, the matrix elements (I) are usually calculated in leading order only — it is not known how to append a parton cascade in full generality and consistently to a calculation beyond leading order[2]. The virtue of a generator is that it produces a final state as it is seen in an experiment, and that hadronization effects are included. The main drawback is that generator results usually depend on a large number of tuned parameters.

A conceptually cleaner approach is to rely on a fixed-order QCD calculation with a suitable observable that can be evaluated both for parton and hadron final states. This is presently the standard procedure for data analysis of DIS hadronic final states. Generators are used to determine correction factors from the hadron to the parton level. The spread from various fragmentation models gives a contribution to the systematic error of a measurement. Observables are chosen according to the criterion of small hadronization corrections. Another important property for suitable observables is *infrared-safety*. Together with the factorization theorems of QCD, this property makes sure that expectation values of observables calculated in perturbative QCD are finite. Infrared-safe observables behave well in soft and collinear limits: collinear parton splitting and radiation of soft partons do not change the value of an observable. Incidentally, this is also a property that good experimental observables are supposed to have. Calorimetric measurements are insensitive to modifications of the final state of this kind. For more details, I wish to refer the reader to Refs. [6,7].

Cancellation of infrared singularities occurs between real and virtual corrections, a mechanism originally discovered in QED. The main problem in next-to-leading-order calculations is to perform the calculation of the real corrections without knowing the specific form of the observable to be calculated. There are two basic methods to do this: the phase-space-slicing method and the subtraction method. Here I do not have enough space to describe these methods in detail, descriptions can be found in Refs. [8,6] and the references cited therein. I just mention that for phase-space slicing a small technical cut-off parameter $y_{\min}$ has to be introduced. Partons are considered to be technically unresolvable if a quantity such as their scaled invariant mass is smaller than this cut-off. Infrared singularities from unresolved configurations of this kind cancel explicitly. The method is observable-independent, because infrared-safe observables are insensitive to unresolved partons. Contrary to this, the cancellation of IR singularities in the subtraction method is done locally in phase space. No technical cut-off is required, but the construction of the required subtraction terms is cumbersome.

In QCD, there are two distinct types of IR singularities:

- Two massless partons may be collinear (Fig. 3). The real cross section σ_{real} in the collinear limit may be approximated by the expression

$$\int \frac{\mathrm{d}\varphi}{2\pi} \sigma_{\mathrm{real}} \sim \frac{\alpha_s}{2\pi} \frac{1}{2p_j p_k} \hat{P}_{kj\leftarrow i}(u)\, \sigma_{\mathrm{Born}}. \tag{1}$$

It is proportional to the Born cross section σ_{Born} and an unsubtracted splitting function $\hat{P}_{kj\leftarrow i}(u)$.

---

[2] There is a recent proposal for a matching procedure, see Ref. [5].

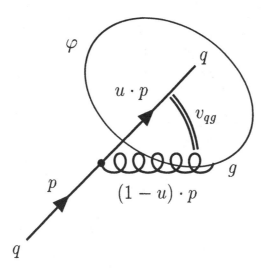

Fig. 3. Collinear limit: splitting of a quark into a quark and a gluon. The incident quark's momentum fraction carried by the outgoing quark in the collinear limit is u, and the azimuthal angle of the gluon with respect to a fixed external axis is φ.

- A soft gluon is radiated (Fig. 4). An eikonal approximation yields an expression of the form

$$\sigma_{\text{real}} \sim \frac{\alpha_s}{2\pi} \sum_{i,j\neq k} C_{ijk} \frac{p_i p_j}{(p_i p_k)(p_j p_k)} \sigma_{\text{Born}} \tag{2}$$

with fixed coefficients C_{ijk}.

After a separation of overlapping singularities, these expressions are used to generate an approximation term for the subtraction method. For collinear singularities in the case of the splitting $g \to gg$, the subtraction term is actually dependent on the angle φ.

3   Calculations of Jet Cross Sections in DIS

By now there are several calculations for (2+1)-jet processes available with corresponding weighted Monte-Carlo programs. The two early programs PROJET [10–12] and DISJET [13,15,14] were restricted to the modifed JADE clustering scheme. Subsequent calculations were universal in the sense that in principle arbitrary infrared-safe observables could be calculated:

- MEPJET [16]: This is a program for the calculation of arbitrary observables which uses the phase-space-slicing method. The corresponding calculation [17] uses the Giele–Glover formalism [18] for the analytical calculation of

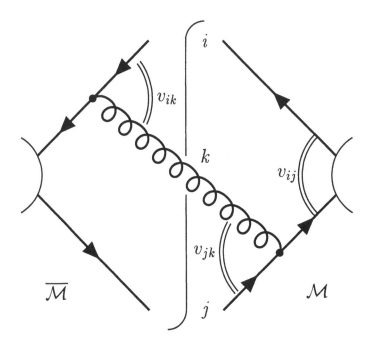

Fig. 4. Soft limit: the gluon attaches to two different quark lines i and j from the matrix element $\mathcal{M}$ and its complex conjugate $\overline{\mathcal{M}}$.

the IR-singular integrals of the real corrections, and the crossing-function technique [19] to handle initial-state singularities. The latter requires the calculation of "crossing functions" for each set of parton densities.

- **DISENT** [20]: This program is based on the subtraction method. The subtraction term is defined by means of the dipole formalism[3] [21,22].
- **DISASTER++** [23]: This is a **C++** class library[4] . The subtraction method is employed, and the construction of the subtraction term resembles the method of Ref. [8], i.e. it is obtained by the evaluation of the residues of the cross section in the soft and collinear limits. Double counting of soft and collinear singularities is avoided by means of a general partial fractions method.
- **JetViP** [24]: This program implements the calculation of [25], which extends the previous calculations into the photoproduction limit $Q^2 \to 0$. The cal-

---

[3] The subtraction term is written as a sum over dipoles (an "emitter" formed from two of the original partons and a "spectator" parton). Besides the factorization theorems of perturbative QCD, the main ingredient is an exact factorization formula for the three-particle phase space, which allows for a smooth mapping of an arbitrary 3-parton configuration onto the various singular contributions.

[4] The acronym stands for "Deeply Inelastic Scattering: All Subtractions Through Evaluated Residues". The program is written in **C++**. A **FORTRAN** interface is available; thus there is no problem to interface the class library to existing **FORTRAN** code.

culation has been done by means of the phase space slicing method. Up to now, the polarization of the virtual photon is restricted to be longitudinal or transverse.

These four programs have been compared in detail during the recent workshop on Monte Carlo generators for HERA physics [26–28], both for jet cross sections and event shape variables. It turns out that DISENT and DISASTER++ agree well for jet cross sections [27]. The results for event shapes are in agreement as well, except for the current jet broadening variable B [28]. It is not yet known where this residual discrepancy comes from; jet broadening is an infrared-safe shape variable, but it regularizes the (1+1)-jet singularity poorly. It may well be that the different results come from a numerical instability in one of the programs. The MEPJET program shows systematic differences with respect to DISASTER++/DISENT for jet cross sections (no comparison of shape variables has been done) even for physical setups of integration parameters. The results are stable against variations of the phase-space slicing parameter. The JetViP results show deviations from the DISASTER++/DISENT results as well, but the differences are smaller than for MEPJET. The JetViP results depend on the technical phase-space slicing cut-off parameter [27], which is a sign that the final result is not completely stable.

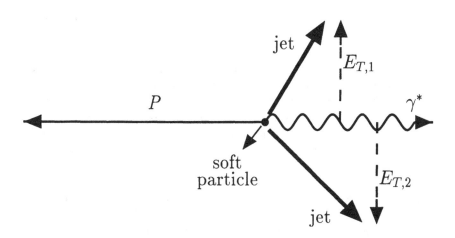

Fig. 5. A (2+1)-jet event in the Breit frame. The transverse momentum is balanced. An additional soft parton of an event close to the transverse momentum cut-off drives the transverse momentum of one of the jets below threshold.

A problem with transverse momentum cuts which has recently been observed is illustrated in Fig. 5. A cut of $E_{T,\min}$ on the transverse momenta of both jets leads to a quantity which is sensitive to the emission of soft particles. If the event is close to the threshold, the emission of a particle with a very small energy removes it from the event sample. This in itself is not critical; after all, this is what cuts are supposed to do. However, what is affected is the cancellation of infrared singularities. In a next-to-leading-order calculation, a process with two high-E_T partons and a soft gluon contributes to the cancellation of IR singularities from the virtual corrections. The singularities themselves will of course cancel, but the cross section will become very sensitive to other cuts, which is, of course, undesirable. A possible cure for this is to impose asymmetric cuts on the transverse momenta, i.e. conditions such as $E_{T,1} > E_{T,\min}$, $E_{T,2} > E_{T,\min} + \Delta$ for $E_{T,1} > E_{T,2}$, and $\Delta = \mathcal{O}(\text{few GeV})$.

4   Comparison with Data: $\alpha_s(\mu^2)$ and $f_{g/P}(\xi, \mu_f^2)$

Both H1 and ZEUS have published measurements of the strong coupling constant $\alpha_s(\mu^2)$ from jet rates in DIS [1,2]. The H1 results are shown in Fig. 6. Experimental jet cross sections for the inclusive $k_\perp$ algorithm vs. NLO theory (DISENT) for CTEQ5M parton densities and $\alpha_s(m_Z^2) = 0.118$ are displayed in Fig. 6(a). This impressive agreement suggests that the strong coupling constant may be extracted from hadronic final states in DIS. This is illustrated in Fig. 6(b). The chosen renormalization scale in this case is the transverse momentum E_T. The final result of a combined fit of all data points is $\alpha_s(m_Z^2) = 0.1181 \pm 0.0030(\text{exp.})^{+0.0039}_{-0.0046}(\text{th.})$, which is consistent with the world average value. Moreover, the plots show the running of the coupling constant. A measurement using Q as the renormalization scale (not shown here) leads to the result $\alpha_s(m_Z^2) = 0.1221 \pm 0.0034^{+0.0054}_{-0.0059}$, which is compatible with the result for $\mu = E_T$. According to Ref. [1], the larger theoretical error comes from the larger renormalization scale dependence of the choice $\mu = Q$.

The direct extraction of the gluon density by H1 is shown in Fig. 7 [29][5]. The procedure was to do a fit of quark and gluon densities to a combined data set of the total DIS cross section and the inclusive DIS jet cross section for a given value of α_s; the former constraining the quark densities, and the latter constraining the gluon density. The result for $f_{g/P}(\xi)$ is compatible with the results from global fits.

There is, of course, the question of the right scale choice. Scale dependence comes in because the perturbative series is truncated. Physical results to all orders would be scale-independent. The choice of a scale essentially corresponds to the choice of a renormalization prescription. To a certain degree, perturbation theory takes care of itself: scale variations of a quantity calculated to order α_s^n in perturbation theory have a residual scale dependence of order α_s^{n+1}. This residual scale dependence may be large, which means that the coefficient of this term is large. In practice, there are two approaches for the choice of a scale. The first approach, a procedure employed frequently in experimental analyses, is to choose some central value μ equal to one of the energy scales of the process, and to take the spread of the results in the interval $\mu/2$ to 2μ to be the uncertainty from the theoretical prediction. This is justified by the reasoning that the scale

---

[5] See also Ref. [30].

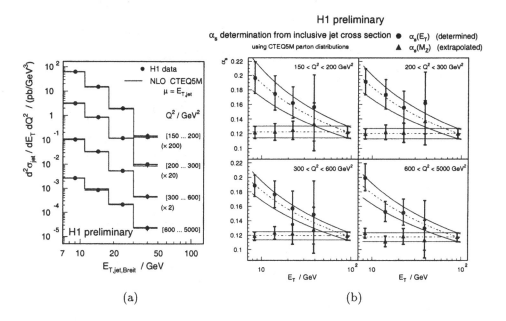

Fig. 6. Preliminary H1 results [1] for jet cross sections and the strong coupling constant: (a) Comparison of data and NLO results (DISENT). (b) Extracted value of $\alpha_s(E_T^2)$ in various bins of Q^2.

dependence of the finite-order result is exactly identical to the scale dependence of the uncalculated higher-order terms. Thus, the scale dependence reflects the size of these terms, and consequently is an estimate of the error made. There are two criticisms: there are scale-independent contributions from higher orders, which are not captured by this procedure, and, in any case, the variation does not determine the absolute size. The second approach invokes some higher principle (such as BLM [31], PMS [32] and FAC [33]), justified in turn either by a physical principle (as for BLM), by an assumption on the behaviour of the perturbative series (for PMS and FAC), or even by a fitted parameter ρ, with, for example, $\mu = \rho Q$. Not all of these scales can be "correct", because they have different values, and what comes out is (necessarily) a typical scale of the process multiplied by a numerical factor. Admittedly, this numerical factor can be different from one. In a certain sense, what is done in the first approach is to estimate the spread of the scale dependence from various *a priori* scenarios.

Typically, there are several physical scales which are relevant in a process, in the case at hand the photon virtuality Q and the transverse momentum E_T of the jets in the Breit frame. Fixed-order perturbation theory breaks down if these scales are very different, because large logarithms of ratios of these scales lead to large coefficients, which in turn spoil the perturbative expansion. The proper way to handle this problem would be resummation. So far, no resummed

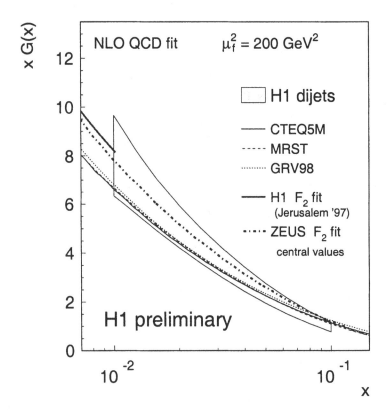

Fig. 7. Preliminary H1 result for the direct extraction of the gluon density [29].

calculation for jet cross sections in DIS is available. Since none of the available scales is to be preferred, the safest procedure at present is to employ all of them in turn, and quote the resulting differences as a theoretical error.

5   Summary

I have given a brief review of the techniques employed in next-to-leading-order QCD calculations. In particular, I have discussed the singularity structure of the real corrections. From a principal point of view, fixed-order perturbation theory is the cleanest approach to theoretical predictions: it depends on the smallest number of input parameters (essentially the strong coupling constant and parton densities), and it is not fragmentation-model dependent.

The situation concerning next-to-leading-order calculations for deeply inelastic processes has improved. Several independent calculations for (2+1)-jet-like quantities are available, with different methods used to extract infrared singularities. The programs have been compared in some detail in the framework of the recent workshop on Monte Carlo generators for HERA. Two of them, DISENT and DISASTER++, agree well, with the exception of the jet broadening shape variable.

Measurements of α_s and $f_{g/P}$ are by now well established. Compared to earlier analyses, the present measurements are more reliable, mainly due to larger event samples and as a consequence the possibility of stricter cuts on transverse momenta of jets, which reduces the problems with jets in the forward direction.

Acknowledgements

I would like to thank the organizers of the workshop for the invitation to Ringberg castle and Doreen Wackeroth for a critical reading of the manuscript.

References

1. H1 Collaboration, *Determination of the Strong Coupling Constant from Inclusive Jet Cross Sections in Deep-Inelastic Positron-Proton Collisions at HERA*, in: Proceedings of the International Europhysics Conference on High Energy Physics, Tampere, Finnland (1999)
2. ZEUS Collaboration, *Measurement of Differential Cross Sections for Dijet Production in Neutral Current DIS at High Q^2 and Determination of α_s*, in: Proceedings of the International Europhysics Conference on High Energy Physics, Tampere, Finnland (1999)
3. J.C. Collins, D.E. Soper, and G. Sterman, in: *Perturbative Quantum Chromodynamics*, ed. A.H. Mueller, World Scientific (Singapore, 1989)
4. B. Andersson, G. Gustafson, and C. Peterson, Z. Phys. **C1** (1979) 105; B. Andersson, G. Gustafson, and B. Söderberg, Z. Phys. **C20** (1983) 317; T. Sjöstrand, Nucl. Phys. **B248** (1984) 469
5. C. Friberg and T. Sjöstrand, *Some thoughts on how to match leading-log parton showers with NLO matrix elements*, to be published in: Proceedings of the workshop on Monte Carlo generators for HERA physics (1998/99)
6. D. Graudenz, *Deeply Inelastic Hadronic Final States: QCD Corrections*, in: Proceedings of the workshop on new trends in HERA physics, Ringberg castle, May 1997
7. D. Graudenz, *Jets and Fragmentation*, in: Proceedings of the Durham workshop on HERA physics 1998
8. Z. Kunszt and D.E. Soper, Phys. Rev. **D46** (1992) 192
9. D. Graudenz, *The three-jet cross section at order α_s^2 in deep-inelastic electron-proton scattering*, Doctoral thesis (September 1990)
10. D. Graudenz, Phys. Lett. **B256** (1991) 518
11. D. Graudenz, Phys. Rev. **D49** (1994) 3291
12. D. Graudenz, Comput. Phys. Commun. **92** (1995) 65
13. T. Brodkorb and J.G. Körner, Z. Phys. **C54** (1992) 519
14. T. Brodkorb and E. Mirkes, Z. Phys. **C66** (1995) 141
15. T. Brodkorb and E. Mirkes, DISJET *program manual*, preprint MAD/PH/821 (Madison 1994)

16. E. Mirkes and D. Zeppenfeld, MEPJET 2.0 *program manual*, unpublished (1997)

17. E. Mirkes and D. Zeppenfeld, Phys. Lett. **B380** (1996) 205

18. W.T. Giele and E.W.N. Glover, Phys. Rev. **D46** (1992) 1980

19. W.T. Giele, E.W.N. Glover and D.A. Kosower, Nucl. Phys. **B403** (1993) 633

20. S. Catani and M. Seymour, DISENT 0.1 *program manual*, unpublished (1997)

21. S. Catani and M. Seymour, in: Proceedings of the workshop on future physics at HERA, eds. G. Ingelman, A. De Roeck, R. Klanner (Hamburg 1995/96)

22. S. Catani and M. Seymour, Nucl. Phys. **B485** (1997) 291

23. D. Graudenz, DISASTER++ *version 1.0.1 program manual*, preprint hep-ph/9710244 (December 1997)

24. B. Pötter, JetViP 1.1 *program manual*, Comput. Phys. Commun. **119** (1999) 45

25. B. Pötter, Nucl. Phys. **B540** (1999) 382

26. D. Graudenz and M. Weber, *NLO Programs for DIS and Photoproduction: Report from Working Group 20*, to be published in: Proceedings of the workshop on Monte Carlo generators for HERA physics (1998/99)

27. C. Duprel, Th. Hadig, N. Kauer, and M. Wobisch, *Comparison of next-to-leading order calculations for jet cross sections in deep-inelastic scattering*, to be published in: Proceedings of the workshop on Monte Carlo generators for HERA physics (1998/99)

28. G. McCance, *NLO program comparison for event shapes* , to be published in: Proceedings of the workshop on Monte Carlo generators for HERA physics (1998/99)

29. H1 Collaboration, *Direct Determination of the Gluon Density in the Proton from Jet Cross Sections in Deep-Inelastic Scattering*, in: Proceedings of the 7th International Workshop on Deep Inelastic Scattering and QCD, Zeuthen, Germany (1999)

30. M. Hampel, *Measurement of the Gluon Density from Jet Rates in Deep Inelastic Scattering*, Doctoral thesis, preprint PITHA 97/13 (Aachen Univ., February 1997)

31. S.J. Brodsky, G.P. Lepage, and P.B. Mackenzie, Phys. Rev. **D28** (1983) 228

32. P.M. Stevenson, Phys. Rev. **D23** (1981) 2916

33. R. Seznec and J. Zinn-Justin, J. Math. Phys. **20** (1979) 1398

Hadronic Final States
in Deep Inelastic Scattering at HERA

Nick Brook

Dept. of Physics & Astronomy, University of Glasgow,
Glasgow, G12 8QQ. United Kingdom.

(On behalf of the H1 and ZEUS Collaborations)

Abstract. Results on the analysis of the hadronic final state in neutral current deep inelastic scattering at HERA are presented; recent results on inclusive single particle distributions, particle correlations and event shapes are highlighted.

1 DIS kinematics

The event kinematics of deep inelastic scattering, DIS, are determined by the negative square of the four-momentum transfer at the lepton vertex, $Q^2 \equiv -q^2$, and the Bjorken scaling variable, $x = Q^2/2P \cdot q$, where P is the four-momentum of the proton. In the quark parton model (QPM), the interacting quark from the proton carries the four-momentum xP. The variable y, the fractional energy transfer to the proton in the proton rest frame, is related to x and Q^2 by $y \simeq Q^2/xs$, where $\sqrt{s}$ is the positron-proton centre of mass energy.

Neutral current (NC) DIS occurs when an uncharged boson (γ, Z^0) is exchanged between the lepton and proton. In QPM there is a 1+1 parton configuration which consists of a single struck quark and the proton remnant, denoted by "+1". At HERA energies there are significant higher-order quantum chromodynamic (QCD) corrections: to leading order in the strong coupling constant, α_s, these are QCD-Compton scattering (QCDC), where a gluon is radiated by the scattered quark and boson-gluon-fusion (BGF), where the virtual boson and a gluon fuse to form a quark-antiquark pair. Both processes have 2+1 partons in the final state. There also exist calculations for the higher, next-to-leading (NLO) processes.

A natural frame in which to study the dynamics of the hadronic final state in DIS is the Breit frame [1]. In this frame, the exchanged virtual boson (γ^*) is completely space-like and has a four-momentum $q = (0, 0, 0, -Q = -2xP^{Breit}) \equiv (E, p_x, p_y, p_z)$, where P^{Breit} is the momentum of the proton in the Breit frame. The particles produced in the interaction can be assigned to one of two regions: the current region if their z-momentum in the Breit frame is negative, and the target region if their z-momentum is positive. The main advantage of this frame is that it gives a maximal separation of the incoming and outgoing partons in the QPM. In this model the maximum momentum a particle can have in the current region is $Q/2$, while in the target region the maximum is $\approx Q(1-x)/2x$. In the Breit frame, unlike the hadronic centre of mass $(\gamma^* p)$ frame, the two regions are asymmetric, particularly at low x, where the target region occupies most of the available phase space.

2   Current Fragmentation Region

The current region in the ep Breit frame is analogous to a single hemisphere of e^+e^- annihilation. In $e^+e^- \rightarrow q\bar{q}$ annihilation the two quarks are produced with equal and opposite momenta, $\pm\sqrt{s_{ee}}/2$. The fragmentation of these quarks can be compared to that of the quark struck from the proton; this quark has an outgoing momentum $-Q/2$ in the Breit frame. In the direction of this struck quark the scaled momentum spectra of the particles, expressed in terms of $x_p = 2p^{Breit}/Q$, are expected to have a dependence on Q similar to that observed in e^+e^- annihilation [2–4] at energy $\sqrt{s_{ee}} = Q$.

2.1   Evolution of $\ln(1/x_p)$ Distributions

Within the framework of the modified leading-log approximation (MLLA) there are predictions of how the higher order moments of the parton momentum spectra should evolve with the energy scale [5,6]. These parton level predictions in practice depend on two free parameters, a running strong coupling, governed by a QCD scale Λ, and an energy cut-off, Q_0, below which the parton evolution is truncated. In this case Λ is an effective scale parameter and is not to be identified with the standard QCD scale, e.g. $\Lambda_{\overline{MS}}$. In particular, predictions can be made at $Q_0 = \Lambda$ yielding the so-called limiting spectrum. The hypothesis of local parton hadron duality (LPHD) [7], which relates the observed hadron distributions to the calculated parton distributions via a constant of proportionality, is used in conjunction with the parton predictions of the MLLA to allow the calculation to be directly compared to data.

The moments of the $\ln(1/x_p)$ distributions have been investigated up to the 4th order [8]; the mean (l), width (w), skewness (s) and kurtosis (k) were extracted from each distribution by fitting a distorted Gaussian of the following form:

$$\frac{1}{\sigma_{tot}}\frac{d\sigma}{d\ln(1/x_p)} \propto \exp\left(\frac{1}{8}k - \frac{1}{2}s\delta - \frac{1}{4}(2+k)\delta^2 + \frac{1}{6}s\delta^3 + \frac{1}{24}k\delta^4\right), \quad (1)$$

where $\delta = (\ln(1/x_p) - l)/w$, over a range of ± 1.5 units (for $Q^2 < 160$ GeV^2) or ± 2 units (for $Q^2 \geq 160$ GeV^2) in $\ln(1/x_p)$ around the mean. The equation is motivated by the expression used for the MLLA predictions of the spectra [5].

Figure 1 shows the moments of the $\ln(1/x_p)$ spectra as a function of Q^2. It is evident that the mean and width increase with increasing Q^2, whereas the skewness and kurtosis decrease. Similar fits performed on e^+e^- data [9] show a reasonable agreement with our results, consistent with the universality of fragmentation for this distribution at large Q^2.

The data are compared to the MLLA predictions of Ref. [6], using a value of $\Lambda = 175$ MeV, for different values of Q_0. A comparison is also made with

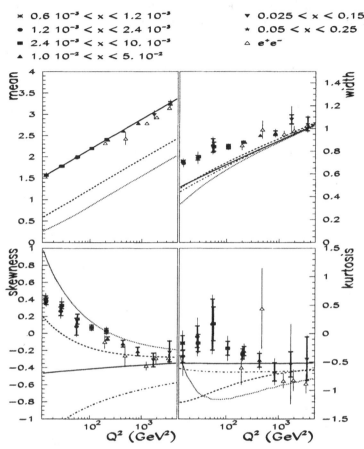

ZEUS 1994–1997

✳ 0.6 10^{-3} < x < 1.2 10^{-3}	▼ 0.025 < x < 0.15
● 1.2 10^{-3} < x < 2.4 10^{-3}	★ 0.05 < x < 0.25
■ 2.4 10^{-3} < x < 10. 10^{-3}	△ e^+e^-
▲ 1.0 10^{-2} < x < 5. 10^{-2}	

Fig. 1. Evolution of the mean, width, skewness and kurtosis of the $\ln(1/x_p)$ distribution in the current fragmentation region with Q^2. Data from e^+e^- and ep are shown together with the MLLA predictions of Dokshitzer *et al* [6] (the full line is $Q_0 = \Lambda$, the dashed $Q_0 = 2\Lambda$, and the dotted $Q_0 = 3\Lambda$) and the limiting spectrum predictions of Fong and Webber [5] (dash-dotted line where available.) The overlapping points are different x ranges in the same Q^2 range. The inner error bars are the statistical error and the outer error bars are the systematic and statistical errors added in quadrature.

the predictions of Ref. [5] for the limiting spectrum ($Q_0 = \Lambda$). The MLLA predictions of the limiting spectrum in Ref. [6] describe the mean well. However both of the MLLA calculations predict a negative skewness which tends towards zero with increasing Q^2 in the case of the limiting spectra. This is contrary to the measurements. The qualitative description of the behaviour of the skewness with Q^2 can be achieved for a truncated cascade ($Q_0 > \Lambda$), but a consistent description of the mean, width, skewness and kurtosis cannot be achieved.

It can be concluded that the MLLA predictions, assuming LPHD, do not describe the data. It should be noted however that a moments analysis has been performed [10], taking into consideration a mass to account for the fact the MLLA predictions are for scaled energy not momentum; this yields good agreement between the limiting case of the MLLA [6] and e^+e^- data over a large energy range, $3.0 < \sqrt{s_{ee}} < 133.0$ GeV.

2.2   Evolution of the x_p Distributions

Scaling violations are predicted in the fragmentation functions, which represent the probability for a parton to fragment into a particular hadron carrying a given fraction of the parton's energy. Fragmentation functions, like parton densities, cannot be calculated in perturbative QCD but can be evolved with the hard-process scale, using the DGLAP evolution [11] equations, from a starting distribution at a defined energy scale; this starting distribution can be derived from a fit to data. If the fragmentation functions are combined with the cross sections for the inclusive production of each parton type in the given physical process, predictions can be made for scaling violations, expressed as the Q^2 evolution of the x_p spectra of final state hadrons [12]. The NLO calculations (CYCLOPS) [13] of the scaled momentum distribution exist for DIS.

The inclusive charged particle distribution, $1/\sigma_{tot}\, d\sigma/dx_p$, in the current fragmentation region of the Breit frame is shown in bins of x_p and Q^2 in Fig. 2. The ZEUS [8] and H1 [14] data are in good agreement. The fall-off as Q^2 increases for $x_p > 0.3$ (corresponding to the production of more particles with a smaller fractional momentum) is indicative of scaling violations in the fragmentation function. The distributions rise with Q^2 for $x_p < 0.1$ and are discussed in more detail below. The data are compared to e^+e^- data [15] (divided by two to account for the production of a $q\bar{q}$ pair) at $Q^2 = s_{ee}$. For the higher Q^2 values shown there is a good agreement between the measurements in the current region of the Breit frame in DIS and the e^+e^- results; this again supports the universality of fragmentation. The fall-off observed in the HERA data at low x_p and low Q^2 is greater than that observed in e^+e^- data at SPEAR [16]; this can be attributed to processes not present in e^+e^- (e.g. scattering off a sea quark and/or boson gluon fusion (BGF)) which depopulate the current region [17,18].

A kinematic correction has recently been suggested [19] to the NLO calculation [13] of the inclusive charged particle distribution which has the form, $1/(1+(m_{\rm eff}/(Qx_p))^2)$, where $m_{\rm eff}$ is an effective mass to account for the assumption of massless hadrons in the fragmentation functions. It is expected to lie in the range 0.1 GeV $< m_{\rm eff} < 1.0$ GeV. The x_p data are compared to the CYCLOPS NLO QCD calculation incorporating this correction in Fig. 2. This calculation combines a full next-to-leading order matrix element with the MRSA$'$ parton densities (with $\Lambda_{\rm QCD} = 230$ MeV) and NLO fragmentation functions derived from fits to e^+e^- data [20]. The kinematic correction allows a more legitimate theoretical comparison to lower Q^2 and x_p than was possible in earlier publications [21]. The bands represent the uncertainty in the predictions by taking the extreme cases of $m_{\rm eff} = 0.1$ GeV and $m_{\rm eff} = 1.0$ GeV. These uncertainties are

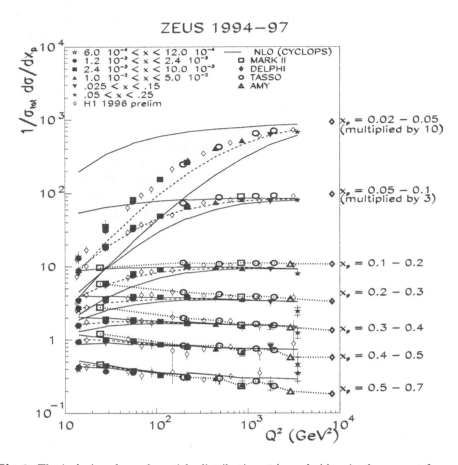

Fig. 2. The inclusive charged particle distribution, $1/\sigma_{tot}\, d\sigma/dx_p$, in the current fragmentation region of the Breit frame. The inner error bar is the statistical and the outer error bar shows the systematic and statistical errors added in quadrature. The open points represent data from e^+e^- experiments divided by two to account for q and $\bar{q}$ production (also corrected for contributions to the charged multiplicity from K_S^0 and Λ decays). The low energy MARK II data has been offset slightly to the left for the sake of clarity. NLO predictions [13] multiplied by the kinematic correction described in the text. The shaded area represents the extreme cases 0.1 GeV $< m_{eff} <$ 1.0 GeV. The upper band corresponds to $m_{eff} = 0.1$ GeV and the lower band $m_{eff} = 1.0$ GeV.

large at low Q^2 and low x_p, becoming smaller as Q^2 and x_p increase. Within these theoretical uncertainties there is good agreement throughout the selected kinematic range. ZEUS found the kinematic correction describes the general trend of the data but it was not possible to achieve a good χ^2 fit for m_{eff} over the whole x_p and Q^2 range. In contrast H1 reported at DIS'99 [14] that a good description of the data could be achieved with a value of $m_{eff} = 0.6$ GeV. The dashed line shows the ZEUS NLO calculation from CYCLOPS multiplied by the kinematic correction using $m_{\text{eff}} = 0.6$ GeV. There is not a good description of the data, therefore it is concluded the distribution also depends strongly on the parameters used to generate the NLO predictions. The uncertainties introduced by this kinematic correction restrict to high Q^2 and high x_p the kinematic range that may be used to extract α_s from the observed scaling violations.

3   Target Fragmentation

DIS at low x allows a study of fragmentation in the target region following the initial scattering off a sea quark (or antiquark). The description based on MLLA [22] is shown schematically in Fig. 3, where the quark box at the top of the gluon ladder represents the scattered sea quark plus its antiquark partner. The MLLA predictions are made up of a number of contributions. Contribution C, the top leg of the quark box, corresponds to fragmentation of the outgoing quark in the current region. Three further contributions (T1, T2 and T3), which are sources of soft gluons, are considered in these analytical calculations to be associated with the target region. It is predicted [22] that the contribution T1(the bottom leg of the quark box) behaves in the same way as the current quark C and so should have no x dependence. The contribution T2 is due to the colour field between the remnant and the struck quark, and the contribution T3 corresponds to the fragmentation of the rungs in the gluon ladder. Both T2 and T3 are predicted to have x and Q^2 dependences which differ from T1. Both the T1 and T2 contributions have been calculated and give particles of momenta $< Q/2$. The collinear gluons T3, on the other hand, generally fragment to particles with momentum $\gtrsim Q/2$. For values of the scaled momentum $x_p < 1.0$, the region of phase space is analogous to the current region and has contributions mainly from T1 and T2. The parton momentum spectra predicted by MLLA, over a range of Q^2 and x, are shown in more detail in Ref. [3]. In the target region these spectra are approximately Gaussian for $x_p < 1$; they peak at a value of $x_p \sim 0.1 - 0.2$ in the range of x and Q^2 measured by ZEUS [8], falling to a plateau region for $1 < x_p < (1 - x)/x$ (the maximum value of x_p in the target region).

The distributions in $\ln(1/x_p)$ are shown for both the target and current regions in Fig. 4. The fitted curves shown are two-piece normal distributions [23] to guide the eye. In contrast to the current region, the target region distribution does not fall to zero as $\ln(1/x_p)$ tends to zero. Although the magnitude of the single particle density at the peak position of the current region distribution grows by a factor of about three over the Q^2 range shown, the single particle density of the target distribution, at the x_p value corresponding to the peak of the current distribution (contribution C is equivalent to contribution T1), depends less strongly on Q^2 and increases by only about 30%. In addition the $\ln(1/x_p)$ distribution shows no significant dependence on x when Q^2 is kept constant. In the target region the peak position of the $\ln(1/x_p)$ distribution increases more rapidly with Q^2 than in the current region; this is consistent with the behaviour expected from cylindrical phase space. The approximate Gaussian distribution of the MLLA predictions peaking at $\ln(1/x_p) \sim 1.5 - 2.5$ [3] is not observed.

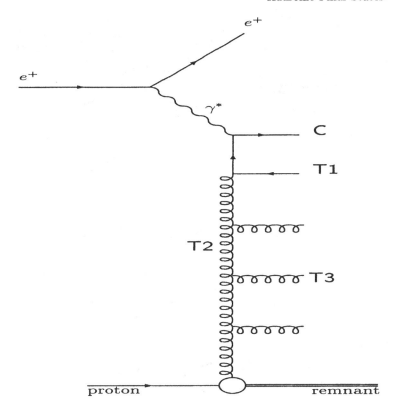

Fig. 3. A schematic of DIS scattering at low x within the MLLA framework. Quark C represents the struck sea quark in the current fragmentation region. T1 is the other half of the quark box which is in the target region. T2 is the t-channel gluon exchange and T3 the rungs of the gluon ladder.

The results strongly suggest that the target distributions are inconsistent with the MLLA predictions when used in conjunction with LPHD.

4   Rapidity Distributions

There are predictions, based on LPHD, for the rapidity distribution of charged particles in the Breit frame [24].
(Rapidity is defined as $Y = \ln\left((E + p_z)/(E - p_z)\right)$, where E is the energy of the particle and p_z is the longitudinal component of its momentum.) It is predicted that there is a sharp rise in the charged particle density followed by a plateau with a width proportional to $\ln(Q)$ as one moves from the current to the target fragmentation region. It is also predicted that there will be another increase in the particle density and an appearance of a second plateau, with the ratio of the two plateaux being 9/4. This ratio reflects the change in colour charge from the

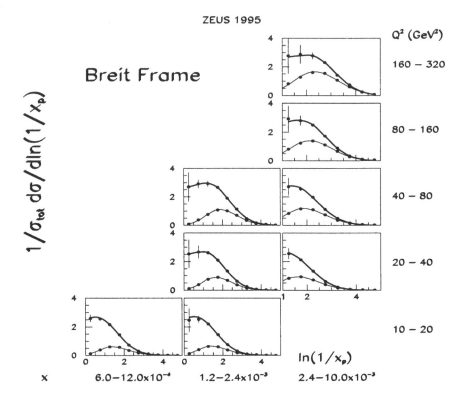

Fig. 4. The corrected $\ln(1/x_p)$ distributions for the target and current regions for the 1995 data. Fitted two-piece normal distributions are shown to guide the eye. The heavy line corresponds to the target region, the light line to the current region. The error bars are the sum of the statistical and systematic errors in quadrature.

dominant quark one, in the current region, to the dominant gluonic one in the target fragmentation region.

The rapidity distributions for charged particles are shown in Fig. 5 [14]. A flat plateau is observed at low transverse momentum, p_t. As p_t increases, QCD effects gradually evolve the flat plateau into an approximate Gaussian, peaking near zero. This illustrates the nature of the Breit frame in separating the current and target fragmentation region. The expected step between the current and target region of the rapidity spectra is not observed. Also shown in Fig. 5 is a comparsion of the data with predictions from LO Monte Carlo models. In general the data are well described. ARIADNE [25] agrees well with the data but LEPTO [26] has problems in describing the high p_t data, and the introduction of soft colour interactions [27] destroys the agreement with the data.

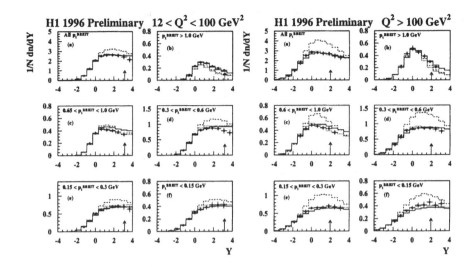

Fig. 5. The rapidity distributions of charged tracks in intervals of transverse momentum and Q^2 in the Breit frame. The error bars are the sum of the statistical and systematic errors added in quadrature. The arrow indicates the position of the origin of the hadronic centre of mass system for the $\langle Q \rangle$ of the data. The histograms show the predictions of LO Monte Carlo models, the solid line is the ARIADNE [25] Monte Carlo and the dashed (dotted) line the LEPTO ME+PS [26] prediction with (without) soft colour interactions.

5   Current-Target Correlations

The correlation coefficient κ:

$$\kappa = \sigma_c^{-1}\sigma_t^{-1}\mathrm{cov}(n_c, n_t) \qquad \mathrm{cov}(n_c, n_t) = \langle n_c n_t \rangle - \langle n_c \rangle \langle n_t \rangle, \qquad (2)$$

is used to measure the dependence between charged particle production in the current region, n_c, and production in the target region, n_t, where σ_c and σ_t are the standard deviations of the multiplicity distributions in the current and target regions respectively. For positive correlations, κ is positive whilst for anti-correlations it is negative. At low Q^2 these correlations are sensitive to the BGF process which depends on the gluon density of the proton [28].

Figure 6 shows the behaviour of the correlation coefficient κ as a function of the average values of Q^2 and x. Anti-correlations are observed for all values of x and Q^2 [29]. The magnitude of κ decreases with increasing $\langle Q^2 \rangle$ from 0.35 to 0.1. According to the analytic results of [28] these observed anti-correlations can be due to the $\mathcal{O}(\alpha_s)$ effects (QCDC and BGF). The $\mathcal{O}(\alpha_s)$ kinematics in the Breit frame can reduce the particle multiplicity in the current region and increase it in the target region. The magnitude of the anti-correlations increases with decreasing $\langle x \rangle$. According to [28] this can be due to an increase of the fraction of events with one or two jets produced in the target region. This behaviour is driven

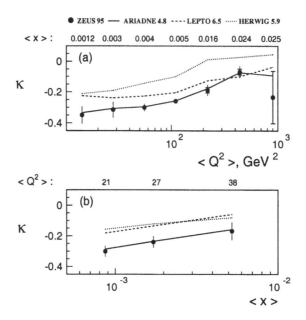

Fig. 6. (a) represents the evolution of the coefficient of correlations κ with predominant variation in Q^2 for corrected data and MC predictions; **(b)** shows the same quantity where predominantly x varies. The corrected values of $\langle Q^2 \rangle$ and $\langle x \rangle$ are indicated for each plot. The inner error bars on the data show the statistical uncertainties. The full error bars include the systematic uncertainties.

by an increase of the BGF rate, due to an increase in the gluon density inside the proton. These observations are, qualitatively, consistent with the depopulation effects discussed in section 2.2.

In addition, Fig. 6 shows a comparison of the data with various LO Monte Carlo models. The ARIADNE model agrees well with the data. The LEPTO and HERWIG [30] predictions show the same trend as the data but do not reproduce the magnitude of the correlations.

6   Event Shapes

The event shape dependence on Q can be due to the logarithmic change of the strong coupling constant $\alpha_s(Q) \propto 1/\ln Q$, and/or power corrections (hadronisation effects) which are expected to behave like $1/Q$. Recent theoretical developments suggest that $1/Q$ corrections are not necessarily related to hadronisation, but may instead be a universal soft gluon phenomenon associated with the behaviour of the running coupling at small momentum scales [31]. These nonperturbative corrections are governed by a parameter $\bar{\alpha}_0$. The scale dependence of any event shape mean $\langle F \rangle$ can be written as the sum of two terms: one associated with the perturbative contribution, $\langle F \rangle^{\text{pert}}$, and the other related to the power corrections, $\langle F \rangle^{\mathrm{pow}}$. The perturbative contribution to an event shape can be calculated from NLO programs, such as DISENT [32].

At HERA a number of infrared-safe event-shape variables have been investigated [33]. Their definitions are given below, where the sums extend over all hadrons h (being a calorimetric cluster in the detector or a parton in the QCD calculations) with four-momentum $p_h = \{E_h, \mathbf{p}_h\}$ in the current hemisphere of the Breit frame. The current hemisphere axis $\mathbf{n} = \{0, 0, -1\}$ coincides with the virtual boson direction.

- **Thrust T_c**

$$T_c = 1 - \tau_c = \max \frac{\sum_h |\mathbf{p}_h \cdot \mathbf{n}_T|}{\sum_h |\mathbf{p}_h|} \qquad \mathbf{n}_T \equiv \text{thrust axis},$$

- **Thrust T_z**

$$T = 1 - \tau = \frac{\sum_h |\mathbf{p}_h \cdot \mathbf{n}|}{\sum_h |\mathbf{p}_h|} = \frac{\sum_h |p_{zh}|}{\sum_h |\mathbf{p}_h|} \qquad \mathbf{n} \equiv \text{hemisphere axis},$$

- **Jet Broadening B_c**

$$B_c = \frac{\sum_h |\mathbf{p}_h \times \mathbf{n}|}{2 \sum_h |\mathbf{p}_h|} = \frac{\sum_h |\mathbf{p}_{\perp h}|}{2 \sum_h |\mathbf{p}_h|} \qquad \mathbf{n} \equiv \text{hemisphere axis},$$

- **Scaled Jet Mass ρ_c**

$$\rho_c = \frac{M^2}{Q^2} = \frac{(\sum_h p_h)^2}{Q^2}.$$

- **C Parameter**

$$C = 3(\lambda_1 \lambda_2 + \lambda_2 \lambda_3 + \lambda_3 \lambda_1)$$

with λ_i being the eigenvalues of the momentum tensor

$$\Theta_{jk} = \frac{\sum_h \frac{p_{jh} p_{kh}}{|\mathbf{p}_h|}}{\sum_h |\mathbf{p}_h|}$$

Also investigated was the variable y_{fJ} (over the whole of phase space). This variable represents the transition value for $(2+1) \rightarrow (1+1)$ jets of the factorizable JADE jet algorithm for a particular event.

A common feature of all the mean event shape values, illustrated in Fig. 7, is the fact that they exhibit a decrease with rising Q. This is due to the fact that the energy flow becomes more collimated along the event shape axis as Q increases, a phenomenon also observed in e^+e^- annihilation experiments.

A simple ansatz for the power correction would be $\langle F \rangle^{\mathrm{pow}} = \Gamma/Q$. However the fits using Γ alone are poor and support the more detailed approach outlined in [31]. In this approach $\langle F \rangle^{\mathrm{pow}}$ is parameterised as follows:

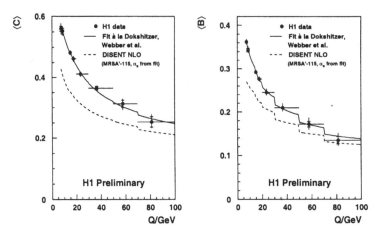

Fig. 7. The points are the corrected mean values of C and B as a function of Q. The inner error bars on the data show the statistical uncertainties. The full error bars include the systematic uncertainties. The full line corresponds to a power correction fit according to the approach in [31]. The dashed line is the perturbative (NLO) prediction from DISENT using the value of α_s found from the full fit.

$$\langle F \rangle^{\mathrm{pow}} = a_F \frac{32}{3\pi^2} \mathcal{M}\left(\frac{\mu_I}{Q}\right)$$
$$\left[\bar{\alpha}_0(\mu_I) - \alpha_s(Q) - \frac{\beta_0}{2\pi}\left(\ln\frac{Q}{\mu_I} + \frac{K}{\beta_0} + 1\right)\alpha_s^2(Q)\right], \qquad (3)$$

where β_0 and K are constants dependent on the number of flavours, a_F is a calculable coefficient dependent on the observable F, μ_I is an 'infra-red' matching scale ($\mu_I = 2$ GeV), $\frac{2}{\pi}\mathcal{M} \approx 1.14$ is a 2-loop correction (known as the Milano factor) and $\bar{\alpha}_0$ is an universal, non-perturbative effective strong coupling below μ_I.

The results of the fit are shown in Fig. 6. The parameter $\bar{\alpha}_0$ is observed to be ≈ 0.5 for all event shapes (except for the jet rate parameter y_{fJ}, not shown), consistent with theoretical expectation. However there is a large spread in the values of α_s. The theoretically calculated parameter for the power corrections for y_{fJ} was $a_{y_{fJ}} = 1$. This is contrary to the observed need for small negative hadronisation corrections [33]. A reasonable fit for y_{fJ} was achieved by using a value of $a_{y_{fJ}} = -0.25$. The extended analysis of the mean event shapes in DIS is consistent with the application of power corrections according to [31] though there is still need for further understanding.

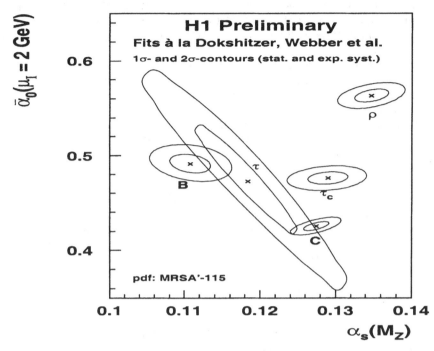

Fig. 8. Results of the fit to $\bar{\alpha}_0$ and $and\alpha_s$ i for the means of τ, B, τ_C, ρ and C. The ellipses illustrate the 1σ and 2σ contours including both statistical and systematic uncertainties.

7   Summary

To understand the underlying QCD processes in DIS it is necessary to study the hadronic final state. At the current level of understanding, QCD works well and describes the HERA data. As the precision of the HERA data improves and further NLO QCD calculations become available the framework of QCD is being tested more thoroughly.

References

1. R. P. Feynman: *Photon-Hadron Interactions.* (Benjamin, New York 1972).
2. Yu. Dokshitzer et al.: Rev. Mod. Phys. **60**, 373 (1988).
3. A. V. Anisovich et al.: Il Nuovo Cimento A **106**, 547 (1993).
4. K. Charchuła: J. Phys. **G19**, 1587 (1993).
5. C. P. Fong and B. R. Webber: Phys. Lett. B **229** 289 (1989); C. P. Fong and B. R. Webber: Nucl. Phys. B **355** 54 (1991).
6. Yu. Dokshitzer, V. Khoze and S. Troyan: Int. J. Mod. Phys. A **7** 1875 (1992).
7. Ya. Azimov et al.: Z. Phys. C **27** 65 (1985) .
8. ZEUS Collab., J. Breitweg et al.: DESY–99–041, to appear in Eur. Phys. J. C.

180   N. H. Brook

9. OPAL Collab., M.Akrway et al.: Phys. Lett. B **247** 617 (1990); TASSO Collab., W. Braunschweig et al.: Z. Phys. C **47** 187 (1990); TASSO Collab., W. Braunschweig et al.: Z. Phys. C **22** 307 (1984); TOPAZ Collab., R. Itoh et al.: Phys. Lett. B **345** 335 (1995).

10. S. Lupia and W. Ochs: Eur. Phys. J. C **2** 307 (1998).

11. V.N. Gribov and L.N. Lipatov: Sov. J. Nucl. Phys. **15** 438 and 675 (1972); Yu.L. Dokshitzer: Sov. Phys. JETP **46** 641 (1977); G. Altarelli and G. Parisi: Nucl. Phys. B **126** 298 (1977).

12. G. Altarelli et al.: Nucl. Phys. B **160** 301 (1979); P. Nason and B. R. Webber: Nucl. Phys. B **421** 473 (1994).

13. D. Graudenz, CERN–TH/96–52; D. Graudenz, CYCLOPS program and private communication.

14. D. Kant: 'Fragmentation functions and rapidity spectra in the Breit frame at H1', to appear in the proceedings of DIS'99, Nucl. Phys. B(Proc. Suppl.) **79** (1999).

15. TASSO Collab., W. Braunschweig et al.: Z. Phys. C **47** 187 (1990); MARK II Collab., A. Petersen et al.: Phys. Rev. D **37** 1 (1988); AMY Collab., Y. K. Li et al.: Phys. Rev. D **41** (1990) 2675 (1990); DELPHI Collab., P. Abreu et al.: Phys. Lett. B **311**, 408 (1993).

16. MARK II Collab., J. F. Patrick et al.: Phys. Rev. Lett. **49** 1232 (1982).

17. K.H. Streng, T.F. Walsh and P.M. Zerwas: Z. Phys. C **2** 237 (1979).

18. ZEUS Collab., M. Derrick et al.: Phys. Lett. B **338** (1994) (1994).

19. Yu. Dokshitzer and B. R. Webber, discussion at Third UK Phenomenology Workshop on HERA Physics, Durham, UK, 20-25 Sept. 1998.

20. J. Binnewies et al.: Z. Phys. C **65** 471 (1995).

21. ZEUS Collab., M. Derrick et al.: Phys. Lett. B **414** 428 (1997); H1 Collab., C. Adloff et al.: Nucl. Phys. B **504** 3 (1997).

22. Yu. Dokshitzer et al.: Sov. Phys. JETP **68** 1303 (1988).

23. Bank of England Quarterly Bulletin , February 1998, and references therein.

24. W. Ochs: 'Recent Tests of Parton Hadron Duality in Multiparticle Final States.'. In:*New Trends in HERA Physics, Tegernsee, Germany, May 25–30, 1997* ed. by B. Kniehl, G. Kramer and A. Wagner (World Scientific, 1998).

25. L.Lönnblad: Comp. Phys. Comm. **71** 15 (1992).

26. G. Ingelman, A. Edin and J. Rathsman: Comp. Phys. Comm. **101** 108 (1997).

27. A. Edin, G. Ingelman and J. Rathsman: Phys. Lett. B **366** 371 (1996).

28. S. V. Chekanov: J. Phys. G **25** 59 (1999).

29. ZEUS Collab., J. Breitweg et al.: DESY–99–063, submitted to Eur. Phys. J. C.

30. G. Marchesini et al.: Comp. Phys. Comm. **67** 465 (1992).

31. Yu. Dokshitzer and B. R. Webber: Phys. Lett. B **352** 451 (1995); B. R. Webber: 'Hadronic Final States'. In:*DIS'95, 5th International workshop on Deep Inelastic Scattering, Paris, France, April 24–28, 1995*, ed. by J.F. Laporte, Y. Sirois (Ecole Polytechnique, 1995).

32. S. Catani and M. H. Seymour: Nucl. Phys. B **485** 291 (1997), erratum-ibid B **510** 503 (1997).

33. K. Rabbertz: 'Event shapes and power corrections in *ep* DIS ', to appear in the proceedings of DIS'99, Nucl. Phys. B(Proc. Suppl.) **79** (1999).

Power Corrections to the Parton Model in DIS

Valentin Zakharov

Max-Planck Institut für Physik, Föhringer Ring 6, 80805 München, Germany.

Abstract. We review the techniques to evaluate the $1/Q^2$ corrections to the Deep Inelastic structure functions. The importance of the power corrections is that they indicate the kinematical region beyond which the leading-twist calculations are no longer valid.

1 Introduction

This Workshop is to discuss the HERA physics, which is primarily the physics of the Deep Inelastic scattering at small x. We approach the DIS at small x from the pure perturbative side and represent a generic structure function F_i as

$$F_i(x, Q^2) = F_i^{p.th}(x, Q^2) + \frac{\Lambda_{QCD}^2}{Q^2} h_i(x) \tag{1}$$

where $F_i^{p.th}$ is the leading twist contribution. Note that the $1/Q^2$ term is considered to be a correction to the leading twist. However, if the power correction reaches, say, 10% at some "moderate" Q^2, it blows up fast at smaller Q^2. Note also that we do not reserve for a possible $\ln Q^2$ dependence of the power correction but this is just for simplicity.

One of the basic questions addressed by the theory of the DIS at small x is when one is to stop using the leading order perturbative calculation which might extend much further than one would expect naively (for a recent review and further references see [1]). There do exist estimates of the higher twist effects which show that the $1/Q^2$ corrections are pretty large at small x [3,4]:

$$\frac{\delta F_2(x \sim 10^{-3}, Q^2)}{F_2(x \sim 10^{-3}, Q^2)} \approx -\frac{1 \; GeV^2}{Q^2} \tag{2}$$

where we picked up $x \sim 10^{-3}$ for the sake of definiteness. The result has been found for any x [3,4].

If true, the estimate (2) is an important piece of information. There has been not much discussion yet of this estimate in the literature. Thus, we are going to provide an overview of the assumptions made to derive (2). Responding to the request of the organizers, we will try to make the presentation understandable to a broader audience.

2   General Remarks on the Power Corrections

We consider QCD and processes determined by physics at short distances. Which means that there is a generic large mass scale, $Q \gg \Lambda_{QCD}$ where Λ_{QCD} is the position of the Landau pole in the coupling:

$$\alpha_s(Q^2) \approx \frac{1}{b_0 ln Q^2/\Lambda_{QCD}^2}. \tag{3}$$

In case of DIS, Q is the 4-momentum of the virtual photon.

Moreover, we consider power corrections $(\Lambda_{QCD}^2/Q^2)^k$ to the parton model. Note first of all that the power corrections appear to be a pure non-perturbative effect. Indeed, on one hand we have

$$\left(\frac{\Lambda_{QCD}}{Q}\right)^k = exp(-k/2b_0\alpha_s(Q^2)). \tag{4}$$

On the other hand, the function $exp(-const/\alpha)$ with a positive $const$ is a classical example from the math courses of a function which has a trivial Taylor expansion at $\alpha = 0$:

$$exp(-const/\alpha)|_{\alpha=0} = 0 + 0 \cdot \alpha + 0 \cdot \alpha^2 + ... \tag{5}$$

since the function itself and all its derivatives vanish at $\alpha = 0$. Thus, this function, being a non-zero, vanishes identically as a perturbative expansion, which is the expansion at $\alpha = 0$.

It could be quite well that the power corrections, unlike pure perturbative contributions, are sensitive to the mechanism of confinement (for a recent review and further references see, e.g., [5,6]). However, here we would not go into this issue since the interest in the estimates like (2) is rather pragmatic than theoretical.

The only point we would like to mention is that we are assuming the validity of the standard picture for the non-perturbative effects. While calculating the perturbative expansions is a well defined procedure in QCD, at least as a matter of principle, the definition of the non-perturbative terms, at first sight is close to saying that these are unknown terms, the rest of the amplitudes upon subtraction of the perturbative contributions. In other words, working with the power corrections relies to a great extent on intuition and heuristic models. In particular, we would not like to use any concrete form of the non-perturbative fluctuations since very little is known about this. Nevertheless, there is one basic assumption underlying all further analysis. Namely, we will assume here that the non-perturbative fields are soft. In other words, the typical size of the non-perturbative fluctuations is of order Λ_{QCD}^{-1}. One could challenge this assumption as well [5] but so far there were no alternative models for the power corrections in DIS developed.

Finally, let us mention that actually working with an infinite perturbative expansion would be awfully difficult in practice. Thus, in reality one is always relying on a kind of a *truncated* series, keeping, say, only the leading perturbative plus power corrections. The assumption behind this truncation is that the power corrections are somehow *enhanced* numerically. There is no proof of this enhancement but it is an indispensable ingredient of any phenomenology based on the power corrections. Historically, this assumption was made first in case of the so called QCD sum rules and worked well. It might fail, however, in other cases.

If there exist high-precision data one may try to check this assumptions varying the number of terms in the perturbative expansion kept explicit and determining the corresponding variation in the fitted values of the power corrections. So far, this procedure was implemented in the most careful way in case of the DIS at large x [7]. The analysis certainly indicates that the fitted value of h_i diminishes quite dramatically with inclusion of higher terms in the perturbative expansion. A similar conclusion is reached in a very detailed model study of power corrections to the thrust [8]. Although such conclusions might seem quite disturbing, we would like to treat them again rather pragmatically. Namely, we would understand the estimate (2) as a correction to the leading log approximation and leave open the question whether it would be absorbed into higher order perturbative corrections.

3   Operator Product Expansion

The basic theoretical tool which allows to exploit the idea that the non-perturbative fields are soft is the Operator Product expansion (OPE). In particular, it allows for a systematic treatment of the power corrections in DIS [9]. The most important outcome of the analysis is that the power corrections to the structure functions start with Q^{-2} terms and we will outline briefly the derivation.

The OPE corresponds to a particular way of preparing ordinary Feynman graphs [10]. The basic ingredient is to separate hard and soft lines. By hard lines we understand the lines which carry virtual momentum of order Q while soft lines carry momentum of order Λ_{QCD}. Then, the hard lines are treated according to the standard Feynman rules. Moreover, in a space-time picture of the interaction all the points along these lines are close to each other, $\Delta x \sim 1/Q$ and, in the first approximation can be reduced to a single point. Note that it is true only in the Euclidean space, while in the Minkowskian space there could be large cancellations between space- and time- components of the space-time interval. The relation of the results obtained within the OPE to the physical processes (measurable in the Minkowski space) is provided by dispersion relations.

On the other hand, soft lines are not integrated out at all and their effect is absorbed into matrix elements of the corresponding operators. The logic behind this step is that the soft lines with $p \sim \Lambda_{QCD}$ are modified drastically by confinement effects. Note that in the approximation that all the hard lines are reduced to a single point the soft lines originate and end up at the same point. Thus, instead of propagators of soft particles which are unknown functions of soft momenta we have now only matrix elements of local operators, that is numbers, not functions.

In this way, the route of the flow of the large momentum Q becomes crucial. For a given order of perturbation theory all the possibilities for this flow should be tried. Examples of the OPE approach to some Feynman graphs in the case of DIS is given in Fig. 1.

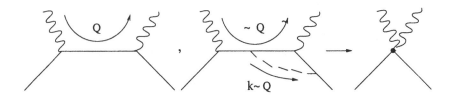

Fig. 1. Examples of applying the OPE to Feynman graphs. One follows the route of the large momentum Q flow. All the points along these, hard lines are reduced to a single point indicates by a black blob. The large momentum Q is brought in and taken away by the virtual photons, denoted by wavy lines. The dashed line denotes gluon and solid lines are quarks.

Applying furthermore the standard dispersion relations for the forward scattering of the virtual photon off the nucleon one arrives to the well-known predictions for the moments:

$$M_n(Q^2) \equiv \int F_i(x, Q^2) x^N dx = M_n(Q_0^2) \left(\frac{\alpha_s(Q^2)}{\alpha_s(Q_0^2)} \right)^{\gamma_N} \tag{6}$$

where γ_N are calculable and reflect the effect of the hard lines while $M_n(Q_0^2)$ are to be borrowed once from experiment for some Q_0^2, as a reflection of our inability to evaluate the effect of the soft quark lines. Moreover, for each moment we have a new matrix operator and the corresponding matrix element.

Note that radiative corrections are dominated by virtual momenta $k \sim Q$, as indicated in Fig. 1. However, there is also a possibility that the gluon momentum k is small, $k^2 \ll Q^2$. Then the effect of the gluon exchange is to be included into new matrix elements, as indicated in Fig. 2

Restricting momenta of virtual gluons to the infrared region costs suppression of the corresponding contributions by powers of Q^{-2}, due to phase space (and gauge invariance). Applying the OPE allows to immediately find out this power. Indeed, the leading twist (see Fig. 1) is related to the matrix elements of operators bilinear in the quark fields, which can generically be represented as $\bar{q}\Gamma q$ where Γ is some matrix in the spinor and color spaces. If a gluon line is added, see Fig. 2, then the corresponding operator becomes $\bar{q}F\Gamma q$ where F is the gluon field strength tensor (and we suppress all Lorentz and color indices). The dimension of this extra factor, F, is $d = 2$ and it is to be compensated by

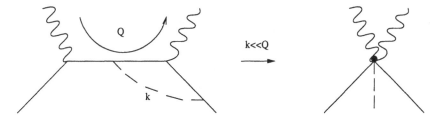

Fig. 2. Emergence of higher twists. If the virtual gluon is soft, $k \sim \Lambda_{QCD}$ its effect cannot be evaluated but is rather to be included into matrix elements of operators containing an extra power of the gluon field strength tensor.

the corresponding factor Q^{-2}:

$$\delta M_n \sim M_n^{(0)} \frac{\beta_n \Lambda_{QCD}^2}{Q^2}. \tag{7}$$

To summarize, the OPE allows to parameterize in a systematic way the infrared sensitive contributions to the ordinary Feynman graphs. In particular, in case of the DIS one readily concludes that the leading power corrections are of order Q^{-2}. These corrections are associated with matrix elements of operators bilinear in quark fields and linear in the gluon field. Borrowing the matrix elements from experiment amounts to introducing new structure functions. Without this input, the use of the OPE does not allow for any concrete numerical predictions for the power corrections.

4   Renormalons or Infrared Sensitive Gluon Mass

Renormalons (for review see, e.g., [11]) represent a particular model for the power corrections which avoids the introduction of a new structure function. Qualitatively, the basic idea can be explained as follows. The general formulation of the OPE outlined above relates the Q^{-2} correction to an admixture of gluons in the nucleon wave function. One could argue, however, that the nucleon wave function is build on the valence quarks alone and all the gluon effects can be understood in terms of the gluon bremsstrahlung from the valence quarks.

In fact this idea, to reduce the gluon structure functions to the quark structure functions has been tried first in the leading twist case [12]. Namely, one assumes that at some $Q_0^2 \sim \Lambda_{QCD}^2$ there are only quark distributions, with no gluons. Gluons evolve at larger Q^2, as a result of bremsstrahlung. The model is not without a success although one cannot go too far because of the matching point at low Q_0^2.

Now, we need technical means to realize the idea that the soft gluons responsible for the power-like corrections (see Fig. 2 and discussion of it) are produced exclusively through bremsstrahlung. Introduce a non-vanishing gluon mass squared λ^2 and try first to expand in λ^2, see Fig. 3. The expansion amounts

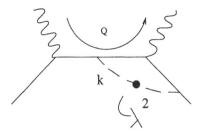

Fig. 3. Introduction of a (fictitious) gluon mass $\lambda^2 \to 0$ into the radiative corrections to the parton model. One evaluates contribution proportional to $\lambda^2 \ln \lambda^2$.

to replacing the gluon propagator by $(k^2+\lambda^2)^{-1} \to k^{-2}\lambda^2 k^{-2}$. Upon this replacement the integral over the momentum of the gluon k becomes logarithmically divergent in the infrared and as a result we come to terms of order $\lambda^2 \ln \lambda^2$. Indeed,

$$\lambda^2 \int_\lambda \frac{d^4 k}{k^4} ~\sim~ \lambda^2 \ln \lambda^2, \tag{8}$$

where the lower bound of the integration $k \sim \lambda$ is introduced because at such k the expansion in λ^2 is no longer valid.

This estimate illustrates a general point that terms non-analytical in λ^2, like $\lambda^2 \ln \lambda^2$ can develop only through contributions of soft gluons with momenta $k \sim \lambda$. Thus, non-analyticity in λ^2 can be used to mark the infrared sensitive part of the Feynman integrals [13]. To make connection with QCD one replaces

$$\lambda^2 \ln \lambda^2 ~\to~ \Lambda^2_{QCD}, \tag{9}$$

with the overall coefficients in this relation left as a fitting parameter. Indeed, for the sake of parameterization of the contribution of momenta $k \sim \Lambda_{QCD}$ we can simply adjust $\lambda \sim \Lambda_{QCD}$. In the particular case of the DIS, detailed calculations along these lines can be found in [14].

Note that so far we have not mentioned renormalons. Actually, this is another way, consistent with the gauge invariance, to isolate contributions of soft gluons. Namely, one inserts fermionic bubbles into the gluon line, see Fig. 4. It can be readily seen that if the number of fermionic bubbles is large, the virtual momentum of the gluon, k_{eff} dominating the Feynman integrals is small. Indeed, each bubble is proportional to $b_0 \ln(k^2/Q^2)$ where b_0 is the fermionic contribution to the first coefficient in the β-function and the log factor vanishes at the gluon momentum $k = Q$ because of normalizing the running coupling at $k = Q$. Moreover, assume that the number of the fermionic bubbles, n is large, $n \gg 1$. Then the integral over the gluon line takes the form

$$\int \frac{d^4 k}{k^2} \left(b_0 \ln(k^2/Q^2) \right)^n ~\sim~ b_0^n n! ~. \tag{10}$$

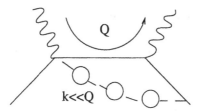

Fig. 4. Renormalon chain inserted into a gluon line. The number of the fermionic bubbles is assumed to be large large, $n \gg 1$.

and is large at large n. What is even more important, it is dominated by low momenta

$$(k^2)_{eff} \sim Q^2 \cdot exp(-n).$$

The type of graphs represented in Fig. 4 is called renormalon chain. The motivation to use the renormalon chain is exactly the same as to isolate the terms of order $\lambda^2 \ln \lambda^2$, see above. Namely, the renormalon chain enhances greatly infrared sensitive contributions to the Feynman graphs and allows for their parameterization. Using the renormalons one comes to the same results for the power corrections as by using the infinitesimal gluon mass, see above.

To summarize, renormalons allow to parameterize contribution of soft gluons to the radiative corrections in DIS. The method is equivalent to introducing an infinitesimal gluon mass λ and following the terms non-analytical in λ^2. The model has much greater predictive power than the general OPE since no independent quark-gluon structure function is introduced (see Fig. 2). Instead, the power correction to a structure function is proportional to the structure function itself (see Figs. 3,4).

5   Reservations, Successes, Predictions

The great simplification brought by the renormalon model comes not without a price tag. The basic new assumption is that external quark lines are considered hard with respect to the gluon lines even if both are in fact soft. We called this assumption a valence quark approximation, and it might be valid only for rough estimates.

There are some further caveats as well. In particular, the effect of, say, two soft gluon emission is not suppressed compared to a single soft gluon considered so far. Indeed, the emission of soft gluons is governed by $\alpha_s(\Lambda^2)$ which is not small. Moreover, in case of some observables one can prove that the multiple gluon emission is important numerically [15].

Thus, the renormalon approach is of openly heuristic and approximate nature and can be justified only by its success, if any, in the description of the data. On the other hand renormalons do fit nicely the x dependence of the power corrections at large x, see, e.g., [2,16]. This is especially true if one truncates both the perturbative expansion and expansion in powers of Q^{-2} to their first respective terms.

This agreement of the theory and data allows to fix the unknown overall coefficient in front of the power corrections and extend the calculation to the region of small x as well. Since small x are dominated by the singlet structure functions one has to include the effect of the infinitesimal gluon mass in graphs of the type represented in Fig. 5 [3,4]. The prediction for the x dependence of

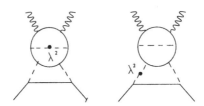

Fig. 5. Power corrections to the singlet structure functions. The black blob indicates insertion of the infinitesimal gluon mass.

the power corrections appears to be highly non-trivial. The correction practically dies away at moderate x but blows up again, with a negative sign at small x [3,4]. I would expect that there should be a simple kinematical reason for such a behaviour but I am not in the position to identify it at the moment.

Thus, we come back to the estimate (2) and may conclude that despite many potential uncertainties the estimate looks quite conservative and might well be realistic.

6   Acknowledgments

I would like to thank the organizers of the workshop, and B. Kniehl in particular for the invitation to the workshop with its enjoyable atmosphere. I am gratefully acknowledging numerous discussions of the power corrections in DIS with R. Akhoury.

References

1. R.D. Ball, S. Forte, hep-ph/9906222.
2. S. Forte, hep-ph/9812382.
3. E. Stein, M. Maul, L. Mankiewicz, A. Schäfer, *Nucl. Phys.* **B536**, 318 (1998) (hep-ph/9803342).
4. G.E. Smye, hep-ph/9812251.
5. F.V. Gubarev, M.I. Polikarpov, V.I. Zakharov, hep-ph/9908292.
6. Yu. L. Dokshitzer, hep-ph/9812252.
7. A.L. Kataev, G. Parente, A.V. Sidorov, hep-ph/9907310.
8. E. Gardi, G. Grunberg, hep-ph/9908458.

9. E.V. Shuryak, A.I. Vainshtein, *Nucl. Phys.* **B201**, 141 (1982).

10. M.A. Shifman, A.I. Vainshtein, V.I. Zakharov, *Nucl. Phys.* **B147**, 385 (1978);
    V.A. Novikov, M.A. Shifman, A.I. Vainshtein, V.I. Zakharov, *Fortsch. Phys.* **32** 585 (1985).

11. R. Akhoury, V.I. Zakharov, *Nucl. Phys. Proc. Suppl.* **54A** 217 (1997) (hep-ph/9610492);
    B.R. Webber, *Nucl. Phys. Proc. Suppl.* **71** 66 (1999) (hep-ph/97122360;
    M. Beneke, hep-ph/9810263.

12. V.A. Novikov, M.A. Shifman, A.I. Vainshtein, V.I. Zakharov, *Annals Phys.* **105**, 276 (1977);
    M. Gluck, E. Reya, *Phys. Lett.* **B270** 65 (1991).

13. K.G. Chetyrkin, V.P. Spiridonov, *Sov. J. Nucl. Phys.* **47**, 522 (1988);
    M. Beneke, V.M. Braun, V.I. Zakharov, *Phys. Rev. Lett.* **73**, 3058 (1994).

14. Yu.L. Dokshitzer, G. Marchesini, B.R. Webber, *Nucl. Phys.* **B469**, 93 (1996).

15. V.I. Zakharov, *Prog. Theor. Phys. Suppl.* **131**, 107 (1998) (hep-ph/9802416).

16. S. Liuti, *Nucl. Phys. Proc. Suppl.* **74**, 380 (1999) (hep-ph/9809248).

Soft Limits of Multiparticle Observables and Parton Hadron Duality

Wolfgang Ochs

Max-Planck-Institut für Physik (Werner Heisenberg-Institut), Föhringer Ring 6, D-80805 München, Germany

Abstract. We discuss observables in multiparticle production for three kinds of limits of decreasing kinematical scales: 1. the transition jet → hadron (limit $y_{cut} \to 0$ of the resolution parameter y_{cut}); 2. single particle inclusive distributions normalized at threshold $\sqrt{s} \to 0$ and 3. particle densities in the limit of low momentum $p, p_T \to 0$. The observables show a smooth behaviour in these limits and follow the perturbative QCD predictions, originally designed for large scales, whereby a simple prescription is supplemented to take into account mass effects. A corresponding physical picture is described.

1 Introduction

A successful description of multiparticle production based on perturbative QCD has been established for "hard" processes which are initiated by an interaction of elementary quanta (quarks, leptons, gauge bosons, ...) at large momentum transfers $Q^2 \gg \Lambda^2$, whereby the characteristic scale in QCD is $\Lambda \sim$ few 100 MeV. In this kinematic regime the running coupling constant $\alpha_s(Q^2)$ is small and the lowest order terms of the perturbative expansion provide the desired accuracy. The coloured quarks and gluons which emerge from the primary hard process cannot escape towards large distances because of the confinement of the colour fields. Rather, they "fragment" into particle jets which may consist of many stable and unstable hadrons.

Here we are interested in the emergence of the hadronic final states and jet structure. The partons participating in the hard process generate parton cascades through gluon Bremsstrahlung and quark antiquark pair production processes which can be treated again perturbatively, at least approximately. The singular behaviour of the gluon Bremsstrahlung in the angle Θ and momentum k

$$\frac{dn}{dk d\Theta} \propto \alpha_s(k_T/\Lambda)\frac{1}{k\Theta}, \quad k_T > Q_0. \tag{1}$$

(in lowest order and for small angles) leads to the collimation of the partons and the jet structure. The transverse momentum k_T is taken as characteristic scale for the coupling $\alpha_s \sim 1/\ln(k_T/\Lambda)$, so it will rise with decreasing scale during jet evolution and one expects the perturbation theory to loose its valididy below a limiting scale Q_0.

The transition to the hadronic final state, finally, proceeds at small momentum transfers $k_T \sim Q_0$ by non-perturbative processes. There have been different approaches to obtain predictions on the hadronic final states:

1. "Microscopic" Monte Carlo models

In a first step a parton final state is generated perturbatively corresponding to a cut-off scale like the above Q_0. Then, according to a non-perturbative model intermediate hadronic systems (clusters, strings, ...) are formed which decay, partly through intermediate resonances, into the final hadrons of any flavour composition. Depending on the considered complexity a larger number of adjustable parameters are allowed for in addition to the QCD scale and cut-off parameters. Because of the complexity of these models only Monte Carlo methods are available for their analysis. They are able to reproduce many very detailed properties of the final state successfully.

2. Parton Hadron Duality approaches

One compares the perturbative QCD result for particular observables directly with the corresponding result for hadrons. The idea is that the effects of hadronization are averaged out for sufficiently inclusive observables. In this case analytical results are aimed for which are closer to a direct physical interpretation than the MC results (for reviews, see [1,2]). This general idea comes in various realizations, we emphasize three kinds of observables:

Jet cross sections: Jets are defined with respect to a certain resolution criterion (parameter y_{cut}), then the cross sections for hadron and parton jets are compared directly at the same resolution. This phenomenological ansatz has turned out to be extremely successful in the physics of energetic jets. A priori, it is non-trivial that an energetic hadron jet with dozens of hadrons should be compared directly to a parton jet with only very few (1-3) partons.

Infrared and collinear safe observables: The value of such an observable is not changed if a soft particle with $k \to 0$ or a collinear particle ($\Theta \to 0$) is added to the final state. It is then expected that the observables are less sensitive to the kinematic region $k_T \sim Q_0$ in (1). Especially, event shape observables like "Thrust" or energy flow patterns belong into this category. Perturbative calculations with all order resummations have been generally successful. In recent years perturbative calculations to $O(\alpha_s^2)$ in combination with power corrections $\sim 1/Q^q$ have found considerable interest.

Infrared sensitive observables: Global particle multiplicities as well as inclusive particle distributions and correlations belong into this category; these observables are divergent for $Q_0 \to 0$ and therefore are particularly sensitive to the transition region from partons to hadrons. Q_0 plays the role of a nonperturbative hadronization parameter.

In this report we will be concerned with the last class of observables to learn about the soft phenomena and ultimately about the colour confinement mechanisms. Specific questions concerning the role of perturbative QCD are

- What is the limiting value of Q_0 for which perturbative QCD can be applied successfully. Especially, can Q_0 be of the order of $\Lambda \sim$ few 100 MeV?
- Is there any evidence for the strong rise of the coupling constant α_s towards small scales below 1 GeV?
- Is there evidence for characteristic QCD coherence effects at small scales which are expected for soft gluons, evidence for the colour factors C_A, C_F?

2 Theoretical approach

2.1 Partons

The evolution of a parton jet is described in terms of a multiparticle generating functional $Z_A(P, \Theta; \{u(k)\})$ with momentum test functions $u(k)$ for a primary parton A $(A = q, g)$ of momentum P and jet opening angle Θ. This functional fulfils a differential-integral equation [1]

$$\frac{d}{d \ln \Theta} Z_A(P, \Theta) = \frac{1}{2} \sum_{B,C} \int_0^1 dz$$

$$\times \frac{\alpha_s(k_T)}{2\pi} \Phi_A^{BC}(z) \left[Z_B(zP, \Theta) Z_C((1-z)P, \Theta) - Z_A(P, \Theta)\right] \tag{2}$$

and has to be solved with the constraint $k_T > Q_0$ and with the initial condition

$$Z_A(P, \Theta; \{u\})|_{P\Theta=Q_0} = u_A(k = P). \tag{3}$$

which means that at threshold $P\Theta = Q_0$ there is only one particle in the jet. From the functional Z_A one can obtain the inclusive n-parton momentum distributions by functional differentiation with the functions $u(k_i)$, $i = 1 \ldots n$, at $u = 1$ and then one finds the corresponding evolution equations as in (2). This "Master Equation" includes the following features: the splitting functions $\Phi_A^{BC}(z)$ of partons $A \to BC$; evolution in angle Θ yielding a sequential angular ordering which limits the phase space of soft emission as a consequence of colour coherence; the running coupling $\alpha_s(k_T)$. For large momentum fractions z the equation approaches the usual DGLAP evolution equations.

The solution of the evolution equations can be found by iteration and then generates an all order perturbation series; it is complete in leading order ("Double Logarithmic Approximation – DLA) and in the next to leading order ("Modified Leading Log Approximation" – MLLA), i.e. in the terms $\alpha_s^n \log^{2n}(y)$ and $\alpha_s^n \log^{2n-1}(y)$. The logarithmic terms of lower order are not complete, but it makes sense to include them as well as they are important for taking into acount energy conservation and the correct behaviour near threshold (3). The complete partonic final state of a reaction may be constructed by matching with an exact matrix element result for the primary hard process.

2.2 Hadrons

We investigate here the possibility that the parton cascade resembles the hadronic final state for sufficiently inclusive quantities. One motivation is "preconfinement" [3], the preparation of colour neutral clusters of limited mass within the perturbative cascade. If the cascade is evolved towards a low scale $Q_0 \sim \Lambda$, a successful description of inclusive single particle distributions has been obtained ("Local Parton Hadron Duality"-LPHD [4]). More generally, one could test relations between parton and hadron observables of the type

$$O(x_1, x_2, \ldots)|_{hadrons} = K\, O(x_1, x_2, \ldots; Q_0, \Lambda)|_{partons} \tag{4}$$

where the nonperturbative cut-off Q_0 and an arbitrary factor K are to be determined by experiment (for a review, see [2]). In comparing differential parton and hadron distributions there can be a mismatch near the soft limit because of mass effects, especially, the (massless) partons are restricted by $k_T > Q_0$ in (2) but hadrons are not. This mismatch can be avoided by an appropriate choice of energy and momentum variables. In a simple model [5,6] one compares partons and hadrons at the same energy (or transverse mass) using an effective mass Q_0 for the hadrons, i.e.

$$E_{T,parton} = k_{T,parton} \qquad \Leftrightarrow \qquad E_{T,hadron} = \sqrt{k_{T,hadron}^2 + Q_0^2}, \qquad (5)$$

then, the corresponding lower limits are $k_{T,parton} \to Q_0$ and $k_{T,hadron} \to 0$.

3   From Jets to Hadrons, the Limit $y_{cut} \to 0$

We turn now to the discussion of several observables and their behaviour in the limit of a small scale. First, we consider the transition from jets to hadrons by decreasing the resolution scale of jets. Jet physics is a standard testing ground for perturbative QCD, the transition to hadrons therefore corresponds to the transition from the known to the unknown territory.

The jets are defined in the multiparticle final state by a cluster-algorithm. Popular is the "Durham algorithm" [8] which allows the all order summation in the perturbative analysis. For a given resolution parameter $y_{cut} = (Q_{cut}/Q)^2$ in a final state with total energy Q particles are successively combined into clusters until all relative transverse momenta are above the resolution parameter $y_{ij} = k_T^2/Q^2 > y_{cut}$.[1] We study now the mean jet multiplicity N_{jet} in the event as function of y_{cut}. In e^+e^--annihilation for $y_{cut} \to 1$ all particles are combined into two jets and therefore $N_{jet} = 2$, on the other hand, for $y_{cut} \to 0$ all hadrons are resolved and $N_{jet} \to N_{had}$.

Results on jet multiplicities are shown in Fig. 1. The jet multiplicity rises only slowly with decreasing y_{cut}. For $y_{cut} \gtrsim 0.01$ the data are well described by the complete matrix element calculations to $O(\alpha_s^2)$ (first results of this kind in [9]) and allow the precise determination of the coupling or, equivalently, of the QCD scale parameter $\Lambda_{\overline{MS}}$ [10,11]. In the region above $y_{cut} > 10^{-3}$ the resummation of the higher orders in α_s becomes important [12] and the MLLA calculation describes the data well. The lower curve shown in Fig. 1 is obtained [7] from a full (numerical) solution of the evolution equations corresponding to (2), matched with the $O(\alpha_s)$ matrix element, and describes the data obtained at LEP-1 [10,11] down to 10^{-4}.

The theoretical curve diverges for small cut-off $Q_{cut} \to \Lambda$ as in this case the coupling $\alpha_s(k_T)$ diverges. In the duality picture discussed above the parton final state corresponds to a hadron final state at the resolution $k_T \sim Q_0$ according to (4) and this limit is reached for $Q_{cut} \to Q_0$. The calculation meets the hadron

---

[1] more precisely, the distance is defined by $y_{ij} = 2(1 - \cos \Theta_{ij}) \min(E_i^2 E_j^2)/Q^2 > y_{cut}$.

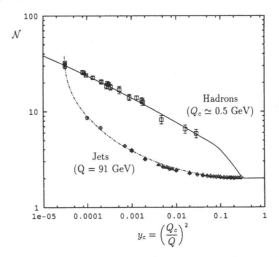

Fig. 1. Data on the average jet multiplicity $\mathcal{N}$ at $Q = 91$ GeV for different resolution parameters y_c (lower set) and the average hadron multiplicity (assuming $\mathcal{N} = \frac{3}{2}\mathcal{N}_{ch}$) at different *cms* energies between $Q = 3$ and $Q = 91$ GeV using $Q_c = Q_0 = 0.508$ GeV in the parameter y_c (upper set). The curves follow from the evolution equation (2) with $\Lambda = 0.5$ GeV; the upper curve for hadrons is based on the duality picture (4) with $K = 1$ and parameter Q_0 (Fig. from [7])

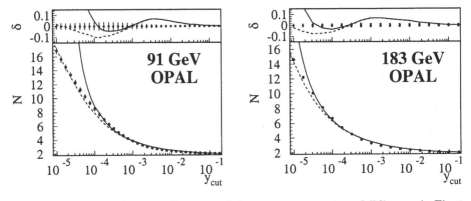

Fig. 2. Jet multiplicities extending towards lower y_{cut} parameters; full lines as in Fig. 1 for jets, dashed lines the same predictions but shifted $y_{cut} \to y_{cut} - Q_0^2/Q^2$ according to the different kinematical boundaries as in (5), with parameters as in Fig. 1 (preliminary data from OPAL [13])

multiplicity data for the cut-off parameter $Q_0 \simeq 0.5$ GeV. If this calculation is done for lower *cms* energies, agreement with all hadron multiplicity data down to $Q = 3$ GeV is obtained with the same parameter Q_0 as seen in Fig. 1 by the upper set of data and the theoretical curve. Moreover, the normalization constant in (4) can be chosen as $K = 1$ whereas in previous approximate calculations $K \approx 2$ (see, e.g. [5]). This result implies that the hadrons, in the duality picture, correspond to very narrow jets with resolution $Q_0 \simeq 0.5$ GeV.

In this unified description of hadron and jet multiplicities the running of the coupling plays a crucial role. Namely, for constant α_s both curves for hadrons and jets in Fig. 1 would coincide, as only one scale Q_{cut}/Q were available. With running $\alpha_s(k_T/\Lambda)$ the absolute scale of Q_{cut} matters: α_s varies most strongly for $Q_{cut} \to \Lambda$ for jets at small y_{cut} in the transition to hadrons and for hadrons near the threshold of the process at large y_{cut} where $\alpha_s > 1$. It appears that the final stage of hadronization in the jet evolution can be well represented by the parton cascade with the strongly rising coupling.

Preliminary results on jet multiplicities at very small y_{cut} have been obtained recently by OPAL [13] and examples are shown in Fig. 2. Whereas in the theoretical calculation all hadrons (partons in the duality picture) are resolved for $Q_{cut} \to Q_0$, for the experimental quantities this limit occurs for $Q_{cut} \to 0$.

This is an example of the kinematical mismatch between experimental and theoretical quantities discussed above and can be taken into account [7] by a shift in y_{cut} according to (5). The shifted (dashed) curves in Fig. 2 describe the data rather well (also at intermediate *cms* energies) whereby the Q_0 parameter has been taken from the fit to the hadron multiplicity before; the predictions fall a bit below the data at lower energies like 35 GeV. The nonperturbative Q_0 correction becomes negligible for $Q_c \gtrsim 1.5$ GeV.

We conclude that in case of this simple global observable the perturbative QCD calculation provides a good description of hard and soft phenomena in terms of one non-perturbative parameter $Q_0 \sim \Lambda$ (from fit [7] $Q_0 \approx 1.015\Lambda$). Multiplicity moments are described very well in this approach also [16].

4   Shape of Energy Spectrum, the Limit $\sqrt{s} \to 0$

A standard procedure in perturbative QCD is the derivation of the Q^2 evolution of the inclusive distributions – either of the structure functions in DIS ($Q^2 < 0$) or of the hadron momentum distributions ("fragmentation functions", $Q^2 \equiv s > 0$). One starts from an input function at an initial scale Q_i^2 and predicts the change of shape with Q^2.

In the LPHD picture one derives the parton distribution from the evolution equation (2) with initial condition (3) at threshold, here the spectrum is simply

$$D(x, Q_0) = \delta(x - 1). \tag{6}$$

If we start from this initial condition the further QCD evolution predicts the absolute shape of the particle energy distribution at any higher *cms* energy $\sqrt{s}$. Within certain high energy approximations one can let $Q_0 \to \Lambda$ and obtains an explicit analytical expression for the spectrum in the variable $\xi = \ln(1/x)$, the so-called "limiting spectrum" [4] which has been found to agree well with the data in the sense of (4) – disregarding the very soft region $p \lesssim Q_0$ (see, e.g. the review [2]). In the more general case $Q_0 \neq \Lambda$ the cumulant moments κ_q of the ξ distribution have been calculated as well [14,15]; they are defined by

$\kappa_1 = <\xi> = \bar{\xi}, \kappa_2 \equiv \sigma^2 = <(\xi - \bar{\xi})^2>, \kappa_3 = <(\xi - \bar{\xi})^3>, \kappa_4 = <(\xi - \bar{\xi})^4> - 3\sigma^4,$

... ; also one introduces the reduced cumulants $k_q \equiv \kappa_q/\sigma^q$, in particular the skewness $s = k_3$ and the kurtosis $k = k_4$.

In the comparison with data some attention has to be paid again to the soft region. The experimental data are usually presented in terms of the momentum fraction $x_p = 2p/\sqrt{s}$, then $\xi_p \to \infty$ for $p \to 0$. On the other hand, the theoretical distribution, because of $p > p_T > Q_0$, is limited to the interval $0 < \xi < Y$, $Y = \ln(\sqrt{s}/2Q_0)$. Therefore, in this region near and beyond the boundary the two distributions cannot agree. A consistent description can be obtained if theoretical and experimental distributions are compared at the same energy as in (5), then both ξ spectra have the same upper limit Y. With a corresponding "transformation" of $E\frac{d^3n}{d^3p}$ the spectra are well described by the appropriate theoretical formula near the boundary [5].

The cumulant moments of the energy spectrum of hadrons determined in this way have been compared [5] with the theoretical calculation based on the MLLA evolution equation [15]. As seen in Fig. 3 the data agree well with the limiting spectrum result ($Q_0 = \Lambda$), both in their energy dependence and their absolute normalization at threshold (the moments vanish because of (6)). This suggests that perturbative calculations are realistic even down to threshold if a treatment of kinematic mass effects is supplemented.

Recently, results on cumulant moments have been presented by the ZEUS group at HERA (see talk by N. Brook [17]). The moments have been determined directly from the momentum distribution of particles in the Breit frame. The ξ_p distributions are seen to extend beyond the theoretical limit Y. The cumulant moments of order $q \geq 2$ determined from this distribution show large deviations from the MLLA predictions at low energies Q^2. The kinematic effects become less important at higher energies and at $Q^2 \gtrsim 1000 \text{ GeV}^2$ the agreement with the predictions using $Q_0 = \Lambda$ is restored. These results demonstrate the importance of the soft region in the analysis of the ξ-moments.

5 Particle Spectra: the limit of small momenta p, $p_T \to 0$

In this limit simple expectations follow from the coherence of the soft gluon emission. If a soft gluon is emitted from a $q\bar{q}$ two jet system then it cannot resolve with its large wave length all individual partons but only "sees" the total charge of the primary partons $q\bar{q}$. Consequently, in the analytical treatment, the soft gluon radiation is determined by the Born term of $O(\alpha_s)$ and one expects a nearly energy independent soft particle spectrum [4]. The consequences and further predictions have been studied recently in more detail.

5.1 Energy Independence

The limit of small momenta p and p_T has been considered in [6]. The behaviour of the inclusive spectrum in rapidity and for small p_T is given by

$$\frac{dn}{dydp_T^2} \sim C_{A,F}\frac{\alpha_s(p_T)}{p_T^2}\left(1 + O\left(\ln\frac{\ln(p_T/\Lambda)}{\ln(Q_0/\Lambda)}\ \ln\frac{\ln(p_T/(x\Lambda))}{\ln(p_T/\Lambda)}\right)\right) \qquad (7)$$

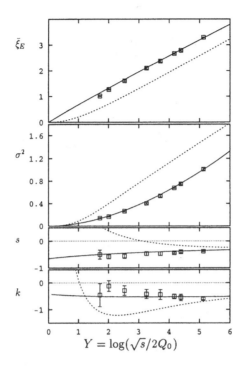

Fig. 3. The first four cumulant moments of charged particles' energy spectra i.e., the average value $\bar{\xi}_E$, the dispersion σ^2, the skewness s and the kurtosis k, are shown as a function of cms energy $\sqrt{s}$ for $Q_0 = 270$ MeV and $n_f = 3$, in comparison with MLLA predictions of the "limiting spectrum" (i.e. $Q_0 = \Lambda$) for running α_s (full line) and for fixed α_s (dashed line) (from [5])

where the second term is known within MLLA and vanishes for $p_T \to Q_0$. Again, the limit $p_T \to Q_0$ at the parton level corresponds to $p_T \to 0$ at the hadron level. Only the first term (the Born term) is energy independent. The approach to energy independence for the soft particles at $p \to 0$ is seen from e^+e^- data [5,6] and also from DIS [18], see Fig. 4. Although the detailed behaviour depends a bit on the specific implementation of the kinematic relations between partons and hadrons the approach towards energy independence in the limit $p \to 0$ is universal and this expectation is nicely supported by the data.

5.2 Colour Factors C_A and C_F

A crucial test of this interpretation is the dependence of the soft particle density on the colour of the primary partons in (7): The particle density in gluon and quark jets should approach the ratio $R(g/q) = C_A/C_F = 9/4$ in the soft limit. This factor has been originally considered for the overall event multiplicity in colour triplet and octet systems but is approached there only at asymptotically

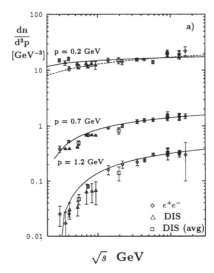

Fig. 4. Particle density at fixed momentum p as function of cms energy, from [6]

high energies [19]. On the other hand, the prediction (7) for the soft particles applies already at finite energies [6].

In practice, it is difficult to obtain gg jet systems for this test. An interesting possibility is the study of 3-jet events in e^+e^- annihilation with the gluon jet recoiling against a $q\bar{q}$ jet pair with relative opening angle of $\sim 90°$ [20]. For such "inclusive gluon jets" the densities of soft particles in comparison to quark jets approach a ratio $R(g/q) \sim 1.8$ for $p \lesssim 1$ GeV which is above the overall multiplicity ratio ~ 1.5 in the quark and gluon jets but still below the ratio $C_A/C_F = 9/4$ (see Fig. 5). This difference may be attributed to the deviation of the events from exact collinearity. If the analysis is performed as function of p_T of the particles the ratio becomes consistent with $9/4$ but not for small $p_T < 1$ GeV [20]. This behaviour indicates the transition from the very soft emission which is coherent from all primary partons to the semisoft emission from the parton closest in angle (q or g) which yields directly the ratio C_A/C_F.

In order to test the role of the colour of the primary partons further in realistic processes it has been proposed [6] to study the soft particle emission perpendicular to the primary partons in 3-jet events in e^+e^- annihilation or in 2-jet production either in pp or in ep collisions, in particular in photoproduction. In these cases, for special limiting configurations of the primary partons, the particle density is either proportional to C_F or to C_A, but it is also known for all intermediate configurations. A first result of this kind of analysis has been presented by DELPHI [21] which shows the variation of the density by about 50% in good agreement with the prediction. The findings by OPAL [20] (Fig. 5) and DELPHI [21] are hints that also the soft particles indeed reflect the colour charges of the primary partons.

Important tests are possible at HERA with two-jet production from direct and resolved photons. The former process corresponds to quark exchange, the

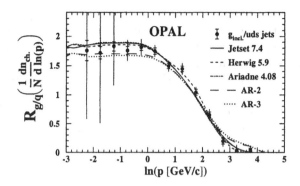

Fig. 5. Ratio of particle densities at small momenta p in inclusive gluon jets and quark jets [20]

latter to gluon exchange. The associated soft perpendicular radiation again reflects the different flow of the primary colour charges: At small scattering angles $\Theta_s \to 0$ in the di-jet *cms* the ratio $R_\perp$ of the soft particles approaches the limits

$$\text{direct } \gamma p \text{ production (q exchange)} : \qquad R_\perp \to 1 \qquad (8)$$

$$\text{indirect } \gamma p \text{ production (g exchange)} : \qquad R_\perp \to C_A/C_F. \qquad (9)$$

In a feasibility study [22] using the event generator HERWIG these ratios have been studied as a function of the particle p_T and the angle Θ_s. With an assumed luminosity of 4.5 pb^{-1} significant results can be obtained. In the MC the predicted ratios are approached for small $p_T \lesssim 0.5$ GeV but deviate considerably for larger p_T. A study towards small angles Θ_s appears feasible. It would be clearly interesting to carry out such an analysis.

5.3 Rapidity Plateaux

Another consequence of the lowest order approximation (7) is the flat distribution in rapidity y at fixed (small) p_T. An interesting possibility appears in DIS where the soft gluon in the current hemisphere is emitted from the quark, in the target hemisphere from a gluon. This would lead one to expect a step in rapidity by factor ~2 between both hemispheres at high energies [6,23].

This problem has been studied recently by the H1 group [24]. They observed a considerable change of the rapidity spectrum with the p_T cut: for large $p_T > 1$ GeV the spectrum was peaked near $y = 0$ in the Breit frame – as expected from maximal perturbative gluon radiation – whereas for small $p_T < 0.3$ GeV a plateau develops in the target hemisphere. On the other hand, no plateau is observed in the current direction at all. A MC study of the e^+e^- hadronic final state did not reveal a clear sign of a flat plateau at small p_T either.

The reason for the failure seeing the flat plateau is apparently the angular recoil of the primary parton which is neglected in the result (7); this introduces an uncertainty in the definition of p_T, especially for the higher momenta. We have investigated this hypothesis further by studying the rapidity distribution in selected MC events where all particles are limited in transverse momentum $p_T < p_T^{max}$. Then the events are more collimated and the jet axis is better defined. The MC results in Table 1 show that the rapidity density gets flatter if the transverse size of the jet decreases with the p_T^{max} cut which is in support of the above hypothesis. This selection, however, considerably reduces the event sample. A step in the rapidity height of DIS events should therefore be expected only in events with strong collimation of particles.

Table 1. Density of particles with $p_T < 0.15$ GeV in rapidity y, normalized at $y = -1$, in events with $p_T < p_T^{max}$ selection. Results obtained from the ARIADNE MC [26] (parameters $\Lambda = 0.2$ GeV, $\ln(Q_0/\Lambda) = 0.015$ as in [25])

p_T^{max}	$y = 0$	$y = -1$	$y = -2$	$y = -3$	$y = -4$	fraction of events
no cut	1.03	1.0	0.69	0.34	0.052	100 %
0.5	0.9	1.0	0.78	0.50	0.13	9 %
0.3	0.9	1.0	0.90	0.84	0.37	0.7 %

5.4   Multiplicity Distributions of Soft Particles and Poissonian Limit

The considerations on the inclusive single particle distributions can be generalized to multiparticle distributions [25]. Interesting predictions apply for the multiplicity distributions of particles which are restricted in either the transverse momentum $p_t < p_T^{cut}$ or in total momentum $p < p^{cut}$.

In close similarity to QED the soft particles are independently emitted in rapidity for limited p_T: because of the soft gluon coherence the secondary emissions at small angles are suppressed. This is less so for the spherical cut. For small values of the cut parameters one finds the following limiting behaviour of the normalized factorial multiplicity moments

$$\text{cylinder}: \quad F^{(q)}(X_\perp, Y) \simeq 1 + \frac{q(q-1)}{6} \frac{X_\perp}{Y} \tag{10}$$

$$\text{sphere}: \quad F^{(q)}(X, Y) \simeq \text{const} \tag{11}$$

where we used the logarithmic variables $X_\perp = \ln(p_T^{cut}/Q_0)$, $X = \ln(p^{cut}/Q_0)$ and $Y = \ln(P/Q_0)$ at jet energy P. Both cuts act quite differently and for small cylindrical cut p_T^{cut} the multiplicity distribution approaches a Poisson distribution (all moments $F^{(q)} \to 1$).

This prediction is verified by the ARIADNE MC at the parton level. Interestingly, the predictions from the full hadronic final state after string hadronization yield factorial moments rising at small $p_T^{cut} < 1$ GeV. These predictions provide a novel test of soft gluon coherence in multiparticle production.

6   Conclusions and Physical Picture

The simple idea to derive hadronic multiparticle phenomena directly from the partonic final state works surprisingly well also for the soft phenomena discussed here which do not belong to the standard repertoire of perturbative QCD. Nevertheless, some clear QCD effects can be noticed in the soft phenomena and the three questions at the end of the introduction can be answered positively. A description with small cut-off $Q_0 \sim \Lambda$ is possible for various inclusive quantities. The coupling is running by more than an order of magnitude at small scales as is seen, in particular, in the transition from jets to hadrons. Also, coherence effects from soft gluons are reflected in the behaviour of soft particles. These effects for the soft particles need further comparison with quantitative predictions. Especially worthwhile are the tests on soft particle flows as function of the primary emitter configuration. Predictions exist also for nontrivial limits of multiparticle soft correlations.

The different threshold behaviour of partons and hadrons can be taken into account by appropriate relations between the respective kinematical variables. Some apparent discrepancies between MLLA predictions and observations can be related to such mass effects.

Finally, we remark on the physical picture which is supported by these results (Fig. 6). The partons in the perturbative cascade are accompagnied by ultrasoft partons with $p_T \lesssim Q_0 \sim \Lambda$ as in very narrow jets; they cannot be further resolved because of confinement and therefore the perturbative partons resolved with $p_T \geq Q_0$ correspond to single final hadrons. This is consistent with the finding of normalization unity ($K = 1$ in (4)) in the transition jet $\to$ hadron ($y_{cut} \to 0$). Colour at each perturbative vertex can be neutralized by the (non-perturbative) emission of one (or several) soft quark pairs; in this way the partons in the perturbative cascade evolve as colour neutral systems outside a volume with confinement radius $R \sim Q_0^{-1}$. In the timelike cascade there is only parton splitting, no parton recombination into massive colour singlets as in the preconfinement model. Such a picture can only serve as a rough guide, it can certainly not be complete as is exemplified by the existence of resonances. Nevertheless, its intrinsic simplicity with only one non-perturbative parameter Q_0 besides the QCD scale Λ makes it attractive as a guide into a further more detailed analysis.

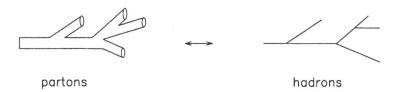

partons   hadrons

Fig. 6. Dual picture of parton and hadron cascades. Ultrasoft partons are confined to narrow tubes with $p_T < Q_0 \sim \Lambda$ around the partons in the perturbative cascade.

References

1. Yu. L. Dokshitzer, V. A. Khoze, A. H. Mueller and S. I. Troyan, *Basics of Per-turbative QCD*, ed. by J. Tran Thanh Van (Editions Frontiéres, Gif-sur-Yvette, 1991)
2. V.A. Khoze, W. Ochs, Int. J. Mod. Phys. A **12**, 2949 (1997)
3. D. Amati, G. Veneziano, Phys. Lett. B **83**, 87 (1979)
4. Ya. I. Azimov, Yu. L. Dokshitzer, V. A. Khoze and S. I. Troyan, Z. Phys. C **27**, 65 (1985); C **31**, 213 (1986)
5. S. Lupia, W. Ochs, Phys. Lett. B **365**, 339 (1996); Eur. Phys. J. C **2**, 307 (1998)
6. V. A. Khoze, S. Lupia, W. Ochs, Phys. Lett. B **386**, 451 (1996); Eur. Phys. J. C **5**, 77 (1998)
7. S. Lupia, W. Ochs, Phys. Lett. B **418**, 214 (1998)
8. Yu. L. Dokshitzer, Proc. Durham Workshop, see W. J. Stirling, J. Phys. G **17**, 1567 (1991)
9. G. Kramer, B. Lampe, Z. Phys. C **34**, 497 (1987); C **39** (1988)
10. L3 Collaboration: O. Adriani et al., Phys. Lett. B **284**, 471 (1992)
11. OPAL Collaboration: R. Acton et al., Z. Phys. C **59**, 1 (1993)
12. S. Catani, Yu. L. Dokshitzer, M. Olsson, G. Turnock B. R. Webber, Phys. Lett. B **269**, 432 (1991)
13. P. Pfeifenschneider, thesis, Technical University Aachen, 1999; OPAL preliminary, Physics note PN403, July 1999.
14. C. P. Fong, B. R. Webber, Nucl. Phys. B **355**, 54 (1991)
15. Yu. L. Dokshitzer, V. A. Khoze, S. I. Troyan, Int. J. Mod. Phys. A **7**, 1875 (1992)
16. S. Lupia, Phys. Lett. B **439**, 150 (1998)
17. ZEUS Collaboration: J. Breitweg et al., hep-ex/9903056; N. Brook, these proceedings
18. H1 Collaboration: C. Adloff et al., Nucl. Phys. B **504**, 3 (1997)
19. S. J. Brodsky, J. F. Gunion, Phys. Rev. Lett. **37**, 402 (1976); K. Konishi, A. Ukawa and G. Veneziano, Phys. Lett. B **78**, 243 (1978)
20. OPAL Collaboration: Abbiendi et al., hep-ex/9903027, subm. Eur. Phys. J. C.; J. W. Gary, Phys. Rev. D **49**, 4503 (1994)
21. K. Hamacher, O. Klapp, P. Langefeld, M. Siebel, DELPHI 99-115 CONF 302, subm. to the HEP'99 Conference, Tampere, Finland, July 1999
22. J. M. Butterworth, V. A. Khoze and W. Ochs, J. Phys. G **25**, 1457 (1999)
23. W. Ochs, in *Proc. Ringberg Workshop 'New Trends in HERA Physics'*, Tegernsee, Germany 1997, ed. by B. A. Kniehl, G. Kramer, A. Wagner (World Scientific, Singapore, 1998) p. 173
24. H1 Collaboration: C. Adloff et al. in: *29th Int. Conf. on High Energy Physics, Vancouver, Canada, July 1998*, paper 531; K. T. Donovan, D. Kant, G. Thompson, J. Phys. G **25**, 1448 (1999)
25. S. Lupia, W. Ochs, J. Wosiek, Nucl. Phys. B **540**, 405 (1999)
26. L. Lönnblad, Comp. Phys. Comm. **71**, 15 (1998)

Theory and Phenomenology of Instantons at HERA

Andreas Ringwald and Fridger Schrempp

Deutsches Elektronen-Synchrotron DESY, D-22603 Hamburg, Germany

Abstract. We review our on-going theoretical and phenomenological investigation of the prospects to discover QCD-instantons in deep-inelastic scattering at HERA.

1 Introduction

It is a remarkable fact that non-Abelian gauge fields in four Euclidean space-time dimensions carry an integer topological charge. Instantons [1] (anti-instantons) are classical solutions of the Euclidean Yang-Mills equations and also represent the simplest non-perturbative fluctuations of gauge fields with topological charge $+1$ (-1). In QCD, instantons are widely believed to play an essential rôle at long distance: They provide a solution of the axial $U(1)$ problem [2], and there seems to be some evidence that they induce chiral symmetry breaking and affect the light hadron spectrum [3]. Nevertheless, a direct experimental observation of instanton-induced effects is still lacking up to now.

Deep-inelastic scattering at HERA offers a unique window to discover QCD-instanton induced events directly through their characteristic final-state signature [4–7] and a sizeable rate, calculable within instanton perturbation theory [8–10]. It is the purpose of the present contribution to review our theoretical and phenomenological investigation of the prospects to trace QCD-instantons at HERA.

The outline of this review is as follows:

We start in Sect. 2 with a short introduction to instanton physics, contentrating especially on two important building blocks of instanton perturbation theory, namely the instanton size distribution and the instanton-anti-instanton interaction. A recent comparison [11] of the perturbative predictions of these quantities with their non-perturbative measurements on the lattice [12] is emphasized. It allows to extract important information about the range of validity of instanton perturbation theory. The special rôle of deep-inelastic scattering in instanton physics is outlined in Sect. 3: The Bjorken variables of instanton induced hard scattering processes probe the instanton size distribution and the instanton-anti-instanton interaction [8,9]. By final state cuts in these variables it is therefore possible to stay within the region of applicability of instanton perturbation theory, inferred from our comparison with the lattice above. Moreover, within this fiducial kinematical region, one is able to predict the rate and the (partonic) final state. We discuss the properties of the latter as inferred from our Monte Carlo generator QCDINS [5,13]. In Sect. 4, we report on a possible search strategy for instanton-induced processes in deep-inelastic scattering at HERA [7].

2   Instantons in the QCD Vacuum

In this section let us start with a short introduction to instantons and their properties, both in the perturbative as well as in the non-perturbative regime. We shall concentrate on those aspects that will be important for the description of instanton-induced scattering processes in deep-inelastic scattering in Sect. 3. In particular, we shall report on our recent determination of the region of applicability of instanton perturbation theory for the instanton size distribution and the instanton-anti-instanton interaction [11]. Furthermore, we elucidate the connection of instantons with the axial anomaly.

Instantons [1], being solutions of the Yang-Mills equations in Euclidean space, are minima of the Euclidean action S. Therefore, they appear naturally as generalized saddle-points in the Euclidean path integral formulation of QCD, according to which the expectation value of an observable $\mathcal{O}$ is given by

$$\langle \mathcal{O}[A, \psi, \overline{\psi}] \rangle = \frac{1}{Z} \int [dA][d\psi][d\overline{\psi}] \, \mathcal{O}[A, \psi, \overline{\psi}] \, e^{-S[A,\psi,\overline{\psi}]}, \tag{1}$$

where the normalization,

$$Z = \int [dA][d\psi][d\overline{\psi}] \, e^{-S[A,\psi,\overline{\psi}]}, \tag{2}$$

denotes the partition function. Physical observables (e.g. S-matrix elements) are obtained from the Euclidean expectation values (1) by analytical continuation to Minkowski space-time. In particular, the partition function (2) corresponds physically to the vacuum-to-vacuum amplitude.

Instanton perturbation theory results from the generalized saddle-point expansion of the path integral (1) about non-trivial minima of the Euclidean action[1]. It can be shown that these non-trivial solutions have integer topological charge,

$$Q \equiv \frac{\alpha_s}{2\pi} \int d^4x \, \frac{1}{2} \mathrm{tr}(F_{\mu\nu} \tilde{F}_{\mu\nu}) = \pm 1, \pm 2, \ldots, \tag{3}$$

and that their action is a multiple of $2\pi/\alpha_s$,

$$S \equiv \int d^4x \, \frac{1}{2} \mathrm{tr}(F_{\mu\nu} F_{\mu\nu}) = \frac{2\pi}{\alpha_s} |Q| = \frac{2\pi}{\alpha_s} \cdot (1, 2, \ldots). \tag{4}$$

In the weak coupling regime, $\alpha_s \ll 1$, the dominant saddle-point has $|Q| = 1$. The solution corresponding to $Q = 1$ is given by[2] [1] (singular gauge)

$$A_\mu^{(I)}(x; \rho, U, x_0) = -\frac{\mathrm{i}}{g} \frac{\rho^2}{(x - x_0)^2} U \frac{\sigma_\mu (\overline{x} - \overline{x_0}) - (x_\mu - x_{0\mu})}{(x - x_0)^2 + \rho^2} U^\dagger, \tag{5}$$

---

[1] Perturbative QCD is obtained from an expansion about the perturbative vacuum solution, i.e. vanishing gluon field and vanishing quark fields and thus vanishing Euclidean action.

[2] In Eq. (5) and throughout the paper we use the abbreviations, $v \equiv v_\mu \sigma^\mu$, $\overline{v} \equiv v_\mu \overline{\sigma}^\mu$ for any four-vector v_μ.

where the "collective coordinates" ρ, x_0 and U denote the size, position and colour orientation of the solution. The solution (5) has been called "instanton" (I), since it is localized in Euclidean space and time ("instantaneous"), as can be seen from its Lagrangian density,

$$\mathcal{L}\left(A_\mu^{(I)}(x;\rho,U,x_0)\right) = \frac{12}{\pi\alpha_s} \cdot \frac{\rho^4}{((x-x_0)^2+\rho^2)^4} \Rightarrow S\left[A_\mu^{(I)}\right] = \frac{2\pi}{\alpha_s}. \qquad (6)$$

It appears as a spherically symmetric bump of size ρ centred at x_0.

Lagrange Density   **Topological Charge Density**

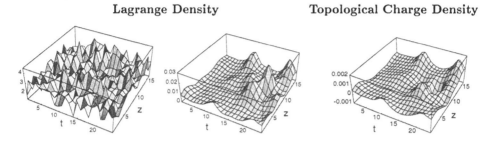

Fig. 1. Instanton content of a typical slice of a gluon configuration on the lattice at fixed x, y as a function of z and t [17]. Lagrange density before "cooling", with fluctuations of short wavelength $\mathcal{O}(a)$ dominating (left). After "cooling" by 25 steps, 3 I's and 2 $\bar{I}$'s may be clearly identified as bumps in the Lagrange density (middle) and the topological charge density (right).

The natural starting point of instanton perturbation theory is the evaluation of the instanton contribution to the partition function (2) [2], by expanding the path integral about the instanton (5). Since the action is independent of the collective coordinates, one has to integrate over them and obtains the I-contribution $Z^{(I)}$, normalized to the topologically trivial perturbative contribution $Z^{(0)}$, in the form[3]

$$\frac{1}{Z^{(0)}}\frac{dZ^{(I)}}{d^4x} = \int_0^\infty d\rho\, D_m(\rho) \int dU. \qquad (7)$$

The size distribution $D_m(\rho)$ is known in the framework of I-perturbation theory for small $\alpha_s(\mu_r)\ln(\rho\,\mu_r)$ and small $\rho\, m_i(\mu_r)$, where $m_i(\mu_r)$ are the running quark masses and μ_r denotes the renormalization scale. After its pioneering evaluation at 1-loop [2] for $N_c = 2$ and its generalization [14] to arbitrary N_c, it is meanwhile available [15] in 2-loop renormalization-group (RG) invariant form, i.e. $D^{-1}\, dD/d\ln(\mu_r) = \mathcal{O}(\alpha_s^2)$,

$$\frac{dn_I}{d^4x\, d\rho} = D_m(\rho) = D(\rho) \prod_{i=1}^{n_f} (\rho\, m_i(\mu_r))\, (\rho\,\mu_r)^{n_f\,\gamma_0\frac{\alpha_{\overline{MS}}(\mu_r)}{4\pi}}, \qquad (8)$$

---

[3] For notational simplicity, we call the I-position in the following x (instead of x_0).

with the reduced size distribution

$$D(\rho) = \frac{d_{\overline{MS}}}{\rho^5} \left(\frac{2\pi}{\alpha_{\overline{MS}}(\mu_r)}\right)^{2N_c} \exp\left(-\frac{2\pi}{\alpha_{\overline{MS}}(\mu_r)}\right)(\rho\,\mu_r)^{\beta_0 + (\beta_1 - 4N_c\beta_0)\frac{\alpha_{\overline{MS}}(\mu_r)}{4\pi}}. \quad (9)$$

Here, γ_0 is the leading anomalous dimension coefficient, β_i ($i = 0, 1$) denote the leading and next-to-leading β-function coefficients and $d_{\overline{MS}}$ is a known [16] constant.

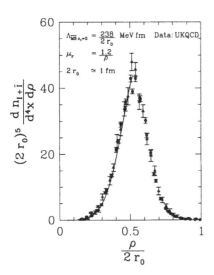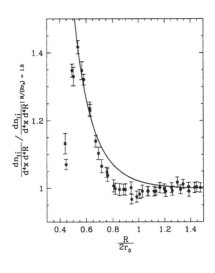

Fig. 2. Continuum limit [11] of "equivalent" UKQCD data [12,18] for the $(I + \overline{I})$-size distribution (left) and the normalized $I\overline{I}$-distance distribution (right) along with the respective predictions from I-perturbation theory and the valley form of the $I\overline{I}$-interaction [11]. The 3-loop form of $\alpha_{\overline{MS}}$ with $\Lambda_{\overline{MS}}^{(0)}$ from ALPHA [19] was used.

The power-law behaviour of the (reduced) I-size distribution,

$$D(\rho) \sim \rho^{\beta_0 - 5 + \mathcal{O}(\alpha_s)}, \quad (10)$$

generically causes the dominant contributions to the I-size integrals (e.g. Eq. (7)) to originate from the infrared (IR) regime (large ρ) and thus often spoils the applicability of I-perturbation theory. Since the I-size distribution not only appears in the vacuum-to-vacuum amplitude (7), but also in generic instanton-induced scattering amplitudes (c.f. Sect. 3) and matrix elements, it is extremely important to know the region of validity of the perturbative result (9).

Crucial information on the range of validity comes [11] from a recent high-quality lattice investigation [12] on the topological structure of the QCD vacuum (for $n_f = 0$). In order to make I-effects visible in lattice simulations with given lattice spacing a, the raw data have to be "cooled" first. This procedure is designed to filter out (dominating) fluctuations of short wavelength $\mathcal{O}(a)$, while affecting the topological fluctuations of much longer wavelength $\rho \gg a$ comparatively little. After cooling, an ensemble of I's and $\bar{I}$'s can clearly be seen (and studied) as bumps in the Lagrange density and in the topological charge density (c.f. Fig. 1).

Figure 2 (left) illustrates the striking agreement in shape and normalization [11] of $2\,D(\rho)$ with the continuum limit of the UKQCD lattice data [12] for $dn_{I+\bar{I}}/d^4x\,d\rho$, for $\rho \lesssim 0.3 - 0.35$ fm. The predicted normalization of $D(\rho)$ is very sensitive to $\Lambda^{(0)}_{\overline{MS}}$ for which we took the most accurate (non-perturbative) result from ALPHA [19]. The theoretically favoured choice $\mu_r \rho = \mathcal{O}(1)$ in Fig. 2 (left) optimizes the range of agreement, extending right up to the peak around $\rho \simeq 0.5$ fm. However, due to its two-loop renormalization-group invariance, $D(\rho)$ is almost independent of μ_r for $\rho \lesssim 0.3$ fm over a wide μ_r range. Hence, for $\rho \lesssim 0.3$ fm, there is effectively no free parameter involved.

Turning back to the perturbative size distribution (8) in QCD with $n_f \neq 0$ light quark flavours, we would like to comment on the appearent suppression of the instanton-induced vacuum-to-vacuum amplitude (7) for small quark masses, $\rho m_i \ll 1$. It is related [2] to the axial anomaly [20] according to which any gauge field fluctuation with topological charge Q must be accompanied by a corresponding change in chirality,

$$\Delta Q_{5i} = 2\,Q\,; \qquad i = 1, \ldots, n_f\,. \tag{11}$$

Thus, pure vacuum-to-vacuum transitions induced by instantons are expected to be rare. On the other hand, scattering amplitudes or Green's functions corresponding to anomalous chirality violation (c.f. Fig. 3) are expected to receive their main contribution due to instantons and do not suffer from any mass suppression.

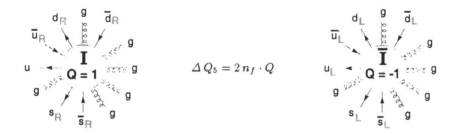

Fig. 3. Instantons and anti-instantons induce chirality violating amplitudes.

Let us illustrate this by the simplest example of one light flavour ($n_f = 1$): The instanton contribution to the fermionic two-point function can be written as

$$\langle \psi(x_1)\overline{\psi}(x_2)\rangle^{(I)} \simeq \int d^4x \int_0^\infty d\rho\, D(\rho) \int dU\,(\rho\,m)\,S^{(I)}(x_1, x_2; x, \rho, U)\,. \tag{12}$$

Expressing the quark propagator in the I-background, $S^{(I)}$, in terms of the spectrum of the Dirac operator in the I-background, which has exactly one right-handed zero mode[4] κ_0 [2],

$$-i\,\not{D}^{(I)}\kappa_n \quad = \lambda_n \kappa_n; \qquad \text{with } \lambda_0 = 0 \text{ and } \lambda_n \neq 0 \text{ for } n \neq 0, \qquad (13)$$

$$S^{(I)}(x_1, x_2; \ldots) = \frac{\kappa_0(x_1; \ldots)\,\kappa_0^\dagger(x_2; \ldots)}{m} + \sum_{n \neq 0} \frac{\kappa_n(x_1; \ldots)\,\kappa_n^\dagger(x_2; \ldots)}{m + i\lambda_n}, \quad (14)$$

we see that for $m \to 0$ only the zero mode contribution survives in Eq. (12),

$$\langle\psi(x_1)\overline\psi(x_2)\rangle^{(I)} \simeq \int d^4x \int_0^\infty d\rho\, D(\rho) \int dU\, \rho\, \kappa_0(x_1; x, \rho, U)\, \kappa_0^\dagger(x_2; x, \rho, U). \, (15)$$

Note that $\kappa_0 \kappa_0^\dagger$ has $Q_5 = 2$, exactly as required by the anomaly (11). For the realistic case of three light flavours ($n_f = 3$), the generalization of Eq. (15) leads to non-vanishing, chirality violating six-point functions corresponding to the anomalous processes shown in Fig. 3.

Finally, let us turn to the interaction between instantons and anti-instantons. In the instanton-anti-instanton ($I\bar I$) valley approach [22] it is determined in the following way: Starting from the infinitely separated ($R \to \infty$) $I\bar I$-pair,

$$A_\mu^{(I\bar I)}(x; \rho, \overline\rho, U, R) \overset{R \to \infty}{=} A_\mu^{(I)}(x; \rho, 1) + A_\mu^{(\bar I)}(x - R; \overline\rho, U) \qquad (16)$$

one looks for a constraint solution, which is the minimum of the action for fixed collective coordinates, $\rho, \overline\rho, U, R$. The valley equations have meanwhile been solved for arbitrary separation R [23] and arbitrary relative color orientation U [24]. Due to classical conformal invariance, the $I\bar I$-action $S^{(I\bar I)}$ and the interaction Ω,

$$S[A_\mu^{(I\bar I)}] = \frac{4\pi}{\alpha_s}\, S^{(I\bar I)}(\xi, U) = \frac{4\pi}{\alpha_s}\,(1 + \Omega(\xi, U)) \qquad (17)$$

depend on the sizes and the separation only through the "conformal separation",

$$\xi = \frac{R^2}{\rho\overline\rho} + \frac{\overline\rho}{\rho} + \frac{\rho}{\overline\rho}. \qquad (18)$$

Because of the smaller action, the most attractive relative orientation (c.f. Fig. 4) dominates in the weak coupling regime. Thus, in this regime, nothing prevents instantons and anti-instantons from approaching each other and annihilating.

From a perturbative expansion of the path integral about the $I\bar I$-valley, one obtains the contribution of the $I\bar I$-valley to the partition function (2) in the form

$$\frac{1}{Z^{(0)}}\frac{dZ^{(I\bar I)}}{d^4x} = \int d^4R \int_0^\infty d\rho \int_0^\infty d\overline\rho\, D_{I\bar I}(R, \rho, \overline\rho), \qquad (19)$$

---

[4] According to an index theorem [21], the number $n_{R/L}$ of right/left-handed zero modes of the Dirac operator in the background of a gauge field with topological charge Q satisfies $n_R - n_L = Q$. For the instanton: $n_R = Q = 1$; $n_L = 0$.

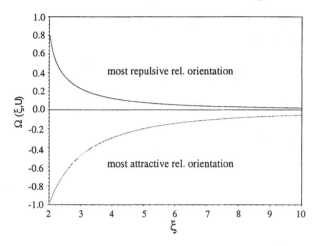

Fig. 4. The instanton-anti-instanton interaction as a function of the conformal separation ξ, for the most attractive and the most repulsive relative orientation, respectively.

where the group-averaged distribution of $I\bar{I}$-pairs, $D_{I\bar{I}}(R, \rho, \bar{\rho})$, is known, for small α_s, m_i, and for sufficiently large R [9–11],

$$\frac{dn_{I\bar{I}}}{d^4x \, d^4R \, d\rho \, d\bar{\rho}} \simeq D_{I\bar{I}}(R, \rho, \bar{\rho}) = \tag{20}$$

$$D(\rho) \, D(\bar{\rho}) \int dU \, \exp\left[-\frac{4\pi}{\alpha_{\overline{MS}}(s_{I\bar{I}}/\sqrt{\rho\bar{\rho}})} \, \Omega\left(\frac{R^2}{\rho\bar{\rho}}, \frac{\bar{\rho}}{\rho}, U\right)\right] \omega(\xi, U)^{2n_f}.$$

Here, the scale factor $s_{I\bar{I}} = \mathcal{O}(1)$ parametrizes the residual scheme dependence and

$$\omega = \int d^4x \, \kappa^\dagger_{0\,I}(x; \dots) \, [\mathrm{i} \, \not{D}^{(I\bar{I})}] \, \kappa_{0\,\bar{I}}(x - R; \dots) \tag{21}$$

denotes the fermionic interaction induced by the quark zero modes.

We will see below in Sect. 3 that the distribution (20) is a crucial input for instanton-induced scattering cross sections. Thus, it is extremely welcome that the range of validity of (20) can be inferred from a comparison with recent lattice data. Fig. 2 (right) displays the continuum limit [11] of the UKQCD data [12,18] for the distance distribution of $I\bar{I}$-pairs, $dn_{I\bar{I}}/d^4x \, d^4R$, along with the theoretical prediction [11]. The latter involves (numerical) integrations of $\exp(-4\pi/\alpha_s \cdot \Omega)$ over the $I\bar{I}$ relative color orientation (U), as well as ρ and $\bar{\rho}$. For the respective weight $D(\rho)D(\bar{\rho})$, a Gaussian fit to the lattice data was used in order to avoid convergence problems at large $\rho, \bar{\rho}$. We note a good agreement with the lattice data down to $I\bar{I}$-distances $R/\langle\rho\rangle \simeq 1$. These results imply first direct support for the validity of the "valley"-form of the interaction Ω between $I\bar{I}$-pairs.

In summary: The striking agreement of the UKQCD lattice data with I-perturbation theory is a very interesting result by itself. The extracted lattice constraints on the range of validity of I-perturbation theory can be directly translated into a "fiducial" kinematical region for our predictions [9,11] in deep-inelastic scattering, as shall be discussed in the next section.

3   Instantons in Deep-Inelastic Scattering

In this section we shall elucidate the special rôle of deep-inelastic scattering for instanton physics. We shall outline that only small size instantons, which are theoretically under contrôl, are probed in deep-inelastic scattering [8]. Furthermore, we shall show that suitable cuts in the Bjorken variables of instanton-induced scattering processes[5] allow us to stay within the range of validity of instanton perturbation theory, as inferred from the lattice [9,10]. We review the basic theoretical inputs to QCDINS, a Monte Carlo generator for instanton-induced processes in deep-inelastic scattering [5,13]. Finally, we discuss the final state characteristics of instanton-induced events.

Let us consider a generic I-induced process in deep-inelastic scattering (DIS),

$$\gamma^* + g \Rightarrow \sum_{\substack{n_f \\ \text{flavours}}} [\bar{q}_R + q_R] + n_g\, g\,, \tag{22}$$

which violates chirality according to the anomaly (11). The corresponding scattering amplitude is calculated as follows [8]: The respective Green's function is first set up according to instanton perturbation theory in Euclidean position space, then Fourier transformed to momentum space, LSZ amputated, and finally continued to Minkowski space where the actual on-shell limits are taken. Again, the amplitude appears in the form of an integral over the collective coordinates [8],

$$\mathcal{T}_\mu^{(I)\,(2n_f + n_g)} = \int_0^\infty d\rho\, D(\rho) \int dU\, \mathcal{A}_\mu^{(I)\,(2n_f + n_g)}(\rho, U)\,. \tag{23}$$

In leading order, the momentum dependence of the amplitude for fixed ρ and U,

$$\mathcal{A}_\mu^{(I)\,(2n_f + n_g)}(q, p; k_1, k_2, \ldots, k_{2n_f}, p_1, \ldots, p_{n_g}; \rho, U)\,, \tag{24}$$

factorizes, as illustrated in Fig. 5 for the case $n_f = 1$: The amplitude decomposes into a product of Fourier transforms of classical fields (instanton gauge fields; quark zero modes, e.g. as in Eq. (15)) and effective photon-quark "vertices" $\mathcal{V}_\mu^{(t(u))}(q, -k_{1(2)}; \rho, U)$, involving the (non-zero mode) quark propagator [26] in the instanton background. These vertices are most important in the following argumentation since they are the only place where the space-like virtuality $-q^2 = Q^2 > 0$ of the photon enters.

---

[5] Our approach, focussing on the I-induced final state, differs substantially from an exploratory paper [25] on the I-contribution to the (inclusive) parton structure functions. Ref. [25] involves implicit integrations over the Bjorken variables of the I-induced scattering process. Unlike our approach, the calculations in Ref. [25] are therefore bound to break down in the interesting domain of smaller $x_{\text{Bj}} \lesssim 0.3$, where most of the data are located.

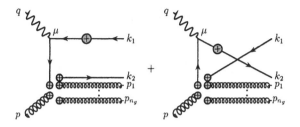

Fig. 5. Instanton-induced chirality-violating process, $\gamma^*(q) + g(p) \to \bar{q}_R(k_1) + q_R(k_2) + g(p_1) + \ldots + g(p_{n_g})$, for $n_f = 1$ in leading order of I-perturbation theory. The corresponding Green's function involves the products of the appropriate classical fields (lines ending at blobs) as well as the (non-zero mode) quark propagator in the instanton background (quark line with central blob).

After a long and tedious calculation one finds [8] for these vertices,

$$V_\mu^{(t)}(q, -k_1; \rho, U) = 2\pi i \rho^{3/2} \left[\epsilon \sigma_\mu \overline{V}(q, k_1; \rho) U^\dagger \right] , \tag{25}$$

$$V_\mu^{(u)}(q, -k_2; \rho, U) = 2\pi i \rho^{3/2} \left[U V(q, k_2; \rho) \overline{\sigma}_\mu \epsilon \right] , \tag{26}$$

where

$$V(q, k; \rho) = \left[\frac{(q-k)}{-(q-k)^2} + \frac{k}{2q \cdot k} \right] \rho \sqrt{-(q-k)^2} \, K_1 \left(\rho \sqrt{-(q-k)^2} \right) \tag{27}$$
$$- \frac{k}{2q \cdot k} \rho \sqrt{-q^2} \, K_1 \left(\rho \sqrt{-q^2} \right) .$$

Here comes the crucial observation: Due to the (large) space-like virtualities $Q^2 = -q^2 > 0$ and $Q'^2 = -(q-k)^2 \geq 0$ in DIS and the exponential decrease of the Bessel K-function for large arguments in Eq. (27), the I-size integration in our perturbative expression (23) for the amplitude is effectively cut off. Only small size instantons, $\rho \sim 1/Q$, are probed in DIS and the predictivity of I-perturbation theory is retained for sufficiently large $\mathcal{Q} = \min(Q, Q')$.

The leading[6] instanton-induced process in the DIS regime of $e^{\pm} P$ scattering for large photon virtuality Q^2 is illustrated in Fig. 6. The inclusive I-induced cross section can be expressed as a convolution [4,9], involving integrations over the target-gluon density, f_g, the virtual photon flux, P_{γ^*}, and the known [9,10] flux $P_{q'}^{(I)}$ of the virtual quark q' in the I-background (c.f. Fig. 6).

The crucial instanton-dynamics resides in the so-called instanton-subprocess (c.f. dashed box in Fig. 6) with its associated total cross section $\sigma_{q'g}^{(I)}(Q', x')$,

---

[6] I-induced processes initiated by a quark from the proton are suppressed by a factor of α_s^2 with respect to the gluon initiated process [9]. This fact, together with the high gluon density in the relevant kinematical domain at HERA, justifies to neglect quark initiated processes.

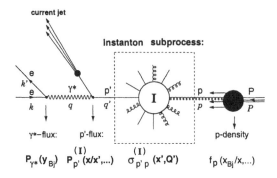

Fig. 6. The leading instanton-induced process in the deep-inelastic regime of $e^\pm P$ scattering ($n_f = 3$).

depending on its own Bjorken variables,

$$Q'^2 = -q'^2 \geq 0; \quad x' = \frac{Q'^2}{2p \cdot q'} \leq 1. \tag{28}$$

The cross section is obtained [9,10] in the form of an integral over $I\bar{I}$ collective coordinates[7],

$$\sigma_{q'g}^{(I)} \sim \int d^4R \int_0^\infty d\rho \int_0^\infty d\bar\rho \, D(\rho)D(\bar\rho) \int dU e^{-\frac{4\pi}{\alpha_s}\Omega\left(\frac{R^2}{\rho\bar\rho},\frac{\bar\rho}{\rho},U\right)} \omega \left(\frac{R^2}{\rho\bar\rho},\frac{\bar\rho}{\rho},U\right)^{2n_f-1}$$

$$\times e^{-Q'(\rho+\bar\rho)} e^{i(p+q')\cdot R} \{\ldots\}. \tag{29}$$

Thus, as anticipated in Sect. 2, the group averaged distribution of $I\bar{I}$-pairs (20) is closely related to the instanton-induced cross section. The lattice constraints on this quantity are therefore extremely useful.

Again, the quark virtuality Q'^2 cuts off large instantons. Hence, the integrals in (29) are finite. In fact, they are dominated by a unique saddle-point [9,10],

$$U^* = \text{most attractive relative orientation};$$

$$\rho^* = \bar\rho^* \sim 1/Q'; \quad R^{*2} \sim 1/(p+q')^2 \quad \Rightarrow \quad \frac{R^*}{\rho^*} \sim \sqrt{\frac{x'}{1-x'}}, \tag{30}$$

from which it becomes apparent (c.f. Fig. 7) that the virtuality Q' contrôls the effective I-size, while x' determines the effective $I\bar{I}$-distance (in units of the size ρ). By means of the discussed saddle-point correspondence (30), the lattice constraints may be converted into a "fiducial" region for our cross section

---

[7] Both an instanton and an anti-instanton enter here, since cross sections result from taking the modulus squared of an amplitude in the single I-background. In the present context, the $I\bar{I}$-interaction Ω takes into account the exponentiation of final state gluons [9].

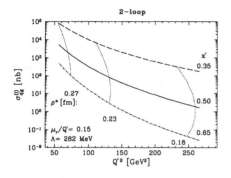
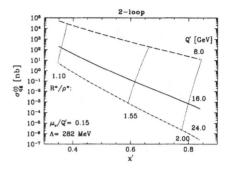

Fig. 7. I-subprocess cross section [9] displayed versus the Bjorken variable Q'^2 with x' fixed (left) and versus x' with Q'^2 fixed (right) for $n_f = 3$. The dotted lines indicate the corresponding effective I-sizes ρ^* [fm] (left) and $I\bar{I}$-distances R^* in units of ρ^* (right), respectively.

predictions in DIS [9],

$$\left.\begin{array}{l} \rho^* \lesssim 0.3 - 0.35 \text{ fm;} \\ \frac{R^*}{\rho^*} \gtrsim 1 \end{array}\right\} \Rightarrow \left\{\begin{array}{ll} Q'/\Lambda_{\overline{\text{MS}}}^{(n_f)} \gtrsim 30.8; \\ x' \gtrsim 0.35. \end{array}\right. \tag{31}$$

As illustrated in Fig. 7, $\sigma_{q'g}^{(I)}(Q', x')$ is very steeply growing for decreasing values of Q'^2 and x', respectively. The constraints (31) from lattice simulations are extremely valuable for making concrete predictions. Note that the fiducial region (31) and thus all our predictions for HERA never involve values of the $I\bar{I}$-interaction Ω smaller than -0.5 (c.f. Fig. 4), a value often advocated as a lower reliability bound [27].

Let us present an update of our published prediction [9] of the I-induced cross section at HERA. For the following *modified* standard cuts,

$$\mathcal{C}_{\text{std}} = x' \geq 0.35, \ Q' \geq 30.8 \, \Lambda_{\overline{\text{MS}}}^{(n_f)}, \ x_{\text{Bj}} \geq 10^{-3}, \tag{32}$$

$$0.1 \leq y_{\text{Bj}} \leq 0.9, \ Q \geq 30.8 \, \Lambda_{\overline{\text{MS}}}^{(n_f)},$$

involving the minimal cuts (31) extracted from lattice simulations, and an update of $\Lambda_{\overline{\text{MS}}}$ to the 1998 world average [28], we obtain

$$\sigma_{\text{HERA}}^{(I)}(\mathcal{C}_{\text{std}}) = 29.2^{+9.9}_{-8.1} \text{ pb.} \tag{33}$$

Note that the quoted errors in the cross section (33) only reflect the uncertainty in $\Lambda_{\overline{\text{MS}}}^{(5)} = 219^{+25}_{-23}$ MeV [28], on which $\sigma^{(I)}$ is known to depend very strongly [9]. We have also used now the 3-loop formalism [28] to perform the flavour reduction of $\Lambda_{\overline{\text{MS}}}^{(n_f)}$ from 5 to 3 light flavours. Finally, the value of $\sigma^{(I)}$ is substantially reduced compared to the one in Ref. [9], since we preferred to introduce a further cut in Q^2, with $Q^2_{\min} = Q'^2_{\min}$, in order to insure the smallness of the I-size ρ in contributions associated with the second term in Eq. (27).

Based on the predictions of I-perturbation theory, a Monte Carlo generator for simulating QCD-instanton induced scattering processes in DIS, QCDINS, has been developed [5,13]. It is designed as an "add-on" hard process generator interfaced by default to the Monte Carlo generator HERWIG [29]. Optionally, an interface to JETSET [30] is also available for the final hadronization step.

QCDINS incorporates the essential characteristics that have been derived theoretically for the hadronic final state of I-induced processes: notably, the isotropic production of the partonic final state in the I-rest system ($q'g$ center of mass system in Fig. 6), flavour "democracy", energy weight factors different for gluons and quarks, and a high average multiplicity $2n_f + \mathcal{O}(1/\alpha_s)$ of produced partons with a (approximate) Poisson distribution of the gluon multiplicity.

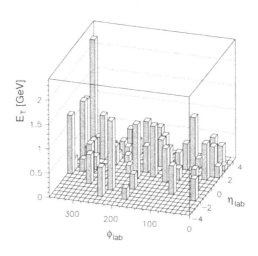

Fig. 8. Lego plot of a typical instanton-induced event from QCDINS.

The characteristic features of the I-induced final state are illustrated in Fig. 8 displaying the lego plot of a typical event from QCDINS (c.f. also Fig. 6): Besides a single (not very hard) current jet, one expects an accompanying densely populated "hadronic band". For $x_{\mathrm{Bj\,min}} \simeq 10^{-3}$, say, it is centered around $\bar{\eta} \simeq 2$ and has a width of $\Delta\eta \simeq \pm 1$. The band directly reflects the isotropic production of an I-induced "fireball" of $\mathcal{O}(10)$ partons in the I-rest system. Both the total transverse energy $\langle E_T \rangle \simeq 15$ GeV and the charged particle multiplicity $\langle n_c \rangle \simeq 13$ in the band are far higher than in normal DIS events. Finally, each I-induced event has to contain strangeness such that the number of K^0's amounts to $\simeq 2.2/$event.

4   Search Strategies

In a recent detailed study [7], based on QCDINS and standard DIS event generators, a number of basic (experimental) questions have been investigated: How to

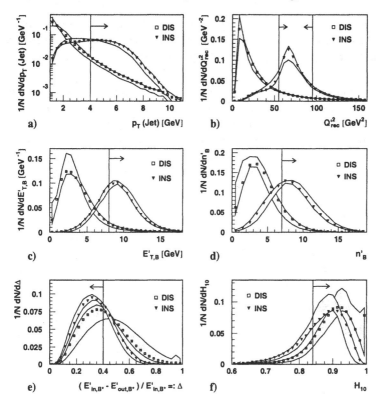

Fig. 9. Distributions of various observables for normal DIS and I-induced processes [7]. Shown are the distributions for the "reference Monte Carlos" (INS markers = QCDINS + HERWIG, DIS markers = ARIADNE [31], including Pomeron exchange) and their variations (shaded band) resulting from the choice of different models or the variation of parameters of a model (c.f. Fig. 10). The lines and the corresponding arrows show the cut applied in each of the observables, with the arrows pointing in the direction of the allowed region.

isolate an I-enriched data sample by means of cuts to a set of observables? How large are the dependencies on Monte-Carlo models, both for I-induced (INS) and normal DIS events? Can the Bjorken-variables (Q', x') of the I-subprocess be reconstructed?

All the studies presented in Ref. [7] were performed in the hadronic center of mass frame, which is a suitable frame of reference in view of a good distinction between I-induced and normal DIS events (c.f. Ref. [6]). The results are based on a study of the hadronic final state, with typical acceptance cuts of a HERA detector being applied.

Let us briefly summarize the main results of Ref. [7]. While the "I-separation power"= $\mathrm{INS}_{\mathrm{eff(iciency)}}/\mathrm{DIS}_{\mathrm{eff(iciency)}}$ typically does not exceed $\mathcal{O}(20)$ for single observable cuts, a set of six observables (among ~ 30 investigated in Ref. [6]) with much improved joint I-separation power = $\mathcal{O}(130)$ could be found, see

Fig. 9. These are (a) the p_T of the current jet, (b) Q'^2 as reconstructed from the final state, (c) the transverse energy and (d) the number of charged particles in the I-band region[8], and (e,f) two shape observables that are sensitive to the event isotropy.

The systematics induced by varying the modelling of I-induced events remains surprisingly small (Fig. 10). In contrast, the modelling of normal DIS events in the relevant region of phase space turns out to depend quite strongly on the used generators and parameters [7]. Despite a relatively high expected rate for I-events in the fiducial DIS region [9], a better understanding of the tails of distributions for normal DIS events turns out to be quite important.

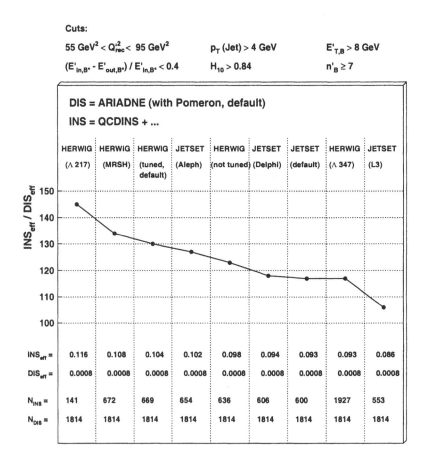

Fig. 10. I-separation power ($\mathrm{INS}_{\mathrm{eff}}/\mathrm{DIS}_{\mathrm{eff}}$) of a multidimensional cut-scenario depending on the variation of MC models and parameters used to simulate I-induced events [7]. The efficiencies and remaining event numbers for an integrated luminosity $\mathcal{L} \simeq 30$ pb^{-1} and corresponding to the cross section from QCDINS 1.6.0 are listed.

---

[8] With the prime in Fig. 9 (c,d,e) indicating that the hadrons from the current jet have been subtracted.

References

1. A. Belavin, A. Polyakov, A. Schwarz, Yu. Tyupkin: Phys. Lett. B **59**, 85 (1975)
2. G. 't Hooft: Phys. Rev. Lett. **37**, 8 (1976); Phys. Rev. D **14**, 3432 (1976); Phys. Rev. D **18**, 2199 (1978) (Erratum); Phys. Rep. **142**, 357 (1986)
3. T. Schäfer, E. Shuryak: Rev. Mod. Phys. **70**, 323 (1998)
4. A. Ringwald, F. Schrempp: 'Towards the Phenomenology of QCD-Instanton Induced Particle Production at HERA', hep-ph/9411217. In: *Quarks '94, Proc. 8th Int. Seminar, Vladimir, Russia, May 11–18, 1994*, ed. by D. Grigoriev et al. (World Scientific, Singapore 1995) pp. 170–193
5. M. Gibbs, A. Ringwald, F. Schrempp: 'QCD-Instanton Induced Final States in Deep Inelastic Scattering', hep-ph/9506392. In: *Workshop on Deep Inelastic Scattering and QCD (DIS 95), Paris, France, April 24–28, 1995*, ed. by J.F. Laporte, Y. Sirois (Ecole Polytechnique, Paris 1995) pp. 341-344
6. J. Gerigk: 'QCD-Instanton-induzierte Prozesse in tiefunelastischer $e^{\pm}p$-Streuung', Dipl. Thesis (in German), University of Hamburg (unpublished) and MPI-PhE/98-20, Nov. 1998
7. T. Carli, J. Gerigk, A. Ringwald, F. Schrempp: 'QCD Instanton-Induced Processes in Deep-Inelastic Scattering – Search Stragegies and Model Dependencies', hep-ph/9906441. To appear in: *Proc. DESY Workshop 1998/1999 on Monte Carlo Generators for HERA Physics*
8. S. Moch, A. Ringwald, F. Schrempp: Nucl. Phys. B **507**, 134 (1997)
9. A. Ringwald, F. Schrempp: Phys. Lett. B **438**, 217 (1998)
10. S. Moch, A. Ringwald, F. Schrempp: in preparation
11. A. Ringwald, F. Schrempp: Phys. Lett. B **459**, 249 (1999)
12. D.A. Smith, M.J. Teper (UKQCD): Phys. Rev. D **58**, 014505 (1998)
13. A. Ringwald, F. Schrempp: in preparation
14. C. Bernard: Phys. Rev. D **19**, 3013 (1979)
15. T. Morris, D. Ross, C. Sachrajda: Nucl. Phys. B **255**, 115 (1985)
16. A. Hasenfratz, P. Hasenfratz: Nucl. Phys. B **193**, 210 (1981)
    M. Lüscher: Nucl. Phys. B **205**, 483 (1982)
17. M.-C. Chu, J.M. Grandy, S. Huang, J.W. Negele: Phys. Rev. D **49**, 6039 (1994)
18. M. Teper: private communication
19. S. Capitani, M. Lüscher, R. Sommer, H. Wittig: Nucl. Phys. B **544**, 669 (1999)
20. S. Adler: Phys. Rev. **177**, 2426 (1969)
    J. Bell, R. Jackiw: Nuovo Cimento **51**, 47 (1969)
    W. Bardeen: Phys. Rev. **184**, 1848 (1969)
21. M. Atiyah, I. Singer: Ann. Math. **87**, 484 (1968)
22. A. Yung: Nucl. Phys. B **297**, 47 (1988)
23. V.V. Khoze, A. Ringwald: Phys. Lett. B **259**, 106 (1991)
24. J. Verbaarschot: Nucl. Phys. B **362**, 33 (1991)
25. I. Balitsky, V. Braun: Phys. Lett. B **314**, 237 (1993)
26. L. Brown, R. Carlitz, D. Creamer, C. Lee: Phys. Rev. D **17**, 1583 (1978)
27. V. Zakharov: Nucl. Phys. B **353**, 683 (1991)
    M. Maggiore, M. Shifman: Nucl. Phys. B **365**, 161 (1991); *ibid.* **371**, 177 (1991)
    G. Veneziano: Mod. Phys. Lett. A **7**, 1661 (1992)
28. C. Caso et al. (Particle Data Group): Eur. Phys. J. C **3**, 1 (1998)
29. G. Marchesini et al.: Comp. Phys. Commun. **67**, 465 (1992)
30. T. Sjöstrand: Comp. Phys. Commun. **82**, 74 (1994)
31. L. Lönnblad: Comp. Phys. Commun. **71**, 15 (1992)

Part III

Photon Structure and Photoproduction of Jets and Hadrons

Experimental Results on Two-Photon Physics from LEP

Richard Nisius

CERN, CH-1211 Genève 23, Switzerland

Abstract. This review covers selected results from the LEP experiments on the structure of quasi-real and virtual photons. The topics discussed are the total hadronic cross-section for photon-photon scattering, hadron production, jet cross-sections, heavy quark production for photon-photon scattering, photon structure functions, and cross-sections for the exchange of two virtual photons.

1 Introduction

The photon structure has been investigated in detail at LEP based on the scattering of two electrons[1] proceeding via the exchange of two photons, as shown in Figure 1. The reactions are classified depending on the virtualities of the pho-

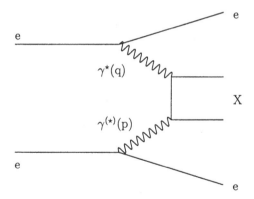

Fig. 1. A diagram of the reaction ee $\rightarrow$ eeX, proceeding via the exchange of two photons.

tons, with $-q^2 = Q^2$ and $-p^2 = P^2$, and on the nature of the final state X. If a photon has small virtuality, and the corresponding electron is not observed in the experiment, it is called a quasi-real photon, γ. If the electron is observed,

---

[1] Fermions and anti-fermions are not distinguished, for example, electrons and positrons are referred to as electrons. The natural system of units, which means, $c = \hbar = 1$ is used.

the photon is far off-shell, and the virtual photon is denoted with $\gamma^\star$. As the two photons can either be quasi-real or virtual the reactions are classified as $\gamma\gamma$ scattering (photon-photon scattering or anti-tagged events), $\gamma\gamma^\star$ scattering (deep inelastic electron-photon scattering or single tag events), and $\gamma^\star\gamma^\star$ scattering (double tag events). In each of the classes, different aspects of the photon structure can be investigated. Due to space limitations not all results can be reviewed here, only a personal selection has been chosen driven by the relevance of the different topics in the context of this workshop on HERA physics. The main results not covered here concern resonance production and glueball searches, which are described in Ref. [1].

2   Results from $\gamma\gamma$ Scattering

The $\gamma\gamma$ scattering reaction has the largest hadronic cross-section at LEP2 energies. The main topics studied are the total hadronic cross-section for photon-photon scattering, $\sigma_{\gamma\gamma}$, and more exclusively, hadron production, jet cross-sections and the production of heavy quarks.

2.1   The Total Photon-Photon Cross-Section

The measurement of $\sigma_{\gamma\gamma}$ is both, interesting and challenging. It is interesting, because in the framework of Regge theory $\sigma_{\gamma\gamma}$ can be related to the total hadronic cross-sections for photon-proton and hadron-hadron scattering, $\sigma_{\gamma p}$ and σ_{hh}, and a slow rise with the photon-photon center-of-mass energy, $s = W^2$, is predicted. It is challenging, firstly, because experimentally the determination of the hadronic invariant mass, W, is very difficult due to limited acceptance and resolution for the hadrons created in the reaction and secondly, because the composition of different event classes, for example, diffractive and quasi-elastic processes, is rather uncertain, which affects the overall acceptance of the events. The first problem is dealt with by determining W from the visible hadronic invariant mass using unfolding programs. The second uncertainty is taken into account by using two models, namely PHOJET [2] and PYTHIA [3], for the description of the hadronic final state and for the correction from the accepted cross-section to $\sigma_{\gamma\gamma}$, leading to the largest uncertainty of the result.

The published measurements of $\sigma_{\gamma\gamma}$ by L3 [4] and by OPAL [5] are shown in Figure 2(left). Both results show a clear rise as a function of W. The cross-section $\sigma_{\gamma\gamma}$ is interpreted within the framework of Regge theory, motivated by the fact that $\sigma_{\gamma p}$ and σ_{hh} are well described by Regge parametrisations using terms to account for pomeron and reggeon exchanges. The originally proposed form of the Regge parametrisation for $\sigma_{\gamma\gamma}$ is

$$\sigma_{\gamma\gamma}(s) = X_{1\gamma\gamma}s^{\epsilon_1} + Y_{1\gamma\gamma}s^{-\eta_1} , \tag{1}$$

where s is taken in units of GeV^2. The first term in the equation is due to soft pomeron exchange and the second term is due to reggeon exchange. The

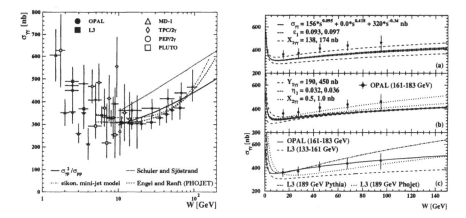

Fig. 2. The published results on $\sigma_{\gamma\gamma}$ as a function of W (right), and an illustration of the spread of the fit results to various data (left).

exponents ϵ_1 and η_1 are assumed to be universal. The presently used values of $\epsilon_1 = 0.095 \pm 0.002$ and $\eta_1 = 0.034 \pm 0.02$ are taken from Ref. [6]. The parameters were obtained by a fit to the total hadronic cross-sections of pp, $p\bar{p}$, $\pi^{\pm}p$, $K^{\pm}p$, γp and $\gamma\gamma$ scattering reactions. The coefficients $X_{1\gamma\gamma}$ and $Y_{1\gamma\gamma}$ have to be extracted from the $\gamma\gamma$ data. The values obtained in Ref. [6] by a fit to older $\gamma\gamma$ data, including those of L3 from Ref. [4], are $X_{1\gamma\gamma} = (156 \pm 18)$ nb and $Y_{1\gamma\gamma} = (320 \pm 130)$ nb. Recently an additional hard pomeron component has been suggested in Ref. [7] leading to

$$\sigma_{\gamma\gamma}(s) = X_{1\gamma\gamma}s^{\epsilon_1} + X_{2\gamma\gamma}s^{\epsilon_2} + Y_{1\gamma\gamma}s^{-\eta_1}, \tag{2}$$

with a proposed value of $\epsilon_2 = 0.418$ and an expected uncertainty of ϵ_2 of about ± 0.05. Different fits to the data have been performed by the experiments.

The interpretation of the results is very difficult, because, firstly the parameters are highly correlated, secondly, the main region of sensitivity to the reggeon term is not covered by the OPAL measurement and thirdly, different assumptions have been made when performing the fits. The correlation of the parameters of Eq. (2) can be clearly seen in Figure 2(right a,b), where the theoretical predictions are shown, exploring the uncertainties for the soft pomeron term in (a) and for the reggeon as well as for the hard pomeron term in (b), using the central values and errors quoted in Ref. [6]. It is clear from Figure 2(right a,b) that by changing different parameters in (a) and (b) a very similar effect on the rise of the total-cross section can be achieved. Figure 2(right c) shows the spread of the best fit curves for various data and various fit assumptions explained below. In Figure 2(right a-c) in addition the results from Ref. [5] are shown to illustrate the size of the experimental uncertainties.

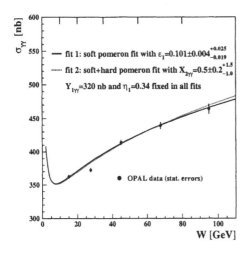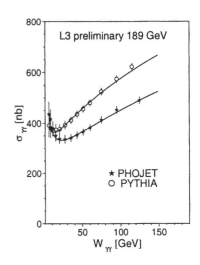

Fig. 3. Fits to the total hadronic cross-section for photon-photon scattering as a function of W for OPAL data at $\sqrt{s}_{ee} = 161 - 189$ GeV (left), and for L3 data at $\sqrt{s}_{ee} = 189$ GeV using two different Monte Carlo models for correcting the data (right).

Examples of different fits are shown in Figure 3 taken from Refs. [5,8]. They yield the following results:

• The OPAL data, within the present range of W, can be accounted for without the presence of the hard pomeron term. When fixing all exponents and $Y_{1\gamma\gamma}$ to the values listed above the fit yields $X_{2\gamma\gamma} = (0.5 \pm 0.2^{+1.5}_{-1.0})$ nb, which is not significantly different from zero, and $X_{1\gamma\gamma} = (182 \pm 3 \pm 22)$ nb, which is consistent with the values from Ref. [6]. Using $X_{2\gamma\gamma} = 0$ and leaving only ϵ_1 and $X_{1\gamma\gamma}$ as free parameters results in $\epsilon_1 = 0.101 \pm 0.004^{+0.025}_{-0.019}$ and $X_{1\gamma\gamma} = (180 \pm 5^{+30}_{-32})$ nb, Figure 2(right, c, full), again consistent with Ref. [6].

• In all fits performed by L3 the hard pomeron term is set to zero. The L3 data from Ref. [4] can be fitted using the old values for the exponents of $\epsilon_1 = 0.0790 \pm 0.0011$ and $\eta_1 = 0.4678 \pm 0.0059$ from Ref. [9] leading to $X_{1\gamma\gamma} = (173 \pm 7)$ nb and $Y_{1\gamma\gamma} = (519 \pm 125)$ nb, Figure 2(right, c, dash). The L3 data at $\sqrt{s}_{ee} = 189$ GeV indicate a faster rise with energy. Using $\epsilon_1 = 0.95$ and $\eta_1 = 0.34$, and the PHOJET Monte Carlo for correcting the data, leads to $X_{1\gamma\gamma} = (172 \pm 3)$ nb and $Y_{1\gamma\gamma} = (325 \pm 65)$ nb, but the confidence level of the fit is only 0.000034 [8]. Fixing only the reggeon exponent to $\eta_1 = 0.34$ leads to $\epsilon_1 = 0.222 \pm 0.019/0.206 \pm 0.013$, $X_{1\gamma\gamma} = (50 \pm 9)/(78 \pm 10)$ nb and $Y_{1\gamma\gamma} = (1153 \pm 114)/(753 \pm 116)$ nb, when using PHOJET/PYTHIA, Figure 2(right, c, dot/dot-dash).

In summary, the situation is unclear at the moment with OPAL being consistent with the universal Regge prediction, whereas L3 indicating a faster rise with s for the data at $\sqrt{s}_{ee} = 189$ GeV. In addition, the L3 data taken at different center-of-mass energies show a different behaviour of the measured cross-section, with the data taken at $\sqrt{s_{ee}} = 133 - 161$ GeV being lower, especially for $W < 30$ GeV.

2.2   The Production of Charged Hadrons

The production of charged hadrons is sensitive to the structure of the photon-photon interactions without theoretical and experimental problems related to the definition and reconstruction of jets. The two main results from the study of hadron production at LEP are shown in Figure 4.

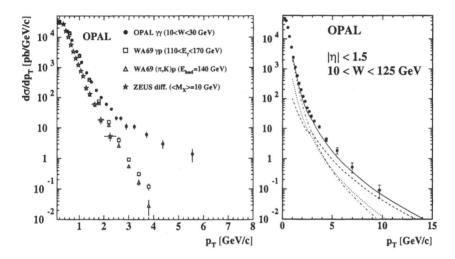

Fig. 4. The differential single particle inclusive cross-section for hadron production in photon-photon scattering at $\sqrt{s_{ee}} = 161 - 172$ GeV compared to other experiments for $10 < W < 30$ GeV (left), and compared to next-to-leading order calculations for $10 < W < 125$ GeV (right).

In Figure 4(left) the differential single particle inclusive cross-section $d\sigma/dp_T$ for charged hadrons for $\gamma\gamma$ scattering as obtained by OPAL [10], with $10 < W < 30$ GeV, is shown, together with results from γp, πp and Kp scattering from WA69 with a hadronic invariant mass of 16 GeV. The WA69 data are normalised to the $\gamma\gamma$ data at $p_T \approx 0.2$ GeV. In addition, ZEUS data from Ref. [11] on charged particle production in γp scattering with a diffractively dissociated photon are shown. These data have an average invariant mass of the diffractive system of 10 GeV, and again they are normalised to the OPAL data. In Figure 4(right) the differential single particle inclusive cross-section for $10 < W < 125$ GeV is compared to next-to-leading order QCD predictions. The main findings are:
- The spectrum of transverse momentum of charged hadrons in photon-photon scattering is much harder than in the case of photon-proton, hadron-proton and 'photon-Pomeron' interactions. This can be attributed to the direct component of the photon-photon interactions.

• The production of charged hadrons is found to be well described by the next-to-leading order QCD predictions from Ref. [12] over a wide range of W. These next-to-leading order calculations are based on the QCD partonic cross-sections, the next-to-leading order GRV parametrisation of the parton distribution functions and on fragmentation functions fitted to e^+e^- data. The renormalisation and factorisation scales are set equal to p_T.

2.3   Jet Production

Jet production is the classical way to study the partonic structure of particle interactions. At LEP the di-jet cross-section in $\gamma\gamma$ scattering was studied in Ref. [13] at $\sqrt{s_{ee}} = 161 - 172$ GeV using the cone jet finding algorithm with $R = 1$. Three event classes are defined, direct, single-resolved and double-resolved interactions. Here, direct means that the photons as a whole take part in the hard interaction, whereas resolved means that a parton of a hadronic fluctuation of the photon participates in the hard scattering reaction. Experimentally, direct and double-resolved interactions can be clearly separated using the quantity

$$x_\gamma^\pm = \frac{\sum_{\text{jets}=1,2}(E \pm p_z)}{\sum_{\text{hadrons}}(E \pm p_z)}, \qquad (3)$$

whereas a selection of single-resolved events cannot be achieved with high purity. Ideally, in leading order direct interactions have $x_\gamma^\pm = 1$, however, due to resolution and higher order corrections the measured values of $x_\gamma^\pm$ are smaller. Experimentally, samples containing large fractions of direct events can be selected by requiring $x_\gamma^\pm > 0.8$, and samples containing large fractions of double-resolved events by using $x_\gamma^\pm < 0.8$.

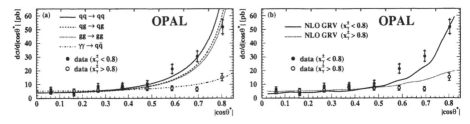

Fig. 5. The angular dependence of di-jet production at $\sqrt{s_{ee}} = 161-172$ GeV compared to leading order matrix elements (left) and to next-to-leading order (NLO) predictions (right).

The measurement of the distribution of $\cos\theta^\star$, the cosine of the scattering angle in the photon-photon centre-of-mass system, allows for a test of the different matrix elements contributing to the reaction. The scattering angle is calculated from the jet rapidities in the laboratory frame using

$$\cos\theta^\star = \tanh\frac{\eta^{\text{jet1}} - \eta^{\text{jet2}}}{2}. \qquad (4)$$

In leading order the direct contribution $\gamma\gamma \to q\bar{q}$ leads to an angular dependence of the form $(1 - \cos^2\theta^\star)^{-1}$, whereas double-resolved events, which are dominated by gluon induced reactions, are expected to behave approximately as $(1 - \cos^2\theta^\star)^{-2}$. The steeper angular dependence of the double-resolved interactions can be clearly seen in Figure 5(left), where the shape of the di-jet cross-section, for events with di-jet masses above 12 GeV and average rapidities of $|(\eta_1 + \eta_2)/2| < 1$, is compared to leading order predictions. In addition, the shape of the angular distribution observed in the data is roughly described by the next-to-leading order prediction from Refs. [14], Figure 5(right). In both cases the theoretical predictions are normalised to the data in the first three bins.

These next-to-leading order calculations well account for the observed inclusive differential di-jet cross-section, $d\sigma/dE_T^{\text{jet}}$, as a function of jet transverse energy, E_T^{jet}, for di-jet events with pseudorapidities $|\eta^{\text{jet}}| < 2$. As expected the

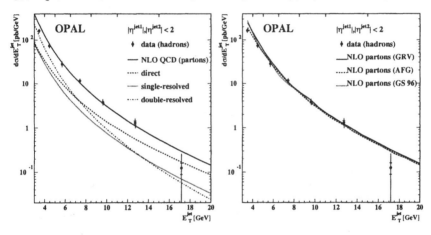

Fig. 6. The E_T^{jet} dependence of di-jet production at $\sqrt{s}_{ee} = 161 - 172$ GeV compared to next-to-leading order (NLO) predictions for different event classes (left) and for different parametrisations of the parton distribution functions of the photon (right).

direct component can account for most of the cross-section at large E_T^{jet}, whereas the region of low E_T^{jet} is dominated by the double-resolved contribution, shown in Figure 6(left). The calculations from Refs. [15] for three different next-to-leading order parametrisations of the parton distribution functions of the photon are in good agreement with the data shown in Figure 6(right), except in the first bin, where theoretical as well as experimental uncertainties are large. Unfortunately, this is the region which shows the largest sensitivity to the differences of the parton distribution functions of the photon.

2.4   Heavy Quark Production

The production of heavy quarks in photon-photon scattering is dominated by charm quark production, as the bottom quarks are much heavier and have a smaller electric charge. Due to the large scale of the process provided by the charm quark mass, the production of charm quarks can be predicted in next-to-leading order perturbative QCD. In QCD the production of charm quarks at LEP2 energies receives contributions of about equal size from the direct production mechanism and from the single-resolved contribution, shown in Figure 7. In contrast, the double-resolved contribution is expected to be very small, see Ref. [16] for details.

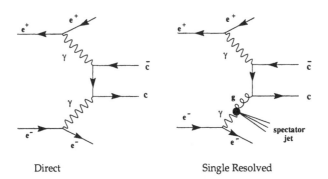

Direct   Single Resolved

Fig. 7. The direct (left) and single-resolved (right) contributions to charm quark production in photon-photon collisions.

In photon-photon collisions the charm quarks have been tagged using standard techniques, either based on the observation of semileptonic decays of charm quarks using identified electrons and muons in Ref. [17], or by the measurement of $D^\star$ production in Refs. [18–20] using the decay $D^\star \to D^0 \pi$, where the pion has very low energy, followed by the D^0 decay observed in one of the decay channels, $D^0 \to K\pi, K\pi\pi^0, K\pi\pi\pi$. The leptons as well as the $D^\star$ can be clearly separated from background processes, as shown in Figure 8. Using the ratio of the transverse energy of the electron measured in the calorimeter and the transverse momentum measured in the tracking chamber the electrons can be well separated from other charged particles, Figure 8(left) from Ref. [17]. Utilising the low energy of the slow pion a clear peak can be observed in the mass difference, ΔM, between the mass of the $D^\star$ and the mass of the D^0 candidate, as shown in Figure 8(right) from Ref. [19]. However, due to the small branching ratios and selection inefficiencies the selected event samples are small and the measurements are limited mainly by the statistical error.

Based on these tagging methods differential cross-sections for charm quark production and $D^\star$ production in restricted kinematical regions have been obtained, examples of which are shown in Figure 9. Figure 9(left) shows the differential cross-section for charm quark production, with semileptonic decays into electrons fulfilling $|\cos\theta_e| < 0.9$ and $E_e > 0.6$ GeV and for $W > 3$ GeV. The data are compared to the leading order prediction from PYTHIA, normalised to the number of data events observed. The shape of the distribution is well reproduced by the leading order prediction. Figure 9(right) shows the differential

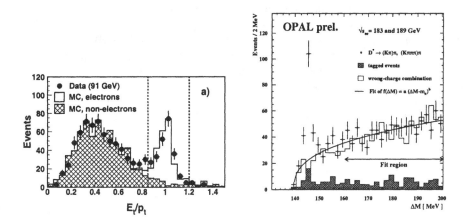

Fig. 8. Charm quark tagging via electrons from semileptonic decays (left), and via the mass difference between the mass of the $D^\star$ and the mass of the D^0 candidate (right).

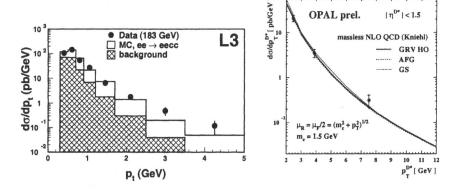

Fig. 9. Differential cross-sections for charm quark production with semileptonic decays into electrons (left), and for $D^\star$ production (right), both determined in restricted kinematical regions.

cross-sections for $D^\star$ production as a function of the transverse momentum of the $D^\star$, for $|\eta^{D^\star}| < 1.5$ compared to the next-to-leading order predictions from Ref. [21] calculated in the massless approach. The differential cross-sections as functions of the transverse momentum and rapidity of the $D^\star$ are well reproduced by the next-to-leading order perturbative QCD predictions, both for the OPAL results [19] and for the L3 results [20]. The shape of the OPAL data can be reproduced by the NLO calculations from Ref. [16], however, the theoretical predictions are somewhat lower than the data, especially at low values of transverse momentum of the $D^\star$.

Based on the observed cross-sections in the restricted ranges in phase space the total charm quark production cross-section is derived, very much relying on the Monte Carlo predictions for the unseen part of the cross-section. Two issues are addressed, firstly the relative contribution of the direct and single-resolved processes, and secondly the total charm quark production cross-section. The direct and single-resolved events, for example, as predicted by the PYTHIA Monte Carlo, show a different distribution as a function of the transverse momentum of the D^* meson, $p_T^{D^*}$, normalised to the visible hadronic invariant mass, W_{vis}, as can be seen in Figure 10(left) from Ref. [19]. This feature has been used to experimentally determine the relative contribution of direct and single-resolved events, which were found to contribute about equally to the cross-section.

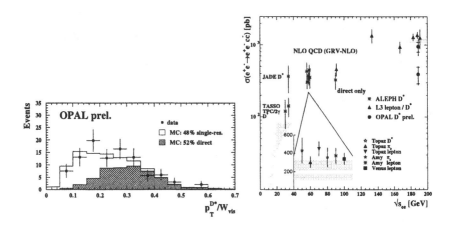

Fig. 10. The separation of the D^* production into direct and single-resolved contributions (left) and the total cross-section for charm quark production (right).

The total cross-section for the production of charm quarks is shown in Figure 10(right). The LEP results are consistent with each other and the theoretical predictions are in agreement with the data. The measurements suffer from additional errors due to the assumptions made in the extrapolation from the accepted to the total cross-section, which are avoided by only measuring cross-sections in restricted ranges in phase space. It has been shown in Ref. [16] that the NLO calculations are flexible enough to account for the phase space restrictions of an experimental analysis and that the predicted cross-sections in restricted ranges in phase space are less sensitive to variations of the charm mass and to alterations of the renormalisation as well as the factorisation scale. Given this, more insight into several aspects of charm quark production may be gained by comparing experimental results and theoretical predictions for cross-sections in restricted ranges in phase space.

In addition to the measurements of the charm quark production cross-sections, a preliminary measurement of the cross-section for bottom quark production has been reported in Ref. [22].

3   Results from $\gamma\gamma^\star$ Scattering

In this kinematical region the reaction can be described as deep-inelastic electron-photon scattering and allows for measurements of photon structure functions, similarly to measurements of proton structure functions in the case of electron-proton scattering at HERA. The measurements of photon structure functions have been discussed in detail in the literature and the reader is referred to the most recent review, Ref. [23], and to references therein. Only the main results from the LEP experiments are shortly mentioned here.

- The QED structure function $F_{2,\mathrm{QED}}^\gamma$ has been precisely measured using data in the approximate range of average virtualities $\langle Q^2 \rangle$ of $1.5 - 130$ GeV^2. The LEP data are so precise that the effect of the small virtuality P^2 of the quasi-real photon can clearly be established.

- The structure functions $F_{A,\mathrm{QED}}^\gamma$ and $F_{B,\mathrm{QED}}^\gamma$ give more insight into the helicity structure of the $\gamma\gamma^\star$ interaction. They were obtained from the shape of the distribution of the azimuthal angle between the plane defined by the momentum vectors of the muons and the plane defined by the momentum vectors of the incoming and the deeply inelastically scattered electron. Both structure functions were found to be significantly different from zero, and the recent theoretical predictions from Ref. [24], which take into account the important mass corrections up to $\mathcal{O}(m_\mu^2/W^2)$, are consistent with the measurements.

- The hadronic structure function F_2^γ has been measured using data in the approximate range of average virtualities of $\langle Q^2 \rangle$ of $1.9 - 400$ GeV^2. The general features of the measurements can be described by several parametrisations of F_2^γ. However, the data are precise enough to disfavour those parametrisations which predict a fast rise of F_2^γ at low values of x, driven by large gluon distribution functions.

- The evolution of F_2^γ with Q^2 has been studied in bins of x. The measurements are consistent with each other and a clear rise of F_2^γ with Q^2 is observed. The general trend of the data is followed by the predictions of several parametrisations of F_2^γ. It is an interesting fact that at medium values of x this rise can also be described reasonably well ($\mathcal{O}(15\%)$ accuracy) by the leading order augmented asymptotic prediction detailed in Ref. [25], which uses the asymptotic solution from Ref. [26] for F_2^γ for the light flavour contribution as predicted by perturbative QCD for $\alpha_s(M_Z) = 0.128$.

4   Results from $\gamma^\star\gamma^\star$ Scattering

The QED and the hadronic structure of virtual photons have been studied at LEP. The structure functions of virtual photons can be determined for the situation where one photon has a much larger virtuality than the other, $Q^2 \gg P^2$, by measuring the cross-section for events where both electrons are observed. For the situation where both photons have similar virtualities, $Q^2 \approx P^2$, the structure function picture is no longer applicable and differential cross-sections for the exchange of two highly-virtual photons have been measured instead. The main results from the LEP experiments are shortly mentioned here, for a more detailed discussion the reader is referred to Ref. [23].

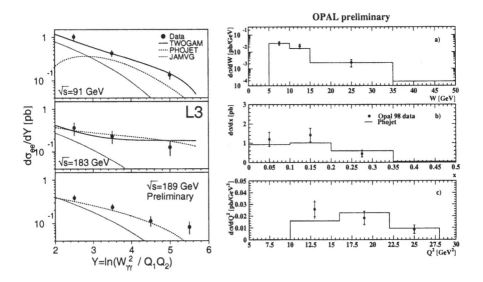

Fig. 11. The differential cross-sections for the exchange of two highly virtual photons as functions of various variables from L3 (left) and OPAL (right).

• The effective hadronic structure function [27] has been measured by L3 [28] for average virtualities of $\langle Q^2 \rangle = 120 \text{ GeV}^2$ and $\langle P^2 \rangle = 3.7 \text{ GeV}^2$. A consistent picture is found for the effective structure function between the older PLUTO result from Ref. [27] and the L3 data, and the general features of both measurements are described by the next-to-leading order predictions from Ref. [29].

• The cross-section for the exchange of two highly virtual photons with muon final states has been measured in Ref. [30]. There is good agreement between the measured cross-section and the QED prediction. The measurement shows that the interference terms, which are usually neglected in investigations of the hadronic structure of the photon, are present in the data in the kinematical region of the analysis, mainly at $x > 0.1$, and that the corresponding contributions to the cross-section are negative.

• The cross-section for the exchange of two highly virtual photons with hadronic final states has been measured in Refs. [31–33], and the main results are shown in Figure 11. The differential cross-sections as functions of various variables are well described by leading order Monte Carlo models. Much larger cross-sections are predicted in the framework of BFKL calculations. These predictions are strongly disfavoured by the data.

Acknowledgement

I am grateful to the organisers for inviting me to this inspiring location and for the fruitful atmosphere they created throughout the meeting. I wish to thank Stefan Söldner-Rembold and Bernd Surrow for carefully reading the manuscript and Jochen Patt for providing me with the figure of combined results for the charm quark production cross-section.

References

1. L3 Collaboration, M. Acciarri et al., Phys. Lett. **B363**, 118–126 (1995);
   L3 Collaboration, M. Acciarri et al., Phys. Lett. **B413**, 147–158 (1997);
   L3 Collaboration, M. Acciarri et al., Phys. Lett. **B418**, 399–410 (1998);
   L3 Collaboration, M. Acciarri et al., Phys. Lett. **B453**, 73–82 (1999);
   L3 Collaboration, M. Acciarri et al., CERN-EP/99-072;
   OPAL Collaboration, K. Ackerstaff et al., Phys. Lett. **B439**, 197–208 (1998);
   ALEPH Collaboration, contributed paper to HEP99 Tampere.
2. R. Engel, Z. Phys. **C66**, 203–214 (1995);
   R. Engel and J. Ranft, Phys. Rev. **D54**, 4244–4262 (1996).
3. T. Sjöstrand, Comp. Phys. Comm. **82**, 74–89 (1994).
4. L3 Collaboration, M. Acciarri et al., Phys. Lett. **B408**, 450–464 (1997).
5. OPAL Collaboration, G. Abbiendi et al., CERN-EP/99-076.
6. Particle Data Group, Eur. Phys. J. **C3**, 1–794 (1998).
7. A. Donnachie and P.V. Landshoff, Phys. Lett. **B437**, 408–416 (1998).
8. L3 Collaboration, A. Csilling, in *Proceedings of Photon '99*.
9. Particle Data Group, Phys. Rev. **D54**, 1– (1996).
10. OPAL Collaboration, K. Ackerstaff et al., Eur. Phys. J. **C6**, 253–264 (1999).
11. ZEUS Collaboration, M. Derrick et al., Z. Phys. **C67**, 227–238 (1995).
12. J. Binnewies, B.A. Kniehl, and G. Kramer, Phys. Rev. **D53**, 6110–6119 (1996).
13. OPAL Collaboration, G. Abbiendi et al., CERN-EP/98-113.
14. M. Klasen, T. Kleinwort, and G. Kramer, Eur. Phys. J. **C1**, 1–105 (1998).
15. M. Klasen, private communication.
16. S. Frixione, M. Krämer, and E. Laenen, hep-ph/9908483.
17. L3 Collaboration, M. Acciarri et al., Phys. Lett. **B453**, 83–93 (1999).
18. ALEPH Collaboration, D. Buskulic et al., Phys. Lett. **B355**, 595–605 (1995).
19. OPAL Collaboration, J. Patt, in *Proceedings of Photon '99*.
20. L3 Collaboration, M. Acciarri et al., CERN-EP/99-106.
21. J. Binnewies, B.A. Kniehl, and G. Kramer, Phys. Rev. **D58**, 014014 (1998).
22. L3 Collaboration, R.R. McNeil, in *Proceedings of Photon '99*.
23. R. Nisius, in *Proceedings of Photon '99*, hep-ex/9907012.
24. R. Nisius and M.H. Seymour, Phys. Lett. **B452**, 409–413 (1999).
25. OPAL Collaboration, K. Ackerstaff et al., Phys. Lett. **B411**, 387–401 (1997).
26. E. Witten, Nucl. Phys. **B120**, 189–202 (1977).
27. PLUTO Collaboration, C. Berger et al., Phys. Lett. **142B**, 119–124 (1984).
28. L3 Collaboration, F.C. Erné, in *Proceedings of Photon '99*.
29. M. Glück, E. Reya, and M. Stratmann, Phys. Rev. **D54**, 5515–5522 (1996).
30. OPAL Collaboration, G. Abbiendi et al., CERN-EP/99-010.
31. L3 Collaboration, M. Acciarri et al., Phys. Lett. **B453**, 333–342 (1999).
32. L3 Collaboration, P. Achard, in *Proceedings of Photon '99*.
33. OPAL Collaboration, M. Przybycień, in *Proceedings of Photon '99*.

Hard Photoproduction and the Structure of the Photon

Laurel Sinclair

Department of Physics and Astronomy, University of Glasgow, G12 8QQ, UK

(For the ZEUS and H1 Collaborations)

Abstract. A pedagogical introduction to the experimental results on hard photoproduction at HERA is provided. Then the latest results in this field from ZEUS and H1 are reviewed.

1 Introduction

It has now been firmly established that the photon can interact strongly, as though it were a hadron. Indeed, two of the first physics results to be published by the HERA experiments H1 and ZEUS were the measurements of the total photoproduction cross section [1,2]. These confirmed the cross section of the photon to be of order 100 μb, close to the area of a typical hadron.

Of course the total cross section is dominated by peripheral collisions which cannot be described in perturbative QCD. In the presence of a hard scale the photon proton cross section factorizes into terms describing the photon and proton structures, and a hard QCD subprocess. Then a perturbative expansion may be applied to determine the subprocess cross section (for an introduction to the theory of hard photoproduction see Michael Klasen's contribution to this proceedings).

With subsequent data sets H1 and ZEUS measured the photoproduction cross section for events in the perturbatively calculable regime, i.e. with jets of at least $E_T^{\text{jet}} > 5$ GeV [3,4]. This cross section is naturally much smaller, of the order of 10 nb, however the hadronic nature of the photon is still prevalent. The HERA experiments clearly established two classes of contribution to the photoproduction of jets: the direct process in which the photon itself participates in the hard subprocess and the resolved process in which the photon fluctuates into a hadronic object and one of its partonic constituents participates in the hard subprocess. For instance, H1 found an excess of energy in the rear direction, over what was expected for direct photoproduction processes [5]. This energy could be attributed to the presence of a photon remnant jet in the resolved photon process.

An unambiguous distinction between resolved and direct processes exists only at leading order (LO). In Figures 1(a) and (b) examples of direct diagrams in LO and next-to-leading order (NLO), respectively, are shown. Figure 1(c) shows an example of a resolved diagram at leading order. Clearly, if the outgoing quark

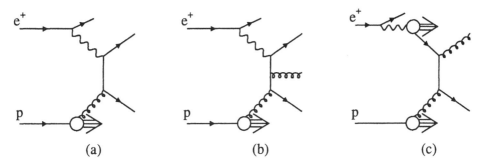

Fig. 1. Illustration of diagrams for (a) LO direct (b) NLO direct and (c) LO resolved photoproduction processes

line in the NLO direct diagram 1(b) has small transverse momentum then this process could as well be represented by the LO resolved diagram 1(c). Therefore some prescription must be introduced in order to make a well-defined distinction between direct and resolved processes.

To this end the observable quantity x_γ^{OBS} has been defined:

$$x_\gamma^{OBS} \equiv \frac{\sum_{\text{jets}} E_T^{\text{jet}} e^{-\eta^{\text{jet}}}}{2yE_e},$$

where E_T^{jet} and η^{jet} are the jet transverse energy and pseudorapidity respectively, y is the fraction of the electron's energy carried by the incoming photon, E_e is the incoming electron energy, and the sum runs over the two highest E_T^{jet} jets within the η^{jet} acceptance. At leading order $x_\gamma^{OBS} = 1$ for direct processes and $x_\gamma^{OBS} < 1$ for resolved processes. However, x_γ^{OBS} is well defined theoretically to any order of perturbation theory (provided, of course, that the jet-finding algorithm is well-behaved). Therefore, a quantitative confrontation of a measurement of a "direct" or "resolved" photoproduction cross section with a pQCD calculation can be made provided the separation between the direct and resolved regions is made in terms of x_γ^{OBS}.

In Figure 2 a measured x_γ^{OBS} distribution is shown for photoproduction events containing two jets of $E_T^{\text{jet1, jet2}} > 11, 14$ GeV within $-1 < \eta^{\text{jet}} < 2$ in the HERA lab frame [6]. The data show a peak at high x_γ^{OBS} values with a broad tail extending to $x_\gamma^{OBS} \sim 0$. The data are compared with the predictions of two parton shower Monte Carlo programs, HERWIG 5.9 [7,8] and PYTHIA 5.7 [9,10]. The Monte Carlo calculations implement the QCD matrix elements at leading order only. The effect of higher order processes is approximated through initial and final-state parton showers. The programs differ in the choice of evolution variable for the parton shower calculation and also in the technique chosen to convert the final partonic configuration into colourless hadrons. Both are able to provide a good description of the x_γ^{OBS} distribution. The distribution of

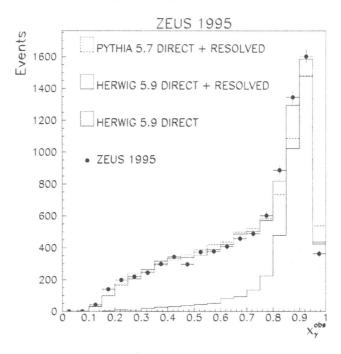

Fig. 2. The x_γ^{OBS} spectrum for events with $E_T^{jet1, \, jet2} > 11, 14$ GeV, compared to the HERWIG 5.9 and the PYTHIA 5.7 Monte Carlo predictions. The direct component from the HERWIG Monte Carlo is shown separately as the shaded histogram. Only statistical uncertainties are plotted

the HERWIG events with LO direct photoproduction subprocesses is shown separately as the shaded histogram. The parton showering and hadronizaton phases smear the x_γ^{OBS} values such that this distribution is peaked just below one and can extend to the lowest available x_γ^{OBS} values. Nevertheless it is plain to see that a sample of events with $x_\gamma^{OBS} > 0.75$ is essentially of the direct photoproduction type. ZEUS and H1 have published several papers in which direct and resolved photoproduction regions are defined based on the x_γ^{OBS} observable [4,6,11–19]

An important confirmation of the presence of the direct and resolved contributions to photoproduction was provided by the ZEUS measurement of dijet scattering angles [13]. A two jet final state can be completely specified in its centre-of-mass frame (up to an arbitrary azimuthal rotation) by the dijet invariant mass, M_{2J}, and the dijet scattering angle, ϑ^*. Direct photoproduction processes should proceed predominantly through quark exchange (the diagram shown at leading order in Figure 1(a)). As the exchanged parton is of spin 1/2, at leading order the dijet scattering angle is distributed according to $1/(1-|\cos \vartheta^*|)$. In contrast, resolved processes are dominated by the exchange of the integer spin gluon which gives rise to the steeper angular dependence, $1/(1-|\cos \vartheta^*|)^2$. ZEUS observed a steeper angular dependence for the events with $x_\gamma^{\mathrm{OBS}} < 0.75$ than for those with $x_\gamma^{\mathrm{OBS}} \geq 0.75$, as shown in Figure 3.

ZEUS 1994

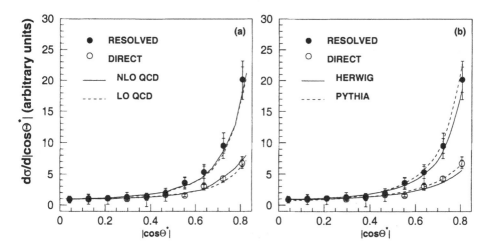

Fig. 3. $d\sigma/d|\cos\vartheta^*|$ normalized to one at $\cos\vartheta^* = 0$ for $x_\gamma^{\mathrm{OBS}} < 0.75$ (*black dots*) and $x_\gamma^{\mathrm{OBS}} \geq 0.75$ (*open circles*) photoproduction. In (a), the ZEUS data are compared to NLO predictions (*solid line*) and LO predictions (*broken line*). In (b), the broken line is the PYTHIA distribution and solid line is the HERWIG distribution. The inner error bars are the statistical errors, the outer error bars are the sum in quadrature of the statistical and systematic uncertainties

This is a compelling observation. An invariant mass cut of $M_{2\mathrm{J}} > 23$ GeV has been applied to ensure that the $E_T^{\mathrm{jet}} > 6$ GeV requirement does not bias the angular distribution. Moreover, the dijet scattering angle is defined in the centre-of-mass frame of the two jets so the fact that the lower x_γ^{OBS} events are boosted more in the proton direction is not responsible for the differences observed in the $\cos\vartheta^*$ distributions. The different $\cos\vartheta^*$ distributions for the high and low x_γ^{OBS} samples are therefore an unambiguous demonstration of the differing underlying QCD subprocess dynamics. There is a greater contribution from gluon exchange processes contributing to the $x_\gamma^{\mathrm{OBS}} < 0.75$ sample, than there is contributing to the $x_\gamma^{\mathrm{OBS}} \geq 0.75$ sample. This is consistent with the expectation that more resolved photon processes contribute to the $x_\gamma^{\mathrm{OBS}} < 0.75$ sample.

Thus it has been established that both direct and resolved processes contribute to photoproduction at HERA. The study of these processes is now providing a fruitful forum for the investigation of strong interactions. Photoproduction processes access the physics involved in the structures of the photon and proton, in the dynamics of hard subprocesses, and in the fragmentation and hadronization of the final state partons.

The structure of the photon is probed in deep inelastic $e\gamma$ experiments at e^+e^- colliders in just the same way as the structure of the proton is probed in ep interactions at HERA. These measurements of the photon's structure function, F_2^γ, constrain the quark densities at intermediate probing energy scales. However, the gluon density is not directly constrained in these experiments, nor do the quark density constraints extend to the high energy scale region accessible at HERA. It is in these two areas that the HERA experiments have concentrated their efforts.

Phenomenological ansätze for the parton densities of the photon exist. These generally involve the assumption that at low probing virtualities, the photon undergoes a quantum fluctuation into a vector meson or an unbound $q\bar{q}$ state and this provides the photon's structure. A DGLAP evolution is then invoked in order to evolve the parton densities to arbitrary scale, using the $e\gamma$ F_2^γ data as a constraint. Thus, studies of photoproduction sensitive to the photon's structure test fundamental physical assumptions, in addition to providing a means to investigate the universality of photon structure data obtained through different processes. In sections 2 and 3 the latest results from H1 and ZEUS pertaining to the structure of the photon are presented.

Perturbative QCD governs the behaviour of the partons emerging from the hard subprocess. As the distribution of the jets of hadrons in the final state bears a close correspondence to the distribution of the underlying partons, measurements may be designed which are sensitive to the subprocess dynamics. The differing dijet angular distributions for direct and resolved photoproduction have already been discussed. As the available luminosity delivered by HERA has increased, it has become possible to look for unusual dynamical signatures in high E_T^{jet} dijet processes, and also to investigate the underlying mechanism of three jet production. Studies of the matrix element dynamics are presented in section 4.

Our knowledge is limited about the physics of hadronization, whereby the partons resulting from a collision are converted to colourless hadrons by the inexorable confinement force of QCD. The hadronization occurs at low momentum transfers where the QCD coupling is much too strong for a perturbative expansion to be relevant. However, there are experimental results which have provided some information about this interesting area of physics. For instance, a universality of the hadronization of quarks is supported by the measurements of jet shapes in ep and e^+e^- collisions [20]. Recently, a procedure for measuring jet structure in hadronic collisions has been proposed which is valid for an all-orders calculation in perturbative QCD [21]. In this way, a well-defined comparison of theory and data can be undertaken which begins well within the regime where a perturbative approach should be valid and then approaches the mysterious realm of the very strong hadron producing force. A measurement of jet substructure using this algorithm is presented in Section 5.

2   Real Photon Structure

A measurement which is sensitive to the gluon density of the photon has been made by the H1 collaboration [22]. The gluonic component of the photon dominates at low x_γ. Therefore the events must be selected keeping E_T^{jet} as low as possible and allowing η^{jet} to extend as far forward into the incoming proton direction as possible. This is a difficult kinematic region experimentally, as the energy-scale uncertainty and angular resolution of the calorimeter are worst here. There are also theoretical limitations as the contribution from events in which there is a secondary scatter, and the smearing between a partonic and hadronic distribution, increase as E_T^{jet} lowers. Nevertheless it has been possible to make the measurement with sufficient precision to illustrate the sensitivity of the data to the distribution of gluons in the photon. Figure 4 shows $d\sigma/dx_\gamma^{jets}$ [1] for events containing two jets of $E_T^{jet} > 6$ GeV within $-0.5 < \eta^{jet} < 2.5$. The data are compared with the predictions of the leading order plus parton shower Monte Carlo program PHOJET [23,24] where the predictions including and neglecting the gluons in the photon are shown separately. Although there is a large systematic uncertainty affecting the measurement, within the PHOJET model

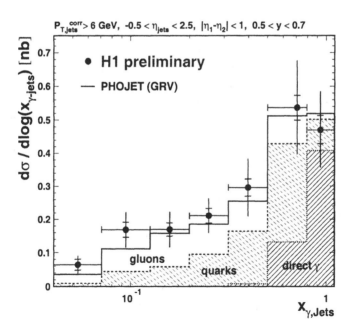

Fig. 4. Dijet cross section in photproduction as a function of x_γ^{jets}. The data are compared to the prediction of PHOJET. The direct photon prediction, and the resolved photon predictions for either quarks are gluons, are shown separately

---

[1] x_γ^{jets} is essentially the same quantity as has been called x_γ^{OBS}.

a significant gluonic contribution at low x_γ^{jets} is required by the data assuming the GRV-LO [25,26] parton densities for the photon and proton.

H1 have gone on to subtract the influence of secondary parton scattering and unfold the data to a leading order effective parton density, using the Monte Carlo model. From this the modelled quark density has been subtracted to yield the gluon density as a function of x_γ where now x_γ refers to the fraction of the momentum of the photon which enters the leading order hard subprocess. This is necessarily a model dependent result; however it can serve to provide insight into photon structure. H1 finds that the gluon density rises as x_γ decreases.

The ZEUS analysis of photon structure has concentrated on limiting the data to that kinematic regime in which perturbative QCD should be applicable, without the need for additional model parameters. Dijet events have been selected with $E_T^{\text{jet1, jet2}} > 11, 14$ GeV within $-1 < \eta^{\text{jet}} < 2$ [6]. The x_γ^{OBS} distribution for this selection is shown in Figure 2. The PYTHIA and HERWIG predictions which are compared with the data in this figure contain no simulation of soft underlying events or of secondary parton scatterings and provide a good description of the data. Also, studies have indicated that hadronization effects in this kinematic regime should be small, at most around 10%. Therefore a strong interpretation of this data within perturbative QCD can be made.

In Figure 5 the cross section for the $E_T^{\text{jet1, jet2}} > 11, 14$ GeV selection, $d\sigma/d\eta_1^{\text{jet}}$, is shown in bins of η_2^{jet}. The kinematic region is restricted to a narrow y range, in order to improve the sensitivity to the photon's structure. The cross section is shown separately for $x_\gamma^{\text{OBS}} \geq 0.75$, indicating that the direct process dominates when the jets tend toward the incoming photon direction. NLO perturbative QCD predictions with the three available photon parton parametrizations [25,27–29] are compared with the data in this figure. (Note that the calculations have been checked by several different groups of theorists as reported in [30].) The predictions underestimate the data in the central rapidity region where experimental and theoretical uncertainties are expected to be particularly small. As previously mentioned, the parton density of the photon is not well constrained by $e\gamma$ data at these high energy scales. Therefore it is expected that the parton density of the photon may be underestimated in this kinematic regime in the currently available parton density functions of the photon.

In another interesting analysis by H1 the photon remnant jet has been tagged in low x_γ^{jet} events by running a clustering algorithm, requiring exactly four jets in the event and defining the photon remnant jet as that jet closest to the incoming photon direction [31]. The E_T^{jet} of the remnant jet was found to be correlated with the E_T^{jet} of the highest transverse energy jets. This points to the presence of the anomalous component of the photon's structure whereby the struck parton arises from a $q\bar{q}$ splitting of the photon rather than as a constituent of a fluctuation into a vector meson. The "remnant jet" can then, as in Figure 1(b), be viewed instead as one of the outgoing partons from a next-to-leading order hard subprocess.

Another promising process for constraining the parton density of the photon is prompt photon production, in which an outgoing quark from the hard subpro-

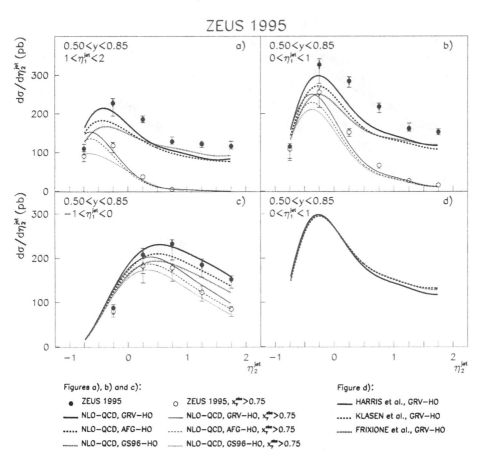

Fig. 5. Figures (a), (b) and (c) show the dijet cross section as a function of η_2^{jet} in bins of η_1^{jet}. The filled circles correspond to the entire x_γ^{OBS} range while the open circles correspond to events with $x_\gamma^{OBS} > 0.75$. The shaded band indicates the uncertainty related to the energy scale. The thick error bar indicates the statistical uncertainty and the thin error bar indicates the systematic and statistical uncertainties added in quadrature. The full, dotted and dashed curves correspond to NLO-QCD calculations, using the GRV-HO, GS96-HO and the AFG-HO parameterizations for the photon structure, respectively. In (d) the NLO-QCD results for the cross section when $0 < \eta_1^{jet} < 1$ and for a particular parameterization of the photon structure are compared

cess is balanced not by a gluon, but by a photon (see [30] for a complete discussion of the contributing diagrams). ZEUS has published a measurement of the cross section for prompt photon production in association with a jet [32]. A more inclusive measurement, whereby only the prompt photon is tagged without the jet requirement, is free of complications due to the matching of the jet definition in theory and experiment and relatively free of hadronization corrections. This then, like high E_T^{jet} production, is an area in which a strong interpretation can be made from the comparision of the data with the predictions of perturbative QCD. Of course the cross section for prompt photon production is suppressed with respect to jet production due to the smallness of the electromagnetic coupling constant and currently the statistics are limited. Nevertheless the technique of prompt photon identification has been refined and a first comparison with the theory indicates a rough agreement [30].

3   Virtual Photon Structure

It is expected that as the virtuality of the incoming photon increases, less time will be available for it to develop a complex hadronic structure. To test this assumption, H1 has measured the dijet triple differential cross section, $d\sigma_{ep}/(dQ^2 d\bar{E}_t^2 dx_\gamma^{\text{jets}})$ where $\bar{E}_t$ is the average jet transverse energy and Q^2 is the negative of the square of the momentum transfer at the scattered lepton vertex [14]. The cross section is presented in Figure 6 as a function of x_γ^{jets} in bins of $\bar{E}_t^2$ in the range $30 < \bar{E}_t^2 < 300$ GeV^2 and in bins of Q^2 in the range $1.6 < Q^2 < 25$ GeV^2. The data are concentrated near $x_\gamma^{\text{jets}} = 1$ with a small tail to lower values. Compared with the data are predictions from the HERWIG model, where the events with a leading order direct subprocess are shown separately by the shaded histogram. Looking at fixed $\bar{E}_t^2$, there is clear evidence for the expected suppression of the resolved processes as Q^2 increases. However, wherever $\bar{E}_t^2 \gg Q^2$, the direct processes alone are insufficient to account for the low x_γ^{jets} events. Thus there is evidence for resolved photon processes, even well into the deep inelastic scattering regime, $Q^2 > 8$ GeV^2.

H1 have extended the analysis in their publication [14] and unfolded the data to a leading order effective parton density, $f_{\text{eff}} \equiv \sum_i^{N_f} (q_i + \bar{q}_i) + \frac{9}{4} g$, relying on the Monte Carlo simulation to correct the data for hadronization and higher order effects. The behaviour of this parton density is consistent with a logarithmic rise with the probing resolution, P_t^2, where P_t is the transverse momentum of the two outgoing partons at leading order. This is in contrast with the approximate scaling behaviour which has been observed for hadrons and reflects the presence of the anomalous, or $q\bar{q}$ splitting term, unique to the structure of the photon.

The effective parton density is also found to exhibit a dependence on the photon's virtuality Q^2 which is consistent with the expected logarithmic suppression implemented in the existing virtual photon parton density functions. Incorporating an earlier measurement of f_{eff} at $Q^2 = 0$ with these data, a drop in f_{eff} with Q^2 is indicated. (To compare the earlier measurement with this one the data have been evolved using the GRV LO parton densities to the same P_t^2 and x_γ range.)

The ZEUS collaboration has measured the dijet cross section $d\sigma/dx_\gamma^{\text{OBS}}$ for jets of $E_T^{\text{jet}} > 5.5$ GeV in the three different photon virtuality ranges $Q^2 \sim 0$

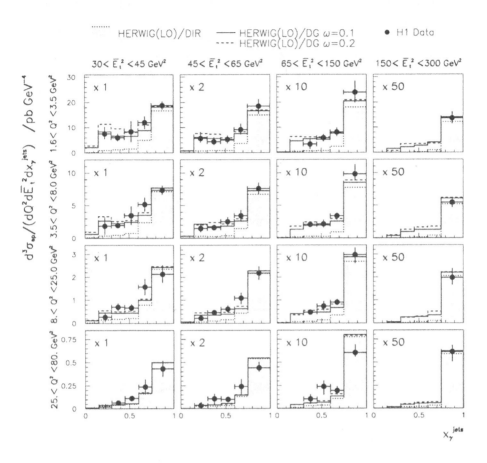

Fig. 6. The differential dijet cross section $d\sigma_{ep}/(dQ^2 d\bar{E}_t^2 dx_\gamma^{jets})$ shown as a function of x_γ^{jets} for different regions of $\bar{E}_t$ and Q^2. The scale factors applied to the cross sections are indicated. The error bar shows the quadratic sum of systematic and statistical errors. Also shown is the HERWIG model where the direct component is shown as the shaded histogram

($Q^2 < 1$ GeV^2), $0.1 < Q^2 < 0.55$ GeV^2 and $1.5 < Q^2 < 4.5$ GeV^2 [33]. This measurement is complementary to the H1 analysis in that the $Q^2 \sim 0$ data are analyzed together with the deep inelastic data at $1.5 < Q^2 < 4.5$ GeV^2 in a consistent way. Also, making use of an auxiliary small angle electron tagger, the ZEUS measurement includes data in the important transition region between photoproduction and deep inelastic scattering, $0.1 < Q^2 < 0.55$ GeV^2. From the $d\sigma/dx_\gamma^{OBS}$ measurements, similar conclusions can be drawn as in the H1 analysis: the population of the x_γ^{OBS} distribution is suppressed at low values as

Q^2 increases yet there is evidence for a resolved component of the photon even in the deep inelastic scattering regime, $Q^2 > 1.5$ GeV^2.

In order to make a precise statement concerning the evolution of the virtual photon parton densities with Q^2, ZEUS has measured the ratio of the dijet cross section for $x_\gamma^{\mathrm{OBS}} < 0.75$ to that for $x_\gamma^{\mathrm{OBS}} \geq 0.75$ in bins of Q^2 [33]. The result is shown in Figure 7. The ratio of resolved to direct cross sections is found to fall with Q^2. In comparison, this ratio within the HERWIG model is flat for a

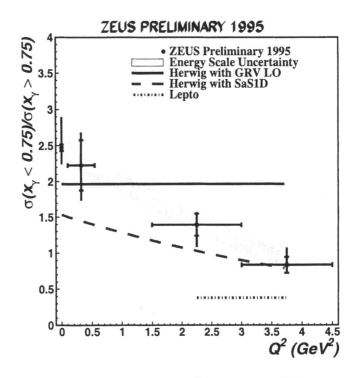

Fig. 7. The ratio of dijet cross sections, $\sigma(x_\gamma^{\mathrm{OBS}} < 0.75)/\sigma(x_\gamma^{\mathrm{OBS}}) \geq 0.75)$, as a function of photon virtuality, Q^2. The inner error bar represents the statistical error and the outer the statistical and systematic uncertainties added in quadrature. The band represents the systematic uncertainty due to the jet energy scale. Also shown are the predictions of HERWIG for two different choices for photon parton densities: GRV for real photons (*full line*) and SaS 1D (*dashed line*) which includes a suppression of the photon parton density with increasing photon virtuality. The LEPTO predictions are shown for $Q^2 > 1.5$ GeV^2 (*dot-dashed line*)

photon parton density which does not fall with Q^2 (GRV LO) and falling for a photon parton density which does (SaS 1D [34]). Therefore the data indicate that the photon parton density is suppressed with Q^2 in this LO description. Furthermore, as Q^2 increases the data tend toward a leading order prediction which does not include any resolved photon contribution (Lepto 6.5.1 [35]).

4   QCD Matrix Elements

The influence of the dominant QCD subprocesses on dijet angular distributions for events with $M_{2J} > 23$ GeV has been observed, as already discussed. As the integrated luminosity delivered by HERA continues to increase, it is important to re-measure these distributions in the newly accessible kinematic regimes, in order to check for new contributing processes to dijet production. Using the 1995 to 1997 integrated data set ZEUS has measured the M_{2J} and $\cos \vartheta^*$ distributions in the kinematic regime $M_{2J} > 47$ GeV and $|\cos \vartheta^*| < 0.8$ [36]. The measurement extends to a dijet invariant mass of $M_{2J} \sim 140$ GeV, and the mass and angular distributions are well described by the predictions of perturbative QCD.

For three massless jets, five quantities are necessary to define the system. These are defined in terms of the energies, E_i, and momentum three-vectors, p_i, of the jets in the three-jet centre-of-mass frame and p_B, the beam direction. They are the three-jet invariant mass, M_{3J}; the energy-sharing quantities X_3 and X_4, $X_i \equiv 2E_i/M_{3J}$; the cosine of the scattering angle of the highest energy jet with respect to the beam, $\cos \vartheta_3 \equiv p_B \cdot p_3/(|p_B||p_3|)$; and ψ_3, the angle between the plane containing the highest energy jet and the beam and the plane containing the three jets, $\cos \psi_3 \equiv (p_3 \times p_B) \cdot (p_4 \times p_5)/(|p_3 \times p_B||p_4 \times p_5|)$, where the jets are numbered, 3, 4 and 5 in order of decreasing energy. ZEUS has measured the three-jet cross section in the kinematic regime defined by $M_{3J} > 50$ GeV, $|\cos \vartheta_3| < 0.8$ and $X_3 < 0.95$ [37]. (In fact, the events have been further required to have at least two jets with $E_T^{jet} > 6$ GeV and a third jet with $E_T^{jet} > 5$ GeV, however these cuts are largely irrelevant because high E_T^{jet} values are forced by the energy and angular cuts.) The measured cross section is well described by $\mathcal{O}(\alpha\alpha_s^2)$ perturbative QCD calculations both in normalization and in the shapes of the M_{3J}, X_i, $\cos \vartheta_3$ and ψ_3 distributions.

The measured angular distributions are of particular interest since they differ markedly from the expectation for three jets distributed evenly over the available phase space, and are therefore sensitive to the underlying physical dynamics. The $\cos \vartheta_3$ distribution, since it is primarily determined by the distribution of the highest energy jet, is similar in three jet production to the distribution of $\cos \vartheta^*$ in two jet production. However, ψ_3 is determined by the orientation of the third, or softest, jet. For orientations in which the third jet is radiated close to the plane defined by the highest energy jet and the beam, $\psi_3 \sim 0$ or π. The data are shown in Figure 8. They indicate that configurations in which the third jet is far from the plane containing the highest energy jet and the beam ($\psi_3 \sim \pi/2$) are suppressed. The dip at $\psi_3 = 0$ and π is caused by a loss of phase space at low angles due to the $E_T^{jet\ 3} > 5$ GeV requirement. Taking this into consideration, the data indicate a strong tendency for the three-jet plane to lie close to the plane containing the highest energy jet and the beam.

To aid in the development of a mental picture for three jet production, the data have been compared to the predictions of the PYTHIA and HERWIG models. The hard subprocess is included only at leading order in these models so three-jet events arise from the parton shower phase of the simulation. Parton

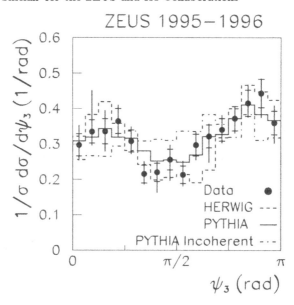

Fig. 8. The area-normalized distribution of ψ_3. The inner error bar shows the statistical error and the outer error bar shows the quadratic sum of the statistical and systematic uncertainties. The solid histogram shows the default PYTHIA prediction. The dashed and dot-dashed histograms show the predictions of HERWIG and of PYTHIA with colour coherence switched off

showers do a remarkably good job of simulating three-jet production as is evident from the agreement of the models with the data in the ψ_3 distribution. Within the PYTHIA model it is possible to switch off the simulation of QCD colour coherence. With no simulation of coherence, PYTHIA predicts a relatively uniform population of the ψ_3 distribution, as shown in Figure 8. Colour coherence in the parton shower model is required to describe the observed suppression of large angle radiation.

5   Jet Substructure

The k_T jet-finding algorithm [38,39] clusters objects into jets based on a distance parameter which is essentially their relative transverse momentum. Subjets may be defined within jets by applying the k_T algorithm to the particles of the jet and counting the subjets as a function of the resolution parameter, y_{cut}. For large values of y_{cut} there is only one subjet, the jet itself, but as y_{cut} decreases more and more subjets are resolved until the subjet multiplicity equals the multiplicity of particles within the jet.

The dependence of the average number of subjets, $\langle n_{\text{subjet}} \rangle$, on y_{cut} has been measured by the ZEUS collaboration for an inclusive sample of jets with $E_T^{\text{jet}} > 15$ GeV [40]. Both the HERWIG and PYTHIA models provide a good description of the evolution of $\langle n_{\text{subjet}} \rangle$ with y_{cut}. Measurements of $\langle n_{\text{subjet}} \rangle$ for $y_{\text{cut}} = 0.01$ have been performed in four different jet pseudorapidity regions, as presented in Figure 9. The average number of subjets increases as the jets

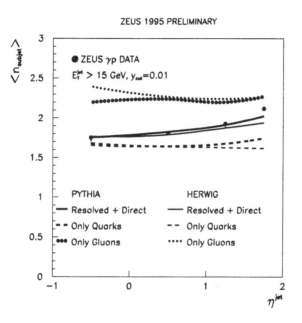

Fig. 9. The mean subjet multiplicity at a fixed value of $y_{\text{cut}} = 0.01$ as a function of η^{jet}. The error bars show the statistical and systematic uncertainties added in quadrature. For comparison, the predictions of PYTHIA including resolved plus direct processes for quark jets (*thick dashed line*), gluon jets (*thick dotted line*), and all jets (*thick solid line*) are shown. The predictions of HERWIG are displayed with thin lines

move toward the incoming proton direction. This behaviour is well described by the PYTHIA and HERWIG models. In the models the dominant leading order direct process is $\gamma g \to q\bar{q}$ while the dominant resolved process is $qg \to qg$. Therefore, relatively more gluon jets are expected for the more forward boosted resolved photon processes. Moreover, the gluon in the $qg \to qg$ subprocess has a tendency to be the more forward parton, further increasing the gluonic content of forward jets. The fundamental expectation of QCD that gluons, which have a higher colour charge than quarks, should yield a higher multiplicity of hadrons, is borne out in the models by a higher average subjet multiplicity for gluon jets than for quark jets. Thus the increase of $\langle n_{\text{subjet}} \rangle$ with η^{jet} may be understood to arise from an increasing admixture of gluon jets in the forward direction. We look forward to an eventual comparison of this data with perturbative QCD calculations.

6   Summary

Hard photoproduction events have been used in a variety of analyses in order to further the understanding of the physics of strong interactions. From studies of the structure of the real photon it has been shown that in a leading order interpretation of the data the gluon density rises as the momentum fraction of the photon accessed in the hard subprocess decreases. There is also an indication that the quark densities of the real photon may be underestimated at high momentum fractions and for high values of the probing energy scale. A global fit of $e\gamma$ and γp measurements should be undertaken to discover whether a parton density can be found which describes all the data. Studies of the structure of the virtual photon have illustrated the expected suppression in the photon's structure as its lifetime decreases. Measurements sensitive to the underlying QCD dynamics have shown that $\mathcal{O}(\alpha\alpha_s^2)$ perturbative QCD matrix elements, and models with $\mathcal{O}(\alpha\alpha_s)$ matrix elements together with parton showers, are successful in explaining the mechanisms of high mass dijet and three-jet production. A measurement of jet substructure has shown the sensitivity of jets produced in hard photoproduction processes to quark and gluon jet differences and revealed their potential to provide insight into the physics of hadronization. Results from photoproduction at HERA have progressed from simple manifestations of the photon's hadronic structure, to detailed investigations of that structure and of QCD in general. As the yearly luminosity deliverable by HERA continues to increase, and as the dialogue between theorists and experimentalists is continuously strengthened, one may look forward to an even greater variety and quality of physics result to emerge from the investigation of hard photoproduction at HERA.

References

1. ZEUS Collaboration, M. Derrick et al.: Phys. Lett. B **293**, 465 (1992)
2. H1 Collaboration, T. Ahmed et al.: Phys. Lett. B **314**, 436 (1993)
3. H1 Collaboration, I. Abt et al.: Phys. Lett. B **322**, 287 (1994)
4. ZEUS Collaboration, M. Derrick et al.: Phys. Lett. B **322**, 287 (1994)
5. H1 Collaboration, T. Ahmed et al.: Phys. Lett. B **297**, 205 (1993)
6. ZEUS Collaboration, J. Breitweg et al.: *DESY 99-057*
7. G. Marchesini et al., Comp. Phys. Commun. **67**, 465 (1992)
8. G. Marchesini et al., hep-hp/9607393
9. H.U. Bengtsson and T. Sjöstrand, Comp. Phys. Commun. **46**, 43 (1987)
10. T. Sjöstrand, Comp. Phys. Commun. **82**, 74 (1994)
11. ZEUS Collaboration, M. Derrick et al.: Phys. Lett. B **348**, 665 (1995)
12. H1 Collaboration, T. Ahmed et al.: Nucl. Phys. B **445**, 195 (1995)
13. ZEUS Collaboration, M. Derrick et al.: Phys. Lett. B **384**, 401 (1996)
14. H1 Collaboration, C. Adloff et al.: *DESY 98-205*
15. H1 Collaboration, C. Adloff et al.: Eur. Phys. J. **C1**, 97 (1998)
16. ZEUS Collaboration, J. Breitweg et al.: Eur. Phys. J. **C1**, 109 (1998)
17. ZEUS Collaboration, J. Breitweg et al.: Eur. Phys. J. **C5**, 41 (1998)
18. H1 Collaboration, C. Adloff et al.: *DESY 98-148* to appear in Eur. Phys. J. C
19. ZEUS Collaboration, J. Breitweg et al.: Eur. Phys. J. **C6**, 67 (1999)

20. ZEUS Collaboration, J. Breitweg et al.: Eur. Phys. J. **C8**, 367 (1999)
21. M.H. Seymour: Nucl. Phys. B **421**, 545 (1994)
22. J. Cvach for the H1 Collaboration: 'Real and Virtual Photon Structure from Di-jet events'. In: *Proc. 7th International Workshop on Deep-Inelastic Scattering and QCD (DIS99) Zeuthen, Germany, 1999*
23. R. Engel: Z. Phys. C **66**, 203 (1995)
24. R. Engel and J. Ranft: Phys. Rev. D **54**, 4244 (1996)
25. M. Glück, E. Reya and A. Vogt: Phys. Rev. D **46**, 1973 (1992)
26. M. Glück, E. Reya and A. Vogt: Z. Phys. C **53**, 127 (1992)
27. M. Glück, E. Reya and A. Vogt: Phys. Rev. D **45**, 3986 (1992)
28. L.E. Gordon and J.K. Storrow: Nucl. Phys. B **489**, 405 (1997)
29. P. Aurenche, J. Guillet and M. Fontannaz: Z. Phys. C **64**, 621 (1994)
30. M. Klasen, *these proceedings*
31. H1 Collaboration: 'Study of the photon remnant in resolved processes at HERA' *Contributed Paper 157j, International Europhysics Conference on High Energy Physics (HEP99), Tampere, Finland, 1999*
32. ZEUS Collaboration, J. Breitweg et al.: Phys. Lett. B **413**, 201 (1997)
33. N. Macdonald for the ZEUS Collaboration: 'Structure of Real and Virtual Photons from ZEUS'. In: *Proc. 7th International Workshop on Deep-Inelastic Scattering and QCD (DIS99) Zeuthen, Germany, 1999*
34. G. Schuler and T. Sjöstrand: Phys. Lett. B **376**, 193 (1996)
35. G. Ingelman, A. Edin and J. Rathsman, Comp. Phys. Commun. **101**, 108 (1997)
36. ZEUS Collaboration: 'High-Mass Dijet Cross Sections in Photoproduction at HERA' *Contributed Paper 805, International Conference on High Energy Physics (ICHEP98), Vancouver, Canada, 1998*
37. ZEUS Collaboration, J. Breitweg et al.: Phys. Lett. B **443**, 394 (1998)
38. S. Catani, Yu.L. Dokshitzer, M.H. Seymour and B.R. Webber: Nucl. Phys. B **406**, 187 (1993)
39. S.D. Ellis and D.E. Soper: Phys. Rev. D **48**, 3160 (1993)
40. ZEUS Collaboration: 'Measurements of Jet Substructure in Photoproduction at HERA' *Contributed Paper 530, International Europhysics Conference on High Energy Physics (HEP99), Tampere, Finland, 1999*

Photon Structure and the Production of Jets, Hadrons, and Prompt Photons

Michael Klasen

HEP Theory Group, Argonne National Laboratory, Argonne, IL 60439, USA

Abstract. We give a pedagogical introduction to hard photoproduction processes at HERA, including the production of jets, hadrons, and prompt photons. Recent theoretical developments in the three areas are reviewed.

1 Introduction

Electron-proton scattering at HERA proceeds dominantly through the exchange of a single photon with small virtuality $Q \simeq 0$. In this photoproduction limit and in the presence of a hard factorization scale $M_{\gamma,p}$, the electron-proton scattering cross section can be decomposed into

$$
\mathrm{d}\sigma_{ep} = \sum_{a,b} \int_0^1 \mathrm{d}y F_{\gamma/e}(y, Q_{\max}^2) \int_0^1 \mathrm{d}x_\gamma F_{a/\gamma}(x_\gamma, M_\gamma^2) \int_0^1 \mathrm{d}x_p F_{b/p}(x_p, M_p^2) \mathrm{d}\sigma_{ab}^{(n)}.
\tag{1}
$$

The Weizsäcker-Williams spectrum of photons in the electron

$$
F_{\gamma/e}(y, Q_{\max}^2) = \frac{\alpha}{2\pi} \left[\frac{1 + (1-y)^2}{y} \ln \left(\frac{Q_{\max}^2}{Q_{\min}^2} \right) + 2 m_e^2 y \left(\frac{1}{Q_{\max}^2} - \frac{1}{Q_{\min}^2} \right) \right]
\tag{2}
$$

is proportional to the electromagnetic coupling constant α. Since the photon virtuality is small, the spectrum can be calculated explicitly by exploiting current conservation and integrating over the unobserved azimuthal angle of the outgoing electron and over the virtuality of the photon. $Q_{\min}^2 = m_e^2 y^2/(1-y)$ depends on the electron mass m_e and the longitudinal momentum fraction y of the photon in the electron. $Q_{\max}^2 = E_e^2(1-y)\theta^2$ is determined experimentally from the incoming electron beam energy E_e, the momentum fraction y, and the scattering angle of the outgoing electron θ.

From deep inelastic scattering (DIS) experiments of virtual photons off protons it is well known that the proton has point-like constituents, quarks q and gluons g, also called partons. As the photon virtuality decreases, the photon itself begins to fluctuate into quark-antiquark ($q\bar{q}$) pairs, which in turn evolve into a vector meson-like structure. At HERA, an almost real photon radiated from the electron can thus interact either directly with the partons in the proton (direct component) or act as a hadronic source of partons which collide with the partons in the proton (resolved component). In the latter case, one does not test the proton structure alone but also the photon structure. Unfortunately, both the photon and the proton structure cannot be calculated theoretically, but have to be determined experimentally.

Before HERA started taking data, information on the parton densities in the photon $F_{a/\gamma}$ came almost exclusively from deep inelastic $\gamma^*\gamma$ scattering at e^+e^- colliders. Whereas the singlet quark densities can be well constrained in this process, it is difficult to obtain information on the gluon density, which is suppressed relative to the quark densities by the strong coupling constant α_s. The gluon density can, however, be constrained from hard photoproduction processes at HERA if the direct contribution is suppressed by measuring the production of jets, hadrons, or prompt photons at low transverse energies E_T. This constitutes an important goal in studies of hard photoproduction.

Among the parton densities in the proton $F_{b/p}$, the quark densities and the gluon density at low x_p are rather well known today from DIS experiments. Of particular interest in photoproduction is the gluon distribution in the proton at large x_p. There, the experimental information from high-E_T jet and prompt photon production in hadron collisions is inconclusive and has considerable potential for improvement from lepton pair production in hadron collisions and from high-E_T photoproduction processes. At large E_T, the direct photoproduction process dominates and uncertainties from the photon structure are suppressed.

In the presence of a hard renormalization scale $\mu \simeq E_T$, $\alpha_s(\mu^2)$ becomes small and the partonic cross section

$$\mathrm{d}\sigma_{ab}^{(n)} = \alpha_s^0(\mu^2)\mathrm{d}\sigma_{ab}^{(0)} + \alpha_s^1(\mu^2)\mathrm{d}\sigma_{ab}^{(1)} + \dots \tag{3}$$

for the scattering of two partons a and b, which can be quarks, gluons, and – in the photon case – also photons, into jets, hadrons, and prompt photons can be calculated in perturbative quantum chromodynamics (QCD). Testing these predictions constitutes a third goal for photoproduction experiments.

Typical leading order (LO) QCD diagrams for the direct and resolved photoproduction of jets are shown in the first line of Fig. 1. Next-to-leading order (NLO) QCD corrections to these graphs arise from virtual loop and real emission diagrams and have been calculated in [1]. The direct and resolved partonic processes are of $\mathcal{O}(\alpha\alpha_s)$ and $\mathcal{O}(\alpha_s^2)$, respectively. Since the parton densities in the photon are of $\mathcal{O}(\alpha/\alpha_s)$ in the asymptotic limit of large M_γ^2, both processes contribute at $\mathcal{O}(\alpha\alpha_s)$. In addition to the goals of photoproduction studies mentioned above, the observation of jets also permits to test and improve jet algorithms of both the cone and cluster types and to study jet profiles and the internal jet structure.

The LO partonic QCD diagrams for photoproduction of hadrons (second line in Fig. 1) are identical to those for the photoproduction of jets, and the NLO corrections for inclusive hadron production have been calculated in [2]. However, in this case the hadronic cross section

$$\mathrm{d}\sigma_{ab\to h}^{(n)} = \int_0^1 \frac{\mathrm{d}z}{z^2} D_{c/h}(z, M_h^2)\mathrm{d}\sigma_{ab\to c}^{(n)} \tag{4}$$

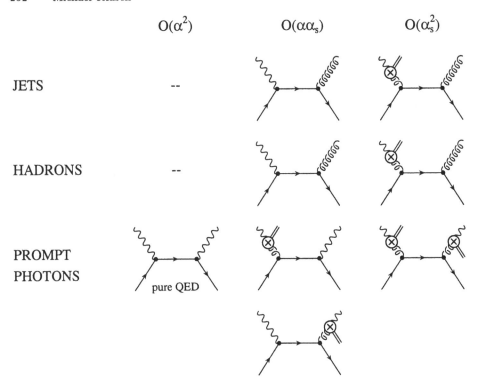

Fig. 1. Typical leading order QCD diagrams for the direct and resolved photoproduction of jets, hadrons, and prompt photons by order of the partonic scattering process. The $\mathcal{O}(\alpha^2)$ process is only present for the photoproduction of prompt photons

is obtained from the partonic cross section $d\sigma_{ab\to c}$ by convoluting it with a fragmentation function $D_{c/h}$. Like parton distribution functions, fragmentation functions cannot be derived from first principles, and photoproduction of hadrons at HERA offers the additional possibility to test hadron fragmentation functions obtained, *e.g.*, in fits to data from e^+e^- colliders.

Like initial state photons, final state (prompt) photons couple to the hard partonic scattering process either directly or through their partonic constituents. The direct (lower left diagram in Fig. 1), single-resolved (center), and double-resolved (right) prompt photon processes contribute at different orders of the hard scattering process, but eventually the photon structure and fragmentation functions compensate for these differences so that all three subprocess types are effectively of $\mathcal{O}(\alpha^2)$. Comparable to hadron fragmentation functions, the partonic content of final state photons is described by a photon fragmentation function $D_{c/\gamma}$. Photoproduction of isolated photons and photons in association with jets has been calculated in NLO QCD in [3]. Since prompt photon production is a purely electromagnetic process in LO, its cross section is smaller than the inclusive jet and hadron cross sections, and data from H1 and ZEUS have only recently become available.

2 Jets

At the first Ringberg workshop on "New Trends in HERA Physics" in 1997, photoproduction of jets had already been studied for five years. Several NLO single jet and the first NLO dijet calculations had been compared to data from H1 and ZEUS [4]. Studies of jet cross sections [5] and jet shapes [6] had demonstrated that the commonly used Snowmass jet cone algorithm [7] suffered from theoretical (double counting, parton distance) and experimental (seed finding, jet overlap) ambiguities, which could only be resolved by the introduction of a phenomenological parton distance parameter R_{sep} in addition to the cone size parameter R.

These ambiguities were found to be absent in the k_T-clustering algorithm in the inclusive mode [8], which has been used by the HERA experiments since then. It combines two hadronic clusters i and j if their distance

$$d_{ij} = \min(E_{T,i}^2, E_{T,j}^2)[(\eta_i - \eta_j)^2 + (\phi_i - \phi_j)^2]/R^2 \tag{5}$$

is smaller than E_T^2. The parameter $R = 1$ now corresponds to the parton distance parameter R_{sep} in NLO QCD. As in the cone algorithm, the transverse energy E_T, rapidity η, and azimuthal angle ϕ of the combined cluster are calculated from the transverse energy weighted sums of the two pre-clusters.

Three different NLO dijet calculations have recently been compared to ZEUS dijet data (see contribution by L. Sinclair to these proceedings) [9]. The definition of the ZEUS dijet cross section has also served as a basis to compare the theoretical predictions among themselves [10]. As can be seen in Fig. 2, the NLO transverse energy distributions agree with each other within the statistical accuracy of the Monte Carlo integration, which is about $\pm 1\%$ at low E_T for the full y range and $\pm 2\%$ for the high y range.

NLO dijet calculations involve the calculation of one-loop $2 \to 2$ and tree-level $2 \to 3$ scattering matrix elements. The latter are then integrated over soft and collinear regions of phase space in order to cancel the infrared divergences arising from the virtual corrections. If used in their original, unintegrated form, the $2 \to 3$ matrix elements can also be used for LO predictions of three jet cross sections. The three NLO dijet predictions mentioned above have also been applied to LO three jet cross sections and found to agree with each other [10]. ZEUS have measured the photoproduction of three jets with transverse energies larger than 6 GeV (two highest jets) and 5 GeV (third jet) and rapidities $|\eta| < 2.4$ [11]. Fig. 3 shows that the total theoretical prediction (full curve) describes the data well in shape and normalization [12]. The agreement in normalization is, however, to some degree coincidental since there is still an uncertainty of a factor of two from the variation of the scales $\mu = M_\gamma = M_p \in [0.5; 2.0] \times \max(E_{T,1}, E_{T,2}, E_{T,3})$ (shaded band). In these LO QCD predictions every parton corresponds to a jet, but a jet definition still has to be implemented for the

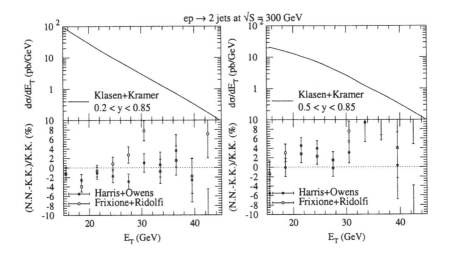

Fig. 2. Comparison of three theoretical predictions for the NLO dijet cross section as a function of the transverse energy E_T of the leading jet for the full (left) and high (right) y range. Both jets lie in a central rapidity range $0 < \eta_{1,2} < 1$

condition that two partons cannot come closer too each other than permitted by the jet algorithm.

H1 have recently analyzed the production of dijets with transverse momentum $p_T > 4$ and 6 GeV, dijet mass $M_{2-\text{jet}} > 12$ GeV, rapidities $-0.5 < \eta_{1,2} < 2.5$, $|\eta_1 - \eta_2| < 1$, and photon energy fraction $0.5 < y < 0.7$ as a function of the variable

$$x_\gamma^{\text{jets}} = \sum_{i=1}^{2} E_{T,i} e^{-\eta_i} / (y\sqrt{S}). \tag{6}$$

This variable can be determined from the two observed jets and is, in LO QCD, directly related to the momentum fraction of the parton in the photon. H1 have used this measurement of the dijet cross section to extract the gluon density in the photon with the result shown in Fig. 4. The 1996 data [13] are consistent with older H1 charged track [14] and jet [15] measurements. They agree very well with the GRV parametrization [16], but rule out both the LAC1 [17] and LAG parametrizations [18] of the gluon density. The systematic (outer) error bars include theoretical uncertainties coming from the variation of different Monte Carlo models, parton densities in the proton, and quark densities in the photon.

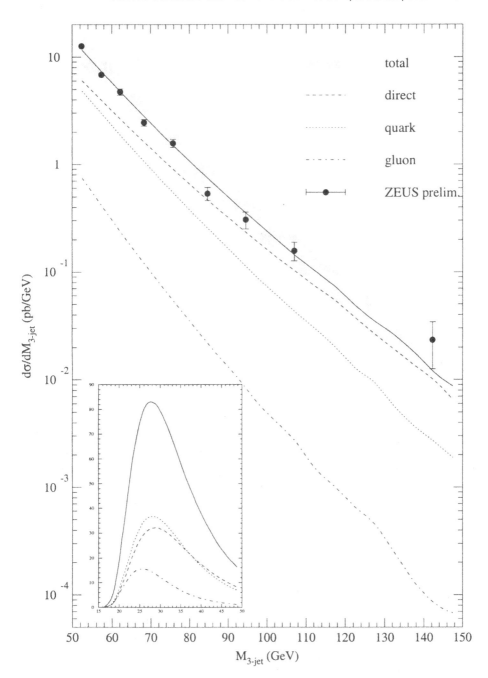

Fig. 3. Total cross section (full curve) for the photoproduction of three jets as a function of the three jet mass M_{3-jet}. We also show the variation of the absolute normalization due to the uncertainty in the scale choice (shaded band) and the contributions from direct photons (dashed), quarks (dotted), and gluons (dot-dashed) in the photon. The ZEUS data [11] agree well with the QCD prediction [12]

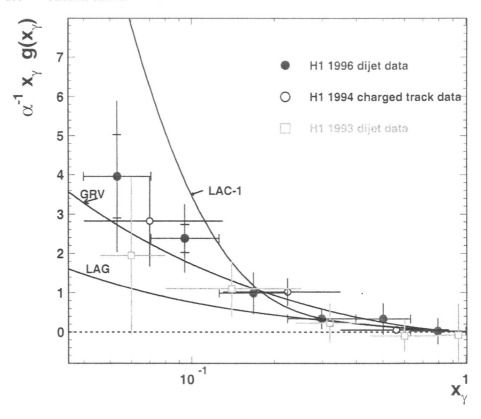

Fig. 4. H1 extraction of the gluon density in the photon from 1996 dijet data [13], 1994 charged track data [14], and 1993 dijet data [15]. The data are consistent with each other and with LO GRV parametrization [16]. The 1996 data rule out both the LAC1 [17] and LAG [18] gluon parametrizations

3 Hadrons

The transformation of quarks and gluons into hadronic final states can be described globally by jet definitions. More detailed experimental information about the hadronization process can be obtained in the production of single hadrons, which is described theoretically by fragmentation functions. Prior to the last Ringberg HERA workshop, several new NLO QCD fits of fragmentations functions for charged pions, charged kaons, and neutral kaons to e^+e^- data from TPC, ALEPH, and Mark II had been performed. They had been applied to a NLO QCD calculation for photoproduction and successfully been compared to HERA data [19]. Since then, fragmentation functions for heavy particles, such as $D^{*\pm}$ and B mesons, have been fitted to ALEPH and OPAL data, and they also compare favorably to HERA data (see contribution by G. Kramer to these proceedings). A new fit of charged pion, charged kaon, and proton fragmentation functions, separately for light flavors, heavy flavors, and gluons, to new LEP and SLC data is currently in progress [20].

In Fig. 5 we compare NLO QCD predictions for the rapidity distribution

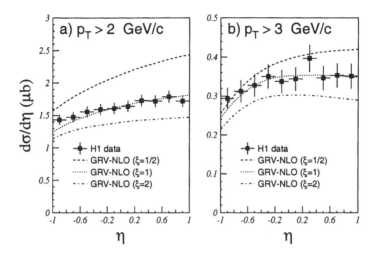

Fig. 5. Differential cross section for the photoproduction of charged hadrons as a function of rapidity for two different transverse momentum cuts. Recent H1 data [14] are compared to NLO QCD predictions [21] with three different scale choices

of charged hadrons [21] to recent H1 data [14]. Good agreement is found for the GRV parton densities in the photon [16] and the proton [22] and a central scale choice of $\mu = M_\gamma = M_p = p_T$, but the theoretical scale uncertainty in NLO QCD is still considerable. It is worth noting that in inclusive hadron production, perturbative QCD works remarkably well down to very low transverse momenta of $p_T > 2$ or 3 GeV. There is no excess in the forward η region as observed in jet photoproduction. Thus, inclusive hadron production offers the potential to extract the gluon density in the photon in low-p_T (and therefore low-x_γ) cross section measurements.

It is also interesting to check the universality of fragmentation functions and the factorization theorem in photoproduction experiments. In Fig. 6 we show a recent ZEUS fragmentation function measurement for charged hadrons [23], measured in photoproduced jets with $E_T > 8$ GeV, $|\eta| < 0.5$, and $R = 1$, which agrees nicely with the NLO QCD fit to e^+e^- data [21]. The error band in the NLO QCD fit comes from a variation of the fragmenting quark flavor, which is unknown in the ZEUS analysis.

4   Prompt Photons

Prompt photon production was not discussed at the 1997 Ringberg HERA workshop – it had yet to be observed. Only recently have H1 and ZEUS reported

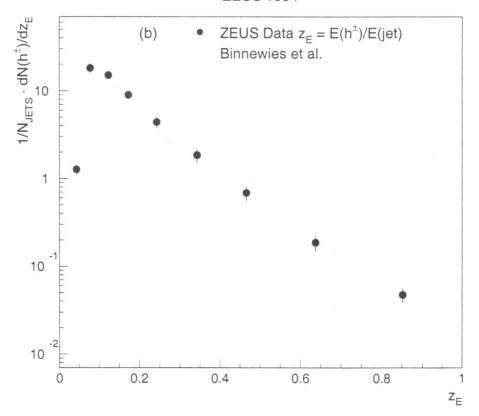

Fig. 6. Fragmentation function of charged particles as a function of their longitudinal momentum fraction. The ZEUS data [23] for particles within photoproduced jets are compared to a NLO fit to e^+e^- data [21]

results on this process [24,25] and have two NLO QCD calculations for isolated photon production and for the production of a photon in association with a jet become available [3,26]. These two calculations differ in their powercounting of the strong coupling constant and in their consideration of higher-order corrections. To illustrate this, we have listed in Fig. 7 the direct, single-resolved, and double-resolved diagrams for photoproduction of prompt photons. The single-resolved process includes either photon structure or photon fragmentation contributions, whereas in the double-resolved case the initial and final state photons contribute both through their partonic constituents. Gordon counts the photon structure and fragmentation functions as $\mathcal{O}(\alpha/\alpha_s)$ from considering the asymptotic limit and calculates the NLO corrections to all three types of subprocesses. Krawczyk *et al.* count the photon structure as $\mathcal{O}(\alpha)$, calculate the NLO corrections only to the direct process, and in addition take into account the box

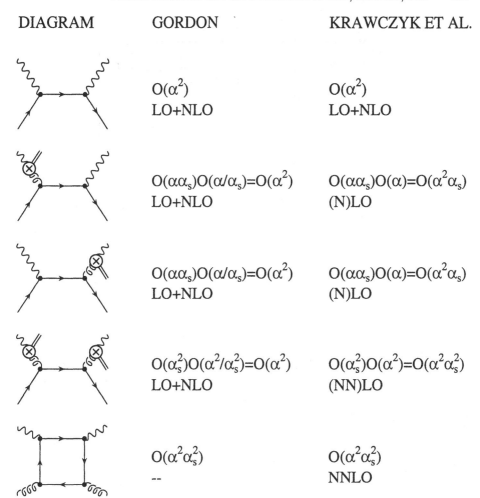

| DIAGRAM | GORDON | KRAWCZYK ET AL. |

$O(\alpha^2)$
LO+NLO

$O(\alpha^2)$
LO+NLO

$O(\alpha\alpha_s)O(\alpha/\alpha_s)=O(\alpha^2)$
LO+NLO

$O(\alpha\alpha_s)O(\alpha)=O(\alpha^2\alpha_s)$
(N)LO

$O(\alpha\alpha_s)O(\alpha/\alpha_s)=O(\alpha^2)$
LO+NLO

$O(\alpha\alpha_s)O(\alpha)=O(\alpha^2\alpha_s)$
(N)LO

$O(\alpha_s^2)O(\alpha^2/\alpha_s^2)=O(\alpha^2)$
LO+NLO

$O(\alpha_s^2)O(\alpha^2)=O(\alpha^2\alpha_s^2)$
(NN)LO

$O(\alpha^2\alpha_s^2)$
--

$O(\alpha^2\alpha_s^2)$
NNLO

Fig. 7. Diagrammatic comparison of two prompt photon QCD calculations. Gordon calculates the NLO corrections to the direct, single-resolved, and double-resolved contributions [3], whereas Krawczyk *et al.* only calculate the corrections to the direct contribution [26]. On the other hand, Krawczyk *et al.* take into account the box diagram that arises at NNLO

diagram in the last line of Fig. 7. Although the box diagram is formally of NNLO, it is known to have a large numerical contribution. This is demonstrated in Fig. 8 where we compare the calculation by Krawczyk *et al.* using GRV photon [16] and proton [22] structure and photon fragmentation functions [27] to ZEUS data for the production of isolated photons with $E_T > 5$ GeV [28]. With higher luminosity and the ensuing better accuracy of the data, prompt photon production at HERA will eventually provide useful tests of photon structure and fragmentation functions.

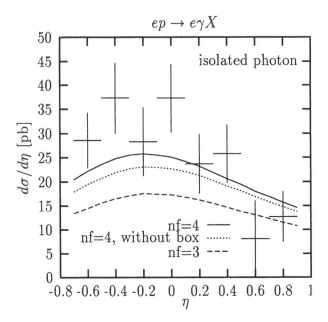

Fig. 8. Differential cross section for the photoproduction of a photon of transverse energy $E_T \in [5; 10]$ GeV, which is isolated in a cone with radius $R = 1$ and hadronic transverse energy fraction $\epsilon = 0.1$, as a function of rapidity. The ZEUS data [28] are compared to QCD predictions with and without the NNLO box diagram and with three and four quark flavors [26]

5 Summary

In summary, hard photoproduction processes can provide very useful information on the hadronic structure of the photon, in particular on the gluon density, which is complimentary to the information coming from deep inelastic photon-photon scattering at electron-positron colliders. Among the different hadronic final states, jets are most easily accessible experimentally and phenomenologically. On the other hand, inclusive hadron production offers the possibility to test the universality of hadron fragmentation functions and measure the photon structure down to very low values of p_T and x_γ. Prompt photon production suffers from a reduced cross section and limited data, but allows for the additional testing of photon fragmentation functions.

Acknowledgment. I would like to thank the organizers of the Ringberg workshop for the kind invitation. This work has been supported by the U.S. Department of Energy under Contract W-31-109-ENG-38.

References

1. M. Klasen, T. Kleinwort and G. Kramer, Eur. Phys. J. Direct **C1** (1998) 1 and references therein
2. B.A. Kniehl and G. Kramer, Z. Phys. **C62** (1994) 53 and references therein
3. L.E. Gordon, Phys. Rev. **D57** (1998) 235 and references therein
4. M. Klasen, hep-ph/9706292 and references therein
5. J.M. Butterworth, L. Feld, M. Klasen and G. Kramer, hep-ph/9608481
6. M. Klasen and G. Kramer, Phys. Rev. **D56** (1997) 2702
7. J.E. Huth *et al.*, *Presented at Summer Study on High Energy Physics, Research Directions for the Decade, Snowmass, CO, Jun 25 - Jul 13, 1990*
8. S.D. Ellis and D.E. Soper, Phys. Rev. **D48** (1993) 3160
9. J. Breitweg *et al.* [ZEUS Collaboration], hep-ex/9905046
10. B.W. Harris, M. Klasen and J. Vossebeld, hep-ph/9905348
11. J. Breitweg *et al.* [ZEUS Collaboration], Phys. Lett. **B443** (1998) 394
12. M. Klasen, Eur. Phys. J. **C7** (1999) 225
13. J. Cvach [H1 Collaboration], hep-ex/9906012
14. C. Adloff *et al.* [H1 Collaboration], hep-ex/9810020
15. T. Ahmed *et al.* [H1 Collaboration], Nucl. Phys. **B445** (1995) 195
16. M. Glück, E. Reya and A. Vogt, Phys. Rev. **D46** (1992) 1973
17. H. Abramowicz, K. Charchula and A. Levy, Phys. Lett. **B269** (1991) 458
18. H. Abramowicz, E. Gurvich and A. Levy, Phys. Lett. **B420** (1998) 104
19. B.A. Kniehl, hep-ph/9709261 and references therein
20. B.A. Kniehl, G. Kramer, and B. Pötter, to be published
21. J. Binnewies, B.A. Kniehl and G. Kramer, Phys. Rev. **D52** (1995) 4947
22. M. Glück, E. Reya and A. Vogt, Z. Phys. **C53** (1992) 127
23. J. Breitweg *et al.* [ZEUS Collaboration], Eur. Phys. J. **C2** (1998) 77
24. J. Breitweg *et al.* [ZEUS Collaboration], Phys. Lett. **B413** (1997) 201
25. K. Muller, *In *Jerusalem 1997, High energy physics* 464-467*
26. M. Krawczyk and A. Zembrzuski, hep-ph/9810253
27. M. Glück, E. Reya and A. Vogt, Phys. Rev. **D48** (1993) 116
28. L.E. Sinclair [ZEUS Collaboration], hep-ex/9810044

Part IV

Heavy-Flavour Production

Measurement of Open b Production at HERA

Daniel Pitzl

DESY, Notkestrasse 85, D-22609 Hamburg, Germany

Abstract. H1 and ZEUS have reported the first measurements of open b photoproduction at HERA. Both experiments use semi-leptonic decays to enrich a two-jet sample with heavy quarks. The beauty fraction is extracted from a fit of the measured p_T^{rel} distribution with contributions from b, c and light quarks according to a leading order QCD simulation. The measured cross section for b quarks is found to be larger than the predictions from next-to-leading order perturbative QCD calculations.

1 Introduction

The production of heavy quarks at HERA is an important testing ground for perturbative QCD calculations, where the heavy quark mass provides a hard scale. The basic mechanism, in leading order, is photon-gluon fusion, where a photon emitted from the electron and a gluon from the proton generate a heavy quark-antiquark pair. The dominant contribution to the cross section is due to the exchange of an almost real photon, whose negative squared four-momentum $Q^2 = -q^2$ is small. In this photoproduction regime, a second production mechanism becomes important, where the photon fluctuates into a hadronic state which is resolved into its partonic composition in the hard collision. In next-to-leading order QCD, however, only the sum of direct and resolved processes is unambiguously defined [1].

Heavy quark production at HERA is dominated by charm, with a total cross section of about $1\,\mu$b. The prediction for open b production is 3.8 nb from the leading order AROMA [2] Monte Carlo event generator, which uses massive matrix elements and considers direct processes only. Predictions from next-to-leading order QCD calculations [3–5] are between 4.7 nb and 10 nb, depending on the mass of the b quark and the parton distribution functions used.

During the years 1994 to 1997 HERA provided positron-proton collisions at a center of mass energy $\sqrt{s} = 300$ GeV. In photoproduction the photon-proton center of mass energy is given by $W_{\gamma p} \approx \sqrt{ys}$, where the proton mass has been neglected, and the inelasticity y is given by $y = q \cdot P/l \cdot P$, with P and l for the four momenta of the incoming proton and positron, respectively.

In the analyses of both H1 and ZEUS heavy quarks are identified by their semileptonic decays, which suppresses the light quark background. The flight directions of the quarks are approximated by reconstructing two jets in the final state. The contributions from b and c quarks are distinguished statistically by exploiting the effects of the higher b quark mass on the kinematics of the final state.

2   H1 muon analysis

The first measurement of the open b photoproduction cross section at HERA was reported by the H1 collaboration [6]. The final analysis [7] uses a data sample corresponding to an integrated luminosity of $6.6\,\mathrm{pb}^{-1}$ collected in 1996. The virtuality of the exchanged photon was restricted to the range $Q^2 < 1\,\mathrm{GeV}^2$ by requiring that no scattered positron candidate was found in the electromagnetic liquid argon or scintillating-fibre calorimeters. The variable y was calculated from a combination of calorimetric and tracking information using the Jacquet-Blondel method [8] and was required to be in the range $0.1 < y < 0.8$, corresponding to $95\,\mathrm{GeV} < W_{\gamma p} < 269\,\mathrm{GeV}$. The event selection required at least two jets in the pseudorapidity range $|\eta| < 2.5$ with transverse energies $E_T > 6\,\mathrm{GeV}$. The jets were identified using a cone algorithm with a radius $r = \sqrt{(\Delta\eta)^2 + (\Delta\phi)^2} < 1$, where η is the pseudorapidity and ϕ is the azimuthal angle of the track or cluster.

At least one muon candidate with a transverse momentum $p_t > 2\,\mathrm{GeV/c}$ had to be identified in the instrumented iron system in the polar angle range $35° < \theta < 130°$. The muon candidate had to be found within one of the jets. This sample contains muons from semi-leptonic decays of b and c hadrons and background from light quark decays and from hadrons faking a muon signature in the detector. A detailed and extensive Monte Carlo simulation was performed to determine the probability $\mathcal{P}_h^\mu(p, \theta)$ that a hadron $h = \pi, K,$ or p produces a muon signature. These probabilities vary with momentum and polar angle but do not exceed 0.6% for pions, 2% for kaons and 0.2% for protons. They were tested using reconstructed decays of K_S^0 and ϕ mesons as sources of charged pions and kaons in the data. Figure 1 shows the distribution of pions from K_S^0 decays that produce a muon signature. It is well described, in shape and magnitude, by the muon fake probability for pions as obtained from the simulation. From the ϕ decays, eight kaons produced a muon signature, compared to an expectation of 7.8.

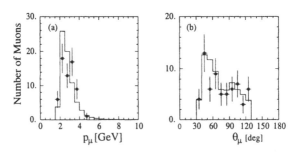

Fig. 1. Pions from K_S^0 decays producing a muon signature in the data (symbols with error bars) and the expectation from the muon fake probability $\mathcal{P}_\pi^\mu(p, \theta)$ (solid histogram).

The decay of a *b* hadron releases an energy of around 3 GeV and can involve decay particles with transverse momenta up to 2.3 GeV/c relative to the *b* hadron direction. The *b* hadron direction was experimentally approximated by the thrust axis of the jet, defined as the axis which maximizes

$$T = \max \frac{\sum_{i \neq \mu} |p_i^L|}{\sum_{i \neq \mu} |p_i|},$$

where the sum is over all particles in the jet except the muon. The transverse momentum of the muon relative to the thrust axis $p_{T,rel}^\mu$ was used to statistically separate *b*, *c* and light quark contributions in the sample. The $p_{T,rel}^\mu$ distribution for the 1996 H1 data is shown in figure 2. It has a maximum at about 0.7 GeV/c and a tail out to 4 GeV/c.

The $p_{T,rel}^\mu$ distribution of hadrons faking a muon signature was determined in shape and absolute normalization from a data sample of similar two-jet events, selected without the requirement of a muon candidate. The muon-fake probabilities $\mathcal{P}_h^\mu(p, \theta)$ were then applied to all particles in the jets, with a pion-kaon-proton ratio as given by the JETSET fragmentation model. The predicted background amounted to 25.9%. The expected shapes of the $p_{T,rel}^\mu$ distributions for *b* and *c* decays were taken from the AROMA Monte Carlo and the normalizations were determined from a fit to the data as $(50.8 \pm 4.9)\%$ for *b* and $(22.4 \pm 5.0)\%$ for *c*, where the errors are statistical only. The sum of the three contributions is in good agreement with the measured distribution.

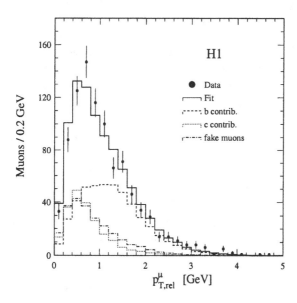

Fig. 2. The $p_{T,rel}^\mu$ distribution as measured by H1 (points with statistical error bars). The solid line is the sum of the fixed hadron background and the beauty and charm contributions from AROMA with normalizations from a fit to the data.

From the number of muons attributed to b events the cross section for open b production in the range $Q^2 < 1\,\mathrm{GeV}^2$, $0.2 < y < 0.8$, $p_T^\mu > 2\,\mathrm{GeV/c}$ and $35^o < \theta^\mu < 130^o$ was determined as

$$\sigma_{vis}(ep \to b\bar{b}X \to \mu Y) = \left(0.176 \pm 0.016\ (stat.)\ {}^{+0.026}_{-0.017}\ (syst.)\right)\ \mathrm{nb}.$$

This cross section contains an extrapolation to the full phase space for the two jets, where the AROMA Monte Carlo gives an acceptance of 42.9% for the requirement of two jets with $E_T > 6\,\mathrm{GeV}$. The systematic error contains contributions due to uncertainties in the muon reconstruction efficiency, the hadronic energy scale in the calorimeter, and the luminosity determination. The systematic error due to the light quark background was determined by varying the fractions for pion, kaon and proton production and by using different data samples. The analysis was also performed using the HERWIG [17] and RAPGAP [10] leading order Monte Carlo Generators, which also contain resolved photon contributions. The effects of fragmentation was studied by varying the parameter ϵ of the Peterson fragmentation function. The variation of the cross section with different simulations amounts to $\pm 7.1\%$ and is included in the systematic error. The cross section predicted by AROMA in leading order and for direct photon processes only is $0.038\,\mathrm{nb}$, where a b-quark mass of $4.75\,\mathrm{GeV/c}^2$ and the MRS-G [9] parton densities for the proton were used.

The measured cross section was compared to a next-to-leading order QCD calculation by the authors of [11]. Their program generates a final state at the parton level. In order to apply the cuts on the muon that define the visible cross section, the fragmentation and weak decay processes had to be added [7]. The b-quark mass was taken as $4.75\,\mathrm{GeV/c}^2$ and the MRS-G [9] and GRV-HO [12] parton densities for the proton and photon, respectively, were used. The cross section from the NLO QCD calculation in the kinematic range defined above was $(0.104 \pm 0.017)\,\mathrm{nb}$, where the uncertainty is due to the fragmentation process and a variation of the renormalization and factorization scales by a factor of two from the default value $\mu_R = \mu_F = \sqrt{m_b^2 + p_{t,b}^2}$. The NLO QCD prediction for the open b visible photoproduction cross section is 2.5 standard deviations below the measured value, where statistical, systematic and theoretical uncertainties were added in quadrature.

Nevertheless, the NLO QCD calculation [11] was used to extrapolate to the full kinematic phase space and to derive the total photoproduction cross section for b-quarks. The total branching ratio for a muon to be produced in the decay chain of b hadrons was applied. The Weizsäcker-Williams approximation was used to determine the equivalent flux of photons emitted by the positron. The average photon-proton center of mass energy is $\langle W_{\gamma p} \rangle \approx 180\,\mathrm{GeV}$. The measured cross section is

$$\sigma(\gamma p \to b\bar{b}X) = (111 \pm 10^{+16}_{-11} \pm 17)\ \mathrm{nb},$$

where the first and second error are the experimental statistical and systematic error and the third error is due to the extrapolation procedure and the uncertainty in the semileptonic branching ratio of B hadrons. The H1 measurement is shown in figure 3 together with the NLO QCD prediction and a measurement by EMC [13] at fixed target energy.

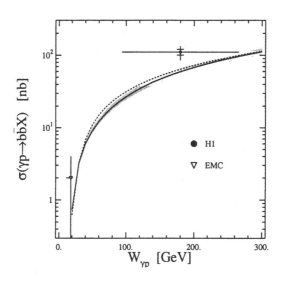

Fig. 3. The total photoproduction cross section $\sigma(\gamma p \rightarrow b\bar{b}X)$. The inner error bar is statistical while the outer error bar includes all systematic errors as explained in the text. The horizontal bar indicates the range of the measurement. The solid curve shows the NLO QCD calculation [11] and the shaded band indicates a change in the factorization scale by a factor of two. The dashed line shows the NLO prediction when the MRST [14] instead of the MRS-G [9] parton densities for the proton are used.

3   ZEUS electron analysis

The ZEUS collaboration has reported a measurement of the b production cross section using decays into electrons [15]. The data were collected during 1996 and 1997 when HERA was running with positrons. The integrated luminosity amounts to $36.9\,\text{pb}^{-1}$. The kinematic range in the analysis was restricted to the region $Q^2 < 1\,\text{GeV}^2$ and $0.2 < y < 0.8$.

The k_T clustering algorithm [16] was used to reconstruct jets and at least two jets with $E_T^{jet1} > 7\,\text{GeV}$ and $E_T^{jet2} > 6\,\text{GeV}$ in the pseudorapidity range $|\eta^{jet}| < 2.4$ were required.

Electrons were identified by using information from the central drift chamber and the uranium-scintillator calorimeter. A transverse momentum $p_T^{e^-} > 1.6\,\text{GeV/c}$ was required and the pseudorapidity was restricted to the range $|\eta^{e^-}| < 1.1$. The specific energy loss dE/dx in the drift chamber is about $1.4\,\text{mip}$

for electrons due to the relativistic rise, while hadrons are minimum ionizing. Tracks were linked to calorimeter clusters, where a cluster is called electron enriched if it has more than 90% of its energy deposited in the electromagnetic section. Conversely, a cluster is called hadron enriched if it has more than 60% of its energy deposited in the hadronic section. The hadron enriched sample contains essentially no electrons and can be used as a background sample.

Figure 4 shows the dE/dx spectra of the electron and hadron enriched samples. The shaded region $dE/dx < 1.1$ mips is used for normalization. The lower part of figure 4 shows the subtraction of the two samples with a clear electron signal. A background of electrons from photon conversions in the material in front of the drift chamber is determined from the data and the detector simulation and is also shown in figure 4.

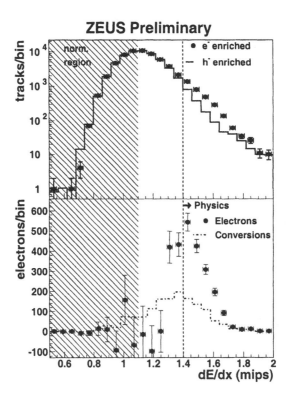

Fig. 4. (top) The dE/dx distributions for the electron enriched and the hadron enriched samples, normalized in the region dE/dx < 1.1 mips. (bottom) The subtraction shows the electron signal. The background from photon conversions is indicated by the dashed-dotted line.

The differential cross section $d\sigma/dp_{T,rel}$, where $p_{T,rel}$ is the transverse momentum of the electron relative to the closest jet, has been measured for the process $e^+ p \rightarrow$ dijet $e^- X$ and is shown in figure 5. The ZEUS data points show the statistical error as the inner error bar and the statistical and systematic errors added in quadrature as the outer bars. The uncertainty due to the ZEUS calorimeter energy scale is shown separately as the shaded band. Figure 5 also shows the prediction from the HERWIG [17] leading order Monte Carlo program for beauty production (17%), for charm and light quarks (83%) and their sum. The overall rate was scaled by a factor of 3.7 in order to match the measured cross section. The HERWIG program was used with the default heavy quark masses ($m_c = 1.55\,\text{GeV}/c^2$ and $m_b = 4.95\,\text{GeV}/c^2$) and with the CTEQ-4D [18] and GRV-LO [19] structure functions for the proton and photon, respectively. Allowing the contributions for beauty and charm plus light flavours to vary, a fraction of $(20 \pm 6\ (stat.)\ ^{+12}_{-7}\ (syst.)\ )\%$ for beauty production was determined. Using this value, the beauty production cross section, measured in the kinematic range $Q^2 < 1\,\text{GeV}^2$, $0.2 < y < 0.8$, $E_T^{jet1} > 7\,\text{GeV}$, $E_T^{jet2} > 6\,\text{GeV}$, $|\eta^{jet}| < 2.4$, $p_T^{e^-} > 1.6\,\text{GeV}/c$, and $|\eta^{e^-}| < 1.1$, was determined as

$$\sigma_{b\bar{b}}^{vis}(e^+\, p \to \text{dijet}\ e^-\ X) = \left(39 \pm 11\ (stat.)\ ^{+23}_{-16}\ (syst.)\ \right) \text{pb, ZEUS preliminary}$$

The measured $b\bar{b}$ cross section is compared to leading order QCD Monte Carlo

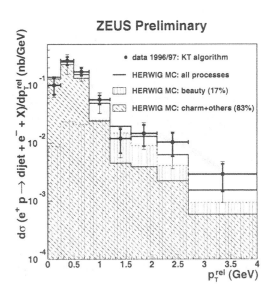

Fig. 5. The differential cross section in $p_{T,rel}$ for dijet events with an electron. The ZEUS data points are compared to the HERWIG Monte Carlo, which was scaled by a factor 3.7 to match the data. A beauty fraction of 17% is the HERWIG prediction.

predictions in figure 6 as a function of the b-quark mass assumed in the simulation. The HERWIG cross section decreases with mass and shows only a small dependence on the choice of the proton and photon structure function. The PYTHIA [20] program contains massless matrix elements and was used with the

GRV94-LO and GRV-LO [19] parton distributions functions for the proton and photon, respectively. It predicts a higher $b\bar{b}$ cross section than HERWIG. Due to the large errors, the present measurement cannot rule out either prediction.

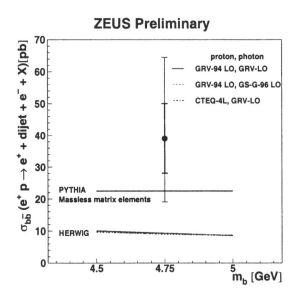

Fig. 6. The measured cross section for $b\bar{b}$ production with two jets and an electron in the final state, in the kinematic range specified in the text, compared to the HERWIG Monte Carlo with massive matrix elements and different proton and photon parton densities and to the PYTHIA Monte Carlo with massless matrix elements.

Further insight in the production mechanism for heavy quarks can be gained by measuring the differential cross section in the variable

$$x_\gamma^{obs} = \frac{\sum_{2jets} E_T^{jet}\, e^{-\eta^{jet}}}{2yE_e},$$

where E_e is the energy of the incoming positron. It is the fraction of the photon's momentum contributing to the production of the two jets. Direct processes involving pointlike photons are observed at $x_\gamma^{obs} \approx 1$, whereas resolved processes involving the partonic structure of the photon lead to smaller values.

Figure 7 shows the measured cross section $d\sigma/dx_\gamma^{obs}$ for events with two jets and an electron in the final state. It has a maximum at high x_γ^{obs} and a significant tail at low x_γ^{obs} which cannot be explained by direct processes and resolution effects alone. The HERWIG Monte Carlo predicts a 40% contribution from resolved photon processes, which, when added to the directed photon processes, provide a good description of the shape of the measured cross section. The overall normalization was scaled by a factor 3.7 to match the data. Fitting the ratio of resolved to direct photon processes yields a resolved contribution of $(35 \pm 6 \ (stat.))\%$ which is consistent with the HERWIG prediction and with the value measured in charm production [21]. A similar result is obtained when the measured cross section is compared to the PYTHIA Monte Carlo generator, which predicts a resolved contribution of 48%.

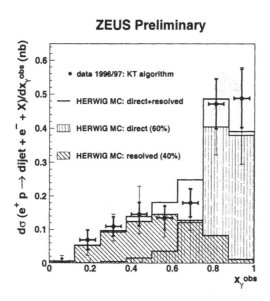

Fig. 7. The differential cross section in x_γ^{obs} for dijet events with an electron. The ZEUS data points are shown with statistical errors and with statistical and systematic errors added in quadrature. The shaded band shows the uncertainty due the calorimeter energy scale separately. The HERWIG prediction for direct and resolved contributions was scaled by a factor 3.7 to match the data.

4 Conclusions

The first measurements of open *b* photoproduction have been reported by H1 and ZEUS. Both experiments find cross sections that are about a factor of 4 larger than the expectations from leading order Monte Carlo Generators with massive matrix elements. The cross sections are measured in a visible range that involves the kinematics of the decay lepton. A comparison to next-to-leading order QCD calculations at the parton level requires an extrapolation to the full kinematic range or the addition of fragmentation and weak decays to the calculations. Both approaches add further systematic errors. The second approach has been followed by H1, which finds a cross section significantly larger than the NLO expectation.

The *b* purity is 50% for the H1 muon analysis and 20% for the ZEUS electron analysis. A higher purity can be obtained by using lifetime information from existing or future silicon vertex detectors. It would allow a study of the resolved photon contribution to b photoproduction.

At higher Q^2 the present data samples should allow the first observation of open b production in deep inelastic scattering. A significant measurement of the structure function $F_2^b(x, Q^2)$, however, requires the increased luminosity expected from the HERA upgrade.

References

1. S. Frixione, M.L. Mangano, P. Nason, G. Ridolfi in *Heavy Flavours II*, ed. by A.J. Buras and M. Lindner, World Scientific, 1998
2. G. Ingelman, J. Rathsman, G.A. Schuler, Comp. Phys. Comm. 101 (1997) 135
3. R.K. Ellis, P. Nason, Nucl. Phys. B312 (1989) 551
4. J. Smith, W.L. van Neerven, Nucl. Phys. B374 (1992) 36
5. S. Frixione et al., Phys. Lett. B348 (1994) 633
6. U. Langenegger, to appear in *QCD and High Energy Hadronic Interactions. Proceedings, 33rd Rencontres de Moriond*, Les Arcs, France, March 1998
   G. Tsipolitis, in *Deep Inelastic Scattering and QCD. Proceedings, 6th International Workshop, DIS'98*, Brussels, Belgium, April 4–8, 1998, ed. by G. Coremans and R. Roosen, World Scientific, 1998
   H1 Collaboration, contributed paper no. 575 to the 29th International Conference on High Energy Physics, ICHEP'98, Vancouver, Canada, July 23-29, 1998
7. H1 Collaboration, C. Adloff et al., Phys. Lett. B467 (1999) 156
8. F. Jacquet, A. Blondel in *Proceddings of the Study of an ep facility for Europe*, ed. by U. Amaldi, DESY 79-48 (1979) 391
9. A.D. Martin, R.G. Roberts, W.J. Stirling, Phys. Lett. B354 (1995) 155
10. H. Jung, Comp. Phys. Comm. 86 (1995) 147
11. S. Frixione, M.L. Mangano, P. Nason, G. Ridolfi, Phys. Lett. B348 (1995) 348
12. M. Glück, E. Reya, A. Vogt, Phys. Rev. D46 (1992) 1973
13. EMC Collaboration, J.-J. Aubert et al., Phys. Lett. B106 (1981) 419
14. A.D. Martin, R.G. Roberts, W.J. Stirling, R.S. Thorne, Eur. Phys. J. C4 (1998) 463
15. M. Hayes, in *34th Rencontres de Moriond on QCD and Hadronic Interactions*, March 20-27, 1999, Les Arcs, France, and hep-ex/9905033
    M. Wing, in *7th International Workshop on Deep Inelastic Scattering and QCD*, April 19-23, 1999 , Zeuthen, Germany, and hep-ex/9905051
    ZEUS Collaboration, contributed paper no. 498 to the International Europhysics Conference on High Energy Physics, Tampere, Finland, July 1999
16. S. Catani, Y.L. Dokshitzer, B.R. Webber, Phys. Lett. B285 (1992) 291
17. G. Marchesini et al., Comp. Phys. Comm. 67 (1992) 465
18. H.L. Lai et al., Phys. Rev. D55 (1997) 1280
19. M. Glück, E. Reya, A. Vogt, Phys. Rev. D45 (1992) 3986
20. H.-U. Bengtsson, T. Sjöstrand, Comp. Phys. Comm. 46 (1987) 43
21. ZEUS Collaboration, J. Breitweg et al., Eur. Phys. J. C6 (1999) 67

Open Heavy-Flavour Photoproduction at NLO

Gustav Kramer

II. Institut für Theoretische Physik, Universität Hamburg, Luruper Chaussee 149, D-22761 Hamburg, Germany

Abstract. We review the construction of non-pertubative fragmentation functions for $D^{*\pm}$ mesons, both at leading and at next-to-leading order in two different subtraction schemes in the so-called massless approach. The fragmentation functions are determined by fitting to OPAL and ALEPH data on inclusive $D^{*\pm}$ production in e^+e^- annihilation. It is shown, how recent data on inclusive $D^{*\pm}$ photoproduction in ep collisions at HERA are reproduced with two sets of fragmentation functions differing in the subtraction schemes. The influence of different assumptions on the charm distribution of the photon is pointed out.

1 Introduction

The production of heavy-flavoured hadrons (with charm or bottom) in the colliding beam experiments with HERA provides new opportunities to study the dynamics of perturbative QCD and to obtain information on the proton and photon structure. Heavy-flavour production at HERA is dominated by photoproduction events, where the electron (or positron) is scattered by a small angle producing photons of almost zero virtuality. In this case the electron scattering is equivalent to the scattering of on-shell photons carrying a fraction of the electron energy distributed according to the Weizsäcker-Williams spectrum [1]. The dynamics of charm and bottom photoproduction can be probed at HERA in a large kinematical region, the available center-of-mass energy being one order of magnitude larger than in fixed target experiments. As is well known, photoproduction at HERA is described with two mechanisms, direct photoproduction, where the photon interacts directly with the quarks coming out of the proton, and via resolved photon interactions, where the photon fluctuates into parton constituents (quarks or gluons), which undergo hard scattering with the partons from the proton. Two types of approaches have been pursued in the calculation of photoproduction cross sections for heavy-flavour production in next-to-leading order (NLO). In the massive approach [2] one assumes that light quarks and gluons are the only active flavours within the ingoing proton and photon. In this approach the heavy quark mass $m_Q \gg \Lambda_{QCD}$ acts as a cutoff for initial and final state singularities and sets the scale for the perturbative calculation. The cross section factorizes into a partonic hard-scattering cross section multiplied by light-quark and gluon densities. In this so-called massive quark scheme, the number of active flavours in the initial state is $n_f = 3$, if charm and bottom are treated as massive, and, for example, $n_f = 4$, if only the bottom quark is considered massive. For the prediction of the total charm or bottom photo-production cross section this is the only possibility. The massive approach is also the correct approach for the production of heavy flavoured hadrons with transverse momenta $p_T^2 \simeq m_Q^2$. If $p_T^2 \gg m_Q^2$, mass terms in the hard scattering cross sections are suppressed by powers of m_Q/p_T, but potentially large terms $\propto \ln(p_T^2/m_Q^2)$ arise from the collinear branching of gluons or photons into heavy quark-antiquark pairs or from the collinear emission of a gluon by a heavy quark at large transverse momentum. For $p_T \gg m_Q$ these terms spoil the convergence of the pertubation series and must be summed by absorbing the collinear $\ln(p_T^2/m_Q^2)$ singular terms into heavy-quark parton density functions (PDF's) of the proton or photon, respectively, or into fragmentation functions (FF's) of the heavy quark into the heavy hadron. An alternative way to the absorption of the large logarithm into parton densities and fragmentation functions is to treat the heavy quark Q as massless from the start and to absorb the collinear singularities into PDF's and FF's in the same way as it is done for the lighter u, d and s quarks using the standard $\overline{MS}$ subtraction scheme. This allows for the consistent usage of the standard NLO parton densities with $n_f = 4$ ($n_f = 5$) flavours for the proton or photon which have been constructed in this scheme. This way of calculating heavy-flavour production has been denoted the massless scheme [3]. The mass of the heavy quark appears only in the starting scale of the c or b quark parton distribution functions of the proton and the photon and the fragmentation functions of the heavy quarks, c or b, into heavy flavoured hadrons. In the case of hadrons containing bottom quarks only the total photoproduction cross section has been measured so far [4], so that the massless approach has not become relevant yet.

For the production of charmed mesons, both H1 [5] and ZEUS [6,7] have presented experimental results for $D^{*\pm}$ photoproduction. This channel is favoured because of its clear experimental signal coming from the $D^* \to D\pi$ decay with subsequent decay of the D into $K\pi$ and $K\pi\pi$ channels. The data cover the p_T range up to $p_T \simeq 10.5 \ GeV$ [5] and $p_T \simeq 12 \ GeV$ [6,7], so that at least for the data in the high p_T range the massless scheme should be applicable. In the medium p_T range it is conceivable that the massive (fixed-order) and the massless (resummed) approach are equally reliable, and only in the low-p_T range $p_T \simeq m_c$ the massive scheme is preferred.

In the massless scheme it is also easy to incorporate information on the fragmentation of c quarks into D^* mesons from other processes, as for example $e^+ + e^- \to D^* + X$, into photoproduction of D^*'s at HERA, since, according to the general factorization theorems, this information is universal and can be used for the prediction of any inclusive D^* production process in terms of universal FF's for $c \to D^*$. This information on the $c \to D^*$ FF's comes from high-energy e^+e^- data from LEP, where several data sets for $e^+e^- \to D^*X$, older [8,9] and newer [10,11] ones, exist, which have been used to determine the FF's of D^*'s employing two different subtraction schemes.

In this report I shall review the influence of different descriptions of the fragmentation process inside the resummed approach, valid for $p_T \gg m_c$, on the predictions of inclusive D^* production in ep collisions at HERA. I shall give a comparison with ZEUS data only in cases relevant for the comparison of these different approaches. A more complete confrontation of the recent H1 [5] and ZEUS [6,7] data with the different theoretical approaches, massive and massless, has been presented elsewhere at this workshop [12] and in the H1 [5] and ZEUS [6,7] papers.

In the next section I shall give a short account of the NLO formalism for massless partons. In sect. 3 the calculation of the fragmentation functions for D^* mesons based on two different schemes is reviewed. Sect. 4 contains a discussion of their consequences for the inclusive photoproduction of D^*'s at HERA. In sect. 5 I finish with some conclusions and an outlook to possible work in the future.

2   NLO Formalism for Massless Partons

The calculation of the NLO inclusive cross section for the production of a hadron h with transverse momentum p_T and rapidity y in the resolved mode starts from the formula

$$\frac{d^3\sigma}{dy\,d^2p_T} = \frac{1}{\pi} \sum_{a,b,c} \int dx_\gamma dx_p \frac{dx_h}{x_h^2} F_a^\gamma(x_\gamma, M_\gamma^2) F_b^p(x_p, M_p^2) D_c^h(x_h, M_h^2)$$

$$\left[\frac{d\sigma_{ab\to c}^0}{dt}(s,t,u,\mu^2)\,\delta\left(1+\frac{t+u}{s}\right)\right.$$

$$\left. + \frac{\alpha_s(\mu^2)}{2\pi} K_{ab\to c}(s,t,u,\mu^2,M_\gamma^2,M_p^2,M_h^2)\,\theta\left(1+\frac{t+u}{s}\right)\right], \quad (1)$$

where $s = (p_a+p_b)^2$, $t = (p_a-p_c)^2$ and $u = (p_b-p_c)^2$. The NLO corrections [13] $K_{ab\to c}$ comprise 16 channels and their $t \to u$ crossed counterparts. $d\sigma_{ab\to c}^0/dt$ stands for the LO hard-scattering cross sections for the transition $ab \to cX$, where $X = d$ is a single parton in LO. F_a^γ and F_b^p are the PDF's of the photon and proton, respectively, which must be calculated at scales M_γ and M_p with the help of the NLO evolution equations derived for the $\overline{MS}$ subtraction scheme. Similarly, D_c^h denotes the FF for $c \to h$ (in our case $h = D^{*\pm}$) evolved up to the scale M_h, also in the $\overline{MS}$ subtraction scheme. We emphasize that the summation over parton labes a, b and c includes also the charm quark. This means, that the charm quark is also present in the initial state inside the proton and particularly inside the photon.

The formula for $\gamma + p \to h + X$ via direct photoproduction can be deduced from (1) by the replacements

$$F_a^\gamma(x_\gamma, M_\gamma^2) \to \delta(1-x_\gamma),$$

$$d\sigma_{ab\to c}^0 \to d\sigma_{\gamma b\to c}^0,$$

$$K_{ab\to c} \to K_{\gamma b\to c}. \quad (2)$$

At LO, only the photon-gluon fusion process $\gamma g \to q\bar{q}$ and the Compton process $\gamma q \to gq$ contribute. NLO corrections [14] $K_{\gamma b\to c}$ include eight channels. It is well known that NLO corrections lead to a reduction of the dependence on the renormalization scale μ and the factorization scales M_γ, M_p and M_h This way, they reduce the theoretical uncertainty and lead in certain regions of phase space to sizable shifts in the cross section due to additional parton emissions in the final state. It should also be mentioned, that in NLO the definition of direct and resolved cross section looses its meaning. At higher orders only their sum is uniquely defined and can be compared to experimental data.

The derivation of these formulas is done in dimensional regularization, which is needed to extract the ultraviolet, infrared and collinear divergencies. Since the charm quark is considered massless as all the other quark flavours, divergencies originating from the cg collinear singularities are encountered, which are absorbed into the bare PDF's and FF of the photon, proton and D^* meson. At this step finite terms in the NLO matrix elements $K_{ab\to c}$ remain, which are fixed by selecting the $\overline{MS}$ subtraction scheme. In the massive scheme these collinear singularities are cut off by the finite quark mass m_c and no subtraction is necessary. In addition, there are no bare PDF's or FF's in this scheme, which could absorb the subtraction terms.

3   Fragmentation Functions for D^*

Several different procedures to describe the fragmentation of c quarks into D^*'s have been developed. They differ whether and how information for the production of massive quarks in e^+e^- annihilation is included in the transition $c \to D^*$. They are described by introducing the function $D_a^C(x, \mu^2, m_c^2)$ for the transition of a massless parton a (including the massless charm quark c) to a massive c quark (denoted C). These functions have been derived by Mele and Nason [15] by matching the cross section $d\sigma(e^+e^- \to C + X)/dx$, where all powers of m_c^2/s have been neglected, in the massive scheme ($m_c \neq 0$) to the cross section $d\sigma(e^+e^- \to a\bar{a} + X)$ in the massless scheme ($m_c = 0$ from the beginning). The result is

$$\frac{d\sigma_{e^+e^-\to C+X}}{dx}(x, s, m_c^2) = \sum_a \int_x^1 \frac{dz}{z} D_a^C\left(\frac{x}{z}, \mu^2, m_c^2\right) \frac{d\sigma_{e^+e^-\to a+X}}{dz}(z, s, \mu^2)(3)$$

where

$$D_a^C(x, \mu^2, m_c^2) = \delta_{aC}\delta(1-x) - \frac{\alpha_s(\mu^2)}{2\pi} d_{a\to C}(x, \mu^2, m_c^2) \tag{4}$$

$$d_{a\to C}(x, \mu^2, m_c^2) = -P_{q\to q}^{(0,T)}(x)\ln\frac{\mu^2}{m_c^2} + \frac{4}{3}\left\{-2\delta(1-x)\right.$$

$$\left. + 2\left(\frac{1}{1-x}\right)_+ + 4\left[\frac{\ln(1-x)}{1-x}\right]_+ - (1+x)[1+2\ln(1-x)]\right\} \tag{5}$$

$$d_{g\to C}(x, \mu^2, m_c^2) = -P_{g\to q}^{(0,T)}(x)\ln\frac{\mu^2}{m_c^2} \tag{6}$$

$$d_{q,\bar{q},\bar{c}\to C}(x, \mu^2, m_c^2) = 0 \tag{7}$$

This formula can be interpreted in many ways. Of course, when $D_a^C(x, \mu^2, m_c^2) = \delta_{aC}\delta(1-x)$, the left hand side of (3) is just the cross section for $e^+e^- \to c\bar{c} + X$ with massless c quarks. If, on the other hand, the remaining terms in D_a^C are moved to the hard scattering factor $d\sigma(e^+e^- \to a + X)/dx$, we recover the formula for the production of massive c quarks at NLO up to the neglected terms proportional to m_c. μ is the factorization scale for the massless cross section, which is usually chosen as $\mu = \sqrt{s}$, where s is the squared center-of-mass energy. When $\mu = \sqrt{s} \simeq 2m_c$ the formula (3) with (4) and (5) can be used directly as it stands. When $\mu = \sqrt{s} \gg m_c$ large logarithms appear which must be summed. To preserve the structure contained in $d_{a\to C}(x, \mu^2, m_c^2)$ for finite m_c the function (4) is assumed to determine the FF for $a \to C$ at the starting scale $\mu_0 \simeq m_c$, and one calculates the FF at larger scales μ via the usual Altarelli- Parisi equations [16]. It follws from (5) that the evolution is given by the well known massless evolution kernel $P_{q\to q}^{(0,T)}(x)$. In this way these so-called perturbative FF's are fully determined concerning the μ dependence and also the initial conditions. Only the starting scale μ_0 has to be chosen appropriately. This approach with pertubative FF's was first applied to the calculation of cross sections for the production of b quarks in hadron-hadron collisions [17] and later to the production of c quarks in γp [18] and $\gamma\gamma$ collisions [19] based on the massless scheme described in the previous section. Later it became clear that the pertubative FF as just described is not sufficient to predict cross sections for the production of heavy flavoured hadrons, like D^* mesons, and $D_a^C(x, \mu^2, m_c^2)$ must be convoluted with a non-pertubative FF $D_C^{D^*}(x, \mu_0^2)$ which accounts for the transition $C \to D^*$ [18]. This FF can not be calculated in pertubation theory. It is asumed to be universal, so that it can be fixed by comparing with one set of experimental data and then used to predict other D^* production cross sections. It has been obtained from inclusive $e^+ + e^- \to D^* + X$ cross sections measured at DORIS and LEP [18]. It is clear that with this procedure the total fragmentation $a \to D^*$, starting from the massless c quark a, is a convolution of the pertubative FF with the non-pertubative FF.

Instead of introducing the concept of a pertubative component for the FF one can incorporate the difference between the massless and massive cross section for $e^+ + e^- \to C + X$ in a different way, namely, by combining the terms in D_a^C proportional to $d_{a\to C}$ with the hard scattering cross section $d\sigma(e^+e^- \to a + X)/dz$ in the sense of changing the final-state factorization scheme. This amounts to substituting in the hard-scattering cross section

$$\alpha_s(\mu^2)P_{a\to c}^{(0,T)}(x)\ln\frac{s}{\mu^2} \to \alpha_s(\mu^2)P_{a\to c}^{(0,T)}(x)\ln\frac{s}{\mu^2} - \alpha_s(\mu_0^2)d_{a\to C}(x, \mu_0^2, m_c^2) \quad (8)$$

One should note that the coupling in front of $d_{a\to C}$ is evaluated at the input scale μ_0. This shift leaves the cross section for massive charm production for scales $\mu \simeq \mu_0 \simeq m_c$ unchanged, when $D_a^C(x, \mu^2, m_c^2) = \delta_{aC}\delta_{aC}(1-x)$ is retained. In [20] it was shown that this way of incorporating the differnce of the massless and massive hard-scattering cross section is consistent with the cross section for $e^+ + e^- \to C + X$ including all terms up to $O(\alpha_s^2)$ in the leading logarithmic approximation. It is clear that when this change of final-state factorization scheme is performed for calculating the $e^+e^- \rightarrow D^*X$ cross section, in order to construct the FF's for $a \rightarrow D^*$ from experimental data, it must be intoduced also in all other hard-scattering cross sections for predicting inclusive D^* production. This change of the final-state factorization has been used for studying the influence of various approximations in the fragmenation of D^* mesons in [21] and somewhat later for predicting photoproduction cross sections with fragmentation information coming from LEP experiments in [22].

The method of combining pertubative and non-pertubative FF's in the analysis of D^* production at DORIS and LEP for the calculation of NLO inclusive D^* photoproduction cross sections has been applied in [18] and reviewed by Cacciari at the 1997 Workshop [23]. As already said, it turned out that the fragmentation of charm or bottom quarks into D^* mesons cannot be calculated from first principles in pertubation theory. In fact, in order to realistically desribe the formation of the D^* meson, a non-pertubative component, which is not known theoretically, is always needed. Hence, it is appropriate to give up the pertubative component of the FF input altogether and to describe the $c \rightarrow D^*$ and $b \rightarrow D^*$ transitions entirely by non-pertubative FF's, as it also done for the fragmentation of u, d and s quarks into light mesons. This is the basis of our recent work [24], which has been compared with the new ZEUS and H1 experimental results in [6,7] and [5] (see also [12]). It should be clear that the parameters of the non-pertubative FF's differ, depending on whether a pertubative component of the FF is introduced (via the initial conditions respectively via a change of the subtraction scheme) or not.

The construction of the non-pertubative FF's for $c \rightarrow D^*$ by fitting large-s data on $e^+ + e^- \rightarrow D^* + X$ is complicated since in hadronic Z decays charmed mesons are expected to be produced either directly through the hadronization of charm quarks in the process $Z \rightarrow c\bar{c}$ or via weak decays of B hadrons produced in $Z \rightarrow b\bar{b}$ with approximately equal rate. Charmed mesons from $Z \rightarrow c\bar{c}$ allow us to measure the c-quark FF. The main task of the experimental analysis was to disentangle these two main sources of $D^{*\pm}$ production at the Z boson resonance. In our first analysis, in which the $d_{a\rightarrow c}$ subtraction scheme was used, we made use of the LEP data from ALEPH [8] and OPAL [9].

The starting scales for the D^* FF's of the gluon and the u, d, s and c quarks and antiquarks was set to $\mu_0 = 2m_c$ with $m_c = 1.5\ GeV$, whereas for bottom it was set to $\mu_0 = 2m_b$ with $m_b = 5\ GeV$. The FF's of the gluon and the first three quark flavours are assumed to be zero at the starting scale. For the parametrization of charm- and bottom quark FF's at their respective starting scales two different forms were employed, the so-called standard (S) and mixed (M) sets [22]. Here we review only the results with the mixed set. The results with the standard set are very similar and lead to almost identical predictions for the photoproduction of D^*'s. In the mixed set the FF of the bottom quark is assumed to be of the form

$$D_b^{D^{*\pm}}(x, \mu_0^2) = N x^\alpha (1-x)^\beta \tag{9}$$

which contains three parameters and is best suited for FF's which peak at smaller x. The charm-quark FF at the initial scale is equal to the Peterson distribution

[25]

$$D_c^{D^{*\pm}}(x,\mu_0^2) = N \frac{x(1-x)^2}{[(1-x)^2 + \epsilon x]^2} \tag{10}$$

which is particularly suitable to describe FF's that peak at large x. It depends only on two free parameters, N and ϵ. Both ALEPH [8] and OPAL [9] extracted the total $D^{*\pm}$ momentum distribution as well as the individual contributions due to $Z \to c\bar{c}$ and $Z \to b\bar{b}$. In the performed LO (NLO) analysis the one-loop (two-loop) formula for α_s with $\Lambda_{\overline{MS}}^{(5)} = 108$ MeV $(227$ $MeV)$ was used. The obtained values of N, α, β and ϵ resulting from the combined LO (NLO) fits to the ALEPH and OPAL data are given in [22]. The ϵ parameter in the LO (NLO) fit is $\epsilon = 0.0856$ (0.0204). In the NLO fit ϵ is very much reduced since part of the fragmentation is described by the pertubative component. In Fig. 1 the fits are compared to the OPAL [9] data. For the fits to the ALEPH data see [22]. Both, the LO and NLO results are shown. Except for very small x, the LO and NLO curves are very similar. At small x large differences occur, indicating that in this region the pertubative treatment ceases to be valid. Also, in this region, the massless approximation is not valid anymore. In any case, due to the finite D^* mass, the D^* can only be produced for $x > x_{min} = 2m(D^*)/M_Z = 0.046$. This is the region, where the NLO results turn negative. Therefore the results are meaningful only for $x > x_{cut} = 0.1$. For the fitting actually only the x bins in the interval $[0.1, 0.9]$ were used. The quality of the fit measured in terms of average χ^2 for all selected data points is quite good. Details are found in [22]. Also the branching fractions of the two transitions $c \to D^*$ and $b \to D^*$ which are calculated from the FF's over x in the interval $x_{cut} < x < 1$ came out in agreement with the measured values: $B_c = 0.259$ and $B_b = 0.252$ [9] with errors of 25%. The corresponding theoretical values are 0.231 (0.221) and 0.215 (0.219) for the LO (NLO) fit. These numbers show that we have the correct normalization. Another quantity of interest is the mean momentum fraction $< x >_Q (\mu)$ where $Q = c, b$. This depends on the scale, where it is evaluated. For $\mu = M_Z$ we have $< x >_c = 0.490$ (0.556) for the LO (NLO) FF, which has to be compared to $< x >_c = 0.495^{+0.010}_{-0.011} \pm 0.007$ [8] and $< x >_c = 0.515^{+0.008}_{-0.005} \pm 0.010$ [9].

Looking at the curves in Fig. 1 we notice that the NLO fits lead to negative cross sections in the vicinity of $x = 1$. This is directly conneted with the subtraction of the pertuabative $d_{a \to C}$ functions. To remove this unphysical behaviour one must resum the pertubative contributions for $x \to 1$ originating from the finite mass cross section. Also due to this extra complication the description in terms of pure non-pertubative FF's is preferred. Negative cross sections at $x \leq 1$ occur also in all the other formulations with pertubative pieces in the FF's [18,26]. With the same ALEPH [8] and OPAL [9] data we performed fits also in the truly massless scheme discarding the pertubative contributions in form of the $d_{a \to C}$ functions. Here the fit could be done in the interval $[0.1, 1]$ in contrast to the fit with the $d_{a \to C}$ functions, which was limited to $[0.1, 0.9]$. The main difference in the NLO fit was a change of the Peterson parameter to

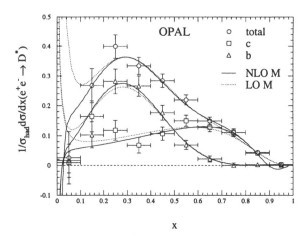

Fig. 1. Comparison of the OPAL data [10] on $e^+ + e^- \to D^{*\pm} + X$ with the LO M (dashed) and NLO M (solid) calculations. The three sets of curves and data correspond to the $Z \to c\bar{c}$ and $Z \to b\bar{b}$ samples as well as their combination.

$\epsilon = 0.116$. The fit was equally good, the χ^2 changed only from 0.96 to 1.01. Of course, also the fragmentation parameters for $b \to D^*$ fragmentation changed somewhat, but much less than the parameters for the c fragmentation.

In 1997 more precise data for $e^+ + e^- \to D^{*\pm} + X$ from ALEPH [10] and OPAL [11] became available. These new samples are compatible with the respective older samples, but not with each other. Both ALEPH [10] and OPAL [11] presented momentum distributions for their full $Z \to D^{*\pm} + X$ sample and for the $Z \to b\bar{b}$ subsample, but not separately for $Z \to c\bar{c}$ as in the previous measurements. We used the new data to perform only fits without the $d_{a \to C}$ subtraction. The fit parameters for the two data sets in LO and NLO and for the two flavours c and b are displayed in Tab. I of [24]. The obtained parameters for ALEPH and OPAL come out different indicating that the two data sets do not overlap inside their errors. Both sets have much larger ϵ parameters in NLO compared to the earlier fits with the massive subtraction. In Fig. 2 the OPAL [11] data are compared with the sets LO O and NLO O (O stands for OPAL). The comparison of the corresponding LO and NLO fits to the ALEPH data can be found in Fig. 1b of [24]. Except at very small x, the LO and NLO results are again very similar. There are large differences between the LO and NLO parameters obtained from the same data. The value $\epsilon = 0.0851$ of set LO O is very similar to $\epsilon = 0.0856$ of the set obtained from the older data [8,9]. Compared to the NLO fit with the $d_{a \to C}$ subtraction shown in Fig. 1, the new NLO fit in Fig. 2 stays positive also for $x \geq 0.9$, coincides there with the LO result, and nicely describes the data.

As with the fits to the older data, we studied also the branching fractions and the mean momentum fractions for the two sets and compared them with the available experimental information. Both quantities turned out to be in

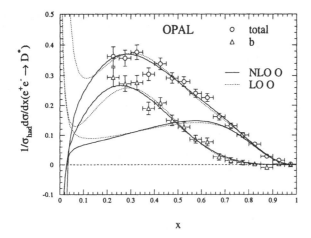

Fig. 2. Comparison of the OPAL data [12] on $e^+e^- \to D^{*\pm} + X$ with the LO O (dashed) and NLO O (solid) calculations. The three sets of curves correspond to the $Z \to c\bar{c}$, $Z \to b\bar{b}$ and full samples.

reasonable agreement with the data, details are found in [24]. We tested also, whether the FF's constructed from LEP data lead to a consistent description of $e^+e^- \to D^*X$-data at lower energies. This presents a direct test of the underlying scaling violations of fragmentation via the time-like Altarelli-Parisi equations. The comparison with data from ARGUS [27] at $\sqrt{s} = 10.49\ GeV$, from HRS [28] at $\sqrt{s} = 29\ GeV$ and from TASSO [29] at $\sqrt{s} = 34.2\ GeV$ calculated with the LO and NLO OPAL and ALEPH sets can be seen in in Fig. 2a,b of [24] together with the OPAL and ALEPH data, respectively. The agreement with all the data is quite good. These comparisons assure us that the $D^{*\pm}$ FF's, although constructed from data at $\sqrt{s} = M_Z$, lead to a valid description of $D^{*\pm}$ fragmentation at lower scales.

Based on the universality of fragmentation we have made predictions for inclusive $D^{*\pm}$ photoproduction at HERA. The results for the FF's based on the older LEP data and empoying the massive charm subtraction scheme are found in [22] as well as the comparison with the ZEUS data [30] published in 1997. The more recent H1 [5] and ZEUS [6,7] have been compared only with the more recent predictions based on FF's obtained with the recent ALEPH and OPAL results and without the massive subtraction. Such comparisons have been shown also at the workshop in Coldewey's contribution [12]. In [24] we compared with preliminary ZEUS data from their 1996 analysis [31]. In the next section I shall present some of these results with particular emphasis on how the two fragmentation schemes, with and without the massive subtraction, influence the predictions. In addition I shall discuss the contribution of the resolved cross section.

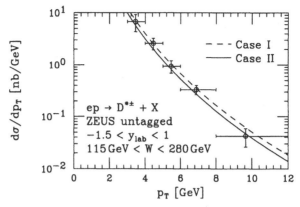

Fig. 3. Comparison of the ZEUS data on $e + p \to D^{*\pm} + X$ with the NLO calculations according to cases I (dashed) and II (solid).

4 $D^{*\pm}$ Production in Low-Q^2 ep Collisions

It is clear that the cross section for inclusive $D^{*\pm}$ production in low-Q^2 ep collisions at HERA depends on input from different sources. First, it depends on the way the FF for the $D^{*\pm}$ is constructed. In [21] several versions of approximating the D^* fragmentation have been investigated. For example, in [21] the ϵ parameter in the Peterson fragmentation formula was varied by ± 0.02. This changed the cross section by $\pm 15\%$ with the branching fraction fixed to its experimental value. This is not a very large change. Since in our more recent work [22,24] the D^* FF was fixed by LEP data this dependence on the FF is largely eliminated. The final prediction, though, may still depend on the scheme used to construct the fragmentation in NLO. i.e. with or without the massive subtraction terms. When we fitted to the LEP data, we also had to take into account $D^{*\pm}$ production by bottom-quark fragmentation. In ep collisions, however, the bottom quark does not contribute to $D^{*\pm}$ production below the threshold at $m_T = \sqrt{p_T^2 + m_c^2} = m_b$. In contrast to e^+e^-- annihilation the bottom quark does not contribute significantly even above threshold. In [22] we found that the relative contribution due to the b-quark FF's is below 5% and decreases with increasing p_T.

Another issue is the problem of the scheme dependence in the construction of the FF's. Above we presented the results for $D^{*\pm}$ FF's obtained from fitting the same LEP data, but with the different schemes, one with and one without the massive subtraction $d_{a \to c}$. As we have seen, these two schemes lead to completely different non-pertubative components in NLO, expressed by two completely different ϵ parameters, $\epsilon = 0.0204$ and $\epsilon = 0.116$, respectively. In Fig. 3 I show the predictions for the $D^{*\pm}$ photoproduction cross section with the two schemes. The dashed curve (case I) is with $d_{a \to c}$ subtraction and the full curve (case II) without. In Fig. 3 the p_T distributions $d\sigma/dp_T$ is compared to the older ZEUS [30] data. We observe, that the two predictions differ approximately by

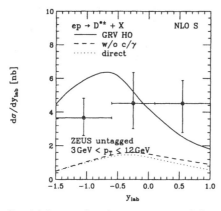

Fig. 4. Sensitivity to the charm content of the photon.

25%. We think that this difference is characteristically the error caused by using two different fragmentation approaches. We note that the cross section with the massive subtraction lies higher than without the subtraction, which is based on the pure $\overline{MS}$ factorization. This is in contrast to the predictions of Cacciari et al. [18], where the $d_{a \to c}$ functions are taken into account via the initial conditions of the FF's. Although these two approaches are equivalent up to $O(\alpha_s^2)$ in the logarithmic terms, they differ in nonlogarithmic terms in the same order. We assume that this difference in the treatment of the $d_{a \to c}$ terms is the explanation for the smaller D^* photoproduction cross sections in [18].

The largest uncertainty in the calculation of the $D^{*\pm}$ photoproduction cross section is related to the charm content of the resolved photon. The most sensitive distributions to study this dependence are the differential rapidity distributions for fixed p_T. To illustrate the sensitivity to the charm PDF of the photon, the NLO analysis (here with the NLO S fragmentation function and with the FF in the massive subtraction scheme from [22]) with this PDF switched off is shown in Fig. 4 and compared to the full calculation. We observe that 73% to 100% (50% to 89%) of the resolved (full) cross section is induced by the charm content of the photon, which is most important in the backward direction. In Fig. 4 the cross section is plotted as a function of y_{lab}, integrated over $3 < p_T < 12 \, GeV$. We emphasize, the lower p_T limit may be somewhat too small to justify the massless scheme. A larger p_T limit should be suited better. However, here the influence of the resolved component diminishes somewhat. The three experimental points are from the first ZEUS measurements [30].

It is well known, that the pure massive scheme [2] contains a contribution from the charm component in the photon. This is hidden in the NLO corrections to the direct cross section. As long as all m_c^2 / p_T^2 terms are negligible this contribution, in a first approximation, should be equivalent in the massless scheme to the LO resolved component. Therefore it is of interest to see the influence of the NLO corrections to the resolved component since this contribution is certainly absent in the massive scheme. This is shown in Fig. 5, where the NLO

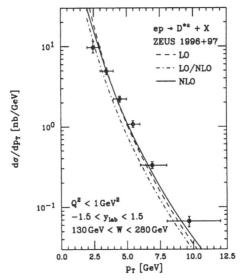

Fig. 5. Comparison of the ZEUS data on $e + p \rightarrow D^{*\pm} + X$ with the LO O (dashed) and NLO O (solid) calculations. The dot-dashed curve emerges from the solid one by switching off the NLO terms in the partonic cross sections of the resolved processes.

O calculation of $d\sigma/dp_T$ is compared to the case, where the NLO terms in the partonic cross sections are switched off. For $p_T \geq 5\ GeV$ we observe an increase of up to 50% due to the NLO corrections. This is almost the difference between the results of the massive scheme and the massless scheme, as can be seen in Fig. 2 of [6]. In Fig. 5 the most recent ZEUS data [6] are compared to the full NLO massless calculation of [24]. The experimental data lie somewhat above the predicted results, in particular for the larger p_T values, where the agreement actually should be better. The difference amounts to approximately 30% (not considering the experimental errors) for $p_T > 5\ GeV$. In Fig. 5 we plotted also the LO O prediction. It is very similar to the NLO O result showing good pertubative stability. We emphasize that the LO curves are obtained with LO PDF's, FF's and the one-loop formula for α_s.

We have seen above that the charm component of the resolved photon makes an important contribution to the $D^{*\pm}$ photoproduction cross section. All the results shown so far are obtained with the parametrization of GRV in LO or NLO [32]. It is interesting to see the results for other photon PDF's. Such comparisons are shown in Fig. 6, where the differential rapidity distributions are plotted after integration over four different p_T ranges $p_{T,min} < p_T < 12\ GeV$ with $p_{T,min} = 2, 3, 4$ and $6\ GeV$. Besides GRV the photon PDF's of Gordon and Storrow (denoted GS92) [33] and two others from Aurenche et al. for massless charm quarks [34] denoted AFG and for massive charm quarks [35] denoted ACFGP(m_c) are employed. We observe a significant variation of the cross sections, especially also in the backward direction. The difference is particularly

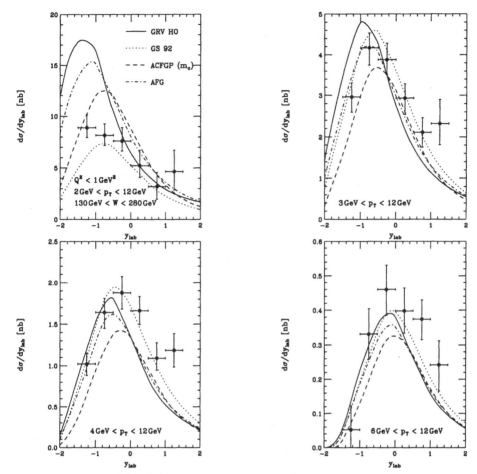

Fig. 6. Comparison of the ZEUS data on $e + p \rightarrow D^{*\pm} + X$ with the NLO O calculation for different photon PDF's.

marked for the curves with $p_{T,min} = 2\ GeV$. For $p_{T,min} = 6\ GeV$ the differences are much less. In this figure the various predictions are also compared to the recent ZEUS data [6]. Of course, for $p_{T,min} = 2\ GeV$ we would not expect that the massless scheme yields reliable predictions, although the GS92 curve comes close to the data. All other PDF's give too large cross sections in this case. Also for the other $p_{T,min} = 3, 4$ and $6\ GeV$ the GS92 prediction agrees reasonably well with the data points, but also GRV and AFG agree for the two largest $p_{T,min}$'s. The GS92 prediction is also better in the forward direction where the deviation between the data and the prediction is largest.

Very recently the ZEUS collaboration also presented data in a smaller total center-of-mass energy region W, $80 < W < 120\ GeV$ [7] compared to the high W region $130 < W < 280\ GeV$ already published [6]. The p_T and y_{lab} intervals of these data are $2 < p_T < 8\ GeV$ and $-1 < y_{lab} < 1.5$. The predicted $d\sigma/dp_T$

distribution based on our fragmentation approach without the massive subtraction and $\epsilon = 0.116$ is in good agreement with the data. The cross section as a function of y_{lab} integrated over the region $3.25 < p_T < 8 \; GeV$ agrees best with the newer Gordon-Storrow photon PDF [36] (see [7]), although some access in the forward region is still evident if compared to the prediction.

5   Conclusions

The study of inclusive $D^{*\pm}$ production in e^+e^- and ep collisions allows for a meaningful test of the QCD-improved parton model. Inside the massless approach, where the charm-quark mass is neglected, we discussed two different subtraction schemes for the fragmentation functions in NLO. Both can be used to fit $D^{*\pm}$ production at LEP and lead to comparable photoproduction cross sections at HERA. It is emphazised that the fit results for the input parameter ϵ of the Peterson et al. parametrization is highly scheme dependent at NLO and must not be naively compared disregarding the theoretical framework which it refers to.

A large part of the $D^{*\pm}$ photoproduction cross section originates from the resolved process, eventually allowing to determine the charm distribution function of the photon more accurately. The reasonable agreement between the NLO masslesss theory and recent $D^{*\pm}$ photoproduction data from ZEUS is quite encouraging. This is also true for the $\gamma\gamma \to D^{*\pm}X$ process, for which new data were shown at this workshop [37]. It is pointed out, that the theoretical cross section is particularly enhanced by the NLO corrections to the resolved component, a contibution, which is absent in any pure massive calculation. It is understood, that the massless-quark approximation is only appropriate for $D^{*\pm}$ production at large enough p_T, say $p_T > 4 \; GeV$. So it should not be expected to yield meaningful predictions for the photoproduction at small-p_T $D^{*\pm}$ mesons at HERA. The next step in theory would be to retain the finite charm-quark mass and to subtract in the NLO corections to the direct process the colllinear logarithmic mass divergences in such a way that the $\overline{MS}$ subtracted massless theory is aproached in the limit $m_c \to 0$. This formulation should be applicable at small and large p_T.

Acknowledgements

This work was supported by Bundesministerium für Forschung und Technologie, Bonn, Germany, under Contract 05 7 HH 92P (0), and by the EU Fourth Framework Program *Training and Mobility of Researchers* through Network *Quantum Chromodynamics and Deep Structure of Elementary Particles* under Contract FMRX–CT98–0194 (DG12 MIHT). I would like to thank J. Binnewies, B. A. Kniehl and M. Spira for their collaboration and B. A. Kniehl for help with the manuscript.

References

1. C. F. v. Weizsäcker, Z. Phys. 88 (1934) 612; E. J. Williams, Phys. Rev. 45 (1934) 729
2. S. Frixione, P. Nason and G. Ridolfi, Nucl. Phys. B454 (1995) 3 and the earlier references given there
3. B. A. Kniehl, M. Krämer, G. Kramer and M. Spira, Phys. Lett. B356 (1995) 539
4. D. Pitzl in these Proceedings
5. C. Adloff et al., H1 Collaboration, Nucl. Phys. B545 (1999) 21
6. J. Breitweg et al., ZEUS Collaboration, Eur. Phys. J. C6 (1999) 67
7. Y. Eisenberg, On behalf of the ZEUS Collaboration, Talk given at DIS99, 19-23 April 1999, Zeuthen, Germany, hep-ex/9905008
8. D. Buskulic et al., ALEPH Collaboration, Z. Phys. C62 (1994) 1
9. R. Akers et al., OPAL Collaboration, Z. Phys. C67 (1995) 27
10. ALEPH Collaboration, Contributed Paper No. 623 to the International Europhysics Conference on High Energy Physics, Jerusalem, Israel, 1997 and CERN/EP 99-094, July 8, 1999
11. K. Ackerstaff et al., OPAL Collaboration, Eur. Phys. J. C1 (1998) 439
12. C. Coldewey in these Proceedings
13. F. Aversa, P. Chiapetta, M. Greco and J.-Ph. Guillet, Nucl. Phys. B327 (1989) 105
14. P. Aurenche, R. Baier, A. Douiri, M. Fontannaz and D. Schiff, Nucl. Phys. B286 (1987) 553; L. E. Gordon, Phys. Rev. D50 (1994) 6753
15. B. Mele and P. Nason, Nucl. Phys. B361 (1991) 626
16. G. Altarelli and G. Parisi, Nucl. Phys. B126 (1977) 298
17. M. Cacciari and M. Greco, Nucl. Phys. B421 (1994) 530
18. M. Cacciari and M. Greco, Z. Phys. C69 (1996) 459; Phys. Rev. D55 (1997) 7134
19. M. Cacciari, M. Greco, B. A. Kniehl, M. Krämer, G. Kramer and M. Spira, Nucl. Phys. B466 (1996) 173
20. P. Nason and C. Oleari, Phys. Lett. B418 (1998) 199
21. B. A. Kniehl, G. Kramer and M. Spira, Z. Phys. C76 (1997) 689
22. J. Binnewies, B. A. Kniehl and G. Kramer, Z. Phys. C76 (1997) 677
23. M. Cacciari, Proceedings of the Ringberg Workshop, New Trends in HERA Physics edited by B A Kniehl, G Kramer, A Wagner, World Scientific, Singapore-New Jersey-London-Hong Kong
24. J. Binnewies, B. A. Kniehl and G. Kramer, Phys. Rev. D58 (1998) 014014
25. C. Peterson, D. Schlatter, I. Schmitt and P. M. Zerwas, Phys. Rev. D27 (1983) 105
26. P. Nason and O. Oleari, Phys. Lett. B447 (1999) 327; hep-ph/9903541, submitted to Nucl. Phys. B
27. H. Albrecht et al., ARGUS Collaboration, Z. Phys. C52 (1991) 353
28. P. Barninger et al., HRS Collaboration, Phys. Lett. B206 (1988) 551
29. W. Braunschweig et al., TASSO Collaboration, Z. Phys. C44 (1989) 365
30. J. Breitweg et al. ZEUS Collaboration, Phys. Lett. B401 (1997) 192
31. ZEUS Collaboration, Contributed Paper No 653 to the International Europhysics Conference on High Enery Physics, Jerusalem, Israel, 1997
32. M. Glück, E. Reya and A. Vogt, Phys. Rev. D46 (1992) 1973
33. L. E. Gordon and J. K. Storrow, Z. Phys. C56 (1992) 307
34. P. Aurenche, J.-Ph. Guillet and M. Fontannaz, Z. Phys. C64 (1994) 621
35. P. Aurenche, P. Chiapetta, M. Fontannaz, J.-Ph. Guillet and E. Pilon, Z. Phys. C56 (1992) 589
36. L. E. Gordon and J. K. Storrow, Nucl. Phys. B489 (1997) 405
37. R. Nisius in these Proceedings

Open Charm Production in Deep Inelastic Scattering at Next-to-Leading Order at HERA

Brian Harris

HEP Theory Group, Argonne National Laboratory, Argonne, IL 60439, USA

Abstract. An introduction and overview of charm production in deep inelastic scattering at HERA is given. The existing next-to-leading order perturbative QCD calculations are then reviewed, and key results are summarized. Finally, comparisons are made with the most recent HERA data, and unresolved issues are highlighted.

1 Introduction

Electromagnetic interactions have long been used to study both hadronic structure and strong interaction dynamics. Examples include deep inelastic lepton-nucleon scattering, hadroproduction of lepton pairs, the production of photons with large transverse momenta, and various photoproduction processes involving scattering of real or very low mass virtual photons from hadrons. In particular, heavy quark production in deep inelastic electron-proton scattering (DIS) is calculable in QCD and provides information on the gluonic content of the proton which is complementary to that obtained in direct photon production or structure function scaling violation measurements. In addition, the scale of the hard scattering may be large relative to the mass of the charm quark, thus allowing one to study whether and when to treat the charm as a parton.

Early measurements of open charm production in neutral current DIS, performed by the Berkeley-Fermilab-Princeton (BFP) [1] and European Muon Collaboration (EMC) [2] experiments, touched upon the topics relevant for HERA today: production mechanism, charm fragmentation, gluon parton distribution function extraction, and charm contribution to the proton structure function. See [3] for a review of these experiments. Activity in this area has increased recently with new data available from the H1 [4] and ZEUS [5] experiments at HERA. In particular, substantial samples of reconstructed D^* hadrons have been obtained, and a semi-lepton decay mode analysis is underway [6]. Considerably more data is anticipated in the next few years.

Interest in the production mechanism is twofold. First, one is concerned with the leading twist-two term in the operator product expansion which incorporates the factorization theorem for hard scattering. Second, there is interest in studies of a higher twist charm component to the proton.

When considering the leading twist-two term the main issue is whether and when to treat the charm as a parton. Near threshold it is well established that charm is produced through photon-gluon fusion. On the other hand, very far above threshold the charm should be resummed into an effective parton distribution. How one interpolates from one region to another is described by a variable flavor number scheme, several of which have been proposed recently [7–9]. One must look at sufficiently inclusive observables in order to build up the logarithms that are to be summed, so predictive power for some observables is lost. In other words, for differential quantities, the fixed flavor number scheme of photon-gluon fusion provides the most appropriate formalism [10].

The idea of a higher twist charm component to the proton was introduced by Brodsky *et al.* [11]. In this scenario the intrinsic charm quark Fock component in the proton wave function, $|uudc\bar{c}>$, is generated by virtual interactions where gluons couple to two or more valence quarks. The probability for $c\bar{c}$ fluctuations to exist in a light hadron scales as $\mathcal{O}(1/m_c^2)$ relative to the twist-two component, and is therefore higher twist. The EMC data [2] marginally support this idea. A full re-analysis of the EMC data has been carried out including both leading twist and intrinsic components at NLO including mass and threshold effects [12]. The result is that one data point contributes to a $0.8 \pm 0.6\%$ normalization of the intrinsic component relative to leading twist.

Fragmentation is the most contentious topic at present. The production and subsequent hadronization of charmed quarks in DIS is not as clean as the much studied case of production in e^+e^- annihilation. In particular, in DIS there are proton beam remnants around which necessarily talk to the charm quark as it hadronizes. If events too close to the beam direction are selected, one can expect deviations from models which do not account for this effect. Presently it is hoped that detailed experimental studies in the Breit frame, wherein one hemisphere resembles e^+e^-, will provide additional information on the fragmentation process.

Because the production method is dominated by the leading twist photon-gluon fusion near threshold, it is possible to extract a gluon parton distribution function (PDF) and compare with existing PDFs, which derive their gluon information from other sources, such as direct photon production or structure function scaling violation measurements. This has been done by both H1 and ZEUS collaborations. The results are completely consistent with comparable PDF sets in the same x and Q^2 range. While unlikely to ever replace structure function scaling violation measurements, charm production does serve as a nice consistency check on the gluon PDF.

One may also wish to take the open charm measurement and extrapolate the cross section over the full phase space and then extract the charm's contribution to the proton structure function. Historically the structure functions have been useful as input into global analyses and a testing ground for the variable flavor number schemes mentioned above. Care must be taken so that conclusions drawn are not artifacts of the extrapolation procedure (model).

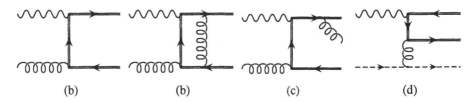

(b)   (b)   (c)   (d)

Fig. 1. Typical Feynman diagrams contributing to the amplitude for neutral current charm production. (a) An order eg_s Born diagram. (b) An order eg_s^3 virtual diagram. (c) An order eg_s^2 gluon-bremsstrahlung diagram. (d) An order eg_s^2 light quark initiated diagram

2   Next-to-leading Order Calculations

The reaction under consideration is charm quark production via neutral-current electron-proton scattering, $e^-(l) + P(p) \to e^-(l') + c(p_1) + X$. When the momentum transfer squared $Q^2 = -q^2 > 0$ ($q = l - l'$) is not too large $Q^2 \ll M_Z^2$, the contribution from Z boson exchange is suppressed and the process is dominated by virtual-photon exchange. After an azimuthal integration, the cross section may be written in terms of structure functions $F_2^c(x, Q^2, m_c)$ and $F_L^c(x, Q^2, m_c)$ as follows:

$$\frac{d^2\sigma}{dydQ^2} = \frac{2\pi\alpha^2}{yQ^4} \left\{ \left[1 + (1-y)^2\right] F_2^c(x, Q^2, m_c) - y^2 F_L^c(x, Q^2, m_c) \right\} \quad (1)$$

where $x = Q^2/2p \cdot q$ and $y = p \cdot q/p \cdot l$ are the usual Bjorken scaling variables, α is the electromagnetic coupling, and m_c is the charm mass. The scaling variables are related to the square of the center of momentum energy of the electron-proton system $S = (l + p)^2$ via $xyS = Q^2$. The total cross section is given by [13]

$$\sigma = \int_{4m_c^2/S}^{1} dy \int_{m_e^2y^2/(1-y)}^{yS-4m_c^2} dQ^2 \left(\frac{d^2\sigma}{dydQ^2} \right) \quad (2)$$

where m_e is the electron mass.

Typical Feynman diagrams for this process are shown in Fig. 1. The interference of the $\mathcal{O}(eg_s)$ Born diagrams with the $\mathcal{O}(eg_s^3)$ one-loop virtual diagrams produces a result that is $\mathcal{O}(\alpha\alpha_s^2)$. The ultraviolet divergences are removed by renormalization in the Collins-Wilczek-Zee scheme [14]. The result is added to the square of the gluon-bremsstrahlung diagrams which is also $\mathcal{O}(\alpha\alpha_s^2)$. Initial state collinear singularities are mass factorized to obtain a finite NLO result. For the light quark initiated subprocess there are no virtual contributions at $\mathcal{O}(\alpha\alpha_s^2)$. One only encounters initial state collinear singularities which are again removed using factorization. The full NLO corrections were first calculated in [15] and may be written in the form

$$F_k^c(x, Q^2, m_c) = \frac{Q^2 \alpha_s(\mu^2)}{4\pi^2 m_c^2} \int_{\xi_{min}}^1 \frac{d\xi}{\xi} \left\{ e_c^2 f_{g/P}(\xi, \mu^2) \right.$$

$$\times \left[c_{k,g}^{(0)} + 4\pi\alpha_s(\mu^2) \left(c_{k,g}^{(1)} + \bar{c}_{k,g}^{(1)} \ln \frac{\mu^2}{m_c^2} \right) \right]$$

$$+ 4\pi\alpha_s(\mu^2) \sum_{i=q,\bar{q}} f_{i/P}(\xi, \mu^2)$$

$$\left. \times \left[e_c^2 \left(c_{k,i}^{(1)} + \bar{c}_{k,i}^{(1)} \ln \frac{\mu^2}{m_c^2} \right) + e_i^2 d_{k,i}^{(1)} + e_i e_c o_{k,i}^{(1)} \right] \right\}$$

$$(3)$$

with $k = 2, L$. The lower limit on the integration over the parton momentum fraction ξ is $\xi_{min} = x(4m_c^2 + Q^2)/Q^2$. The parton momentum distributions in the proton are denoted by $f_{i/P}(\xi, \mu^2)$. The sum is taken over the light quarks, $q = u, d, s$. The mass factorization scale μ_f has been set equal to the renormalization scale μ_r and is denoted by μ. All charges are in units of e. $c_{k,i}^{(0)}$, $c_{k,i}^{(1)}$, $\bar{c}_{k,i}^{(1)}$, $(i = g, q, \bar{q})$, and $d_{k,i}^{(1)}$, $o_{k,i}^{(1)}$, $(i = q, \bar{q})$ are scale independent parton coefficient functions, and are distinguished by their origin. The c-coefficient functions originate from processes involving the virtual photon-heavy quark coupling, the d-coefficient functions arise from processes involving the virtual photon-light quark coupling, and the o-coefficient functions are from the interference between these processes.

In addition to describing the structure functions, the coefficient functions of [15] may also be used to compute the single-inclusive distributions dF_k/dp_t and dF_k/dy [16], where p_t and y are the transverse momentum and rapidity, respectively, of the heavy quark in the virtual photon-proton center of momentum system.

Unfortunately, the analytic expressions of the coefficient functions are too long to be published in journal form. They are however available as computer code, but, for the same reason, the code tends to be slow. Initially this was seen as an impediment to including them in a global fitting program. Then in [17], with the threshold and asymptotic behavior of the coefficient functions removed, it was possible to numerically tabulate grids, with a fast interpolation routine, so that speedy computation of Eq. (1) – (3) became possible. This process was further refined in [18].

The coefficient functions in the large momentum transfer limit are a necessary ingredient for constructing the variable flavor number schemes mentioned in the introduction. Exact analytic formulae for the $d_{k,i}^{(1)}$ together with analytic formulae for the coefficient functions $c_{k,i}^{(0)}$, $c_{k,i}^{(1)}$, $\bar{c}_{k,i}^{(1)}$, $(i = g, q, \bar{q})$, and $d_{k,i}^{(1)}$, $(i = q, \bar{q})$ in the limit $Q^2 \gg m^2$ can be found in [19].

The coefficient functions have also been calculated in a fully differential form [20]. These in turn can be used with Eq. (3) to construct pair-inclusive distributions such as $dF_k/dM_{c\bar{c}}$ [21] where $M_{c\bar{c}}$ is the invariant mass of the produced charm-anticharm system.

The resulting differential structure functions and Eq. (1) have further been used to construct a NLO monte carlo style program, HVQDIS [22,23]. The basic components (in terms of virtual-photon-proton scattering) are the 2 to 2 body squared matrix elements through one-loop order and tree level 2 to 3 body squared matrix elements, for both photon-gluon and photon-light-quark initiated subprocesses, as shown in Fig. 1. It is therefore possible to study fully-, single-, and semi-inclusive production at NLO, and three body final states at leading order. The goal of this NLO calculation is to organize the soft and collinear singularity cancellations without loss of information in terms of observables that can be predicted.

The subtraction method provides a mechanism for this cancellation. It allows one to isolate the soft and collinear singularities of the 2 to 3 body processes within the framework of dimensional regularization without calculating all the phase space integrals in a space-time dimension $n \neq 4$. Expressions for the three-body squared matrix elements in the limit where an emitted gluon is soft appear in a factorized form where poles $\epsilon = 2 - n/2$ multiply leading order squared matrix elements. These soft singularities cancel upon addition of the interference of the leading order diagrams with the renormalized one-loop virtual diagrams. The remaining singularities are initial state collinear in origin. The three-body squared matrix elements appear in a factorized form, where poles in ϵ multiply splitting functions convolved with leading order squared matrix elements. These collinear singularities are removed through mass factorization.

The result of this calculation is an expression that is finite in four-dimensional space time. One can compute all phase space integrations using standard monte carlo integration techniques. The final result is a program which returns parton kinematic configurations and their corresponding weights, accurate to $\mathcal{O}(\alpha\alpha_s^2)$. The user is free to histogram any set of infrared-safe observables and apply cuts, all in a single histogramming subroutine. Additionally, one may study heavy hadrons using the Peterson *et al.* model [24]. Detailed physics results from this program and a description of the necessary cross checks the program satisfies are given in [23]. See also [25] for recent improvements.

3   Results and Current Issues

Charmed meson differential cross sections are measured experimentally [1,2,4,5] within some detector acceptance region, and the corresponding theory predictions can be made using HVQDIS. As we saw above, the cross section is an integral over the structure functions. Therefore, the two share many of the same properties. Before discussing the cross sections, the more salient features of the NLO structure functions will be reviewed. The interested reader can find additional details in the original paper [15] and, more so, in the recent phenomenological analyses [18,26–28].

For moderate $Q^2 \sim 10 \, \text{GeV}^2$ one finds that the charm quark contribution at small $x \sim 10^{-4}$ is approximately 25% of the total structure function (defined as light parton plus heavy quark contributions). In contrast, the contribution from

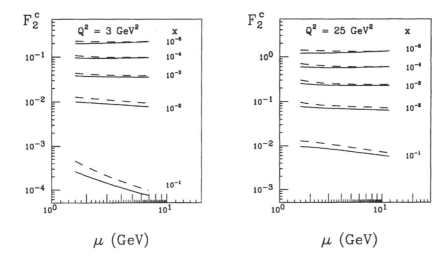

Fig. 2. The renormalization/factorization scale dependence of the structure function $F_2^c(x, Q^2, m_c)$ for $Q^2 = 3$ GeV^2 (left) and $Q^2 = 25$ GeV^2 (right) for various x values. The results for GRV94 (*solid lines*) and CTEQ4F3 (*dashed lines*) parton distribution sets are shown

bottom quarks is only a few percent due to charge and phase space suppression. The structure functions show a marked rise at small x due primarily to the rapidly rising gluon distribution: the gluon initiated contributions comprise most of the structure function. The light quark initiated processes give only a few percent contribution at small x. The scale dependence of the structure functions is very small in the HERA x and Q^2 regions. This is demonstrated in Fig. 2 for various x and Q^2 values. The largest variation comes from our imprecise knowledge of the charm quark mass. For example, a $\pm 10\%$ variation of the charm mass about the central value of 1.5 GeV gives a $\pm 20\%$ variation in the structure function for small x and moderate Q^2.

At moderate Q^2, and x values larger than 0.01, the charm structure function is increasingly dominated by partonic processes near the charm quark pair-production threshold. The large size of the gluon density for small momentum fractions gives relatively large weight to such processes [27]. Although the QCD corrections at presently accessible x values are moderate (about $30 - 40\%$), with an increasing amount of data to be gathered at higher x, it is worthwhile to have a closer look at such near-threshold subprocesses. In this kinematic region, the QCD corrections are dominated by large Sudakov double logarithms. Recently [29], these Sudakov logarithms have been resummed to all orders of perturbation theory, to next-to-leading logarithmic accuracy, and, moreover, in single-particle inclusive kinematics [30]. Let us recall the main results. First, the quality of the approximation for the *next-to*-leading logarithmic threshold resummation was found to be clearly superior to leading logarithmic one. Furthermore, the resum-

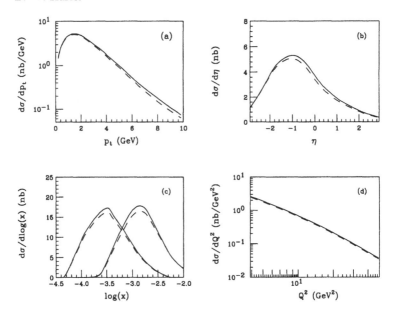

Fig. 3. Next-to-leading order differential cross sections for charm production at $\sqrt{S} =$ 301 GeV at HERA using the GRV94 (*dashed lines*) and CTEQ4F3 (*solid lines*) parton distribution sets. (**a**) Transverse momentum. (**b**) Pseudo-rapidity. (**c**) Bjorken x (*left set*) and parton momentum fraction ξ (*right set*). (**d**) Photon virtuality Q^2

mation provided next-to-next-to-leading order estimates [29], which were found to be sizable for $x \geq 0.05$.

Let us now examine some of the properties of the charm quark cross section. Results are presented in the HERA lab frame with positive rapidity in the proton direction. The proton and electron beam energies are taken to be 820 GeV and 27.6 GeV, respectively. Results are for the kinematic range $3 < Q^2 < 50\,\mathrm{GeV}^2$ and $0.1 < y < 0.7$. The CTEQ4F3 [31] and GRV94 HO [32] proton-parton distribution sets are used. The renormalization and factorization scales have been set equal to μ.

Fig. 3 shows the NLO cross sections differential in transverse momentum p_t, pseudo-rapidity η, Bjorken x, and momentum transfer Q^2 for charm quark production using the GRV94 (dashed) and CTEQ4F3 (solid) parton distribution sets at $\mu = \sqrt{Q^2 + 4m_c^2}$ with $m_c = 1.5\,\mathrm{GeV}$. From Eq. (3) the parton distributions are probed at a momentum fraction ξ which is typically one order of magnitude larger the x. This is illustrated in Fig. 3c where a plot of $d\sigma/d\log(\xi)$ vs. $\log(\xi)$ (right set of curves) is superimposed on the plot of $d\sigma/d\log(x)$ vs. $\log(x)$ (left set of curves). The difference between the curves is approximately 10% at $\xi = 10^{-2.7}$.

The scale dependence of the NLO differential cross sections is shown in Fig. 4. The curves were made using the CTEQ4F3 parton distribution set at $\mu = 2m_c$ (solid) and $\mu = 2\sqrt{Q^2 + 4m_c^2}$ (dashed) with $m_c = 1.5\,\mathrm{GeV}$. The curves show

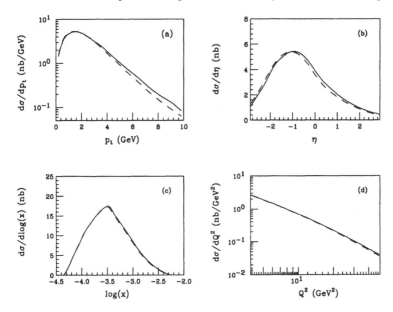

Fig. 4. Same set of distributions as Fig. 3, but this time showing the variation with respect to renormalization/factorization scale, $\mu = 2m_c$ (*solid lines*) and $\mu = 2\sqrt{Q^2 + 4m_c^2}$ (*dashed lines*)

very little scale dependence. This can be anticipated from the results shown in Fig. 2 and the distribution in Bjorken x shown in Fig. 4c. The latter shows the cross section is dominated by $x \sim 10^{-3.5} = 3.2 \times 10^{-4}$ while the former shows that, independent of Q^2, the structure function is very flat in this particular x region. Therefore, the cross section tends to be fairly insensitive to the choice of scale. Other kinematic regions show increased scale dependence.

The largest uncertainty in the structure function calculation is due to the charm quark mass. The same is true for the cross section as shown in Fig. 5. The NLO differential cross sections for charm quark production using the CTEQ4F3 parton distribution set at $\mu = \sqrt{Q^2 + 4m_c^2}$ with $m_c = 1.35\,\text{GeV}$ (solid) and $m_c = 1.65\,\text{GeV}$ (dash) are shown. Mass effects are smaller at the larger transverse mass because they are suppressed by powers of m_c/p_t in the matrix elements. However, if the range is extended much further, large logarithms of the form $\ln(p_t^2/m_c^2)$ appear in the cross section and should be resummed.

Before closing, we compare the NLO calculations described above with the most recent HERA data from H1 [4] and ZEUS [5] collaborations. The measurements make use of a tagging technique wherein the D^* meson kinematics are reconstructed using the tight constrains of the decay $D^{*+} \to D^0 \pi^+_{\text{slow}} \to (K^- \pi^+)\pi^+_{\text{slow}}$. In order to make the comparison, the theory prediction for the charm production cross sections must be converted to those of charmed meson production. This is done using a simple non-evolving Peterson *et al.* model [24] which depends on one parameter ϵ which is taken from $e^+ e^-$ data. The overall

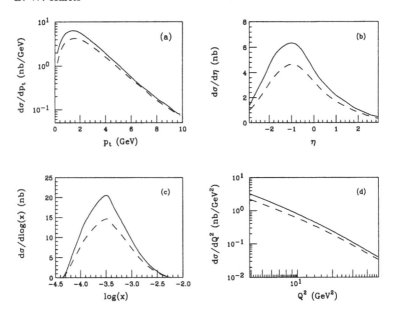

Fig. 5. Same set of distributions as Fig. 3, but this time showing the variation with respect to charm quark mass, $m_c = 1.35$ GeV (*solid lines*) and $m_c = 1.65$ GeV (*dashed lines*)

cross section normalization is set by the hadronization fraction $f(c \to D^*)$ again taken from e^+e^- data. The four-vector for the D^* is constructed from that of the charmed quark by smearing the the charm three-vector in the lab frame. The energy component is then fixed such that the four-vector has the physical D^* mass, 2.01 GeV.

Shown in Fig. 6 are D^* meson cross sections measured by the H1 collaboration [4] differential in transverse momentum $p_\perp$, pseudo-rapidity η, reconstructed parton momentum fraction x_g^{obs}, and momentum transfer Q^2 compared to the NLO calculation described in the Sec. 2 plus the Peterson *et al.* fragmentation model. The shaded band corresponds to varying the charm quark mass of 1.5 GeV by ± 0.2 GeV. Overall the agreement is good, except for the pseudo-rapidity plot in which the theory under (over) estimates the data in the forward (backward) region.

The ZEUS collaboration [5] has also measured D^* meson cross sections differential in momentum transfer Q^2, Bjorken x, hadronic energy W, transverse momentum p_t, pseudo-rapidity η, and D^* momentum fraction $x(D^*) = 2|\mathbf{p}_{\gamma P\mathrm{cms}}|/W$ which are compared with theory in Fig. 7. The boundaries of the bands correspond to varying the charm quark mass of 1.4 GeV by ± 0.1 GeV. Again, the overall agreement is good, but the theory underestimates the data in the forward region and overestimates it in the backward region. Additionally, the D^* momentum fraction data, which is particularly sensitive to the charm

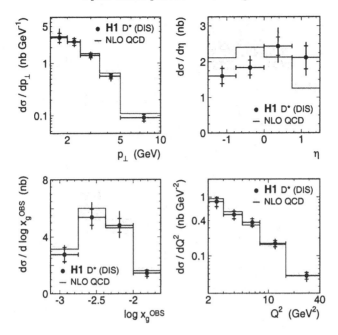

Fig. 6. Various differential cross sections for D^* meson production at HERA as measured by the H1 collaboration [4] compared to the next-to-leading order calculation described in the Sec. 2 plus a Peterson *et al.* fragmentation model. The shaded band corresponds to varying the charm quark mass from 1.3 to 1.7 GeV

hadronization process, is poorly described. Similar effects are seen in the D^* photoproduction data at HERA.

Variations of the parton distribution set, renormalization/factorization scale, charm mass, or fragmentation parameter ϵ are unable to account for the differences between data and theory. It also appears unlikely that an evolving fragmentation function would help; and the momentum transfers are large enough that any photon structure is surely negligible.

One explanation [33] proposed for the photoproduction data appears to work for the DIS data as well. Qualitatively, one is invited to think of a color string connecting the hadronizing charm quark and the proton remnant which pulls (drags) the charmed meson to the forward region. This is made quantitative in the Lund String model modified for heavy flavor production [34], as implemented in Pythia [35]. The shaded band in Fig. 7 shows what happens when the Peterson *et al.* model is replaced by an effective fragmentation model extracted from the Pythia based monte carlo RAPGAP [36]. The agreement is much improved.

Another way to improve the agreement between data and theory is to simply raise the minimum p_t of the events that are selected. Data from a slightly different decay chain, but higher minimum p_t cut are shown as open triangles. Here the Peterson and RAPGAP improved NLO predictions give essentially the same

ZEUS 1996-97

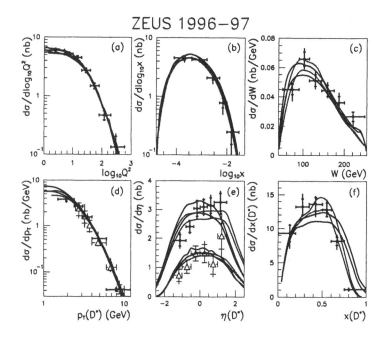

Fig. 7. Various differential cross sections for D^* meson production at HERA as measured by the ZEUS collaboration [5] compared to the next-to-leading order calculation described in the Sec. 2 plus a Peterson *et al.* fragmentation model. For the shaded band the Peterson *et al.* model was replaced by an effective fragmentation model extracted from RAPGAP which includes a drag effect between the proton remnant and the hadronizing charm. The bands result from varying the charm quark mass from 1.3 to 1.5 GeV

results, as expected. However, a seemingly large fluctuation in the forward most data bin somewhat clouds this expectation.

A number of additional studies have been done. For example, H1 [4], using the above cross sections, has extracted a NLO gluon PDF which agrees well with their own gluon PDF, obtained indirectly through structure function scaling violations, and that of CTEQ4F3. In a different approach, ZEUS [5], has extrapolated over the full phase space and extracted the structure function F_2^{charm}.

In closing, the next-to-leading order calculations described herein have been very successful in describing charm production at HERA. A variety of different observables have been studied, and the gluon parton distribution function and the charm contribution to proton structure function have been extracted. The weakest stage of the calculation is, not surprisingly, modeling the hadronization of the produced charm to the observed charmed meson, especially at low transverse momentum. One could take this as an opportunity to study hadronization in the presence of beam remnants.

Acknowledgments. I thank the organizers for the invitation and Z. Sullivan for comments on the text. The work presented herein is the result of collaborations and discussions with J. Smith and E. Laenen, and was supported in part by the U.S. Department of Energy, High Energy Physics Division, Contract No. W-31-109-Eng-38.

References

1. BFP Collaboration, G. D. Gollins *et al.*: Phys. Rev. D24, 559 (1981)
2. EMC Collaboration, J. J. Aubert *et al.*: Nucl. Phys. B213, 31 (1983)
3. M. Strovink: 'Review of multimuon production by muons'. In: *Proceedings of the 1981 International Symposium on Lepton and Photon Interactions at High Energy*, Bonn, 1981, edited by W. Pfeil (Bonn Univ., Phys. Inst. 1981) pp. 594-622
4. H1 Collaboration, C. Adloff *et al.*: Z. Phys. C72, 593 (1996); Nucl. Phys. B545, 21 (1999)
5. ZEUS Collaboration, J. Breitweg *et al.*: Phys. Lett. B407, 402 (1997); DESY 99-101, hep-ex/9908012
6. *Measurement of Charm Production in Deep Inelastic Scattering*, W. Verkerke, Ph.D. Thesis, University of Amsterdam, 1998.
7. M. Aivazis, J.C. Collins, F.I. Olness, and W.K. Tung: Phys. Rev. D50, 3102 (1994)
8. M. Buza, Y. Matiounine, J. Smith and W.L. van Neerven: Phys. Lett. B411, 211 (1997); Eur. Phys. J. C1, 301 (1998)
9. R. S. Thorne, R. G. Roberts: Phys. Rev. D57, 6871 (1998)
10. C. R. Schmidt: 'QCD phenomenology of charm production at HERA'. In: *Proceedings of the 5th International Workshop on Deep Inelastic Scattering and QCD*, Chicago, IL, 1997, edited by J. Repond and D. Krakauer (American Institute of Physics, Woodbury 1997) pp. 381-385
11. S. J. Brodsky, P. Hoyer, C. Peterson, and N. Sakai: Phys. Lett. B93, 451 (1980); S. J. Brodsky, C. Peterson, and N. Sakai: Phys. Rev. D23, 2745 (1981)
12. B. W. Harris, J. Smith, and R. Vogt: Nucl. Phys. B461, 181 (1996)
13. G.A. Schuler: Nucl. Phys. B299, 21 (1988)
14. J.C. Collins, F. Wilczek and A. Zee: Phys. Rev. D18, 42 (1978)
15. E. Laenen, S. Riemersma, J. Smith, and W.L. van Neerven: Nucl. Phys. B392, 162 (1993)
16. E. Laenen, S. Riemersma, J. Smith, and W.L. van Neerven: Nucl. Phys. B392, 229 (1993)
17. S. Riemersma, J. Smith, and W.L. van Neerven: Phys. Lett. B347, 143 (1995)
18. K. Daum, S. Riemersma, B. W. Harris, E. Laenen, and J. Smith: 'The heavy-flavour contribution to proton structure'. In: *Proceedings of the Workshop 1995/96 on Future Physics at HERA*, Hamburg, 1996, edited by G. Ingelman *et al.* (Deutches Elektronen-Synchrotron, Hamburg 1996), p. 89.
19. M. Buza, Y. Matiounine, J. Smith and W.L. van Neerven: Nucl. Phys. B472, 611 (1996)
20. B. W. Harris and J. Smith: Nucl. Phys. B452 109 (1995)
21. B. W. Harris and J. Smith: Phys. Lett. B353, 535 (1995)

22. B. W. Harris: 'Electroproduction of heavy quarks at NLO'. In: *Minneapolis Meeting of the Division of Particles and Fields of the American Physical Society*, Minneapolis, 1996, edited by J.K. Nelson and K. Heller (World Scientific, Singapore 1996) pp. 1019-1021

23. B. W. Harris and J. Smith: Phys. Rev. D57, 2806 (1998)

24. C. Peterson, D. Schlatter, I. Schmitt, and P.M. Zerwas: Phys. Rev. D27, 105 (1983)

25. B. W. Harris, E. Laenen, S. Moch, and J. Smith: 'Heavy quark production in deep-inelastic scattering at HERA'. In: *Proceedings of Workshop on Monte Carlo Generators for HERA Physics*, Hamburg, Germany, 1998, edited by G. Grindhammer, G. Ingelman, H. Jung, and T. Doyle (DESY, Hamburg 1998)

26. M. Glück, E. Reya, and M. Stratmann: Nucl. Phys. B422, 37 (1994)

27. A. Vogt, in *International Workshop on Deep Inelastic Scattering and Related Phenomena (DIS 96)*, Rome, 1996, edited by G. D'Agostini and A. Nigro (World Scientific, Singapore 1997), p. 254.

28. E. Laenen, M. Buza, B. W. Harris, Y. Matiounine, R. Migneron, S. Riemersma, J. Smith, and W. L. van Neerven: 'Deep-inelastic production of heavy quarks'. In: *Proceedings of the Workshop 1995/96 on Future Physics at HERA*, Hamburg, 1996, edited by G. Ingelman *et al.* (Deutches Elektronen-Synchrotron, Hamburg 1996), p. 393.

29. E. Laenen and S. Moch: Phys. Rev. D59, 034027 (1999)

30. E. Laenen, G. Oderda and G. Sterman, Phys. Lett. B438, 173 (1998)

31. H. L. Lai and W. K. Tung: Z. Phys. C74, 463 (1997)

32. M. Glück, E. Reya and A. Vogt: Z. Phys. C67, 433 (1995)

33. E. Norrbin and T. Sjöstrand: 'Drag effects in charm photoproduction'. In: *Proceedings of Workshop on Monte Carlo Generators for HERA Physics*, Hamburg, Germany, 1998, edited by G. Grindhammer, G. Ingelman, H. Jung, and T. Doyle (DESY, Hamburg 1998)

34. M. G. Bowler, Z. Phys. C11, 169 (1981)

35. T. Sjöstrand, Comp. Phys. Comm. 82, 74 (1994)

36. H. Jung: Comp. Phys. Comm. 86, 147 (1995)

Heavy Quarkonium Production at HERA

Christian Kiesling

Max-Planck-Institut für Physik (Werner-Heisenberg-Institut),
Föhringer Ring 6, D-80805 München

Abstract. In this article, recent results on electro- and photoproduction of heavy vector mesons at HERA are reviewed. Various models for the inelastic and elastic production mechanisms are discussed and confronted with the measurements. Elastic production allows comparison of two distinct theoretical concepts, i.e. the Regge ansatz and models using perturbative QCD (pQCD), where the colorless exchange is modeled by multiple gluon exchange. While production of light vector mesons is well described in the Regge picture, the heavy vector meson data show clear deviations from the universal "soft" Pomeron prediction. Perturbative QCD, however, is able to reproduce the trend of the data, as well as a more general Regge fit using an additional "hard" Pomeron.

1 Introduction

One of the most striking early results from the HERA collider was the detection of the strong rise [1,2] of the proton structure function F_2 towards low values of the Bjorken variable x. These findings could be interpreted by a dramatic increase of the gluon density within the proton and provided strong motivation to numerous subsequent measurements of the electron-protron interaction at low x and low momentum transfer Q^2. In the course of these measurements a new class of events were (re-)discovered [3,4], namely the diffractive production of hadrons, both in electroproduction and photoproduction (at HERA the latter are defined as events with an undetected electron in the final state, emitted at very small angles and exhibiting very small Q^2). A surprisingly large fraction of the produced events showed a large rapidity gap between the produced secondary hadrons and the outgoing nucleon, indicative for a color singlet exchange between the two hadronic systems. These events led to a revival of interest in Regge theory, more precisely in Pomeron exchange, which in the past could successfully describe "soft" diffractive processes in hadron-hadron and photon-hadron collisions. The correct prediction of the total cross section for quasi-real photons (γ^*) colliding with protons (p) at HERA, which entered an entirely new kinematic domain, was a clear triumph for the Pomeron exchange interpretation: Pomeron exchange is characterized by a universal trajectory, independent of the composition of the initial hadronic state, and predicts a slow rise of the total cross section with increasing hadronic center-of-mass energy $W = W_{\gamma^* p}$. Other data, such as the slowly rising cross sections for exclusive light vector meson (ρ, ω, ϕ) production, contributed to the strengthening of the "soft" Pomeron interpretation.

A severe blow against this consistent Pomeron picture came through the measurement of J/ψ photoproduction at HERA which, in contrast to ρ photoproduction and the soft Pomeron expectation, was observed to exhibit a much steeper rise[5,6] of the total cross section towards increasing W. The two abovementioned phenomena, i.e. strong rise of F_2 towards low x and strong rise of the cross section for diffractive processes with increasing W, can however be understood within perturbative QCD (pQCD): The rise is caused by an increased gluon density inside the proton as a consequence of the Q^2 evolution predicted in pQCD.

An even more direct path towards the gluon density within the proton is offered by inelastic heavy vector meson electroproduction: At HERA, this process is dominated by photon-gluon fusion, which is tagged as such, e.g., through the presence of a J/ψ meson, originating from the fluctuation of the virtual photon emitted from the electron into a charm-anticharm quark pair. The color singlet nature of the J/ψ in the γg subprocess is ensured by the emission of a hard gluon in the final state (inelastic reaction) or by emission of a soft gluon re-absorbed by the partonic constituents of the proton.

All these processes, involving heavy vector meson production at HERA, are extremely useful in the investigation of the detailed gluonic structure of the nucleon and complement nicely other analyses such as jet production or the Q^2 development of the structure function F_2.

2   Kinematics and Experimental Procedures

Figure 1 shows the lowest order Feynman diagram for inelastic J/ψ production, where also the kinematics of the reaction can be defined. The process is photon-gluon fusion, where the (virtual) photon γ^* with four-momentum q is emitted from the incoming electron (or positron) of four-momentum k, and the gluon is radiated off the constituents of the proton, the four-momentum of which is p_p. Besides the mass squared of the produced vector meson M_V^2, two other scales exist in the process, relevant for the subsequent discussions: The photon virtuality $Q^2 = -q^2$ and the momentum transfer squared $t = (p_p - p_X)^2$ at the lower ("proton") vertex.

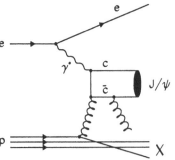

Fig. 1. Feynman diagram for inelastic vector meson (e.g. J/ψ) production as a boson gluon fusion process. The following four-vectors define the kinematics: incoming proton p_p, incoming electron k, scattered electron k', exchanged photon $q = k - k'$, outgoing vector meson p_V. The hadronic center-of-mass energy W is given by the virtual photon, emitted from the electron, and the incoming proton: $W^2 = (p_p + q)^2$

The energy y of the virtual photon, in units of the electron energy in the proton rest system, is given by

$$y = \frac{q \cdot p_p}{k \cdot p_p} . \tag{1}$$

The produced (heavy) vector meson V is emitted with energy fraction z ("elasticity") relative to the incident photon energy fraction y in the proton rest system, which is given by

$$z = \frac{p_p \cdot p_V}{p_p \cdot q} = \frac{E_V}{E_\gamma} . \tag{2}$$

The variable z can be used to separate photon-gluon fusion from resolved photon processes, which dominate for $z < 0.2$ [7]. For elastic reactions, no hard gluon is radiated and the hadronic system X is just the proton. In this limit the vector meson's elasticity z approaches unity: $z \approx 1$.

In order to measure the reaction $\gamma^* p \rightarrow VX$, both H1 and ZEUS require the heavy vector meson V to be detected essentially in the central detector (H1 also uses the backward electromagnetic calorimeter to detect J/ψ mesons decaying into electron-positron pairs). One can distinguish the following reactions producing a heavy vector meson in the final state:

- *inelastic production*: The vector meson is produced accompanied by additional hadrons from the hard gluon (see fig. 1). The initial proton at the lower vertex is transformed to a general hadronic state X. Typically, the elasticity for the J/ψ is small, $z \leq 0.9$ but larger than about 0.2 in order to exclude background where the meson V is produced in the fragmentation chain of a quark- or gluon-initiated jet.
- *dissociative production*: The vector meson is produced "elastically" (no hard gluon in the final state), but the proton dissociates into a state X with $M_X > M_p$ and $z \leq 1$.
- *elastic (exclusive) production*: The vector meson is produced truly elastically and the proton stays intact, $z \approx 1$.

3   Theoretical Concepts of Heavy Quarkonium Production

Electro- and photoproduction of vector mesons $V(= \rho, \omega, \phi, J/\psi, \Upsilon)$, i.e. $ep \rightarrow eXV$, provides a wealth of information on the properties of the strong interaction. As indicated schematically in fig. 1, one of the motivations for the study of inelastic production of vector mesons is the possibility to directly probe the gluon content in the nucleon. Processes such as exclusive vector meson production $(ep \rightarrow epV)$, on the other hand, may be used as sensitive tools to study soft and hard diffractive processes and the hadronic structure of the virtual photon. While perturbative QCD is expected to describe inelastic vector-meson production - given a sufficiently large scale in the scattering process - there are two major theoretical concepts for the description of the diffractive reactions, i.e. Regge phenomenology and, again, perturbative QCD.

3.1 Regge Models

High energy hadron-hadron and photon-hadron scattering display many features characteristic for optical diffraction, among them total cross sections slowly rising with the hadronic center of mass energy W and a strong (exponential) drop of the elastic differential cross section $d\sigma/dt$ away from the forward direction. Donnachie and Landshoff[8] have found an extremely simple parameterization of the energy dependence of all the total and differential cross sections at high energies, based on the exchange of a particular Regge trajectory[9], the so-called Pomeron (P) trajectory α_P, defined by two parameters:

$$\alpha_P(t) = \alpha_P^0 + \alpha_P' \, t \, . \tag{3}$$

The differential cross section $d\sigma/dt$ for elastic scattering processes at low values of $|t|$ is well described by a single exponential in t with a slope parameter, varying in a characteristic manner with the energy W:

$$\frac{d\sigma}{dt} \propto e^{-b|t|} \quad , \quad b = b_0 + 2\alpha_P' \ln\left(\frac{W^2}{W_0^2}\right) \, . \tag{4}$$

Here, b_0 and W_0 are process-dependent constants. Using the optical theorem, the total cross section is then given by

$$\sigma(W) \propto \frac{W^{4(\alpha_P^0-1)}}{b} \sim W^{0.22} \, , \tag{5}$$

where the numerical value in the exponent of W results from $\alpha_P^0 = 1.08$ and $\alpha_P' = 0.25 \text{GeV}^{-2}$, which have been determined from fits to all available total cross section data[8].

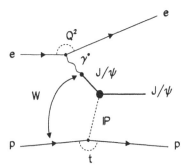

Fig. 2. Feynman diagram for elastic vector meson (e.g. J/ψ) production within the framework of Regge phenomenology, exhibiting vector-meson dominance and Pomeron exchange. In electroproduction, the hadronic center of mass energy W is given by the virtual photon, emitted from the electron, and the incoming proton.

In the framework of Regge phenomenology, electroproduction of vector-mesons is viewed as a two-step process as indicated in fig. 2: First, the (virtual) photon emitted from the electron fluctuates into a vector meson, described by the Vector-Meson Dominance model (VDM[10]), and the vector meson in turn scatters elastically off the proton, exchanging Reggeons like "ordinary" hadron-hadron interactions. Based on this picture, the following predictions are made: The total cross section for $\gamma^* p \to pV$ at high virtual-photon proton energy W rises slowly with W, no additional dependence on the photon virtuality Q^2 is expected concerning the rise of cross section with W. The differential cross section is falling exponentially with t, the slope parameter b increasing logarithmically with W, driven by $\alpha'_P = 0.25$ GeV^{-2}[8]. This feature of the differential cross section is called "shrinkage": due to the increasing b the differential cross section steepens with energy, the diffractive forward peak becomes narrower, or "shrinks". A further prediction of the Regge model is the complete transfer of the photon helicity to the final state vector meson (s-channel helicity conservation, SCHC). SCHC can be tested analyzing the angular distribution of the decay particles in the rest frame of the vector meson.

It is well-known by now that the HERA data on J/ψ photo- and electroproduction are not described by the simple ("soft") Regge model. A natural extension within the Regge phenomenology was to introduce in addition to the soft Pomeron trajectory $\alpha_P(t)$ a "hard" Pomeron trajectory $\alpha_{P'}(t)$ with suitable parameters $\alpha^0_{P'}$ and $\alpha'_{P'}$, so that the data can, again, be correctly described. Such an approach was recently proposed by Donnachie and Landshoff[11].

3.2 Perturbative QCD

Given a sufficiently large scale such as the photon virtuality, the momentum transfer t, or the mass of the produced vector meson, diffractive processes can be viewed as mediated, to lowest order, by an exchange of two gluons (in a color singlet state) between a quark-antiquark pair (emitted from the photon) and the constituents inside the proton (see fig. 3). In models based on pQCD

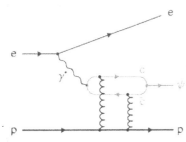

Fig. 3. Feynman diagram for elastic J/ψ production within the framework of perturbative QCD: The virtual photon couples to a $c\bar{c}$ pair and two gluons in a color singlet state are exchanged, in lowest order, between the quark pair and the proton. The $c\bar{c}$ pair forms a J/ψ with a probability derived from the bound-state wave function.

the reaction $\gamma^* p \to Vp$ (the virtual photon γ^* being radiated off an incoming electron or positron) is viewed as a temporal sequence of three sub-processes[12]: First the virtual photon γ^* fluctuates into a $q\bar{q}$ pair, which then scatters off the proton target exchanging, in lowest order, a color singlet two-gluon system or, more generally, a gluon ladder. Finally the $q\bar{q}$ pair turns into a vector meson. Elastic (heavy) vector meson production may thus be used as an alternative way to probe the gluon content of the proton[13]. While in [12] only the case of high Q^2 is considered, where the longitudinal components of the virtual photon and the produced vector meson are supposed to dominate, several authors[14–16] have generalized this model to include also transversely polarized photons and vector mesons, and heavy vector meson photoproduction, the calculations being carried out in leading order (LO) $\alpha_S \ln(Q^2/\Lambda^2_{QCD})$. The calculations of Frankfurt et al.[15] yield the following prediction for the forward differential cross section (the Ryskin formula[16] is similar):

$$\frac{d\sigma^{\gamma^* p}}{dt}\bigg|_{t=0} = \frac{12\pi^3 \Gamma_{ll} M_V^3}{\alpha_{em}(Q^2 + 4m_c^2)^4} \cdot \left|\alpha_S(\tilde{Q}^2)\left(1 + \frac{i\pi}{2}\frac{d}{d\ln x}\right) xg(x, \tilde{Q}^2)\right|^2$$
$$\cdot \left(1 + \epsilon\frac{Q^2}{M_V^2}\right) C(Q^2) . \tag{6}$$

In this expression, ϵ is the flux ratio of longitudinally to transversely polarized incoming virtual photons, given by

$$\epsilon = \frac{\Gamma_L}{\Gamma_T} = \frac{1-y}{1-y+y^2/2} . \tag{7}$$

At HERA, ϵ is very close to unity. The function $C(Q^2)$ is related to the vector meson's wave function and approaches unity from below for $Q^2 \to \infty$. The momentum fraction x_g carried by the gluon is given by

$$x_g \approx \frac{Q^2 + M_V^2}{W^2} . \tag{8}$$

The main uncertainties in these calculations for J/ψ production come from relativistic effects in the vector meson wave function (Fermi motion of the charmed quarks) and from the uncertainty of the charm quark mass. In the model of Ryskin the effective scale is $\tilde{Q}^2 = (Q^2 + M_V^2)/4$, while it is roughly a factor of 2 larger in the Frankfurt et al. model.

As can be seen from eq. 6, pQCD makes a prediction for the forward elastic differential cross section. In order to predict the total elastic cross section, the slope parameter b (see eq. 4) needs to be known, for which no QCD prediction exists. The slope has therefore to be taken from experiment and the total cross section is determined by

$$\sigma^{\gamma^* p} = \frac{1}{b}\frac{d\sigma^{\gamma^* p}}{dt}\bigg|_{t=0} . \tag{9}$$

Concerning inelastic heavy vector meson production, substantial progress has been made during the past years with the introduction of an effective field-theory framework of nonrelativistic QCD (NRQCD)[?]. In this approach the NRQCD matrix elements factorize into short-distance coefficients describing the production of a heavy quark-antiquark system, calculated using ordinary perturbative techniques, and effects from long-distance physics describing the hadronization of the quark-antiquark pair. The long-distance effects are parameterized in terms of color multiplet matrix elements (see, e.g., [18]), which can be determined from experiment or, at least in principle, from non-perturbative calculus such as lattice QCD. In contrast to elastic vector meson production, the inelastic cross section depends linearly on the gluon density in the proton.

Compared to previous work on heavy quarkonium production employing the color singlet model (see, e.g. [19] for an extensive review), the NRQCD formalism allows for the possibility that a $c\bar{c}$ pair produced in a color octet state may evolve non-perturbatively into a J/ψ or a ψ'. Such color octet contributions were in fact successful in explaining the large J/ψ and ψ' production rates observed by the CDF and D0 collaborations[20] at the Tevatron.

4 Heavy Quarkonium Production in Deep Inelastic Scattering

The subsequent discussion of the experimental data on heavy quarkonium production will concentrate mainly on the J/ψ. We will start with inclusive production and then turn to the inelastic and exclusive channels. In order to make meaningful comparisons of the various models introduced above with the experimental data, the separation of elastic and inelastic production of vector mesons is a major experimental issue; the distinction between photoproduction and electroproduction is straightforward through the observation of the scattered electron in the final state.

4.1 Inelastic Production of J/ψ Mesons

In general, the "diffractive" and "inelastic" regimes are separated by a cut in the variable z (see eq.2). The diffractive component is expected at values of z close to unity. However, large non-diffractive contributions, most importantly color octet contributions, are expected[18] at high z as well, motivating a model independent analysis without such a cut. H1 has recently published data on inclusive J/ψ production[21] (see also [22]). The J/ψ was observed in its leptonic decay modes and the kinematic range covered was $2 < Q^2 < 80$ GeV^2 and $40 << 180$ GeV. The di-lepton mass spectra for two different regions of z are shown in fig. 4, where the inelastic component of the events was identified by a cut in the energy deposited close to the proton (forward) direction ($E_{\mathrm{fwd}} > 5$ GeV). This cut excludes the low mass excitations M_X of the nucleon and with it the elastic and most of the quasi-elastic diffractive events. The selection is, however, believed to retain the color octet contributions. It was verified by Monte-Carlo methods that this cut is equivalent to a cut $M_X > 10$ GeV. As expected, the inelastic component dominates at lower values of z (fig. 4a)), while the elastic and quasi-elastic component dominates at high z (see fig. 4b)).

The corrected differential cross sections with respect to several kinematic quantities (Q^2, transverse momentum squared $p_{t,\psi}^2$ of the J/ψ, z, y^* - the rapidity of the J/ψ in the $\gamma^* p$ center of mass system - and W) are shown in fig. 5 for the inclusive as well as the inelastic data. For comparison, the results of a NRQCD calculation[18] are given, which should describe the inelastic part of the cross

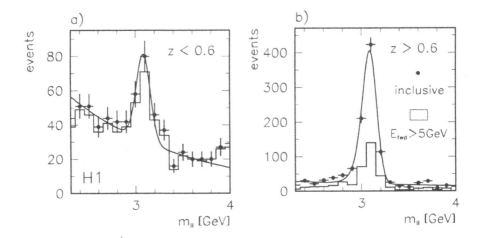

Fig. 4. Di-lepton mass spectra for events from inclusive (points) and inelastic (histogram) electroproduction of J/ψ mesons as measured in the H1 experiment[21]: a) $0.2 < z < 0.6$, b) $0.6 < z$. The curves are from fits to the signal region, convoluting a Gaussian with an exponential tail to account for radiation loss in the case of electronic decay, and an exponentially falling non-resonant background.

sections. The predictions include contributions from the color octet states $^3P_0$ and $^1S_0$, which are of order $\mathcal{O}(\alpha_S)$, and from the color singlet state $^3S_1$, which is of order $\mathcal{O}(\alpha_S^2)$. Note that both contributions are in leading order; the color singlet contribution is shown separately. The color octet contributions are found to dominate for all values of Q^2, the leading order color singlet alone falling below the data by factors 2-3. The observed differences in magnitude between predicted and measured cross sections may be indicative of the necessity to adjust the color multiplet matrix elements, determined so far from the Tevatron data alone. The differences in shape (see e.g. the y^* distribution), might be influenced by a relative adjustment of the matrix elements. Note also that a full prediction of $d\sigma/dz$ is not possible at present[18] and is therefore not shown.

One might be led to the conclusion that, like for the Tevatron data, color octet contributions seem required by the ep data, the color singlet contribution alone not being sufficient to saturate the cross sections. Still, since the calculations are only in leading order, one cannot yet claim the necessity for color octet contributions in the HERA data.

4.2   Exclusive Production of J/ψ Mesons

Both ZEUS[23] and H1[21] have recently published data on exclusive electroproduction of J/ψ mesons, $ep \to eVp$. The Q^2 (W) ranges covered are $2 < Q^2 < 80$ GeV^2 ($25 < W < 180$ GeV) for H1 and $2 < Q^2 < 40$ GeV^2 ($50 < W < 150$ GeV) for ZEUS. Exclusive production of J/ψ mesons is ensured in both experiments by requiring exactly two oppositely charged leptons in the main detector and by

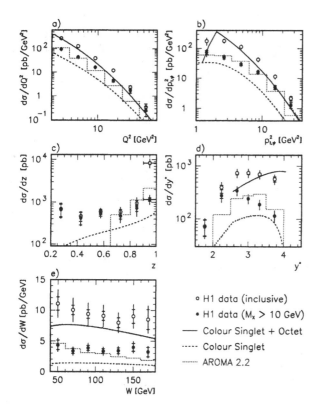

Fig. 5. Differential cross sections for the inclusive (open points) and inelastic ($M_X > 10$ GeV, full symbols) process $ep \to eJ/\psi X$ for $z > 0.2$ from H1[21]. The inner error bars are statistical, the outer bars combined statistical and systematic errors. The curves are leading order predictions for inelastic J/ψ production within the NRQCD factorization approach[18]. The histogram is from a Monte-Carlo simulation incorporating soft color interactions[24].

suppressing quasi-elastic (proton dissociation) events by requiring no activity in the forward (proton direction) parts of the detector. Residual backgrounds in the J/ψ mass region (coming mainly from low mass proton dissociation and from the Bethe-Heitler process $ep \to l^+l^-p$) are of order few percent and have been subtracted using Monte-Carlo simulations. The electroproduction cross section is converted to a virtual photoproduction cross section using

$$\frac{d^2\sigma(ep \to eJ/\psi p)}{dydQ^2} = \Gamma\sigma(\gamma^*p \to J/\psi p) , \tag{10}$$

where the function Γ is the flux of virtual photons[25], defined by

$$\Gamma = \frac{\alpha_{\mathrm{em}}}{2\pi yQ^2}\left(1 + (1 - y)^2\right) . \tag{11}$$

W dependence: Figure 6 shows the resulting cross section measurements for $\gamma^*p \to J/\psi\, p$ as a function of W for various values of the momentum transfer Q^2 in the range $2 < Q^2 < 80$ GeV^2. For comparison, also the photoproduction ($Q^2 = 0$) cross sections are shown (see also below). All cross sections exhibit a steep rise with W, in contrast to the corresponding data on exclusive light vector meson production (e.g. ρ electroproduction[23,26,27]) in a similar (low) Q^2 range. Fits of the form $\sigma \sim W^\delta$ for given Q^2 yield values $\delta \approx 0.8 - 0.9$ with

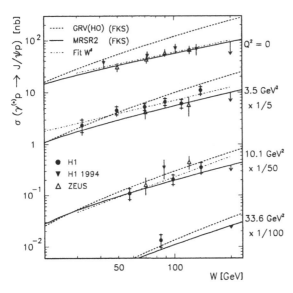

Fig. 6. Exclusive electroproduction of J/ψ mesons from H1[21] and ZEUS[23], converted to virtual photoproduction (see eq. 10) together with real photoproduction data (see discussion below). The full error bars correspond to statistical and systematic errors added in quadrature. The dashed-dotted lines result from fits of the form W^δ. Typical values are $\delta \approx 0.8 - 0.9$, with no significant variation with Q^2. The data are compared to various theoretical models (see text).

typical errors of about 0.2, clearly incompatible with the "soft Pomeron" slope of about 0.22 (see eq. 5). The dash-dotted lines give the results of these fits. Although in the $Q^2 = 10.1$ GeV^2 bin a somewhat large slope ($\delta = 1.3 \pm 0.4$) is measured, no significant variation of the slope with Q^2 is observed within the present precision. In addition to the phenomenological description, the full and dashed lines in the figure are results of calculations based on pQCD[15] using two different parameterizations for the gluon density inside the proton(GRV[28], MRSR2[29]). The absolute normalization of these predictions is sensitive to the mass m_c of the charm quark (see eq. 6). This is indicated in fig. 6 by the length of the little arrows denoting the change in normalization at the point $W = 200$ GeV, when m_c is changed from 1.4 to 1.5 GeV.

Q^2 *dependence*: The Q^2 dependence of the cross section for $\gamma^* p \to J/\psi\, p$ in the W range between 40 and 160 GeV (mean value at $W = 90$ GeV) is shown in fig. 7. At large Q^2, the naive expectation from QCD (see eq. 6) would suggest a Q^6 dependence of the cross section. However, one has to take into account the Q^2 evolution of the gluon distribution which is known to increase strongly as Q^2 grows, and the function $\mathcal{C}(Q^2)$, which both reduce the Q^2 dependence. Parameterizing the Q^2 dependence in the form $\propto (Q^2 + M_V^2)^{-n(V)}$, the J/ψ data are well described[21] by the fit with $n(J/\psi) = 2.38 \pm 0.11$. The dotted line in fig. 7 is from the pQCD model of Frankfurt et al.[15] with the MRSR2 gluon distribution, which was able to describe the W dependence of the total cross section (see fig. 6). This model also reasonably well describes the Q^2 dependence. For comparison, the cross section for $V = \rho$ is also shown in fig. 7, at somewhat lower W. Fitting the ρ data with the same Q^2 dependence yields similar values as for the J/ψ: $n(\rho) = 2.24 \pm 0.09$ from H1[26], in good agreement with ZEUS[23]

$(n(\rho) = 2.32 \pm 0.10)$. While the ρ dominates the cross section at low Q^2, the J/ψ approaches the ρ with increasing Q^2. At sufficiently large Q^2 ($Q^2 \gg M_V^2$), the ratio of cross sections should, according to the additive quark model, only depend on the electric charges of the $q\bar{q}$ state forming the vector meson, $\rho : \omega : \phi : J/\psi = 9 : 1 : 2 : 8$. This limit seems to be reached for $Q^2 > 20$ GeV^2. At very large Q^2 Frankfurt et al.[14] predict an enhancement of the J/ψ cross section, relative to the asymptotic quark model limit, by a factor of about 3.5. Accordingly, the J/ψ cross section would overtake the ρ cross section by such an amount. The presently accessible Q^2 range (up to 50 GeV^2) is obviously not yet high enough to test this prediction.

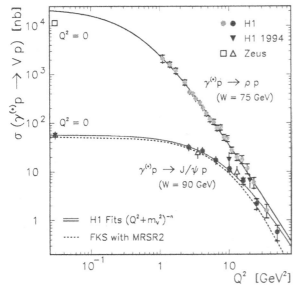

Fig. 7. Exclusive virtual-photoproduction cross sections for J/ψ ($\langle W \rangle = 90$ GeV) and ρ mesons ($\langle W \rangle = 75$ GeV) as function of the momentum transfer Q^2. The full lines are from dipole-like fits (see text) to the measurements. At large Q^2, the cross sections for ρ and J/ψ approach the SU(4) limit of 9:8. The curves are from fits of the form $(Q^2 + M_V^2)^{-n(V)}$, resulting in $n(\rho) = 2.24 \pm 0.09$[26] and $n(J/\psi) = 2.38 \pm 0.11$[21]

t Distribution: Since for electroproduction events the scattered electron is observed in the final state, the momentum transfer t can be determined unambiguously . For $|t|$ values below about 1 GeV^2 the differential cross section for elastic vector meson production can be described by a single exponential fall-off away from $|t| = 0$, characterized by the slope parameter b (see eq. 4). After correcting the experimental $|t|$ distributions for residual backgrounds (mainly from proton dissociation and the Bethe-Heitler process) H1[21] obtain $b = 4.1 \pm 0.3$(stat.)± 0.4(syst.) GeV^{-2} for the kinematic range $2 < Q^2 < 80$ GeV^2, $40 < W < 160$ GeV, in good agreement with the ZEUS[23] value of $b = 5.1 \pm 1.1$(stat.) ± 0.7(syst.) GeV^{-2} for the range $2 < Q^2 < 40$ GeV^2, $55 < W < 125$ GeV. The fits, carried out in the $|t|$ region below 1 GeV^2, are consistent with a single exponential for the elastic t distribution in both experiments. The statistical accuracy of the J/ψ electroproduction data do not yet allow for a determination of b as function of W to test shrinkage.

Note that the slope parameter $b(J/\psi)$ for elastic J/ψ production (average $Q^2 \approx 8$ GeV^2) is substantially smaller than the one for light vector mesons: For elastic ρ production[23,26] at a similar Q^2, e.g., the value $b(\rho) \approx 8$ GeV^{-2} is found. This behavior is qualitatively explained in the optical picture of scattering, where the b parameter can be viewed as the sum of the scattering partner's radii squared: $b \approx R_p^2 + R_V^2$. Due to the large mass of the charm quarks their velocities in a J/ψ bound state is small as is, as a consequence of the uncertainty principle, the J/ψ radius. The slope b is therefore given mainly by the (larger) radius of the proton, which is composed of light quarks. On the basis of this it is expected that the b slope for elastic Υ production will be similar to the one for the J/ψ.

Compared to elastic production, the b slopes for J/ψ production with proton dissociation are smaller[22,31] by typically a factor of 2, both for heavy and light vector mesons. Again the "optical" picture provides a qualitative understanding: The elastic scattering is a coherent effect involving the entire proton, while the dissociative process is thought to proceed as quasi-elastic scattering off the ("smaller") constituent quarks in the proton[32].

Helicity analysis: In order to disentangle the longitudinal and transverse virtual photon components in exclusive vector meson production, the helicity structure of the interaction can be investigated. This is usually done in the s-channel helicity frame, where the meson emission direction in the $\gamma^* p$ rest system serves as the quantization axis. At given values for W and Q^2 the exclusive production and decay of a meson $V \to l^+ l^-$ is described by three angles: θ^* and ϕ, the polar and azimuthal angles of the positively charged lepton in the V rest system, and the angle Φ between the production plane of the meson V and the electron scattering plane. The probability distribution of these three angles can be expressed by the spin density matrix elements or linear combinations of these (see, e.g., [33]). Assuming s-channel helicity conservation (SCHC), the number of independent angles reduces to two, where the polarization angle $\Psi = \phi - \Phi$ absorbs one degree of freedom. The two remaining angular distributions depend on the matrix elements (bilinear combinations of the helicity amplitudes) r_{00}^{04} and $r_{1,-1}^1$, where the lower indices denote the helicities of the vector meson in the bilinear combination and the upper indices are conventional[33] and refer to the photon helicity producing the vector meson. Integrating over the other respective angle, the following distributions are obtained:

$$\frac{d\sigma}{d\cos\theta^*} \propto 1 + r_{00}^{04} + \left(1 - 3r_{00}^{04}\right)\cos^2\theta^* \tag{12}$$

$$\frac{d\sigma}{d\Psi} \propto 1 - \epsilon\, r_{1-1}^1 \cos 2\Psi \ . \tag{13}$$

The acceptance corrected distributions from recent H1 measurements[21] of $\cos\theta^*$ and Ψ are shown in fig. 8 for two ranges of Q^2. The $\cos\theta^*$ distribution nicely supports the expectation of a dominating longitudinal photon polarization as Q^2 is increased: Fits to the data from H1[21] yield the values $r_{00}^{04} = 0.15\pm0.11$ at $\langle Q^2 \rangle = 4$ GeV^2 and $r_{00}^{04} = 0.48\pm0.15$ at $\langle Q^2 \rangle = 16$ GeV^2, suggesting that the longitudinal component of the J/ψ indeed increases with Q^2. The Ψ distributions

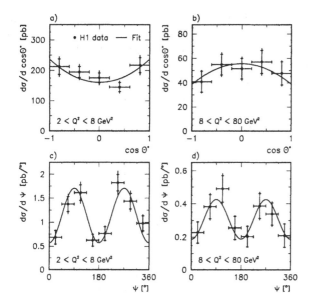

Fig. 8. Distributions of $\cos\theta^*$ (a) and b)) and of Ψ (c) and d)) for the positive lepton from J/ψ decay from the quasi-elastic (elastic plus proton-dissociative) process $ep \to eJ/\psi X$ in the range $40 < W < 160$ GeV, for two intervals in Q^2 from H1[21]. The inner error bars indicate statistical, the outer error bars statistical and systematic errors added in quadrature. The lines are the results of fits described in the text.

provide measurements of $r^1_{1-1} = 0.50 \pm 0.08$ at $\langle Q^2 \rangle = 4$ GeV^2 and $r^1_{1-1} = 0.39 \pm 0.13$ at $\langle Q^2 \rangle = 16$ GeV^2. Assuming natural parity exchange (NPE), the two matrix elements are related by $r^1_{1-1} = \frac{1}{2}(1 - r^{04}_{00})$. Using this relation and the measured values for r^{04}_{00} one obtains agreement within one standard deviation with the direct measurement of r^1_{1-1} from the Ψ distributions. Thus both SCHC and NPE are hypotheses supported by the elastic J/ψ production data at HERA.

Assuming SCHC to hold, the measurements of r^{04}_{00} can be used to determine the ratio R of longitudinal to transverse virtual photon cross sections:

$$R = \frac{\sigma_L}{\sigma_T} = \frac{1}{\epsilon} \frac{r^{04}_{00}}{1 - r^{04}_{00}} . \tag{14}$$

For the two Q^2 regions H1 obtain the results $R = 0.18^{+0.18}_{-0.14}$ at $\langle Q^2 \rangle = 4$ GeV^2 and $R = 0.94^{+0.79}_{-0.43}$ at $\langle Q^2 \rangle = 16$ GeV^2, compatible with the ZEUS result[23] of $R = 0.41^{+0.45}_{-0.52}$ at $\langle Q^2 \rangle = 5.9$ GeV^2. Together with the photoproduction data (see below) the measurements suggest a rise of R, i.e. an increase of the fraction of longitudinally polarized virtual photons, with Q^2.

Within the above-mentioned pQCD models, R is approximately proportional to Q^2/M_V^2. This relation can be tested comparing ρ with J/ψ production with the result that the value for $R(\rho)$ from elastic ρ production is found to be significantly larger[23,26] than $R(J/\psi)$ from elastic J/ψ production. But $R(\rho)$ and $R(J/\psi)$ are of the same order, when the value of Q^2 is reduced appropriately in elastic ρ production to match the values of Q^2/M_V^2 in both reactions.

4.3 Electroproduction of $\psi'(2S)$ Mesons

H1[21] has observed quasi-elastic production of $\psi(2S)$ in deep inelastic scattering using the decay channel $\psi(2S) \to J/\psi\,\pi^+\pi^-$. No separation of elastic and proton dissociation was possible due to reasons of lacking statistics. Qualitatively, the cross section ratio $\sigma(\psi(2S))/\sigma(J/\psi)$ is expected to be significantly smaller than unity due to the smaller wave function at the origin for the $\psi(2S)$ particle (1st radial excitation) compared to the J/ψ. Within the Q^2 range from 1 to 80 GeV^2 the cross section ratio $\sigma(\psi(2S))/\sigma(J/\psi)$ was determined and compared to pQCD models[14,34], which predict a rise of this ratio from ≈ 0.2 (photoproduction, see below) to ≈ 0.4 for large Q^2 ($Q^2 \gg M_V^2$), in agreement with the measurements.

5 Photoproduction of Heavy Quarkonia

Diffractive photoproduction is one of the classical domains for the successful application of Regge theory. Perturbative QCD calculations, on the other hand, are deprived of their "natural" scale Q^2 and cannot be applied here in the usual way, since in photoproduction the photon virtuality does not provide a hard scale. However, due to the Q^2 effect in the emission of photons from the incoming electron or positron suppressing high virtualities, photoproduction events are more abundant at HERA compared to deep inelastically produced events, allowing for selection of data exhibiting an alternative hard scale. In vector meson photoproduction, the mass of a produced heavy quarkonium state and/or the momentum transfer at the proton vertex may serve as such scales. For inelastic reactions, mediated dominantly by photon-gluon fusion, Regge theory cannot be applied but full next-to-leading order (NLO) QCD calculations exist[35].

5.1 Inelastic Photoproduction of Charmonium

Both H1[36] and ZEUS[37] have recently presented analyses on inelastic production of J/ψ in the photoproduction limit. The inelastic component was extracted by a cut in z: ZEUS use $0.55 < z < 0.9$ with a W range of $50 < W < 180$ GeV, while H1 selects $0.3 < z < 0.9$ and $60 < W < 180$ GeV. The data are corrected for the lower z cuts using Monte Carlo techniques. The differential cross sections $d\sigma/dz$ for both experiments (not shown) are well described by the color singlet model in NLO, but are not compatible with a leading order color octet model, as found in the ZEUS analysis. It has been pointed out[38], however, that the theoretical calculations contain considerable uncertainties, so that the color octet model based on the framework on NRQCD[?] cannot be ruled out by the HERA data. This should be compared with the conclusions from the electroproduction data (see above), where the leading order color singlet model was not sufficient to describe the measurements, but, due to missing NLO predictions, octet contributions could still not be claimed.

The ratio of the inelastic $\psi(2S)$ over J/ψ photoproduction cross section was remeasured recently by H1[36] and ZEUS[39] at photon proton center of

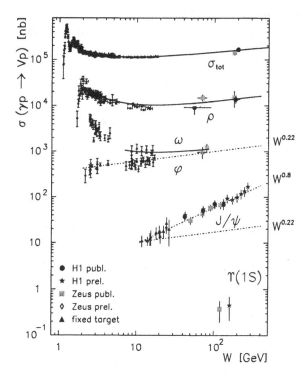

Fig. 9. Data on the total photoproduction cross section and cross sections for elastic vector meson production as function of the photon-proton center of mass energy W, for fixed target experiments and from the HERA collider. The cross sections for light vector mesons $V = \rho, \omega, \phi$ are well described by the soft Pomeron model, while J/ψ cross sections are rising much steeper with W. Statistics for Υ production is not yet sufficient to determine the W dependence of the cross section at HERA.

mass energies $60 < W < 180$ GeV (H1) and $50 < W < 180$ GeV (ZEUS). H1 finds the cross section ratio to be 0.21 ± 0.06 (the error contains statistical and systematic uncertainties), in good agreement with the ZEUS value of $0.242 \pm 0.065 \text{(stat.)} {}^{+0}_{-0.019} \text{(syst.)}$. Comparing to similar measurements from fixed target experiments at typical energies $W \approx 10$ GeV, no dependence of this ratio on W can be observed within the large uncertainties.

5.2   Exclusive Photoproduction of J/ψ Mesons

The total cross section for elastic photoproduction of light vector mesons, $\gamma p \to Vp$, $V = \rho, \omega, \phi$, is well described within the Regge model by soft Pomeron exchange (see, e.g. [27]), predicting a universal, weak dependence $\sigma \sim W^{0.22}$. This model, however, cannot accommodate heavy vector meson photoproduction $(V = J/\psi, \psi(2S), \Upsilon)$, as clearly established with the HERA data [40–45] shown in fig. 9. Fitting the J/ψ cross section to the form $\sigma \sim W^\delta$ yields $\delta \approx 0.80$, resulting from a much steeper rise of the cross section with increasing W compared to the universal expectation.

As is discussed in more detail elsewhere[27], the electroproduction cross sections for the light vector mesons also start to rise faster with W than the soft expectation, when the photon virtuality Q^2 approaches a "hard" scale, typically 10 GeV^2. The fact that the total cross section of elastic photoproduction of J/ψ mesons shows already "hard" behavior suggests that the hard scale is provided by the mass of the heavy vector meson itself ($M^2_{J/\psi} \sim 10$ GeV^2): Perturbative QCD can be applied to the deep inelastic and to the photoproduction processes, since the relevant scale $\tilde{Q}^2$ for these processes can be written[13] as

$$\tilde{Q}^2 = \frac{Q^2 + M^2_V}{4} \, , \tag{15}$$

which is much larger than Λ_{QCD}, even for photoproduction ($Q^2 = 0$) of heavy vector mesons such as the J/ψ.

Figure 10 shows the cross section $\gamma p \to J/\psi\, p$ from the fixed target experiments[46,47] and the results from ZEUS[41] and H1[40,42] together with the leading order pQCD predictions of Frankfurt et al.[15], using various parameterizations of the gluon distribution inside the proton, such as GRV(HO)[28], MRSR2[29], and CTEQ4M[48]. Note the large W range of the preliminary H1 data[42], reaching up to $W = 285$ GeV, close to the kinematic limit of HERA (300 GeV). These measurements have become possible through the use of a novel hardware trigger system at level 2, based on the neural network architecture[49].

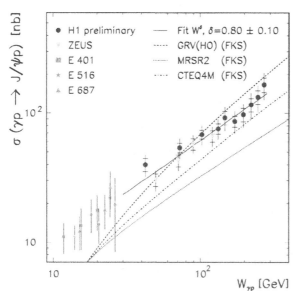

Fig. 10. Total cross section for elastic photoproduction of J/ψ mesons from HERA[40–42] and fixed target experiments[46,47]. The power law fit shown (full line) is for the HERA data alone. The curves are from a pQCD model[15] using various assumptions of the gluon density (see text). The absolute normalization of the pQCD predictions strongly depends on the mass of the charmed quark. Note that these models predict a "convex" shape in this log σ vs log W representation.

While there is a large uncertainty in the absolute value of the prediction, depending strongly on the charm quark mass (see also fig. 6), the W dependence is a "hard" prediction from pQCD: The slope of the data is best described using MRSR2 or CTEQ4, while GRV(HO) is slightly too steep. Fitting a power law ($\sigma \sim W^\delta$) to the combined HERA and fixed target data yields[50] $\delta = 0.83 \pm 0.05$, much steeper than the prediction of the Donnachie and Landshoff model using the soft Pomeron ($\delta \approx 0.22 - 0.32$). A phenomenological solution to this problem was offered by Donnachie and Landshoff[11], introducing a second, "hard", Pomeron trajectory $\alpha_{P'}(t) = 1.4 + 0.1t$. This trajectory exhibits a much stronger W rise (the intercept $\alpha_{P'}(0) = 1.4$, as opposed to 1.08 from the soft Pomeron), and a much reduced W dependence of the b parameter ($\alpha'_{P'} = 0.1$ instead of 0.25 GeV^{-2} for the soft Pomeron). A two-Pomeron fit, adjusting the relative contributions of the two Pomerons, is able to describe the data[50]. A distinctive feature of the two-Pomeron model as opposed to the pQCD prediction is its W dependence: While the pQCD model suggests a "convex" shape in the representation $\log \sigma$ vs $\log W$ (see fig. 10), the two-Pomeron model predicts a "concave" shape with an accelerated rise at the end of the HERA kinematic range. Within the present experimental accuracy, neither of the two shapes can be excluded.

t $distribution$: For photoproduction, the momentum transfer t at the proton vertex can be approximated by the transverse momentum squared, p_T^2, of the J/ψ. The p_T^2-distribution can then be used to fit an exponential $\exp(-b|t|)$ for low values of $|t|$ (typically $|t| < 1$ GeV^2). The resulting b parameters are very close to the ones obtained from electroproduction: H1[50], e.g., find a value of $b = 4.2 \pm 0.2(\text{stat.}) \pm 0.6(\text{syst.})$, in good agreement with the result from ZEUS[51], $b = 4.6 \pm 0.4(\text{stat.})^{+0.4}_{-0.6}(\text{syst.})$. Statistics in photoproduction is larger than in electroproduction so that the W dependence of the slope parameter can be studied[50]: In the range $40 < W < 150$ GeV five W bins were created and $b(W)$ was parameterized according to eq. 4, where the shrinkage parameter α'_P was allowed to freely vary in the fit. The result of the fit to the entire W range is

$$b = (4.5 \pm 0.2) + 4 \cdot (0.18 \pm 0.09) \cdot \ln\left(\frac{W}{90\text{GeV}}\right) \left[\text{GeV}^{-2}\right] . \qquad (16)$$

Shrinkage due to the soft Pomeron model demands $\alpha'_P = 0.25$, compared to 0.18 ± 0.09 from the fit. While the data are clearly compatible with being independent of W, shrinkage cannot be excluded within the present uncertainties.

5.3 Photoproduction of Υ Mesons

An even larger scale in the process $\gamma p \to Vp$ is provided when Υ mesons are produced ($V = \Upsilon$). The naive expectation would suggest a rise of the cross section with W, even steeper than for the J/ψ, driven by the gluon density in the proton which becomes larger as x_g decreases and the relevant scale $\tilde{Q}$ (here the mass of the vector meson V) increases. Both ZEUS[52] and H1[53] have observed photoproduction of Υ mesons, decaying into μ pairs. Due to limited mass resolution, the three radial excitations $\Upsilon(1S)$, $\Upsilon(2S)$ and $\Upsilon(3S)$ cannot be resolved. The main source of background is the non-resonant Bethe-Heitler process, which is subtracted statistically. Experimentally, the sum of the products of the elastic $\Upsilon(iS)$ photoproduction cross sections and their respective branching ratios B_i into $\mu^+\mu^-$ is determined. ZEUS obtain $13.3 \pm 6.0(\text{stat.})^{+2.7}_{-2.3}(\text{syst.})$ pb at a mean center of mass energy of 120 GeV, the preliminary value of H1 is

16 ± 8(stat.)± 4(syst.) pb at a mean center of mass energy of 160 GeV. Both values are in mutual agreement and are well described by a QCD model of Martin et al.[54], using parton-hadron duality. Since Υ photoproduction is reduced by a factor of about 400 relative to the J/ψ (suppressed by $(M_{J/\psi}/M_\Upsilon)^5$, see eq. 6), it is not surprising that statistics is not yet sufficient to study the W dependence of the Υ cross section.

5.4   Photoproduction at High Momentum Transfers

Diffractive processes are characterized by an exponential t dependence, $\sigma/dt \sim e^{-b|t|}$. Most of the above discussions concentrated on elastic processes, exhibiting a steeply falling t behavior and therefore limiting the $|t|$ reach to below ~ 1 GeV^2. When scattering to larger $|t|$ values is considered, an additional hard scale is introduced and improved QCD calculations become possible[55,56]. Calculations concentrating on high momentum transfers can be applied also to the lighter vector mesons.

When $|t|$ is large, the diffractive proton dissociation process is expected to dominate, where a gluon ladder described by the BFKL formalism[57] can be exchanged and predictions can be made for both the t and W dependences[55,56].

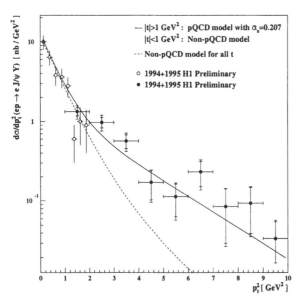

Fig. 11. Differential cross section for J/ψ photoproduction with proton dissociation, as measured by H1[58] (full symbols). The open points are taken from [6], where $Q^2 < 4$ GeV^2. The solid line represents the combination of a pQCD calculation[56] for $|t| > 1$ GeV^2 and a non-perturbative model when $|t| < 1$ GeV^2. The normalization was determined using the measured cross section for $|t| > 1$ GeV^2. The dashed line represents the prediction from the non-perturbative model when the full kinematic range is used.

Both H1[58] and ZEUS[59] have studied quasi-elastic J/ψ photoproduction at large $|t|$. As in the elastic case, the $|t|$ distribution is obtained from the measured p_T^2 of the final state J/ψ, where $t = -p_T^2$ holds to a very good approximation. As an example, the differential cross section $d\sigma/dt$ for J/ψ photoproduction with proton dissociation from H1 is shown in fig. 11(full symbols), the open points are from elastic photoproduction[6]. Both sets of data are restricted to the kinematic range $30 < W < 150$ GeV. A clear decrease of the slope with increasing p_T^2 is observed. The solid line in the figure is the result of a pQCD calculation[56] in the region $|t| > 1$ GeV^2, using a non-perturbative model for $|t| < 1$ GeV^2. The prediction agrees well with the data, implying that the pQCD calculation describes correctly the large $|t|$ behavior.

According to the expectations from perturbative QCD, the diffractive photo-production of vector mesons at a hard scale should be, to a good approximation, flavor-independent and proportional to the electromagnetic coupling of the photon to the mesonic $q\bar{q}$ state. In a ZEUS analysis[59] the large $|t|$ behavior in photoproduction also of the light vector mesons (ρ, ϕ) has been studied. It is found that the cross section ratios $\sigma(J/\psi)/\sigma(\rho)$ and $\sigma(\phi)/\sigma(\rho)$ approach their quark model limits 8:9 and 2:9, respectively, as $|t|$ increases. These results also suggest that $|t|$, just like Q^2 and M_V, can play the role of a hard scale, enabling perturbative QCD calculations. A combination of all the three relevant scales may provide an interesting hard scale $\tilde{Q}^2$: It has in fact been suggested[13] earlier, although limited to heavy quarkonia (J/ψ) production, to use the combined scale $\tilde{Q}^2 = (Q^2 + M_V^2 + |t|)/4$. From the present HERA data it seems that such a restriction to heavy quarkonia may not be necessary.

6 Conclusions

HERA has provided a wealth of precise data on electro- and photoproduction of heavy quarkonia, giving access to the rich field of diffractive phenomena in ep collisions. While the diffractive processes involving light vector mesons (ρ, ω, ϕ) show the typical "soft" behavior, i.e. weak dependence of the cross section on W, well described by Pomeron exchange, the corresponding cross sections for heavy quarkonia (J/ψ) exhibit a steep rise with W, incompatible with the soft picture. Another prediction of the soft Pomeron model , i.e. shrinkage of the differential cross section, cannot be seriously tested with heavy quarkonia at present due to lack of statistics.

The W dependence of the cross sections $\sigma(\gamma^{(*)}p \to Vp, V = J/\psi, \Upsilon)$, however, is explained within pQCD by the steep rise of the gluon density in the proton with decreasing x (= increasing W). Also in contrast to the soft Pomeron model, pQCD does not expect shrinkage of the differential cross section. The hard scale necessary to justify pQCD is apparently provided by the mass of the heavy vector meson. This view is beautifully supported by diffractive photoproduction of J/ψ, where pQCD can successfully describe the observed cross sections, the J/ψ mass being the only hard scale in the reaction.

While the soft, single Pomeron model is unable to account for the cross section rise in heavy quarkonium production, models with an additional "hard" Pomeron (two new parameters which need adjustment, plus the relative strength of the two Pomeron components) are able to describe the HERA data again. Also the question of shrinkage, thought to be a distinctive feature, loses its importance: The hard Pomeron parameters suggest a largely reduced shrinkage effect. As a consequence many of the salient features distinguishing Regge models from pQCD predictions need largely improved statistics, such as shrinkage, the exact shape of the cross section as function of W ("convex" or "concave"), or the open question of SCHC. Future running at HERA with high luminosity will significantly improve the precision of the data.

For future comparisons of theoretical models with, hopefully, statistically improved data one should keep in mind that many of the pQCD predictions for heavy quarkonium production at present are available only in leading order, thus precluding serious confrontation with experiment. One such example is the inelastic production of J/ψ mesons, where the size of possible color octet contributions, apparently necessary for the Tevatron data, cannot yet be quantified in ep collisions: For photoproduction, the NLO color singlet calculation is able to describe the differential cross section $d\sigma/dz$. Additional leading order color octet contributions, especially at high z, would largely exceed the data. But a suppression of such additional color octet contributions in higher order, then being in agreement with the HERA data, is not excluded. On the other hand for electroproduction only the LO color singlet and octet calculations are available. Here, the color singlet prediction alone clearly undershoots the data. Due to a lacking NLO calculation, however, one cannot use this observed discrepancy to claim the necessity of additional octet contributions.

To reach a more complete theoretical understanding along with the expected improvement of the experimental results would be highly desirable: One of the main motivations for studying heavy vector meson production in ep collisions is the possibility to directly extract the gluon density in the proton. This important additional information from the heavy vector meson sector has yet to be supplied and would nicely complement other analyses on the gluonic structure of the nucleon.

Acknowledgments

I am indebted to the organizers for this very pleasant meeting. I would like to thank my colleague Beate Naroska for a critical reading of the manuscript and many helpful suggestions.

References

1. H1 Collaboration: T. Ahmed et al., Nucl. Phys. B439 (1995) 471.
2. ZEUS Collaboration: M. Derrick et al., Z. Phys. C65 (1995) 379.
3. H1 Collaboration: T. Ahmed et al., Nucl. Phys. B429 (1994) 477.
4. ZEUS Collaboration: M. Derrick et al., Phys. Lett. B315 (1993) 481.
5. ZEUS Collaboration: M. Derrick et al., Phys. Lett. B350 (1995) 120.
6. H1 Collaboration: S. Aid et al., Nucl.Phys. B472 (1996) 3.
7. H. Jung, G. A. Schuler and J. Terron, Int. Jour. of Mod. Phys. A7 (1992) 7955.
8. A. Donnachie and P.V. Landshoff, Phys. Lett. B296 (1992) 227.

9. T. Regge, Nuovo Cim. 14 (1959) 951; J.J. Sakurai, Ann. Phys. 11 (1960) 1; see also P.D.B. Collins: An Introduction to Regge Theory and High Energy Physics (Cambridge University Press, 1977).

10. J.J. Sakurai, Phys. Rev. Lett. 22 (1969) 981; H. Fraas and D. Schildknecht, Nucl. Phys. B14 (1969) 543.

11. A. Donnachie and P.V. Landshoff, Phys. Lett. B437 (1998) 408.

12. S.J. Brodsky et al., Phys. Rev. D50 (1994) 3134.

13. M.G. Ryskin, Z. Phys. C57 (1993) 89.

14. L. Frankfurt, W. Koepf and M. Strikman, Phys. Rev. D54 (1996) 3194.

15. L. Frankfurt, W. Koepf and M. Strikman, Phys. Rev. D57 (1998) 512.

16. M.G. Ryskin, R.G. Roberts, A.D. Martin and E.M. Levin, Z. Phys. C76 (1997) 231.

17. G.T. Bodwin, E. Braaten and G.P. Lepage, Phys. Rev. D51 (1995) 1125.

18. S. Fleming and T. Mehen, Phys. Rev. D57 (1998) 1846.

19. G.A. Schuler, CERN-TH-7170-94, hep-ph/9403387.

20. CDF Collaboration: F. Abe et al., Phys. Rev. Lett. 69 (1992) 3704; D0 Collaboration: S. Abachi et al., Phys. Lett. B370 (1996) 239; CDF Collaboration: F. Abe et al., Phys. Rev. Lett. 79 (1997) 572.

21. H1 Collaboration: C. Adloff et el., Eur. Phys. J. C10 (1999) 373.

22. A. Meyer: PhD Thesis, University of Hamburg, 1998.

23. ZEUS Collaboration: J. Breitweg et al., Eur. Phys. J. C6 (1999) 603.

24. G. Ingelman, J. Rathsman and G.A. Schuler, Comp. Phys. Comm. 101 (1997) 135.

25. L.N. Hand, Phys. Rev. 129 (1963) 1834.

26. H1 Collaboration: C. Adloff et al., DESY-99-010, to appear in Eur.Phys.J. C.

27. see also A. Proskuryakov: Experimental Results on the Production of Light Vector Mesons, these proceedings.

28. M. Glück, E. Reya and A. Vogt, Z. Phys. C67 (1995) 433.

29. A.D. Martin, R.G. Roberts and W.J. Stirling, Phys. Lett. B387 (1996) 419.

30. H1 Collaboration: C. Adloff et al., Z. Phys. C75 (1997) 607.

31. H1 Collaboration: Comparison of Elastic and Proton Dissociative ρ Meson Electroproduction from H1, contributed paper 157n to the International Europhysics Conference on High Energy Physics HEP99, Tampere, Finland, July 1999.

32. S.P. Misra, A.R. Panda and B.K. Parida, Phys. Rev. D22 (1980) 1574.

33. K. Schilling and G. Wolf, Nucl. Phys. B61 (1973) 381.

34. J. Nemchik et al., Phys. Lett. B341 (1994) 228 and J. Exp. Theor. Phys. 86 (1998) 1054.

35. M. Krämer et al., Phys. Lett. B348 (1995) 657; M. Krämer, Nucl. Phys. B459 (1996) 3.

36. H1 Collaboration: Inelastic Photoproduction of J/ψ and $\psi(2S)$ at H1, contributed paper 157aj to the International Europhysics Conference on High Energy Physics HEP99, Tampere, Finland, July 1999.

37. ZEUS Collaboration: Measurement of Inelastic J/ψ Photoproduction at HERA, contributed paper 814 to the XXIX International Conference on High Energy Physics ICHEP98, Vancouver, Canada, July 1998.

38. M. Beneke and M. Krämer, Phys. Rev. D55 (1997) 5269; B.A. Kniehl and G. Kramer, Phys. Rev. D56 (1997) 5820; B.A. Kniehl and G. Kramer, Phys. Lett. B413 (1997) 416; M. Beneke, I.Z. Rothstein and M.B. Wise, Phys. Lett. B408 (1997) 373.

39. ZEUS Collaboration: Inelastic Charmonium Photoproduction at HERA, contributed paper 504 to the International Europhysics Conference on High Energy Physics HEP99, Tampere, Finland, July 1999.

40. H1 Collaboration: S. Aid et al., Nucl. Phys. B468 (1996) 3.
41. ZEUS Collaboration: J. Breitweg et al., Z. Phys. C75 (1997) 215.
42. H1 Collaboration: C. Adloff et al.: Elastic Production of J/ψ Mesons in Photo-production and at High Q^2 at HERA, contributed paper 242 to the International Europhysics Conference on High Energy Physics HEP97, Jerusalem, August 1997.
43. ZEUS Collaboration, J. Breitweg et al.: Exclusive Vector Meson Production in Deep Inelastic Scattering at HERA, contributed paper 639 to the International Europhysics Conference on High Energy Physics HEP97, Jerusalem, August 1997.
44. H1 Collaboration: C. Adloff et al., Phys. Lett. B421 (1998) 385.
45. ZEUS Collaboration: J. Breitweg et al., Z. Phys. C76 (1997) 599.
46. E401 Collaboration: M. Brinkley et al., Phys. Rev. Lett. 48 (1982) 73.
47. E516 Collaboration: B.H. Denby et al., Phys. Rev. Lett. 52 (1984) 795.
48. H.L.Lai et al., Phys. Rev. D55 (1997) 1280.
49. J.H. Köhne et al., Realization of a second level neural network trigger for the H1 experiment at HERA, Nucl. Inst. Meth. A389 (1997) 128.
50. H1 Collaboration: Diffractive J/ψ Production, contributed paper 157aj to the International Europhysics Conference on High Energy Physics HEP99, Tampere, Finland, July 1999.
51. ZEUS Collaboration: J. Breitweg et al., kZ. Phys. C75 (1997) 215.
52. ZEUS Collaboration: J. Breitweg et al., Phys. Lett. B437 (1998) 432.
53. H1 Collaboration: Observation of Υ Production at HERA, contributed paper 574 to the XXIX International Conference on High Energy Physics ICHEP98, Vancouver, Canada, July 1998.
54. A.D. Martin, M.G. Ryskin and T. Teubner, Phys. Lett. B454 (1999) 339.
55. J.R. Forshaw and M.G. Ryskin, Z. Phys. C68 (1995) 137.
56. J. Bartels, J.R. Forshaw, H. Lotter and M. Wüsthoff, Phys. Lett. B375 (1996) 301.
57. S.J. Brodsky et al., JETP Lett. 70 (1999) 155.
58. H1 Collaboration: Production of J/ψ Mesons with large $|t|$ at HERA, contributed paper 274 to the International Europhysics Conference on High Energy Physics HEP97, Jerusalem, August 1997.
59. ZEUS Collaboration: Study of Vector Meson Production at Large $|t|$ at HERA and Determination of the Pomeron Trajectory, contributed paper 788 to the XXIX International Conference on High Energy Physics ICHEP98, Vancouver, Canada, July 1998.

Color-Octet Contributions
to Inelastic Heavy-Quarkonium Production

Jungil Lee

II. Institut für Theoretische Physik, Universität Hamburg, 22761 Hamburg, Germany

Abstract. The color-octet mechanism based on the nonrelativistic QCD (NRQCD) factorization approach in inclusive heavy quarkonium production in high energy processes is reviewed. As phenomenological applications, we discuss charmonium production at HERA and Υ production associated with weak gauge bosons at the LHC.

1 Introduction

In high energy experiments, $J^{PC} = 1^{--}$ quarkonium states, such as the charmonium state J/ψ and the bottonium state Υ, have been used as clean probes to study QCD. They provide an enormous suppression of the backgrounds due to their leptonic decay modes. For a long time, the color singlet model (CSM) [1,2] has been a reasonable picture to describe heavy quarkonium production and decay. The CSM allows for a color-singlet heavy quark pair to form a quarkonium system in the same spectroscopic state $^{2S+1}L_J$. Factorization of the short distance process from bound state formation is assumed due to the separation of the two scales governing them. The short distance process is calculated in perturbative QCD, and the wavefunction and its derivatives at the origin, which involve nonperturbative physics, are calculated within potential models or phenomenological analyses.

In the middle of the 1990's, serious discrepancies between several experimental results and the CSM were found, and this naive factorization approach was reexamined. For example, the dominant mechanism of large p_T J/ψ production in $p\bar{p}$ collisions, has been found to be parton fragmentation in the framework of the CSM, by Braaten and Yuan [3]. This fragmentation mechanism has been applied to various studies on prompt charmonium production at the Tevatron [4], and the CDF data on prompt J/ψ production [5] qualitatively meet these theoretical predictions. This fragmentation picture in the CSM was not enough. The ψ' production rate measured by CDF was 30 times larger than the theoretical prediction even after including the parton fragmentation mechanism, which is the so-called ψ' anomaly [6]. Based on NRQCD, Braaten and Fleming proposed the color-octet gluon fragmentation picture as a possible scenario to resolve this problem [7]. They included gluon fragmentation into a color-octet $Q\bar{Q}$ pair in the $^3S_1^{(8)}$ state followed by the soft transition of this pair into physical quarkonium. This subprocess is not allowed in the CSM, where only the color-singlet $Q\bar{Q}$ pair in the same spectroscopic state $(^3S_1^{(1)})$ corresponding to the final-state quarkonium can be hadronized into that quarkonium. As an effective field theory of QCD, NRQCD [8,9] was developed to describe heavy quarkonium production and decay in a rigorous way from the first principles of QCD. In NRQCD, it is allowed to make a quarkonium state from various $Q\overline{Q}$'s with color and spin states which need not be the same as those of the final quarkonium state. Among an infinite number of channels, only a few channels dominate the soft transition process, as may be seen with the help of NRQCD velocity scaling rules [8,9]. Newly introduced nonperturbative NRQCD matrix elements should then be fixed from experimental data. Efforts are being made to extract the NRQCD matrix elements from experiments to test universality.

This talk is organized as follows. We first review the basics of NRQCD in the analysis of the inclusive production of heavy quarkonium. Then we discuss charmonium production at HERA and associate production of Υ and weak gauge bosons at the LHC.

2   NRQCD Basics

2.1   Lagrangian

By use of the mass splitting of heavy quarkonia, one may assume the velocity v of the heavy quark inside the system to be small, and therefore one finds the separation among the three important scales, heavy quark mass M, momentum in meson rest frame Mv and binding energy Mv^2 as

$$M \gg Mv \gg Mv^2 \quad \text{where} \quad v_b^2 \sim 0.1, \quad v_c^2 \sim 0.3. \tag{1}$$

The NRQCD Lagrangian is obtained from full QCD by introducing an ultravio-

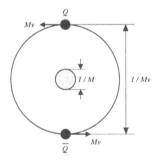

Fig. 1. Typical scales of a heavy quarkonium system.

let momentum cutoff Λ that is of order M, since the important nonperturbative physics involves momenta of order Mv or less. One can compensate the removal of the relativistic states by adding local interactions to the Lagrangian order by order in $1/M$. Due to the decoupling of particle and antiparticle in this region, the dynamics of the heavy quarks can be described in nonrelativistic Schrödinger field theory. Only the heavy quark part is modified from the full QCD Lagrangian [8,9]:

$$\mathcal{L} = \mathcal{L}_{\text{light}} + \mathcal{L}_{\text{heavy}} + \delta\mathcal{L}, \tag{2}$$

$$\mathcal{L}_{\text{heavy}} = \psi^\dagger \left(iD_t + \frac{\mathbf{D}^2}{2M}\right)\psi + \chi^\dagger\left(iD_t - \frac{\mathbf{D}^2}{2M}\right)\chi. \tag{3}$$

In LO ($\delta\mathcal{L} = 0$), there is no spin-flip operator and this leads to *approximate* heavy quark spin symmetry (HQSS) of NRQCD. The LO NRQCD Lagrangian naturally allows for a color-octet term via the chromoelectric dipole (E1) transition which is included in $\mathbf{D}^2/(2M)$. The relativistic effects from full QCD are included in the correction term $\delta\mathcal{L}$. The correction term is composed of bilinear terms

$$\delta\mathcal{L}_{\text{bilinear}} = \frac{c_1}{8M^3}\left(\psi^\dagger(\mathbf{D}^2)^2\psi - \chi^\dagger(\mathbf{D}^2)^2\chi\right) + \cdots$$
$$+ \frac{c_4}{2M}\left(\psi^\dagger(g\mathbf{B}\cdot\sigma)\psi - \chi^\dagger(g\mathbf{B}\cdot\sigma)\chi\right), \tag{4}$$

which describe further details of low energy physics, and four-fermion interaction terms

$$\delta\mathcal{L}_{4-\text{fermion}} = \sum_n\left[\frac{f_n(\Lambda)}{M^{d_n-4}}\psi^\dagger\Gamma_n\chi\chi^\dagger\Gamma'_n\psi + \frac{f_n^{EM}(\Lambda)}{M^{d_n-4}}\psi^\dagger\Gamma_n\chi|0\rangle\langle0|\chi^\dagger\Gamma'_n\psi\right]. \tag{5}$$

The former represent the heavy quark pair annihilation into light hadrons while the latter describe the electromagnetic annihilation in NRQCD. The short distance scale physics is included in the coefficient f_n^{EM}. The bilinear part includes all the higher dimensional operators coming from the expansion in the heavy quark velocity v. Here we find the spin-flip operator σ in the chromomagnetic dipole (M1) transition.

2.2   Velocity Scaling Rules

The velocity scaling rules [8,9] classify the importance of each term in the NRQCD interaction Lagrangian in powers of v_Q^2. For example, the most important channels of J/ψ production can be classified as

$$(c\bar{c})(^3S_1^{(8)}) \to J/\psi : (E1)^2 : v^4 \times (c\bar{c})(^3S_1^{(1)}) \to J/\psi,$$
$$(c\bar{c})(^1S_0^{(8)}) \to J/\psi : (M1)^1 : v^3 \times (c\bar{c})(^3S_1^{(1)}) \to J/\psi, \tag{6}$$
$$(c\bar{c})(^3P_J^{(8)}) \to J/\psi : (E1)^1 : v^4 \times (c\bar{c})(^3S_1^{(1)}) \to J/\psi.$$

Note that $^3P_J^{(8)}$ is further suppressed by v^2 in addition to the E1 transition factor v^2, since the operator itself has a covariant derivative. If we only consider the long distance part, the color singlet matrix element dominates over the color octet ones. Therefore the color octet mechanism dominates over the color singlet channel only when the short distance coefficient for the octet channel is large enough to compensate the suppression in the long distance factor.

2.3   Factorization and Matching in Inclusive Heavy Quarkonium Production

Factorization in the inclusive heavy quarkonium production rate comes from the idea that the probability of forming a bound state from a heavy quark pair produced in a short distance process is significant only if the corresponding production points are separated by a distance of order $1/M$ or less [9]. Then the production rate of the heavy quark pair within a distance scale of order $1/M$ is factored out from the formation of the bound state with distance scale of order $1/(Mv)$ or larger. The latter is accurately described by NRQCD. The vacuum expectation value of a four-quark operator describes the transition rate of the long distance process.

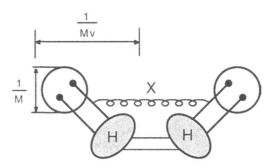

Fig. 2. Diagram representing the NRQCD color octet matrix element.

As shown in Fig.2, the operator creates the heavy quark pair at a spacetime point and propagates into the asymptotic future, where the final state includes the heavy quarkonium state, and then propagates back to the creation point. The factorization formula of inclusive quarkonium H production is then given by [9]

$$\sigma(H) \;=\; \sum_n \frac{F_n(\Lambda)}{M^{d_n-4}} \, \langle 0|\mathcal{O}_n^H(\Lambda)|0\rangle, \tag{7}$$

where F_n is the short distance coefficient, which is infrared and collinear finite, and $\mathcal{O}_n^H$ is the four-quark operator of the inclusive production mentioned above. The factorization scale Λ is of order of the heavy quark mass M. The most important operators in 1^{--} quarkonium production are given by

$$\mathcal{O}_1^H(^3S_1) = \chi^\dagger \sigma^i \psi \left(a_H^\dagger a_H\right) \psi^\dagger \sigma^i \chi,$$

$$\mathcal{O}_8^H(^1S_0) = \chi^\dagger T^a \psi \left(a_H^\dagger a_H\right) \psi^\dagger T^a \chi,$$

$$\mathcal{O}_8^H(^3S_1) = \chi^\dagger \sigma^i T^a \psi \left(a_H^\dagger a_H\right) \psi^\dagger \sigma^i T^a \chi, \tag{8}$$

$$\mathcal{O}_8^H(^3P_0) = \frac{1}{3}\chi^\dagger \sigma^i D^i T^a \psi \left(a_H^\dagger a_H\right) \psi^\dagger \sigma^j D^j T^a \chi,$$

$$a_H^\dagger a_H \quad \equiv \sum_X \sum_{m_J} |H+X\rangle\langle H+X|,$$

where the sums are over the spin states of H and over all other final states. The scale dependence of each matrix element is canceled after summing over all the contributions. Expanding the perturbative QCD amplitude in v, one can find out the partial wave amplitudes with NRQCD operators. In this matching procedure, one can extract the short distance coefficients F_n.

3 Quarkonium Production at HERA

After the color-octet mechanism was applied to high p_T ψ' production via gluon fragmentation at the Tevatron [7], studies of J/ψ production including the color-octet mechanism were activated. Cho and Leibovich considered a large class of color-octet diagrams which can contribute to J/ψ production at hadron colliders [10], below the region where fragmentation is dominant. The same approach has been used in inclusive photoproduction of J/ψ [11,12]. In the region where gluon fragmentation does not dominate, two more NRQCD color-octet matrix elements, $\langle \mathcal{O}_8^{J/\psi}(^1S_0)\rangle$ and $\langle \mathcal{O}_8^{J/\psi}(^3P_0)\rangle$, contribute significantly. Unfortunately, they appear in a linear combination, and this makes it difficult to determine their numerical values separately. Furthermore, the results obtained from the Tevatron data [10] are not appropriate to explain the photoproduction data [11,12]. The energy distribution of the produced J/ψ meson is one of the most important measurable quantities in this process. While the CSM is enough to explain the data in the lower z region [13,14] (z is the J/ψ energy fraction of the initial photon in the proton rest frame), the data in the higher z region region ($z <$ $0.8 \sim 0.9$), where color-octet contribution is suppressed, was not well explained. The diverging behavior of the contribution in the higher z region due to the color-octet mechanism seemed to be troublesome. However, Beneke, Rothstein and Wise showed that the NRQCD expansion may lead to a breakdown near the boundaries of phase space [15]. Further analysis on this end point region dynamics would give a clearer explanation of the J/ψ energy distribution in the high z region. Fleming and Mehen [16] suggested that the leptoproduction analysis could avoid the end point region ambiguity. In the high Q^2 region of leptoproduction, the corrections coming from the non-relativistic approximation of the quarkonium phase space become negligible [16]. They also predicted the polarization of J/ψ in leptoproduction. But the data sample is not sufficient to confirm their predictions yet [17]. Recently, Kniehl and Kramer took into account HO QCD effects [18] due to the multiple emission of gluons, which had been estimated by Monte Carlo techniques [19]. As a consequence, the excess of the NRQCD prediction at z close to unity over the HERA data disappeared. So, the complete NLO analysis on this process is called for.

4 $pp \rightarrow \Upsilon + W/Z + X$

The associated production of a $W^{\pm}$ or Z^0 boson and a J/ψ meson has been studied previously, but only for J/ψ's with large transverse momentum, where the process is dominated by gluon fragmentation into J/ψ [20]. The differential cross sections for $W^{\pm} + J/\psi$ and $Z^0 + J/\psi$ peak at small transverse momentum of the J/ψ. Because the mass of the Υ is larger than that of the J/ψ by a factor of 3, the cross sections for $W^{\pm} + \Upsilon$ and $Z^0 + \Upsilon$ are much smaller than those for $W^{\pm} + J/\psi$ and $Z^0 + J/\psi$. However the larger mass of the Υ makes it possible to observe the leptons from its decay, even if the Υ is produced at small transverse momentum. It is therefore possible to measure the total cross sections for $W^{\pm} + \Upsilon$ and $Z^0 + \Upsilon$.

Let us consider the production of $W^{\pm}/Z + \Upsilon$ at a hadron collider. The $2 \rightarrow 3$ parton processes that produce $W^+/Z + b\bar{b}$ are shown in Fig.3. The cross sections

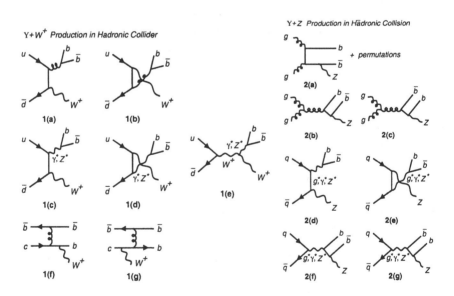

Fig. 3. $2 \rightarrow 3$ parton processes that produce $W^+/Z + b\bar{b}$.

for the production of $W^{\pm} + \Upsilon$ and $Z^0 + \Upsilon$ in pp collisions at the LHC with center-of-mass energy 14 TeV are [21]

$$\sum_{\pm} \sigma(W^{\pm} + \Upsilon) = 10 \text{ fb} \frac{\sum B\langle \mathcal{O}_1(^3S_1)\rangle}{8 \text{ GeV}^3} + 4000 \text{ fb} \frac{\sum B\langle \mathcal{O}_8(^3S_1)\rangle}{0.4 \text{ GeV}^3}, \quad (9)$$

$$\sigma(Z^0 + \Upsilon) \quad = 500 \text{ fb} \frac{\sum B\langle \mathcal{O}_1(^3S_1)\rangle}{8 \text{ GeV}^3} + 1300 \text{ fb} \frac{\sum B\langle \mathcal{O}_8(^3S_1)\rangle}{0.4 \text{ GeV}^3}. \quad (10)$$

where $\sum B\langle \mathcal{O}_{1,8}(^3S_1)\rangle$ includes the feed-down processes from higher bottonium states as well as direct production of Υ. Their numerical values are estimated [21] as shown in the denominator. The leading order GRV-94 parton distributions are used, and the factorization and renormalization scales are set equal. We neglect the masses of the u, d, s, and c quarks, and we use the value 4.7 GeV for the b quark mass. The theoretical error due to unknown higher order perturbative corrections can be estimated by varying the factorization and renormalization scales by a factor of 2 and is roughly 35%. The theoretical error due to the uncertainty in the b quark mass can be estimated by varying m_b by 0.25 GeV and is roughly 25%. With an integrated luminosity of 10 fb^{-1}, the number of events in the purely leptonic decay channels should be about 440 for $W^{\pm} + \Upsilon$ and about 70 for $Z^0 + \Upsilon$. Even after allowing for detector acceptances and efficiencies, there should be enough events to make these processes observable. In Fig. 4, we plot the invariant mass distributions $d\sigma/dM_{W\Upsilon}$ (summed over $W^{\pm}$) and $d\sigma/dM_{Z\Upsilon}$. They peak at only a few GeV above the thresholds at 89.8 GeV for $W^{\pm} + \Upsilon$ and 100.6 GeV for $Z^0 + \Upsilon$. The cross sections for $W^{\pm} + \Upsilon$ and $Z^0 + \Upsilon$ are both dominated by the $\langle \mathcal{O}_8(^3S_1)\rangle$ term. While the color-singlet matrix element is accurately normalized by the leptonic decay width of the Υ, there is a large uncertainty in the normalization of the color-octet matrix element. The uncertainty in the value of $\sum B\langle \mathcal{O}_8(^3S_1)\rangle$ leads to a substantial uncertainty in the normalizations of the cross sections. However, this uncertainty cancels in the ratio $\sum_{\pm} \sigma(W^{\pm}+\Upsilon)/\sigma(Z^0+\Upsilon)$, which is predicted to be about 3 at the Tevatron and about 2 at the LHC. If a significant deviation from this prediction will be observed at the LHC, this might be evidence for an additional contribution from a heavy particle that decays into $W^{\pm} + b\bar{b}$ or $Z^0 + b\bar{b}$.

The processes of $W^{\pm} + \Upsilon$ and $Z^0 + \Upsilon$ production provide lamp-posts under which one can look for new physics. The most promising possibility is to search for a charged Higgs via the decay $H^+ \rightarrow W^+ + \Upsilon$. The decay rate of the charged Higgs into $W + b\bar{b}$ is enhanced by the Yukawa coupling of the Higgs to a virtual top quark. If the mass of the charged Higgs is in the range between 140 GeV and the $t\bar{b}$ threshold and if the Higgs mixing parameter $\tan\beta$ is small, then $W + b\bar{b}$ may be the largest single decay mode [22]. The decay rate of the charged Higgs into $W + \Upsilon$ was first calculated by Grifols, Gunion, and Mendez [23]. For small $\tan\beta$, the branching fraction $B(H^+ \rightarrow W^+ + \Upsilon)$ ranges from about 10^{-4} if the Higgs mass is just above the $W^+ + \Upsilon$ threshold to about 10^{-3} if the Higgs mass is just below the $t\bar{b}$ threshold.

In a hadron collider, most of the standard production mechanisms for a charged Higgs in the mass range below the $t\bar{b}$ threshold involve the production of an additional very massive particle [24]. The standard production mechanisms

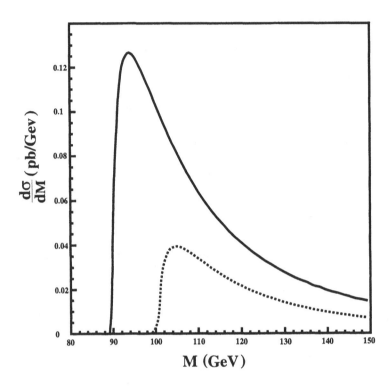

Fig. 4. The invariant mass distributions $d\sigma/dM_{W\Upsilon}$ (summed over $W^{\pm}$) and $d\sigma/dM_{Z\Upsilon}$ of $pp \to \Upsilon + W/Z + X$ at the LHC.

for H^+ include $t\bar{t}$ production followed by the decay $t \to H^+ b$, $\bar{t}H^+$ production, $W^- H^+$ production, and $H^- H^+$ production. Because of the additional very massive particle, events in which a charged Higgs decays into $W + \Upsilon$ will be easily distinguished from $W + \Upsilon$ events produced by Standard Model processes. There is one potentially significant production process for a charged Higgs that could result in events that resemble Standard Model $W + \Upsilon$ events. That process is $qb \to q'bH^+$, which proceeds through a Feynman diagram that involves a virtual W boson and a virtual top quark [25].

5   Conclusion

The introduction of the color-octet mechanism made it possible to explain the enormous surplus in unpolarized $\psi(1S, 2S)$ production at the Tevatron. At HERA, one still cannot definitely say whether NRQCD is working or not since NRQCD loses its predictive power in some kinematic region. High Q^2 leptoproduction of J/ψ proposed by Fleming and Mehen might give a more concrete answer to this. Associate production of Υ and weak gauge bosons at the LHC might be a lamp-post under which one can probe a charged heavy resonance.

Within NRQCD, one can also predict the polarization [26] of heavy quarkonia. Polarization predictions for direct ψ and ψ' production at the Tevatron are already known [27]. But the first experimental result for polarized production at the Tevatron would be that of prompt ψ, which includes cascade effects so that one needs to include the feed-down effect. Further test of polarized quarkonium production at the Tevatron will be one of the most important tests of NRQCD in the near future.

Acknowledgments. The author thanks the organizers of this workshop for the invitation, E. Braaten and S. Fleming for their enjoyable collaborations on the subject discussed in Sec. 4, and B. A. Kniehl for his reading this manuscript. This work was supported in part by the Alexander von Humboldt Foundation.

References

1. E. L. Berger and D. Jones, Phys. Rev. D **23**, 1521 (1981).
2. R. Baier and R. Rückl, Z. Phys. C **19**, 251 (1983).
3. E. Braaten and T. C. Yuan, Phys. Rev. Lett. **71**, 1673 (1993).
4. E. Braaten and M. A. Doncheski, S. Fleming and M. L. Mangano, Phys. Lett. B **333**, 548 (1994);
   D. P. Roy and K. Sridhar, Phys. Lett. B **339**, 141 (1994);
   M. Cacciari and M. Greco, Phys. Rev. Lett. **73**, 1586 (1994).
5. The CDF Collaboration, Fermilab-Conf-94/136-E (1994).
6. The CDF Collaboration, M. Mangano, in *Proceedings of the 27th International Conference on High Energy Physics*, Glasgow, Scotland, 1994, edited by P. J. Bussey and I. G. Knowles (IOP, London, 1995), and references therein; The CDF Collaboration, Phys. Rev. Lett. **79**, 572 (1997).
7. E. Braaten and S. Fleming, Phys. Rev. Lett. **74**, 3327 (1995).
8. G. P. Lepage, L. Magnea, C. Nakhleh, U. Magnea and K. Hornbostel, Phys. Rev. D **46**, 4052 (1992).
9. G. T. Bodwin, E. Braaten and G. P. Lepage, Phys. Rev. D **51**, 1125 (1995);**55**, 5853(E) (1997).
10. P. Cho, A. K. Leibovich, Phys. Rev. D **53**, 150 (1996); *ibid.* **53**, 6203 (1996).
11. M. Cacciari and M. Krämer, Phys. Rev. Lett. **76**, 4128(1996);
    J. Amundson, S. Fleming and I. Maksymyk, Phys. Rev. D **56**, 5844 (1997); M. Krämer, Nucl. Phys. B **459**, 3 (1996);
    F. Maltoni, M. L. Mangano, A. Petrelli, Nucl. Phys. B **519**, 361 (1998).
12. P. Ko, Jungil Lee and H. S. Song, Phys. Rev. D **54**, 4312 (1996); **60**, 119902(E) (1999).
13. H. Jung, G. A. Schuler and J. Terron, Int. J. Mod. Phys., A **7**, 7955 (1992);
    M. Krämer, J. Zunft, J. Steegborn, P. M. Zerwas, Phys. Lett. B **348**, 657 (1995).

14. H1 Collaboration, Nucl. Phys. B **472**, 3 (1996); ZEUS Collaboration, Z. Phys. C **76**, 599 (1997).
15. M. Beneke, I. Z. Rothstein and M. B. Wise, Phys. Lett. **B408**, 373 (1997).
16. S. Fleming and T. Mehen, Phys. Rev. D **57**, 1846 (1998).
17. H1 Collaboration, DESY-99-026 (hep-ex/9903008).
18. B. A. Kniehl and G. Kramer, Eur. Phys. J. C **6**, 493 (1999).
19. B. Cano-Coloma and M. A. Sanchis-Lozano, Nucl. Phys. B **508**, 753 (1997).
20. V. Barger, S. Fleming, and R.J.N. Phillips, Phys. Lett. B **371**, 111 (1996).
21. E. Braaten, J. Lee, and S. Fleming, Phys. Rev. D **60**, 091501 (1999).
22. S. Moretti and W.J. Stirling, Phys. Lett. B **347**, 291 (1995); Phys. Lett. B **366**, 451 (1996); E. Barradas, J.L. Diaz-Cruz, A. Gutierrez, and A. Rosado, Phys. Rev. D **53**, 1678 (1996); A. Djouadi, J. Kalinowski, and P.M. Zerwas, Z. Phys. C **70**, 435 (1996); E. Ma, D.P. Roy, and J. Wudka, Phys. Rev. Lett. **80**, 1162 (1998).
23. J.A. Grifols, J.F. Gunion, and A. Mendez, Phys. Lett. B **197**, 266 (1987); R.W. Robinett and L. Weinkauf, Mod. Phys. Lett A **6**, 1575 (1991).
24. J.F. Gunion, H.E. Haber, G.L. Kane, and S. Dawson, *The Higgs Hunter's Guide* (Addison-Wesley, Redwood City, 1990).
25. S. Moretti and K. Odagiri, Phys. Rev. D **55**, 5627 (1997).
26. E. Braaten and Y.-Q. Chen, Phys. Rev. D **54**, 3216 (1996).
27. M. Beneke and I. Z. Rothstein, Phys. Lett. B **372**, 157 (1996); M. Beneke and M. Krämer, Phys. Rev. D **55**, R5269 (1997); A. K. Leibovich, Phys. Rev. D **56**, 4412 (1997).

Part V

Elastic and Diffractive *ep* Scattering

Light Vector Meson Production at HERA

Alexander Proskuryakov

Moscow State University, 119899 Moscow, Russia

(On behalf of the H1 and ZEUS Collaborations)

Abstract. Recent results from HERA on exclusive vector meson production are reviewed. The diffractive reaction $\gamma^* p \to V p$ ($V = \rho^0, \omega.\phi$) has been measured at photon-proton centre-of-mass energies $20 < W < 160$ GeV, for photon virtualities $Q^2 < 60$ GeV^2 and for $|t|$ up to 11 GeV^2, where t is the square of the four-momentum transferred at the proton vertex. The $\gamma^* p \to V p$ cross section as a function of W, Q^2 and t is presented. The polarisation properties of the ρ^0 produced at low and high Q^2 are discussed. The data are compared with perturbative QCD calculations.

1 Introduction

A detailed description of results on the diffractive production of the vector mesons by real and virtual photons is available in [1,2].

1.1 Kinematics

Figure 1 shows a schematic diagram for exclusive vector meson production in the reaction $ep \to epV$, where V stands for a vector meson (ρ^0, ω, ϕ, J/ψ...).

The kinematics of this reaction is described by the following variables:

- the negative of the four-momentum squared of the exchanged photon, $Q^2 = -(k - k')^2$, where $k(k')$ is the four-momentum of the incident (scattered) electron;
- the centre-of-mass energy of the photon-proton system, $W = \sqrt{(q + P)^2}$, where q and P are the four-momenta of the exchanged photon and the incident proton respectively;
- the Bjorken variable $y = pq/pk$;
- the four-momentum transfer squared at the proton vertex, $t = (P - P')^2$, where P' is the four-momentum of the scattered proton;
- the angle between the vector meson production plane and the electron scattering plane, Φ.

The kinematics of the vector meson decay (in the following only two particle decay is considered) is described by the polar angle θ and the azimuthal angle ϕ in the s-channel helicity frame [3], in which the quantisation axis is defined as the direction of the momentum of the final-state proton in the V rest frame.

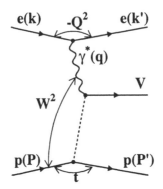

Fig. 1. Schematic diagram for exclusive vector meson production

1.2 Cross section

The cross section of exclusive vector meson production from virtual photons has contributions from longitudinal (helicity 0 state) and transverse (helicity ± 1 state) photons. The differential electron-proton cross section is related to the virtual longitudinal and transverse photon-proton cross sections $\sigma_L^{\gamma^* p}$ and $\sigma_T^{\gamma^* p}$ as:

$$\frac{d\sigma^{ep}}{dydQ^2} = \Gamma_T(y, Q^2)(\sigma_T^{\gamma^* p} + \epsilon \sigma_L^{\gamma^* p}), \tag{1}$$

with the flux of transverse photons given by

$$\Gamma_T = \frac{\alpha_{em}}{2\pi} \frac{1 + (1 - y)^2 - 2(1 - y)Q_{min}^2/Q^2}{yQ^2}, \tag{2}$$

and the polarisation parameter ϵ defined as

$$\epsilon = \frac{2(1 - y)}{1 + (1 - y)^2 - 2(1 - y)Q_{min}^2/Q^2}, \tag{3}$$

where $Q_{min}^2 \simeq m_e^2 y^2/(1 - y)$ is the minimum kinematically allowed value of Q^2. In the kinematic region covered by the HERA data the value of the polarisation parameter ϵ is close to unity. The virtual photon-proton cross section, $\sigma^{\gamma^* p} = \sigma_T^{\gamma^* p} + \epsilon \sigma_L^{\gamma^* p}$, can be related to the total cross section, $\sigma_{TOT}^{\gamma^* p} = \sigma_T^{\gamma^* p} + \sigma_L^{\gamma^* p}$, through the relation:

$$\sigma_{TOT}^{\gamma^* p} = \frac{1 + R}{1 + \epsilon R} \sigma^{\gamma^* p}, \tag{4}$$

where $R = \sigma_L^{\gamma^* p}/\sigma_T^{\gamma^* p}$ is the ratio of the longitudinal and the transverse cross sections.

1.3 Decay angular distributions

The decay angular distributions for the reaction $ep \to e\rho^0 p$ can be written as [3]:

$$
\begin{aligned}
W(\cos\theta, \phi, \Phi) = \frac{3}{4\pi} \Bigg[& \frac{1}{2}(1 - r_{00}^{04}) + \frac{1}{2}(3r_{00}^{04} - 1)\cos^2\theta \\
& -\sqrt{2}\mathrm{Re}\{r_{10}^{04}\}\sin 2\theta \cos\phi - r_{1-1}^{04}\sin^2\theta \cos 2\phi \\
& -\epsilon\cos 2\Phi(r_{11}^{1}\sin^2\theta + r_{00}^{1}\cos^2\theta - \sqrt{2}\mathrm{Re}\{r_{10}^{1}\}\sin 2\theta \cos\phi - r_{1-1}^{1}\sin^2\theta \cos 2\phi) \\
& -\epsilon\sin 2\Phi(\sqrt{2}\mathrm{Im}\{r_{10}^{2}\}\sin 2\theta \sin\phi + \mathrm{Im}\{r_{1-1}^{2}\}\sin^2\theta \sin 2\phi) \\
& +\sqrt{2\epsilon(1 + \epsilon)}\cos\Phi(r_{11}^{5}\sin^2\theta + r_{00}^{5}\cos^2\theta \\
& -\sqrt{2}\mathrm{Re}\{r_{10}^{5}\}\sin 2\theta \cos\phi - r_{1-1}^{5}\sin^2\theta \cos 2\phi) \\
& +\sqrt{2\epsilon(1 + \epsilon)}\sin\Phi(\sqrt{2}\mathrm{Im}\{r_{10}^{6}\}\sin 2\theta \sin\phi + \mathrm{Im}\{r_{1-1}^{6}\}\sin^2\theta \sin 2\phi)\Bigg].
\end{aligned}
\tag{5}
$$

The spin-density matrix elements r_{ij}^{04}, r_{ij}^{α} are the combinations of the helicity amplitudes $T_{\lambda_V \lambda_\gamma}$ (the nucleons helicities are omitted), where λ_V and λ_γ are the helicities of the vector meson and the photon respectively.

The assumption of helicity conservation in the photon-proton centre-of-mass frame (SCHC) reads:

$$
T_{\lambda_V \lambda_\gamma} = T_{\lambda_V \lambda_\gamma}\delta_{\lambda_V \lambda_\gamma}.
\tag{6}
$$

Under this assumption all matrix elements except r_{00}^{04}, r_{1-1}^{1}, $\mathrm{Im}\{r_{1-1}^{2}\}$, $\mathrm{Re}\{r_{10}^{5}\}$ and $\mathrm{Im}\{r_{10}^{6}\}$ are zero. The remaining matrix elements are related via:

$$
r_{1-1}^{1} + \mathrm{Im}\{r_{1-1}^{2}\} = 0,
\tag{7}
$$

$$
\mathrm{Re}\{r_{10}^{5}\} + \mathrm{Im}\{r_{10}^{6}\} = 0.
\tag{8}
$$

The angular distribution in the case of SCHC depends on two angles, θ and $\psi = \phi - \Phi$:

$$
\begin{aligned}
W(\cos\theta, \psi) = \frac{3}{4\pi} \Bigg[& \frac{1}{2}(1 - r_{00}^{04}) + \frac{1}{2}(3r_{00}^{04} - 1)\cos^2\theta \\
& +\epsilon r_{1-1}^{1}\sin^2\theta \cos 2\psi - 2\sqrt{2\epsilon(1 + \epsilon)}\mathrm{Re}\{r_{10}^{5}\}\sin 2\theta \cos\psi \Bigg].
\end{aligned}
\tag{9}
$$

The SCHC hypothesis allows the direct determination of the ratio $R = \sigma_L^{\gamma^* p}/\sigma_T^{\gamma^* p}$ from the value of r_{00}^{04}:

$$
R = \frac{r_{00}^{04}}{\epsilon(1 - r_{00}^{04})}.
\tag{10}
$$

1.4   Theoretical models

Exclusive production of light vector mesons has been studied in a wide range of W and Q^2 in fixed target experiments [1,4] and at HERA [5,6]. For $W > 10$ GeV and at low photon virtuality these reactions have the characteristics of a soft diffractive process: a cross section rising weakly with the centre-of-mass energy, a sharp forward diffractive peak in the differential t distribution and s-channel helicity conservation (SCHC). These features are well described within the framework of Regge phenomenology [7] and the Vector Dominance Model (VDM) [8,9]. In this approach the energy dependence of exclusive vector meson production at high energies is determined by the exchange of a Pomeron trajectory:

$$\frac{d\sigma}{d|t|} \propto e^{-b_0|t|} \left(\frac{W^2}{W_0^2}\right)^{2(\alpha(t)-1)}, \tag{11}$$

where the Pomeron trajectory is parameterised as $\alpha(t) = \alpha(0) + \alpha't = 1.08 + 0.25t$ [10]. The W dependence of cross section in the HERA kinematic region can approximately be parameterised by a simple power behaviour, W^δ, where $\delta \simeq 4(\alpha(0) - \alpha'/b - 1)$. The slope of the differential t distribution depends on the energy as $b = b_0 + 2\alpha' ln(W^2/W_0^2)$.

The same approach fails to describe the energy dependence of the cross section for elastic J/ψ photoproduction measured at HERA [11]. The rapid rise with energy of the elastic J/ψ photoproduction cross section is consistent with a perturbative QCD calculation [12]. The hard scale for this reaction is given by the charm quark mass.

The hard scale can also be provided by the photon virtuality Q^2. Exclusive electroproduction at high values of Q^2 has been the subject of many theoretical investigations (see e.g. [13–18]). The main prediction of the perturbative QCD models is a fast increase of the cross section with W, connected with the Q^2 evolution of the gluon density of the proton $xg(x, Q^2)$. The Q^2 dependence for the longitudinal vector meson production, which dominates at high Q^2, is $\sigma_L^{\gamma^*p} \sim (xg(x, Q^2))^2/Q^6$. The QCD models predict also universal, flavor and energy independent, t dependence for this exclusive diffractive process.

Recent calculations of the helicity amplitudes for the reaction $\gamma^*p \to \rho^0 p$, performed in the framework of perturbative QCD [19–21] predict small violations of SCHC. The largest helicity non-conserving amplitude is expected to be T_{01}, which describes the production of longitudinally polarised vector mesons from transverse photons.

2 Results

2.1 The W dependence of the $\gamma^\star p \to Vp$ cross sections

Figure 2 shows the energy dependence of elastic ρ^0, ω, ϕ and J/ψ photoproduction ($Q^2 \simeq 0$ GeV^2) cross sections measured at HERA and fixed target experiments. The W dependence of the cross section for light vector meson photoproduction is well parameterised at high energy as $\sigma^{\gamma p} \propto W^\delta$ with $\delta \simeq 0.22$. This weak energy dependence is in agreement with the 'soft' Pomeron parameterisation of eq. (11).

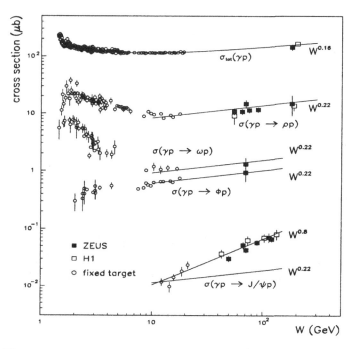

Fig. 2. The energy dependence of the total photoproduction and vector meson photoproduction cross sections for fixed target and HERA experiments.

In contrast, the rapid rise with energy of the elastic J/ψ photoproduction cross section is consistent with a perturbative QCD calculation [12].

The W dependence of the exclusive ρ^0 production cross section at different values of Q^2 is shown in Fig. 3. The lines represent the result of a fit to the data with a function of type $\sigma \propto W^\delta$. The results of the fit are summarised in Fig. 4. As can be seen from Fig. 4, the W dependence of the cross section for the reaction $\gamma^\star p \to Vp$ is consistent with the 'soft' Pomeron approach at low values of Q^2. In the region of high Q^2, where QCD based models predict a steeper cross section dependence on W, there is an indication that the rise of the cross section becomes faster.

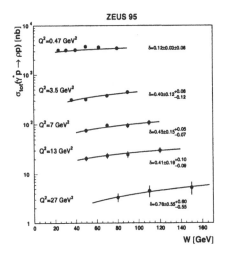

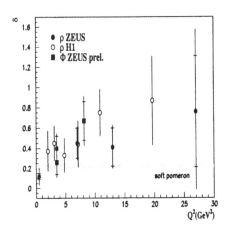

Fig. 3. The energy dependence of exclusive ρ^0 production cross sections for different values of Q^2. The lines represent the function $\sigma \propto W^\delta$ fitted to the data.

Fig. 4. The values of δ as a function of Q^2 for exclusive production of ρ^0 and ϕ mesons.

2.2   The Q^2 dependence of the $\gamma^* p \to V p$ cross sections

The cross section for exclusive ρ^0 production as a function of Q^2 is shown in Fig. 5. The Q^2 dependence of the $\gamma^* p \to \rho^0 p$ cross section is well described in a wide range of Q^2 by the function $d\sigma/dQ^2 \propto 1/(1+Q^2/M_\rho^2)^n$, with $n = 2.28 \pm 0.06$. The curves shown in Fig. 5 represent the predictions of QCD [17,18] and Generalized VDM [9].

Figure 6 shows the ratio of the ϕ and ω production cross sections to that for the ρ^0 meson as a function of Q^2. The ratios in the high Q^2 region are consistent with the prediction of QCD models.

2.3   The t dependence of the $\gamma^* p \to V p$ cross sections

The exponential t dependence of the $\gamma^* p \to V p$ cross section is a typical characteristic of diffractive processes. The slope parameter b, which is related in optical models to the sizes of the interacting particles, is expected to depend on the photon virtuality. Figure 7 shows the slope parameter for exclusive ρ production as a function of Q^2. Despite large statistical errors of the experimental data these results indicate that the slope b decreases with Q^2, approaching the value of $b \simeq 4 - 5 \text{ GeV}^{-2}$.

The slope parameter also depends on the vector meson mass. The value of b is [5,11] $9.8 \pm 0.8(\text{stat}) \pm 1.1(\text{syst}) \text{ GeV}^{-2}$, $7.3 \pm 1.0(\text{stat}) \pm 0.8(\text{syst}) \text{ GeV}^{-2}$ and

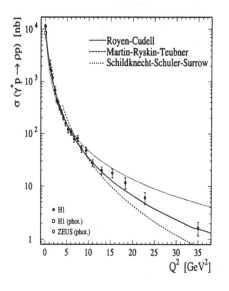

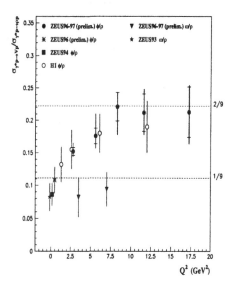

Fig. 5. The Q^2 dependence of the exclusive ρ^0 production cross section. The curves represent the predictions of the QCD models [17,18] and Generalized VDM [9].

Fig. 6. Ratios of ϕ and ω cross sections to that for the ρ^0 meson as a functions of Q^2.

$4.0 \pm 0.4(\text{stat})^{+0.6}_{-0.7}(\text{syst})$ GeV^{-2}, for photoproduction of ρ^0, ϕ and J/ψ mesons respectively.

The differential t distribution at low $|t|$ is well described by Regge phenomenology assuming that the exchange of the Pomeron trajectory dominates at high enough energies. The parameters of the Pomeron trajectory can be measured by studying the energy dependence of $d\sigma/dt$ at different t values (see eq. 11). Figure 8 shows the W dependence for the elastic ρ^0 photoproduction cross section at fixed t values. The resulting Pomeron trajectory, shown in Fig. 9, is $\alpha(t) = (1.096 \pm 0.021) + (0.125 \pm 0.038)t$.

2.4 Decay angular distributions

The analysis of vector meson decay distribution provides important information about helicity properties of the reaction $\gamma^* p \to V p$. In the case of unpolarised beams, the angular distributions are determined by 15 spin density matrix elements (see eq. 5), which are the combinations of the helicity amplitudes $T_{\lambda_V,\lambda_\gamma}$. The full set of spin density matrix elements has been measured for exclusive ρ^0 production. The most significant deviation from the expectation of SCHC is observed for the matrix element r^5_{00}. This matrix element is proportional

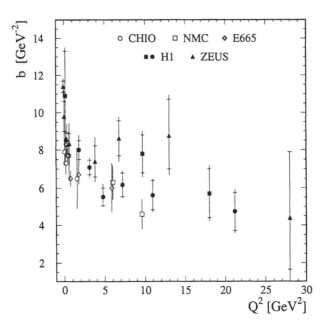

Fig. 7. The slope parameter b as a function of Q^2 for exclusive ρ^0 production.

to the interference between the helicity non-flip amplitude T_{00} and the helicity single-flip amplitude T_{01}. The size of this effect can be quantified by the ratio of the helicity single-flip amplitude to the helicity conserving amplitudes $|T_{01}|/\sqrt{|T_{00}|^2 + |T_{11}|^2}$, which gives $(8 \pm 3)\%$ [6].

Figure 10 shows the 15 spin density matrix elements as a function of Q^2. The results of the QCD based calculations [20,21] shown in Fig. 10 are in good agreement with the data. The ratio $R = \sigma_L^{\gamma^*p}/\sigma_T^{\gamma^*p}$, calculated according to eq. 10 (neglecting small violation of SCHC) , is displayed in Fig. 11 as a function of Q^2 for exclusive ρ^0 production. The curves represent the predictions of the QCD [17,18] and Generalized VDM [9]. Fig. 11 shows that the longitudinal photon-proton cross section dominates at large Q^2 and that the non-linear rise of R can be reproduced by the theoretical models.

3   Conclusions

- The energy dependence of the light vector meson production cross section exhibits a slow rise with W at low values of Q^2. There is an indication that in the region of high Q^2 this rise becomes faster, in agreement with predictions by QCD models. The cross section for exclusive J/ψ production shows a steep rise with W already a low photon virtualities.
- The Q^2 dependence of the $\gamma^*p \to \rho^0 p$ cross section is well described (for $Q^2 > 1$ GeV^2) by the function $d\sigma/dQ^2 \propto 1/(1 + Q^2/M_\rho)^n$, with $n = 2.24 \pm 0.09$.

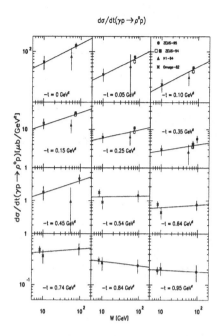

do/dt(γp → ρ⁰p)

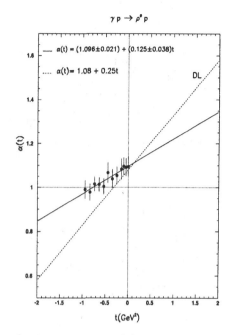

γ p → ρ⁰ p

Fig. 8. Cross section for elastic ρ^0 photoproduction as a function of W at different values of t. The W dependence is fitted by $d\sigma/dt \propto (W^2)^{2\alpha(t)-2}$.

Fig. 9. The Pomeron trajectory determined from the reaction $\gamma p \to \rho^0 p$. The dots are the values of the trajectory at a given t as determined from figure 8. The full line is the result of a linear fit to these values. The Pomeron trajectory given in [10] is shown as the dashed line.

The ratios of the ϕ and ω production cross sections to that for the ρ^0 meson are consistent in the high Q^2 region with the predictions of QCD models.

- The t distribution for the exclusive vector meson production is well parameterised by an exponential dependence e^{bt}. The slope parameter b decreases with Q^2. The Pomeron trajectory, determined from the reaction of elastic ρ^0 photoproduction, is $\alpha(t) = (1.096 \pm 0.021) + (0.125 \pm 0.038)t$.

- The full set of spin density matrix elements measured for exclusive ρ^0 productions shows a small deviation from the SCHC hypothesis. The ratio of the longitudinal and transverse photon-proton cross sections exhibits a nonlinear rise with Q^2. The theoretical models reproduce this behaviour.

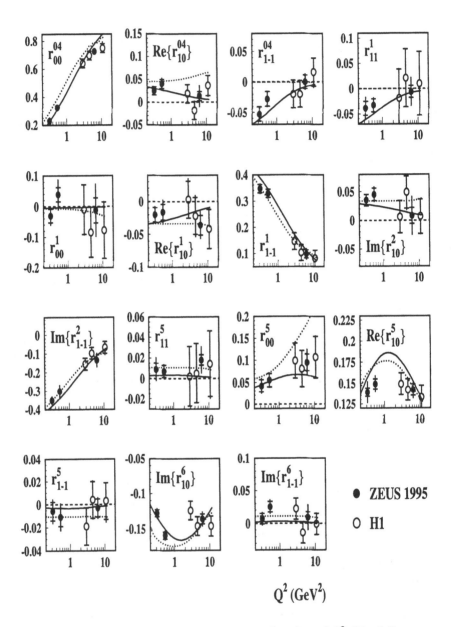

Fig. 10. The 15 spin density matrix elements as a function of Q^2. The full curves are the prediction of Ref. [20], the dotted curves the prediction of Ref. [21]. The dashed lines indicate the expectation of SCHC.

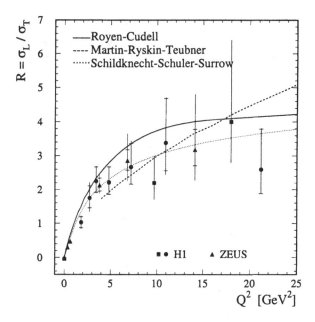

Fig. 11. The ratio of the longitudinal and transverse photon-proton cross sections as a function of Q^2. The curves represent the predictions of the QCD [17,18] and Generalized VDM [9].

References

1. T.H. Bauer et al., Rev. Mod. Phys. 50 (1978) 261.
2. J.A. Crittenden, Exclusive Production of Neutral Vector Mesons at the Electron-Proton Collider HERA, Springer Tracts in Modern Physics, Volume 140 (Springer, Berlin Heidelberg, 1997).
3. K. Schilling and G. Wolf, Nucl. Phys. B61 (1973) 381.
4. R.M. Egloff et al., Phys. Rev. Lett. 43 (1979) 657;
   R.M. Egloff et al., Phys. Rev. Lett. 43 (1979) 1545;
   J. Busenits et al., Phys. Rev. D40 (1989) 1;
   EMC Collab., J.J. Aubert et al., Phys. Lett. B161 (1985) 203;
   NMC Collab., M. Arneodo et al., Nucl. Phys. B429 (1994) 503;
   M.R. Adams et al., Z. Phys., C74 (1997) 237.
5. ZEUS Collab., M Derrick et al., Z. Phys. C69 (1995) 39;
   H1 Collab., S. Aid et al., Nucl. Phys. B463 (1996) 3;
   ZEUS Collab., M. Derrick et al., Phys. Lett. B377 (1996) 259;
   ZEUS Collab., M. Derrick et al., Z. Phys., C73 (1996) 73;
   ZEUS Collab., M. Derrick et al., Z. Phys. C73 (1997) 253;
   ZEUS Collab., J. Breitweg et al., Eur. Phys. J. C2 (1998) 247;
   ZEUS Collab., J. Breitweg et al., Eur. Phys. J. C6 (1999) 603.
6. H1 Collab., C. Adloff et al., DESY Report DESY 99-010, hep-ex/9902019 (1999);
   ZEUS Collab., J. Breitweg et al., DESY Report DESY 99-102.
7. see e.g. P.D.B. Collins, Introduction the Regge Theory and High Energy Physics, Cambridge University Press (1977).
8. J.J. Sakurai, Ann. Phys. 11 (1960) 1;
   J.J. Sakurai, Phys. Rev. Lett. 22 (1969) 981;
   H. Fraas and D. Schildknecht, Nucl. Phys. B14 (1969) 543.
9. D. Schildknecht, G.A. Schuler and B. Surrow, Phys. Lett. B449 (1999) 328.
10. A. Donnachie and P.V. Landshoff, Phys. Lett. B296 (1992) 227.
11. ZEUS Collab., M. Derrick et al., Phys. Lett. B350 (1995) 120;
    H1 Collab., S. Aid et al., Nucl. Phys., B472 (1996) 3;
    ZEUS Collab., J. Breitweg et al., Z. Phys. C75 (1997) 215.
12. M.G. Ryskin, Z. Phys. C57 (1993) 89;
    M.G. Ryskin et al., Z. Phys. C76 (1997) 231.
13. S. Brodsky et al., Phys. Rev. D50 (1994) 3134.
14. J. Nemchik, N.N. Nikolaev and B.G. Zakharov, Phys. Lett. B341 (1994) 228.
15. A. Donnachie and P.V. Landshoff, Phys. Lett. B348 (1995) 213.
16. L. Frankfurt, W. Koepf and M. Strikman, Phys. Rev. D54 (1996) 3194.
17. A.D. Martin, M.G. Ryskin and T. Teubner, Phys. Rev. D55 (1997) 4329.
18. J.R. Cudell and I. Royen, hep-ph/9807294.
19. D.Yu. Ivanov and R. Kirschner, Phys. Rev. D58 (1998) 114025.
20. E.V. Kuraev, N.N. Nikolaev and B.G. Zakharov, JETP Lett. 68 (1998) 696 and Pisma Zh. Eksp. Teor. Fiz. 68 (1998) 667;
    N.N. Nikolaev, Proceedings of the DIS99 workshop, April 19-23, 1999, Zeuthen, Germany, Eds. J. Blumlein and T. Riemann, to apper in Nucl. Phys. B (Proc. Suppl.).
21. I. Royen, Proceedings of the DIS99 workshop, April 19-23, 1999, Zeuthen, Germany, Eds. J. Blumlein and T. Riemann, to apper in Nucl. Phys. B (Proc. Suppl.) and Liege University preprint ULC-PNT-99-1-IR.

Theory of Elastic Vector Meson Production

Thomas Teubner

Deutsches Elektronen-Synchrotron DESY, Notkestrasse 85, D-22607 Hamburg, Germany

Abstract. The elastic production of vector mesons at HERA is discussed from the theoretical point of view. We briefly review different models, their successes and short-comings. Main emphasis is put on recent issues in perturbative QCD calculations. Models including the vector meson wave function are compared with an approach based on parton-hadron duality. We discuss several refinements of these models in some detail, including the important role of off-diagonal parton distributions.

1 Introduction

Why are we interested in elastic vector meson production? First of all the process $\gamma^* p \to V p$ provides us with well distinguishable experimental signals in a wide range of the $\gamma^* p$ c.m. energy W, the virtuality of the photon Q^2, and the mass of the vector meson M_V. Quite some data are already available for $V = \rho, \phi$ and J/Ψ, and even for the heavy Υ first measurements were published recently.[1] In the future the range in Q^2 and W and the precision of the data will increase. This enables us to study vector meson production in detail in the very interesting regime where the transition from soft to hard QCD dynamics is expected (and already seen) to take place. In addition, there is hope to make use of the high sensitivity of this process on the gluon distribution $x\,g(x,\overline{Q}^2)$ in the proton to constrain this quantity at small values of x better than through other processes.

In the following we first sketch the basic picture of elastic vector meson production. In Section 2 we briefly discuss different theoretical models which are not based on the two gluon exchange picture which is then introduced in Section 3. There, starting from the basic leading order result known for a long time, we develop corrections which improve the leading order formula. In Section 4 recent issues in pQCD calculations as off-diagonal parton distributions, the influence of the vector meson wave function and an alternative approach using parton-hadron duality are discussed. We mainly concentrate on diffractive ρ meson electroproduction, but the presented perturbative model is also successful in the case of J/Ψ and Υ. Section 4 contains our conclusions and outlook.

1.1 The basic picture

In Fig. 1 the basic picture for the process $\gamma^* p \to V p$ is shown: first the photon with virtuality $Q^2 = -q^2$ fluctuates into a quark-antiquark pair. This $q\bar{q}$

---

[1] For the discussion of experimental results on the production of light and heavy quarkonia see [1–3] and references therein.

fluctuation then interacts elastically with the proton p, where the zig-zag line represents the (for the moment unspecified) elastic interaction with the proton. The $\gamma^* p$ centre-of-mass energy is denoted by W,

$$W^2 = (q + p)^2 , \tag{1}$$

whereas

$$t = (p - p')^2 \tag{2}$$

is the four-momentum transfer squared. (In the following we will mainly restrict ourselves to the case of small t.) The shaded blob at the right stands for the formation of the vector meson V, which, to leading order, has to form from the $q\bar{q}$ pair with invariant mass squared M_V^2. It is important to note that at high

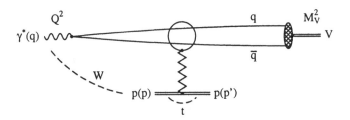

Fig. 1. Diagram for the elastic production of a vector meson V in $\gamma^* p$ collisions

energy W corresponding to small values of x,

$$x = \frac{Q^2 + M_V^2}{Q^2 + W^2} , \tag{3}$$

the timescales involved in the problem are very different:[2] the typical lifetime of the $\gamma^* \to q\bar{q}$ fluctuation as well as the time for the formation of the vector meson V are much longer than the duration of the interaction with the proton, i.e. $\tau_{\gamma^* \to q\bar{q}}, \tau_{q\bar{q} \to V} \gg \tau_i$. Therefore the basic amplitude factorizes, as sketched already in Fig. 1, into the $q\bar{q}$ fluctuation, the interaction amplitude $A_{q\bar{q}+p}$ and the wave function of the vector meson V,

$$A(\gamma^* p \to V p) = \psi^\gamma_{q\bar{q}} \otimes A_{q\bar{q}+p} \otimes \psi^V_{q\bar{q}} , \tag{4}$$

and the process becomes calculable within various models.[3] Formally it has been shown that for Q^2 larger than all other mass scales in the process there is factorization into a hard scattering subprocess, non-perturbative (and off-diagonal, as will be discussed later) parton distributions and the meson wave function

---

[2] This definition for x, which is often called ξ or $x_{I\!P}$, is common in diffractive physics and should not be confused with the ordinary Bjorken-x, $x_{\mathrm{Bj}} = Q^2/(Q^2 + W^2)$.

[3] For a more detailed discussion of the ordering of the timescales see e.g. [4].

[5]. This strict proof of factorization holds for longitudinally polarized photons, whereas meson production through transversely polarized photons is shown to be suppressed by a power of Q.

Let us now turn to the discussion of different models for the $\gamma^* p$ interaction.

2   Some non-perturbative models

The following short section is far from being a review of this rich field, but is intended to give a hint at some non-perturbative models, which contrast the perturbative description of diffractive scattering, which is the main subject of this article.

• We will not cover approaches based on vector meson dominance (see e.g. [6]).

• For Regge-phenomenology-based models of (one or two) Pomeron exchange we refer the reader to [7].

• *The model of the stochastic QCD vacuum*
Dosch, Gusset, Kulzinger and Pirner [8] have developed a model of the interaction with the proton, which is similar to the semi-classical model of Buchmüller discussed in [9] in the context of inclusive diffractive DIS. This model, originally used for hadron-hadron scattering, leads to linear confinement and predicts a dependence of the high-energy scattering on the hadron size. It gives a unified description of low energy and soft high-energy scattering phenomena. Dosch et al. approximate the slowly varying infrared modes of the gluon field of the proton by a stochastic process. Via a path integral method they average over all possible field configurations. For the splitting of the photon into the $q\bar{q}$ pair and for the description of the vector meson they use light cone wavefunctions. Within their model they are able to calculate the Q^2 dependence of the cross section, as well as the dependence on the momentum transfer t, $d\sigma/dt$, and the ratio of the longitudinal to the transverse cross section, L/T, where longitudinal and transverse refer to the polarization of the photon. Their results are in fair agreement with experimental data. There is no prediction for the W dependence of the cross section.

• Rueter has extended the model of Dosch et al. to also describe the W dependence of the cross section [10]. He achieves this by using a phenomenological model based on the exchange of one soft and one hard Pomeron, each being a simple pole in the complex angular momentum plane, similar to the Donnachie-Landshoff model [7]. For the very hard components of the photon fluctuations he treats the interaction perturbatively and achieves a good description of the experimentally observed transition from the soft to the hard regime.

3   The two gluon exchange model

To leading order in QCD the zig-zag line in Fig. 1, which stands for the elastic scattering via the exchange of a colourless object with the quantum numbers of the vacuum, can be described by two gluons. If the scale governing the (transverse) size of the photon fluctuation is large compared to the typical scale of non-perturbative strong interactions, i.e. if

$$Q^2 \gg \Lambda_{\text{QCD}}^2 \quad \text{or} \quad M_V^2 \gg \Lambda_{\text{QCD}}^2, \tag{5}$$

then the coupling of the two gluons to the $q\bar{q}$ fluctuation can be treated reliably within perturbative QCD (pQCD). Another kinematic regime, where pQCD is applicable, is high-t diffraction. There the hard scale which is needed to ensure the validity of the perturbative treatment is given by the large value of the momentum transfer t, and one expects high-t diffraction to be a good place to search for the perturbative Pomeron [11].

It has been shown some time ago that due to the factorization property of the process the coupling of the two gluons to the proton can, in the leading logarithmic approximation, be identified with the ordinary (diagonal) gluon distribution in the proton [12–14]. We will come back to this point later when discussing the importance of off-diagonal gluon distributions.

3.1   The basic formula

The basic leading order formula for diffractive vector meson production is given by [12,13]

$$\frac{d\sigma}{dt} (\gamma^* p \to V p) \Big|_{t=0} = \frac{\Gamma_{ee}^V M_V^3 \pi^3}{48\alpha} \frac{\alpha_s(\overline{Q}^2)^2}{\overline{Q}^8} \left[x\, g(x, \overline{Q}^2) \right]^2 \left(1 + \frac{Q^2}{M_V^2} \right), \tag{6}$$

where α is the electromagnetic coupling and the gluon distribution is sampled at the effective scale

$$\overline{Q}^2 = (Q^2 + M_V^2)/4. \tag{7}$$

In Eq. (6) the non-relativistic approximation for the vector meson wave function is used and the coupling of the vector meson to the photon is encoded in the electronic width Γ_{ee}^V. Note that Eq. (6) is valid for $t = 0$. In the approach discussed in the following there is no prediction for the t dependence of the cross section, which is assumed to be of the exponential form $\exp(-b|t|)$ with an experimentally measured slope-parameter b, which may depend on the vector meson V and on Q^2. On the other hand, Eq. (6) makes predictions for both the Q^2 and the W dependence of the cross section for longitudinally and transversely polarized photons for all sorts of vector mesons, as long as either Q^2 or M_V^2 are large enough to act as the hard scale. It is obvious that the W dependence comes entirely from the gluon distribution $x\, g(x, \overline{Q}^2)$, which enters quadratically in the cross section.

3.2   Improvements beyond the leading order

In the following we will discuss several improvements of the leading order formula.[4]

---

[4] For more detailed discussions see e.g. [15,16] or the recent review [17].

• Eq. (6) contains only the leading imaginary part of the positive-signature amplitude

$$A \propto i\left(x^{-\lambda} + (-x)^{-\lambda}\right).$$ (8)

The real part of the amplitude can be restored using dispersion relations:

$$\mathrm{Re}A = \tan(\pi\lambda/2)\,\mathrm{Im}A,$$ (9)

where λ is given by the logarithmic derivative

$$\lambda = \frac{\partial \log A}{\partial \log(1/x)}.$$ (10)

For the case of ρ production, the contributions from the real part are roughly 15%. For J/Ψ production in the HERA regime they amount to approximately 20% and are even bigger for Υ production [18,19], where larger values of x are probed.

• In Fig. 2 one of four leading order diagrams[5] for the two gluon exchange model is shown with some kinematic variables which will be used below. In the general case the two gluons g_1, g_2 have different x, x' and transverse momenta ℓ_T, ℓ'_T. The leading logarithmic approximation of the ℓ_T^2 loop integral (indicated by the circle in Fig. 2) leads to the identification with the integrated gluon distribution $x\,g(x,\overline{Q}^2)$ at the effective scale $\overline{Q}^2$ defined in Eq. (7). Beyond leading logarithmic accuracy one has to perform the ℓ_T^2 integral over the unintegrated gluon distribution $f(x,\ell_T^2)$. This can lead to numerical results which are, depending on the kinematical regime, twice as big as the result from Eq. (6) [4,15].

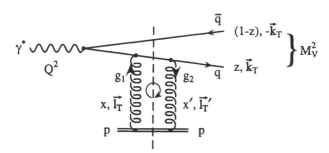

Fig. 2. One of four leading order diagrams for the two gluon exchange model for diffractive vector meson production

Although here we are considering elastic production at small momentum transfer t, the timelike vector meson with mass M_V has to be produced from the

---

[5] There are three similar diagrams: one where both gluons couple to the antiquark, and two where one gluon is attached to the quark, whereas the other couples to the antiquark.

spacelike (or real) photon with virtuality Q^2. This means, that even if there is no transverse momentum transfer, $\ell_T = \ell'_T$, there has to be a difference $x - x' = \left(M_V^2 + Q^2\right) / \left(W^2 + Q^2\right)$ in the longitudinal momentum of the two gluons g_1 and g_2. Therefore the identification with the ordinary diagonal gluon distribution $x\, g(x, \overline{Q}^2)$ is only a good approximation for very small values of x and t, and in the general case the process $\gamma^* p \to V p$ depends on off-diagonal parton distributions [20]. Their importance for diffractive vector meson production will be discussed in the following.

4   Recent issues in pQCD calculations

4.1   Off-diagonal parton distributions

Off-diagonal (also called "skewed" or non-forward) parton distributions[6] are much studied recently.[7] In the case of small t scattering the skewedness comes from the difference between x and x' of the two gluons g_1 and g_2, and the cross section can be shown to be proportional to the square of a skewed gluon distribution,

$$\sigma \propto \left| x' g\left(x, x'; \overline{Q}^2\right)\right|^2 . \tag{11}$$

Here $x = \left(M_{q\bar{q}}^2 + Q^2\right) / \left(W^2 + Q^2\right)$, $x' = \left(M_{q\bar{q}}^2 - M_V^2\right) / \left(W^2 + Q^2\right) \ll x$, and $M_{q\bar{q}}^2$ is the mass squared of the intermediate $q\bar{q}$ pair. (Taking the leading imaginary part of the amplitude corresponds to cutting the amplitude as indicated by the dashed line in Fig. 2 and putting both q and $\bar{q}$ on-shell, which in turn fixes x. x' has to accomodate the difference between $M_{q\bar{q}}$ and M_V and it not fixed due to the integration over all possible quark (and antiquark) momenta. At leading logarithmic order $x' \ll x$, and we can put $x' \simeq 0$.)

For arbitrary kinematics skewed parton distributions are not connected with the diagonal ones and are unknown non-perturbative objects. However, in the case of small x, they are determined completely by the diagonal ones [20,21]. The ratio of skewed to diagonal gluon distribution is given by

$$R_g = \frac{x' g(x, x')}{x\, g(x)} = \frac{2^{2\lambda+3}}{\sqrt{\pi}} \frac{\Gamma\left(\lambda + \frac{5}{2}\right)}{\Gamma\left(\lambda + 4\right)} . \tag{12}$$

Here Γ is the usual Gamma function and the effective power λ can be obtained from the logarithmic derivative of the amplitude A for the $\gamma^* p \to V p$ cross section,

$$\lambda = \frac{\partial \log A}{\partial \log\left(1/x\right)} . \tag{13}$$

As will be shown below, the magnitude of the resulting correction factor for the total cross section, R_g^2, can be sizeable, especially for large Q^2 or M_V^2.

---

[6] These off-diagonal parton distributions are not parton densities in the ordinary probabilistic sense but matrix elements of parton-fields between different initial and final proton states.

[7] See [20] and references therein.

4.2   The vector meson wave function

Another important issue is the treatment of the vector meson wave function. As sketched in Fig. 1 and Eq. (4), it enters the amplitude via a convolution with the scattered $q\bar{q}$ fluctuation. In Eq. (6) the non-relativistic approximation was adopted. This means, that quark and antiquark equally share the longitudinal momentum of the photon, i.e. $z = 1 - z = 1/2$, and that there is no internal (transverse) momentum k_T in the $q\bar{q}$ bound state. Therefore, in this naive approximation,

$$\psi_{q\bar{q}}^V(z, k_T) = \delta^{(2)}(k_T)\,\delta\left(z - 1/2\right)\,, \tag{14}$$

and $M_V = 2m_q$. While this simplification may be suitable for heavy mesons like the Υ, it is clear that the non-relativistic approximation has to break down for light quarks. Various groups have worked on improving this approximation by including the Fermi motion of the quarks in the meson by using a nontrivial wave function [13,15,16,22]. Different models for the meson wave function were used which lead to quite different correction factors: whereas in Gaussian models there is no strong suppression [15], the large k_T tail typical for wave functions from non-relativistic potential models seems to lead to large corrections [16]. On the other hand, considering that within these potential models a big part of the $\mathcal{O}(v^2)$ corrections comes from a regime, where k_T is bigger than the quark mass itself, these large corrections may well be an artefact of the non-relativistic approximation.

Another related problem is the question, which mass for the quarks should be used in the perturbative formulae. Note that Eq. (6) is written in terms of the vector meson mass M_V. However, as discussed in [15], the full expressions used to include higher order (relativistic) corrections contain the quark mass m_q instead of M_V. As the ratio $M_V/(2m_q)$ enters with a high power, this difference is not negligible and should be taken into account in the calculation of the $\mathcal{O}(v^2)$ corrections applied to Eq. (6).

In addition, it is well known that there are other relativistic corrections, which in principle have to be taken into account in a consistent way. As pointed out by Hoodbhoy [23], gauge invariance is only preserved if higher Fock states $(q\bar{q}g, q\bar{q}gg, \dots)$ are included in the wave function. In doing so he arrives at the conclusion, that the relativistic corrections to the quark propagators plus the corrections from the higher Fock states amount to only a few percent for J/Ψ production, in agreement with [15].

After all large relativistic corrections can probably be excluded, but, as different approaches lead to quite different results there remains a considerable uncertainty and the issue a hot topic.

4.3   An alternative approach based on parton-hadron duality

In this section we will discuss an alternative approach, which avoids the meson wave function and leads to results which are in surprisingly good agreement with available data. It was proposed in [24] for ρ meson electroproduction, where the hard scale is provided by Q^2, not by M_ρ.[8] Due to the tiny u and d quark masses, in the case of the ρ non-relativistic approximations cannot be justified, and the wave function is not very well known. Now the crucial problem was that all naive predictions for the ratio of the longitudinal to the transverse cross section, which are based on the perturbative formula (6), lead to

$$\sigma_L/\sigma_T \sim Q^2/M_\rho^2 . \tag{15}$$

This is much too steep and incompatible with experimental data (see below). The inclusion of effects from a light cone wave function for the ρ does not change the picture considerably.[9] These observations indicate that the main effects are not coming from the ρ wave function and lead to the proposal of a different model in [24]: there the cross section for ρ production is predicted via perturbative $u\bar{u}$ and $d\bar{d}$ quark pair electroproduction together with the principle of parton-hadron duality (PHD) [25]. PHD means that the integral of the parton $(q\bar{q})$ production cross section over a mass interval ΔM is approximately equal to the sum over all (corresponding) possible hadron production cross sections in the same mass interval. In the region $M_{q\bar{q}}^2 \approx M_\rho^2$ the production of more complicated partonic configurations (like $q\bar{q} + g$, $q\bar{q} + 2g$, $q\bar{q} + q\bar{q}$, etc.) is heavily suppressed. On the hadronic side the ρ resonance (plus the small admixture of the ω) with its decay into two (three) pions completely saturates the cross section. Therefore we can well approximate the ρ production cross section

$$\gamma^* p \to \rho p \to \pi\pi p$$

by

$$\sigma\left(\gamma^* p \to \rho p\right) \simeq 0.9 \sum_{q=u,d} \int_{M_a^2}^{M_b^2} \frac{\mathrm{d}\sigma\left(\gamma^* p \to (q\bar{q})p\right)}{\mathrm{d}M^2} \tag{16}$$

where M_a and M_b have to be chosen to embrace the ρ resonance appropriately, i.e. $M_b^2 - M_a^2 \sim 1$ GeV^2. The factor 0.9 on the right side of Eq. (16) corrects for the contributions from ω production.

The perturbative formulae for the $q\bar{q}$ production cross section are derived from the amplitudes depicted in Fig. 2 and can be written in terms of the conventional spin rotation matrices $d^J_{\lambda\mu}(\theta)$ (see [24] for details):

$$\left.\frac{\mathrm{d}^2\sigma_L}{\mathrm{d}M^2 dt}\right|_{t=0} = \frac{4\pi^2 e_q^2 \alpha}{3} \frac{Q^2}{(Q^2+M^2)^2} \frac{1}{8} \int_{-1}^{1} \mathrm{d}\cos\theta \left|d^1_{10}(\theta)\right|^2 |I_L|^2 , \tag{17}$$

$$\left.\frac{\mathrm{d}^2\sigma_T}{\mathrm{d}M^2 dt}\right|_{t=0} = \frac{4\pi^2 e_q^2 \alpha}{3} \frac{M^2}{(Q^2+M^2)^2} \frac{1}{4} \int_{-1}^{1} \mathrm{d}\cos\theta \left(\left|d^1_{11}(\theta)\right|^2 + \left|d^1_{1-1}(\theta)\right|^2\right) |I_T|^2$$

[8] Experimentally both the t dependence $\mathrm{d}\sigma/\mathrm{d}t \sim \exp(-bt)$ with $b \simeq 5 - 6$ GeV^{-2} for $Q^2 > 10$ GeV^2 and the W behaviour of the cross section $\sigma \propto W^{0.8}$ indicate that ρ meson electroproduction is not a soft, but mainly a hard process.

[9] One might argue that σ_T receives large contributions from the small k_T region, which is non-perturbative. But those contributions would cause the transverse cross section to fall off even faster with increasing Q^2 and therefore worsen the problem [24].

where e_q is the electric charge of the quark q, α the electromagnetic coupling and θ the polar angle of the quark q in the $q\bar{q}$ rest frame with respect to the proton direction ($k_T = M/2 \sin\theta$). $I_{L,T}$ are integrals over the gluon ℓ_T^2 and given by

$$I_L(K^2) = K^2 \int \frac{d\ell_T^2}{\ell_T^4} \alpha_s(\ell_T^2) f(x, \ell_T^2) \left(\frac{1}{K^2} - \frac{1}{K_\ell^2} \right), \qquad (18)$$

$$I_T(K^2) = \frac{K^2}{2} \int \frac{d\ell_T^2}{\ell_T^4} \alpha_s(\ell_T^2) f(x, \ell_T^2) \left(\frac{1}{K^2} - \frac{1}{2k_T^2} + \frac{K^2 - 2k_T^2 + \ell_T^2}{2k_T^2 K_\ell^2} \right),$$

with f being the unintegrated gluon distribution and

$$K_\ell^2 \equiv \sqrt{(K^2 + \ell_T^2)^2 - 4k_T^2 \ell_T^2}, \qquad K^2 \equiv k_T^2(Q^2 + M^2)/M^2.$$

In Eqs. (17) the different rotation matrices appropriately reflect the different spin states of the $q\bar{q}$ produced from longitudinal and transverse photons, and the integrals $I_{L,T}$ contain the scattering off the proton via the two gluon exchange.[10] In order to pick up only those $u\bar{u}$, $d\bar{d}$ configurations which correspond to the quantum numbers of the ρ, one has to project out the $J^{PC} = 1^{--}$ states. This can be easily done on amplitude level with the same rotation matrices $d^J_{\lambda\mu}(\theta)$, see [24]. (Even higher spin states like the $\rho(3^-)$ can be projected out using the corresponding d function [26].) It is important to note that through the projection on amplitude level the longitudinal and transverse cross sections $\sigma_{L,T}$ are less infrared sensitive than Eqs. (17), and therefore σ_T becomes calculable without a large uncertainty from the treatment of the (non-perturbative) infrared region.

For the complete numerical predictions one also has to include the contributions from the real part of the amplitudes and the skewed gluon distribution as discussed above. Both effects are taken into account on amplitude level. Eqs. (17) give the cross section differential in t for $t = 0$. To arrive at the t integrated total cross section one assumes the exponential form $\exp(-b|t|)$. The slope b can be taken from experiment or theoretical models and depends in general on M^2, W and Q^2. For more details we refer the reader to [27].

To go beyond the leading order prediction in a completely consistent way would require in addition the full set of next-to-leading order gluonic corrections to the $(q\bar{q})$-$2g$ vertex. These corrections are not known yet,[11] but can be estimated by a $\mathcal{K}$ factor [4,24]. Similar to the Drell-Yan process, there are π^2 enhanced terms, which come from the $i\pi$ terms in the double logarithmic Sudakov form factor. Resummation of those leading corrections results in the $\mathcal{K}$ factor $\mathcal{K} = \exp(\pi C_F \alpha_s)$ which leads to a considerable enhancement of the cross section.

---

[10] Here we assume s channel helicity conservation (SCHC), i.e. the produced ρ has the helicity of the virtual photon. However, there are small violations of SCHC (e.g. $\gamma_T^* \to \rho_L$) which can be successfully described in a framework similar to the one discussed here. For a discussion of recent measurements of the 15 spin density matrix elements of ρ production compared to different theoretical predictions see [1].

[11] A first step towards the calculation of the full NLO corrections is provided by [28].

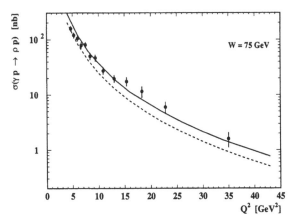

Fig. 3. $\sigma\,(\gamma^*p \to \rho\,p)$ predicted within the PHD model as described in the text compared to recent H1 data [29]. *Continuous line*: using the skewed gluon distribution, *dashed line*: without skewing

In Fig. 3 the complete numerical prediction for $\gamma^*p \to \rho\,p$ using the PHD model[12] is shown as a function of Q^2 together with recent H1 data [29]. The continuous line includes all the effects discussed above, whereas the dashed line does not include the skewed gluon. The importance of the off-diagonal gluon for the Q^2 behaviour of the cross section is obvious and the effect seems to be required to describe the data. Of course the model prediction is not free from uncertainties like the choice of the mass interval $M_b^2 - M_a^2$ in Eqs. (16) or the scale of α_s in the $\mathcal{K}$ factor. These (and other) uncertainties are discussed in detail in [27], but they affect mainly the normalization of the cross section and do not spoil the good agreement with the experimental data. In Fig. 4 the prediction of the PHD model for the ratio L/T is shown as a continuous line. It agrees fairly well with the data points, which show a very modest rise with Q^2 in contrast to the naive prediction from Eq. (15) (dashed line). Thus, in the PHD picture, it is *not* the ρ wave function, but the dynamics of the $q\bar{q}$ pair creation from longitudinal and transverse photons together with the off-diagonal two gluon interaction and the projection onto the 1^- state, that determines the Q^2 dependence and the ratio L/T.

It is important to note that the PHD model also works in the case of massive quarks and heavy mesons. Starting from formulae for diffractive heavy quark production [4] and modifying the projection formalism appropriately, elastic Υ photoproduction was recently predicted using PHD in agreement with first measurements [18]. The same formalism can also be applied to diffractive J/Ψ production [27]. Again, as shown in Fig. 5, there is a surprisingly good agreement between the predicted cross section as a function of Q^2 and the experimental data [33].

---

[12] For the numerical analysis the MRST99 gluon [30] was used, and the scale of α_s in the $\mathcal{K}$ factor was chosen as $2K^2$. For more details see [27].

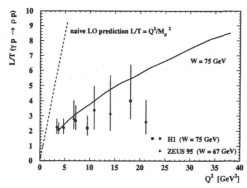

Fig. 4. L/T predicted within the PHD model (*continuous line*) as described in the text compared to experimental data [29,31,32] and the naive prediction (*dashed line*) from Eq. (15)

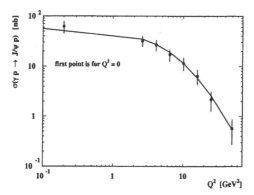

Fig. 5. Cross section for diffractive J/Ψ production as predicted in the PHD model [27] compared to recent H1 data [33]

5   Summary

Elastic vector meson production is a rich field, both from the experimental and theoretical points of view. Different theoretical models describe the data, and more precise data in an increased kinematical range will be needed to clarify the situation. We have briefly discussed some non-perturbative models, but mainly concentrated on perturbative approaches. We have shown that with recent improvements pQCD-based approaches work very well and are in agreement with the data. The fairly large impact of skewed parton distributions on the predictions within these models is supported by the data. There is good hope that in the future we will be able to discriminate between the different models and to understand elastic vector meson production in more detail. By combining different observables from different processes, elastic vector meson production with its high sensitivity to the gluon at small x will finally help to constrain the gluon much better. For this much effort will be needed also from the theoretical side in order to increase the precision of the calculations.

Acknowledgements

I would like to thank G. Grindhammer, B. Kniehl and G. Kramer for the good organization of this stimulating and enjoyable workshop. I also thank Genya Levin, Alan Martin and Misha Ryskin for pleasant collaborations.

References

1. A. Proskuryakov: in these proceedings
2. C. Kiesling: in these proceedings
3. L. Lindemann: in these proceedings
4. E.M. Levin, A.D. Martin, M.G. Ryskin, T. Teubner: Z. Phys. C **74**, 671 (1997)
5. J.C. Collins, L. Frankfurt, M. Strikman: Phys. Rev. D **56**, 2982 (1997)
6. D. Schildknecht, G.A. Schuler, B. Surrow: Phys. Lett. B **449**, 328 (1999)
7. P. Landshoff: in these proceedings;
   A. Donnachie, P.V. Landshoff: Phys. Lett. B **437**, 408 (1998)
8. H.G. Dosch, T. Gousset, G. Kulzinger, H.J. Pirner: Phys. Rev. D **55**, 2602 (1997)
9. W. Buchmüller: in these proceedings
10. M. Rueter: Eur. Phys. J. C **7**, 233 (1999)
11. J.R. Forshaw: talk given at the *7th International Workshop on Deep Inelastic Scattering and QCD (DIS 99), Zeuthen, Germany, 19-23 April 1999*, to be published in the proceedings, hep-ph/9905557
12. M.G. Ryskin: Z. Phys. C **57**, 89 (1993)
13. S.J. Brodsky et al.: Phys. Rev. D **50**, 3134 (1994)
14. J. Bartels, H. Lotter, M. Wüsthoff: Phys. Lett. B **379**, 239 (1996); B **382**, 449(E) (1996)
15. M.G. Ryskin, R.G. Roberts, A.D. Martin, E.M. Levin: Z. Phys. C **76**, 231 (1997)
16. L. Frankfurt, W. Koepf, M. Strikman: Phys. Rev. D **54**, 3194 (1996); Phys. Rev. D **57**, 512 (1998)
17. A.D. Martin, M. Wüsthoff: University of Durham Preprint DTP/99/78, Sep. 1999 and hep-ph/9909362
18. A.D. Martin, M.G. Ryskin, T. Teubner: Phys. Lett. B **454**, 339 (1999)
19. L.L. Frankfurt, M.F. McDermott, M. Strikman: JHEP **9902**, 002 (1999)
20. A.D. Martin: in these proceedings
21. A.G. Shuvaev, K.J. Golec-Biernat, A.D. Martin, M.G. Ryskin: Phys. Rev. D **60** 014015 (1999)
22. J. Nemchik, N.N. Nikolaev, E. Predazzi, B.G. Zakharov: Z. Phys. C **75**, 71 (1997)
23. P. Hoodbhoy: Phys. Rev. D **56**, 388 (1997)
24. A.D. Martin, M.G. Ryskin, T. Teubner: Phys. Rev. D **55**, 4329 (1997)
25. E.C. Poggio, H.R. Quinn, S. Weinberg: Phys. Rev. D **13**, 1958 (1976)
26. A.D. Martin, M.G. Ryskin, T. Teubner: Phys. Rev. D **56**, 3007 (1997)
27. A.D. Martin, M.G. Ryskin, T. Teubner: in preparation
28. V.S. Fadin, A.D. Martin: Preprint BUDKER-INP-99-38, May 1999 and hep-ph/9904505
29. H1 collaboration (C. Adloff et al.): DESY Orange Preprint DESY 99-010, Feb. 1999, to appear in Eur. Phys. J. C
30. A.D. Martin, R.G. Roberts, W.J. Stirling, R.S Thorne: University of Durham Preprint DTP/99/64, July 1999 and hep-ph/9907231
31. H1 collaboration (S. Aid et al.): Nucl. Phys. B **468**, 3 (1996)
32. ZEUS collaboration (J. Breitweg et al.): Eur. Phys. J. C **6**, 603 (1999)
33. H1 collaboration (C. Adloff et al.): DESY Orange Preprint DESY 99-026, March 1999, to appear in Eur. Phys. J. C

Inclusive Diffraction at HERA[*]

Henri Kowalski

Deutsches Elektronen Synchrotron DESY, 22603 Hamburg, Germany

Abstract. Diffractive phenomena observed at HERA are starting to give us a new perspective on the deep inelastic scattering. In this talk the properties of diffractive events observed by the H1 and ZEUS experiments are described. We give also a theoretical interpretation of the diffractive results with special emphasis on the phenomena at the boundary of soft and hard physics. In this region perturbative QCD seems to provide a useful guide. Simple models can be constructed which allow to connect the photoproduction, deep inelastic scattering and diffractive phenomena in a quantitative way.

1 Introduction

One of the most important results of the experiments at HERA is the observation that a substantial amount ($\sim 10\%$) of deep-inelastic electron-proton scattering (DIS) events is of diffractive origin. The usual DIS events are characterized by continuous particle emission between the direction of the struck quark and the proton remnant. In the γ^*-proton rest frame this leads to an almost uniform distribution of emitted particles along the $\gamma^* p$ rapidity axis, The rapidity interval in which particles are radiated has the length $\ln W^2$, where W denotes the $\gamma^* p$ CMS energy in GeV. Diffractive processes are characterized by particle emission in a considerably shorter rapidity interval given by $\ln M_X^2$, where M_X denotes

---

[*] Presented at *New Trends in HERA Physics 1999*, Ringberg Workshop, June 1999

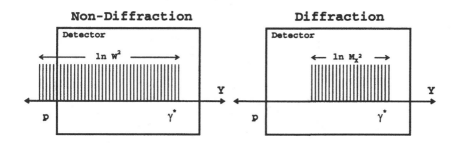

Fig. 1. Particle emission characteristics for non-diffractive (left) and diffractive events (right).

the diffractive mass, see Fig. 1. The mass M_X, which is equal to the invariant mass of all particles emitted in the reaction with exception of the outgoing proton (or the proton dissociated system), is in diffractive reactions considerably smaller than W. Therefore diffractive events exhibit clear rapidity gaps, given by $\Delta Y = \ln W^2 / \ln M_X^2$, [1].

The relatively small amount of hadronic radiation emitted in diffractive reactions is due to the dominance of the colour singlet exchange; this is in contrast to non-diffractive $\gamma^* p$ interactions where coloured gluon exchanges dominate. Diffraction gives new insight into deep inelastic electron proton interactions and gives hope that, because a hard scale is involved, more can be learned about a possible partonic mechanism of diffraction and, eventually, about the transition between the perturbative and nonperturbative regimes.

I will first describe the diffractive measurement and then discuss a possible theoretical interpretation.

2   Diffractive Measurement

2.1   Kinematics

The kinematic quantities used for the description of inclusive DIS, $e(k)+p(P) \rightarrow e(k') + anything$, see Fig. 2, are $Q^2 = -q^2 = -(k - k')^2$, $x = Q^2/(2P \cdot q)$, $y = (P \cdot q)/(P \cdot k)$ and $W^2 = Q^2(1-x)/x + m_p^2 \approx Q^2/x$ for $x \ll 1$. Here k, k' are the four-momenta of the initial and final state positrons; P is the four-momentum of the intial state proton and y is the fractional energy transfer to the proton in its rest frame. For the range of Q^2 and W considered in this talk $W^2 \approx ys$, where $s = 4E_e E_p$ is the square of the ep c.m.s. energy, $\sqrt{s} = 300$ GeV. The scaling variables used to describe DIS diffraction are given by $x_{I\!P} = [(P-N)\cdot q]/(P\cdot q) \approx$

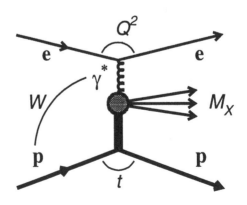

Fig. 2. Diagram showing the diffractive dissociation process in DIS.

$(M_X^2 + Q^2)/(W^2 + Q^2)$ and $\beta = Q^2/[2(P-N)\cdot q] = x/x_{I\!P} \approx Q^2/(M_X^2 + Q^2)$ where N is the four-momentum of the outgoing nucleonic system and M_X is the mass of the system into which the virtual photon dissociated. In models where diffraction is described by the t-channel exchange of a system, for example the pomeron, $x_{I\!P}$ is the momentum fraction of the proton carried by this system and β is the momentum fraction of the struck quark within this system.

2.2 Diffractive Signature

The diffractive and non-diffractive components can be experimentally separated by the analysis of the observed M_X distribution [2,3]. Fig. 3 shows these distributions in terms of $\ln M_X^2$ for various W intervals as observed in the ZEUS detector. The high mass peak is due to non-diffractive events, the plateau-like structure most notably seen at higher W values, is due to diffractive ones. The high mass peak has a steep exponential fall-off, $dN/d\ln M_X^2 \propto \exp(b \ln M_X^2)$, towards smaller $\ln M_X^2$ values. The position of the high mass peak changes proportionally to $\ln W^2$, i.e. shows scaling in $\ln(M_X^2/W^2)$ and the slope, b, of the exponential in $\ln M_X^2$ is approximately independent of W and Q^2. These characteristics are properties of events with uniform, random and uncorrelated particle production along the rapidity axis where, due to the geometrical acceptance of the detector, particles are measured in a limited range of rapidity, see Fig. 1 . In models such as the Feynman gas model or ones dominated by longitudinal phase space [1], the slope b represents the particle multiplicity per unit of rapidity. The exponential in $\ln M_X^2$ and the scaling in $\ln(M_X^2/W^2)$ are directly connected to the exponential suppression of large rapidity gaps by QCD radiation.

These characteristics are also properties of realistic models for particle production in deep inelastic scattering. ARIADNE [4], which gives a good descrip-

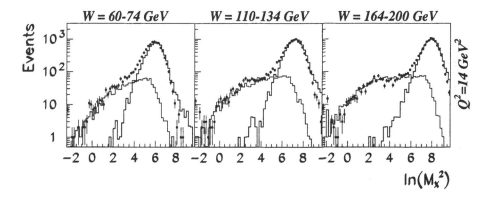

Fig. 3. Distribution of M_X in terms of $\ln M_X^2$. The solid points show the data. The solid histograms show the ARIADNE predictions for non-diffractive reactions. The dashed histograms show the predictions of RAPGAP for diffractive reactions.

tion of particle production by DIS at HERA, also exhibits a pure exponential fall-off with $\ln M_X^2$ and scaling in $\ln M_X^2/W^2$. The distribution predicted by ARIADNE shown in Fig. 3 gives a good account of the high mass peak. The plateau-like part of the $\ln M_X^2$ distribution in Fig. 3 is well described by the diffractive RAPGAP simulation.

2.3 Selection of Diffractive Events

In diffractive events, the system X resulting from the dissociation of the virtual photon is almost fully contained in the detector while the outgoing proton or low mass nucleonic system escapes through the forward beam hole. Diffractive dissociation prefers small M_X values and leads to an event distribution of the form $d\mathcal{N}/dM_X^2 \propto 1/(M_X^2)^{(1+n)}$ corresponding to $d\mathcal{N}/d\ln M_X^2 \propto 1/(M_X^2)^n$, approximately independent of W. At high energies and for large M_X, one expects $n \approx 0$, leading to a constant distribution in $\ln M_X^2$, as discussed below.

In the ZEUS investigation [3] the diffractive contribution was identified as the excess of events at small M_X above the exponential fall-off of the nondiffractive contribution in $\ln M_X^2$. This method is called the M_X method for the determination of the diffractive component and was developed by H. Kowalski and G. Wolf. The exponential fall-off permits the subtraction of the nondiffractive contribution and, therefore, the extraction of the diffractive contribution without assuming the precise M_X dependence of the latter. The $\ln M_X^2$ distribution for both contributions together is expected to be of the form:

$$\frac{d\mathcal{N}}{d\ln M_X^2} = D + c\,\exp(b\,\ln M_X^2), \quad \text{for} \quad \ln M_X^2 \leq \ln W^2 - \eta_0. \tag{1}$$

Here, D denotes the diffractive contribution and the second term the nondiffractive contribution. The diffractive term D is corrected for the detector distortion

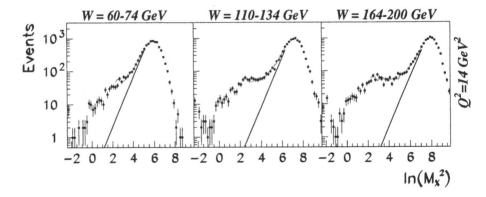

Fig. 4. Distribution of M_X in terms of $\ln M_X^2$. The straight lines give the non-diffractive contribution as obtained from the fits.

effects. The nondiffractive term is not corrected since the detector effects on the exponential fall-off were found to be negligible. The quantity $(\ln W^2 - \eta_0)$ specifies the maximum value of $\ln M_X^2$ up to which the exponential behaviour of the nondiffractive part holds. Its value was determined from the data [3]. The diffractive contribution is not taken from the fit result for D but is determined by subtracting from the observed number of events the nondiffractive contribution found from the fit values of b and c, see Fig. 4. The M_X method selects diffractive events in which the system N, emitted close to the direction of the incoming proton, escapes undetected. Detector acceptance limits the mass of the system N to $M_N < 5.5\,GeV$.

In the H1 investigation [5] the selection of diffractive events was performed by the requirement of a large rapidity gap in the event. This selection makes use of the information from the central and various forward detectors. The particles observed for each event are grouped into two systems, called Y and N, such that the maximum rapidity gap observed is between Y and N. The system N, which includes particles produced in the direction of the incoming proton is required to have a mass $M_N < 1.6\,GeV$. The presence of forward detectors allows H1 to extend the diffractive measurement beyond the region accessible to the ZEUS M_X measurement.

The ZEUS M_X and the H1 rapidity gap methods allow only to measure the diffractive cross section integrated over the four-momentum transfer t. The t dependence of the diffractive cross section was measured by the ZEUS leading proton spectrometer (LPS) [6]. The LPS is designed to measure small angle ($\sim$ 1 mrad) scattered protons which would otherwise escape undetected inside the beam pipe. Isolation of diffractive events with the LPS is rather straight forward: detection of a proton carrying a large fraction of the momentum of the incoming proton, $x_L = p^{LPS}/p_{p_{beam}} > 0.97$, ensures a large rapidity gap between the outgoing proton and the system X. However, the event rate is considerably limited by the acceptance of the LPS.

2.4 Results

The cross section for the virtual photoproduction process $\gamma^* p \rightarrow XN$ can be determined from corresponding cross sections for $ep \rightarrow eXN$ by:

$$\frac{d\sigma_{\gamma^* p \rightarrow XN}^{diff}(M_X, W, Q^2)}{dM_X} \approx \frac{2\pi}{\alpha} \frac{Q^2}{(1-y)^2 + 1} \frac{d\sigma_{ep \rightarrow eXN}^{diff}(M_X, W, Q^2)}{dM_X d\ln W^2 dQ^2} \quad (2)$$

The diffractive cross section determined by ZEUS is shown in Fig. 5 as a function of W for various M_X and Q^2 values [3]. We observe a rapid rise of the cross section with W at all Q^2 values for the M_X bins up to 7.5 GeV.

The differential γp cross section is related to the diffractive structure function, $F_2^{D(3)}(\beta, x_{I\!P}, Q^2)$, by

$$\frac{1}{2M_X} \frac{d\sigma_{\gamma^* p \rightarrow XN}^{diff}(M_X, W, Q^2)}{dM_X} \approx \frac{4\pi^2 \alpha}{Q^2(Q^2 + M_X^2)} x_{I\!P} F_2^{D(3)}(\beta, x_{I\!P}, Q^2). \quad (3)$$

ZEUS 1994

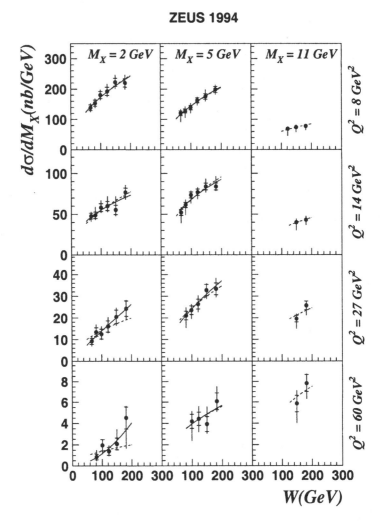

Fig. 5. The differential cross sections $d\sigma^{diff}_{\gamma^*p\to XN}/dM_X$, $M_N < 5.5$ GeV, as a function of W at various M_X values. The inner error bars show the statistical errors and the full bars the statistical and systematic errors added in quadrature. The solid curves show the result from fitting the diffractive cross section for each (W, Q^2) bin separately using the form $d\sigma^{diff}_{\gamma^*p\to XN}/dM_X \propto (W^2)^{a^{diff}}$ where a^{diff} and the normalization constants were treated as free parameters. The dashed curves show the result from the fit where a^{diff} was assumed to be the same for all (W, Q^2) bins.

If $F_2^{D(3)}$ is interpreted in terms of quark densities then it specifies for a diffractive process the probability to find a quark carrying a momentum fraction $x = \beta x_{I\!P}$ of the proton momentum, with x being the Bjorken x.

The H1 results are shown in Fig. 6 in terms of diffractive structure function, $x_{I\!P} F_2^{D(3)}(\beta, x_{I\!P}, Q^2)$ as a function of $x_{I\!P}$ at various β and Q^2 values [5]. The quick rise with W observed in the diffractive cross section measurment corresponds here to the rise of $x_{I\!P} F_2^{D(3)}$ with decreasing $x_{I\!P}$.

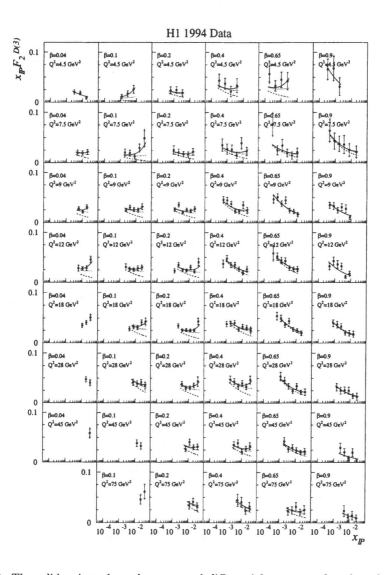

Fig. 6. The solid points show the measured differential structure function plotted as $x_{I\!P} F_2^{D(3)}(\beta, x_{I\!P}, Q^2)$ against $x_{I\!P}$ for various β and Q^2 values. The inner error bars represent the statistical errors and the outer error bars represent the statistical and systematic error added in quadrature. The solid curves show the results of the Regge fit described in text. The dashed curves give the contributions of the pomeron alone and the pomeron with the interfence term.

One of the most important results obtained from the diffractive measurement was the observation that the diffractive cross section rises rapidly with W at all Q^2 values and the measurement of its rate of rise. In the ZEUS case the cross section was fitted for each (M_X, Q^2) bin using the form

$$\frac{d\sigma_{\gamma^* p \to XN}^{diff}(M_X, W, Q^2)}{dM_X} = h \cdot W^{a^{diff}} \quad , \tag{4}$$

where a^{diff} and the normalization constants h were treated as free parameters. In Regge models, a^{diff} is related to the trajectory of the pomeron $\alpha_{I\!P}(t)$, averaged over t: $\overline{\alpha_{I\!P}} = 1 + a^{diff}/4$. The fit value for a^{diff} leads to $\overline{\alpha_{I\!P}} = 1.127 \pm 0.009(stat)^{+0.039}_{-0.012}(syst)$.

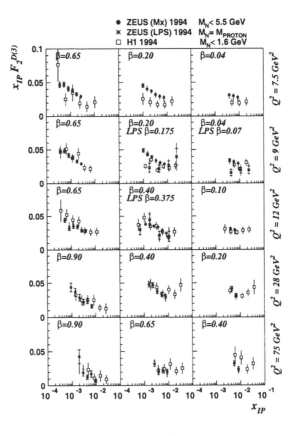

Fig. 7. The diffractive structure function of the proton for $\gamma^* p \to XN$, $M_N < 5.5$ GeV, multiplied by $x_{I\!P}$, $x_{I\!P} F_2^{D(3)}(x_{I\!P}, \beta, Q^2)$, from the M_X analysis (solid points) compared with the results from ZEUS LPS measurement obtained with an identified proton for $\gamma^* p \to Xp$ (stars) and from a subsample of the H1 data (open points) for $\gamma^* p \to XN$, $M_N < 1.6$ GeV.

In the H1 case, due to the larger kinematic region accessed by the measurement, $F_2^{D(3)}(\beta, x_{I\!\!P}, Q^2)$ was fitted using a combined pomeron and reggeon ansatz,

$$F_2^{D(3)}(\beta, x_{I\!\!P}, Q^2) = c_1 (1/x_{I\!\!P})^{2\alpha_{I\!\!P}-1} + c_2 (1/x_{I\!\!P})^{2\alpha_{I\!\!R}-1} \tag{5}$$

H1 [5] has given for the intercept of the pomeron trajectory a value of $\alpha_{I\!\!P}(0) = 1.203 \pm 0.020(stat) \pm 0.013(syst)^{+0.030}_{-0.035}(model)$. Averaging over the t-distribution gives approximately $\overline{\alpha_{I\!\!P}} = \alpha_{I\!\!P}(0) - 0.03$, a value which is in agreement with the ZEUS result.

The $\alpha_{I\!\!P}$ values of both experiments lie above the $\alpha_{I\!\!P}$ values deduced from hadron-hadron scattering, where the intercept of the pomeron trajectory was found to be $\alpha_{I\!\!P}^{soft}(0) = 1.08$ [14]. Averaging over t reduces this value by about 0.02 leading to $\overline{\alpha_{I\!\!P}}^{soft} = 1.06$ [14].

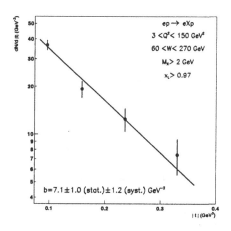

Fig. 8. $|t|$-dependence of the diffractive cross section as measured by the ZEUS collaboration.

In Fig. 7 the data from the ZEUS M_X analysis (solid points) are compared with ZEUS data obtained using the LPS [15] and with those of H1 [5]. For ease of comparison the $x_{I\!\!P} F_2^{D(3)}$ values from this analysis were scaled to the (β, Q^2) values used in the H1 analysis. The LPS data correspond to events of the type $\gamma^* p \to Xp$ with an identified proton. No correction was applied for the contribution from double dissociation which is present in the ZEUS M_X analysis but not in the LPS data. The correction would increase the LPS data by a factor of $1.45^{+0.40}_{-0.26}$. There is consistency between the M_X analysis and the LPS data. The H1 data correspond to $M_N < 1.6$ GeV while those from the M_X analysis are given for $M_N < 5.5$ GeV. No correction was applied. The data from H1 approximately agree with those from the ZEUS M_X analysis. However, for fixed

β, the H1 values have a tendency to rise faster with Q^2 even allowing for an extra scaling factor.

The dependence of the diffractive cross section $d\sigma_{\gamma^*p\to Xp}/dM_X$ on the square of the four-momentum transfer t is shown in Fig. 8 The cross section is steeply falling with $-t$, $d\sigma/dt \propto \exp(bt)$ with $b = 7.1 \pm 1.(stat) \pm 1.2(syst)$. This shows that small momentum transfers between incoming and outgoing proton dominate as expected for diffractive scattering.

3   Theoretical Interpretation

In the parton model the DIS process, $\gamma^*p \to XN$, can be visualized, in the rest frame of the proton, as a fluctuation of the incoming virtual photon into a $q\bar{q}$ pair followed by the interaction of this pair with the incoming proton [7,8]. In a diffractive DIS process the final state will be particularly simple: it consists of a $q\bar{q}$ state plus a proton or debris from the dissociation of the proton, well separated in rapidity.

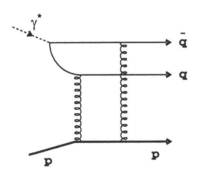

Fig. 9. Feynman diagram describing the fluctuation of $\gamma^* \to q\bar{q}$ with subsequent scattering by two gluon exchange on the proton

Some thirty years ago, J.D. Bjorken has pointed out that configurations with small q (or $\bar{q}$) transverse momenta relative to the direction of the virtual photon will lead to a substantial diffractive cross section in DIS [9]. Therefore, since in general one of the quarks of the pair will have a large logitudinal momentum, the so called aligned jet configuration will dominate. The cross section for this configuration is expected to scale with Q^2 and to have the same W dependence as observed for diffraction in hadron-hadron scattering.

The aligned jet model can be, today, well understood within QCD [10,12]. In diffraction, the interaction of the proton with a quark pair is mediated by the exchange of a colour singlet gluonic system. Such a system cannot resolve a $q\bar{q}$ (neutral) pair when the typical wave length of a gluon is larger than the $q\bar{q}$ pair transverse separation. Since the colour force which binds the proton has a wave length of the order of the proton size only pairs with small transverse momenta will interact with the proton.

The β distribution of the aligned quark configuration from transverse photons was predicted, by A. Donnachie and P.V. Landshoff [13], to be of the form

$$F_{q\bar{q}}^{T} \propto \beta(1-\beta). \qquad (6)$$

N.N. Nikolaev and B.G. Zakharov have shown that the same β dependence was expected in pQCD when the aligned quarks interact with the proton through two-gluon exchange [16].

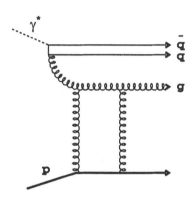

Fig. 10. Feynman diagram describing the fluctuation of $\gamma^{*} \to q\bar{q}g$ with subsequent scattering by two gluon exchange on the proton

The production of a $q\bar{q}g$ system by transverse photons was found by N.N. Nikolaev and B.G. Zakharov to be also of leading twist and was assumed to have a β dependence of the type [16]

$$F_{q\bar{q}g}^{T} \propto (1-\beta)^{\gamma} \qquad (7)$$

with $\gamma = 2$. A later calculation by M. Wüsthoff [17] found $\gamma = 3$.

The contribution to the production of a $q\bar{q}$ system by longitudinal photons was found to be of higher twist and to have a β dependence of the form [17]

$$F_{q\bar{q}}^{L} \propto \beta^{3}(1-2\beta)^{2}. \qquad (8)$$

In pQCD models the $x_{I\!P}$ dependence is expected to be driven by the x dependence of the square of the gluon momentum density of the proton [18], $[x \cdot g(x, \mu^{2})]^{2}$, with $x = x_{I\!P}$ and μ the probing scale.

3.1   BEKW Model

J. Bartels, J. Ellis, H. Kowalski and M. Wüsthoff (BEKW) indentified the theoretical processes leading to the three terms in eqs. 6- 8 as the major contributors
to the diffractive structure function. In [10] the three contributions were calculated in the perturbative region and extended into the soft region. The $x_{I\!P}$
dependence was assumed to be of the form $(1/x_{I\!P})^n$. The power n was allowed
to be different for the transverse (n_T) and the longitudinal (n_L) contributions.

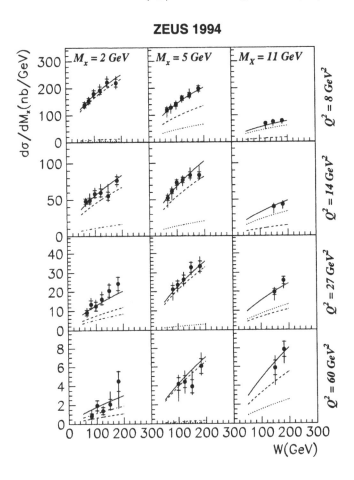

Fig. 11. Comparison of the ZEUS M_X data with the results of the BEKW fit (solid
curve). The transverse $q\bar{q}$ contribution is shown by the dashed curve, the longitudinal
$q\bar{q}$ contribution by the dashed-dotted curve and the transverse $q\bar{q}g$ contribution by the
dotted line.

The normalizations of the three terms were determined from the data:

$$x_{I\!P} F_2^{D(3)}(\beta, x_{I\!P}, Q^2) = c_T \cdot F_{q\bar{q}}^T + c_L \cdot F_{q\bar{q}}^L + c_g \cdot F_{q\bar{q}g}^T \qquad (9)$$

with

$$F_{q\bar{q}}^T = (\frac{x_0}{x_{I\!P}})^{n_T(Q^2)} \cdot \beta(1-\beta) \tag{10}$$

$$F_{q\bar{q}}^L = (\frac{x_0}{x_{I\!P}})^{n_L(Q^2)} \cdot \frac{Q_0^2}{Q^2} \cdot [\ln(\frac{7}{4} + \frac{Q^2}{4\beta Q_0^2})]^2 \cdot \beta^3(1-2\beta)^2 \tag{11}$$

$$F_{q\bar{q}g}^T = (\frac{x_0}{x_{I\!P}})^{n_T(Q^2)} \cdot \ln(1 + \frac{Q^2}{Q_0^2}) \cdot (1-\beta)^\gamma \tag{12}$$

$$n_{T,L}(Q^2) = 0.1 + n_{T,L}^0 \cdot \ln[1 + \ln(\frac{Q^2}{Q_0^2})]. \tag{13}$$

The three terms have different Q^2 dependences. $F_{q\bar{q}}^T$ does almost not depend on Q^2 as a result of the limited quark p_T in the aligned configuration. The term $F_{q\bar{q}}^L$ is higher twist but the power $1/Q^2$ is softened by a logarithmic Q^2 factor; $F_{q\bar{q}g}^T$ grows logarithmically with Q^2 similar to the proton structure function F_2 at low x.

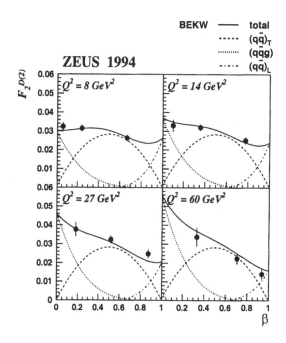

Fig. 12. Comparison of the $F_2^{D(2)}$ with the results of the BEKW fit.

The coefficients c_T, c_L, c_g as well as the parameters n_T^0, n_L^0 and x_0, Q_0^2 were determined from experiment. In the fit the power γ was also considered as a free parameter. Assuming $Q_0^2 = 1$ GeV^2 and $x_0 = 0.0042$ and treating the other constants as free parameters a good fit ($\chi^2/d.o.f. = 56/47 = 1.2$, statistical

errors only) was obtained for the ZEUS M_X data as shown by the solid curves
in Fig. 11. The fit yielded the following parameter values: $n_T^0 = 0.13 \pm 0.03$, $n_L^0 = 0.32 \pm 0.14$, $\gamma = 3.9 \pm 0.9$ [3]. It is instructive to compare the β dependence of
the three components which build up the diffractive structure function using the
results from the BEKW fit as shown in Fig. 12.[1] The ZEUS data suggest that for
$\beta > 0.2$ the colourless system couples predominantly to the quarks in the virtual
photon. The region $\beta \geq 0.8$ is dominated by the contributions from longitudinal
photons. The contribution from coupling of the colourless system to a $q\bar{q}g$ final
state becomes important for $\beta < 0.3$. Figure 13 shows the same quantities as

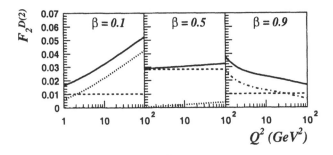

Fig. 13. $F_2^{D(2)}$ obtained from the BEKW fit to the ZEUS M_X data as a function of
Q^2

a function of Q^2 for $\beta = 0.1, 0.5, 0.9$. One finds within the BEKW model and
using ZEUS data, that the gluon term, which dominates at $\beta = 0.1$ rises with Q^2
while the quark term, which is important at $\beta = 0.5$ shows almost no evolution
with Q^2. The contribution from longitudinal photons, which is higher twist and
dominates at $\beta = 0.9$, decreases with Q^2.

In the BEKW model the $x_{I\!P}$-dependence of the quark and gluon contributions
for transverse photons is expected to be close to that given by the soft pomeron,
$n_T \approx 2(\overline{\alpha_{I\!P}}^{soft} - 1)$. However, perturbative admixtures in the diffractive final
state are expected to have a somewhat stronger energy dependence, leading
to an effective $n_T > 2(\overline{\alpha_{I\!P}}^{soft} - 1)$. The $x_{I\!P}$ dependence of the longitudinal
contribution is driven by the square of the proton's gluon momentum density
leading to $n_L > n_T$. The fit results agree with these predictions but the errors
are too large for a definitive statement.

Similar fits were performed to H1 data [10]. In the H1 case the data allowed
two, from the χ^2 point of view, almost equivalent solutions. One with the value
of the parameter $\gamma = 0.28 \pm 0.08$ and another with $\gamma = 8.5 \pm 0.8$. The high γ
solution leads to conclusions which are similar to those obtained from the ZEUS

---

[1] $F_2^{D(2)}$ denotes $F_2^{D(3)}$ at $x_{I\!P} = x_0$

data analysis. In the low γ solution the gluon term is dominant at large β region, which is in agreement with the H1 analysis [5] based on a DGLAP NLO fit.

3.2 Saturation Model

A.H. Mueller in his DIS 98 talk at Brussel [19] drew attention to the fact that the much stronger rise with W of the diffractive cross section in DIS as compared to hadron-hadron scattering combined with the behaviour of the F_2 data suggests the presence of saturation in DIS processes. Motivated by this talk K. Golec-Biernat with M. Wüsthoff developed a model which connects in a quantitative way the low and higher Q^2 total cross section data with the inclusive diffractive process [20,21]. Similar models were also proposed by J.R. Forshaw et al. [11] and W. Buchmüller et al. [12].

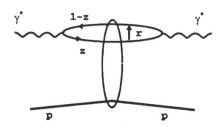

Fig. 14. Diagram of the basic saturation model process

In a frame, where photon and proton are collinear, the total γp and the differential diffractive cross section can be written as [7,8]:

$$\sigma_{T,L}(x,Q^2) = \int d^2r \int dz \, |\Psi_{T,L}(z,r)|^2 \, \hat{\sigma}(x,r^2) \tag{14}$$

$$\frac{d\sigma_{T,L}^{Diff}}{dt}\Big|_{t=0} = \frac{1}{16\pi} \int d^2r \int dz \, |\Psi_{T,L}(z,r)|^2 \, |\hat{\sigma}(x,r^2)|^2 \tag{15}$$

where $\Psi_{T,L}(z,r)|$ denotes the transverse (T) and longitudinally (L) polarized photon wave functions, $\hat{\sigma}(x,r^2)$ the dipole cross section of the $q\bar{q}$ pair with the proton, z the momentum fraction of the photon carried by the quark and r the relative transverse separation between the quarks. The wave functions are solely determined by the coupling of the photon to the quark and therefore are well known in QCD [8]. The cross section of the $q\bar{q}$ pair with the proton is assumed to be a function of the transverse separation r and the saturation radius R_0:

$$\hat{\sigma}(x, r^2) = \sigma_0 \left(1 - \exp[-\frac{r^2}{4R_0^2}]\right) \tag{16}$$

At small r, $r \ll 2\,R_0$, the dipole cross section grows quadratically with r, $\hat{\sigma} \sim \sigma_0 r^2/4R_0^2$, at large r, $r \gg 2\,R_0$, it saturates, $\hat{\sigma} = \sigma_0$.

The wave function, $\Psi(r)$, at small quark transverse separation, $r < 1/Q$, behaves as $Q \cdot 1/(Qr)$, whereas for large quark transverse separation, $r > 1/Q$, it is exponentially suppressed. When the transverse size of the $q\bar{q}$ pairs is much smaller than the saturation radius, $1/Q \ll R_0$, the total cross section behaves as:

$$\sigma_T = \int dz \int_0^{1/Q^2} dr^2 Q^2 \left(\frac{1}{Q^2 r^2}\right) \sigma_0 \frac{r^2}{4R_0^2} \sim \frac{1}{Q^2} \frac{\sigma_0}{R_0^2} \tag{17}$$

When the transverse size of the $q\bar{q}$ pairs exceeds the saturation radius, $1/Q \gg R_0$, the integration over dr^2 splits into two parts:

$$\sigma_T = \int_0^{R_0^2} dr^2 \left(\frac{1}{r^2}\right) \sigma_0 \frac{r^2}{4R_0^2} + \int_{R_0^2}^{1/Q^2} dr^2 \left(\frac{1}{r^2}\right) \sigma_0 \sim \sigma_0 + \sigma_0 \ln\left(\frac{1}{Q^2 R_0^2}\right)$$

and σ_T becomes almost independent of Q^2. From the above approximative arguments it can be seen that the model describes the well known scaling behaviour at higher Q^2, $\sigma_T \sim 1/Q^2$, and the break down of scaling at low Q^2, $\sigma_T \sim \sigma_0$.

In order to describe the W (or x) dependence of the total DIS cross sections measured at HERA, the saturation radius has to be x dependent.

$$R_0^2(x) = \frac{1}{Q_0^2} \left(\frac{x}{x_0}\right)^\lambda \tag{18}$$

with $Q_0 = 1$ GeV. The eqs. 14, 15, 16 and 18 define the model called in the following the saturation model.

The parameters of the model σ_0, x_0 and λ where determined from the fit to the HERA total cross section data. Figure 15 shows the HERA total cross section data in the Q^2 region between 0.1 and 100 GeV^2 and compares them to the fit. Note that a very good description is achieved for the transition between the photoproduction-like region at $Q^2 < 1 GeV^2$ and the scaling region at higher Q^2. The parameters of the model are determined as $\sigma_0 = 23$ (mb), $\lambda = 0.29$, $x_0 = 0.0003$. The value of σ_0 is mainly determined by the behaviour of the data in the photoproduction-like region while λ and x_0 is mainly determined by the DIS region. Note that σ_0, which in the model corresponds to the $q\bar{q}$ proton cross section, is of the same magnitude as the total $\pi^\pm p$ cross section.

The results of the saturation model are compared with the data in Fig. 16. We see a good agreement with data at all Q^2 and M_X values, similar in quality

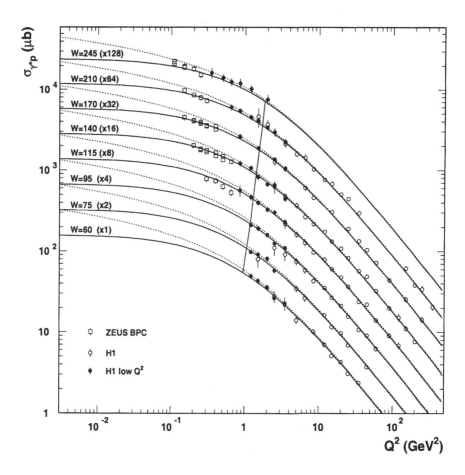

Fig. 15. The $\gamma^* p$ cross section compared to the results of the saturation model fit (solid line). The dotted line shows an alternative fit in which the quark mass was set to zero.

to the BEKW fit as seen in Fig. 11. The difference is that it is now an absolute prediction and not a fit. All parameters of the model were fixed by F_2 data.

It is interesting to compare the data on the ratio of the diffractive and total cross sections with the predictions of the saturation model as shown in Fig. 17. When this ratio was first published the data were difficult to understand since the diffractive and total cross sections show the same rate of growth with W. This is contrary to the naïve expectation that according to the optical theorem the diffractive cross section should grow with a power of W which is twice the power of the total one, $\sigma_{Diff} \sim W^{2a}$ while $\sigma_{TOT} \sim W^a$. This basic property is also present in the saturation model; note that the dipole cross section, $\hat{\sigma}$, appears in quadrature in the equation 15 while only one power of $\hat{\sigma}$ enters equation 14. Therefore, if $\hat{\sigma}$ leads to a growth of the total cross secton with the power λ,

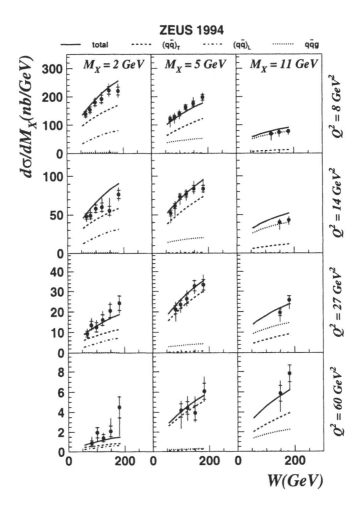

Fig. 16. Comparison of the ZEUS M_X data with the results of the saturation model (solid curve). The transverse $q\bar{q}$ contribution is shown by the dashed curve, the longitudinal $q\bar{q}$ contribution by the dashed-dotted curve and the transverse $q\bar{q}g$ contribution by the dotted line.

$\sigma_T \sim W^{\lambda}$, (see eqs. 17, 18) then the diffractive cross section would, again, be expected to grow with the power 2λ. The saturation model leads, however, to a remarkably good description of the experimentally observed $\sigma_{Diff}/\sigma_{TOT}$ ratio. This is due to the fact that the main contribution to diffraction is coming from $q\bar{q}$ pairs with larger transverse separation, r, than in the inclusive process. In the diffractive case, a similar integration as in eq. 17 leads only to a higher twist contribution. A leading twist contribution can be obtained if the dr^2 integration is extended beyond the $1/Q^2$ limit. However, at these large transverse distances saturation starts to damp the growth of the diffractive cross section [21].

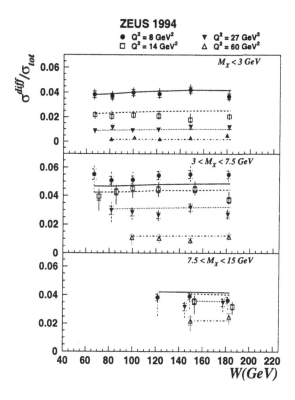

Fig. 17. The ratio of the diffractive and the total cross sections compared with the prediction of the saturation model (solid line).

4 Conclusions

The first round of experimental data on diffraction at HERA has lead to a growing theoretical understanding of relations between the, previously disconnected, phenomena of photoproduction, deep inelastic scattering and diffraction. Perturbative QCD has served as a useful guide into the transition region between hard and soft phenomena.

In the near future HERA experiments can be expected to provide a much larger body of data on diffraction. These should help to understand diffraction in deep inelastic scattering, especially in the transition region between the perturbative and nonperturbative regimes.

5 Acknowledgments

It is a pleasure to thank the organizers, G. Grindhammer, B. Kniehl and G. Kramer for a stimulating and enjoyable workshop. I have considerably benefited from discussions with J. Bartels, E. Levin, A.H. Mueller, G. Wolf and M.

Wüsthoff. I am grateful to G. Wolf for a critical reading of the manuscript and for many comments.

References

1. R.P. Feynmann, "Photon-Hadron Interactions", Benjamin, N.Y. (1972), lectures 50-54.
2. ZEUS Collab., M. Derrick et al., Z. Phys. C70 (1996) 391.
3. ZEUS Collab., J. Breitweg et al., E. Phys. J. C6 (1999) 43.
4. ARIADNE 4.0 Program Manual, L. Lönnblad, DESY-92-046 (1992), Computer Phys. Comm. 71 (1992) 15.
5. H1 Collab., C. Adloff et al., Z. Phys. C76 (1997) 613.
6. ZEUS Collab., J. Breitweg et al., E. Phys. J. C1 (1998) 81.
7. N.N. Nikolaev and B.G. Zakharov, Z. Phys. C49 (1990) 607.
8. J.R. Forshaw and D.A. Ross, *QCD and the Pomeron*, Cambridge University Press, 1997.
9. J.D. Bjorken, Proc. Int. Symp. Electron and Photon Interactions at High Energies, Cornell, 1971, p. 282.
   J.D. Bjorken, J. Kogut and D. Soper, Phys. Rev. D3 (1971) 1382.
   J.D. Bjorken and J. Kogut, Phys. Rev. D8 (1973) 1341.
10. J. Bartels, J. Ellis, H. Kowalski and M. Wüsthoff, E. Phys. J. C7 (1999) 443.
11. J.R. Forshaw, G. Kerley and G. Shaw, CERN-TH/99-58.
12. W. Buchmüller and A. Hebecker, Nucl. Phys. B476 (1996) 203.
    W. Buchmüller, M.F. McDermott, and A. Hebecker, Nucl. Phys. B487 (1997) 283; *ibid.* B500 (1997) 621.
    W. Buchmüller, T. Gehrmann, and A. Hebecker, Nucl. Phys. B537 (1999) 477.
13. A. Donnachie and P.V. Landshoff, Phys. Lett. B191 (1987) 309.
14. A. Donnachie and P.V. Landshoff, Nucl. Phys. B244 (1984) 322; Phys. Lett. B296 (1992) 227.
15. ZEUS Collab., J. Breitweg et al., Eur. Phys. J. C1 (1998) 81.
16. N.N. Nikolaev and B.G. Zakharov, Z. Phys. C53 (1992) 331; M. Genovese, N.N. Nikolaev and B.G. Zakharov, JETP 81 (1995) 625; M. Bertini, M. Genovese, N.N. Nikolaev, A.V.Pronyaev and B.G. Zakharov, Phys. Lett. B422 (1998) 238.
17. M. Wüsthoff, PhD thesis, University of Hamburg, DESY 95-166.
18. M. Ryskin, Sov. J. Nucl. Phys. 52 (1990) 529.
    N.N. Nikolaev and B.G. Zakharov, Phys. Lett. B332 (1994) 177.
    E.M. Levin and M. Wüsthoff, Phys. Rev. D50 (1994) 4306.
    S.J. Brodsky et al., Phys. Rev. D50 (1994) 3134.
    J. Bartels, H. Lotter and M. Wüsthoff, Phys. Lett. B379 (1996) 239; Erratum ibid. B382 (1996) 449.
19. A.H. Mueller, 6th International Workshop on Deep Inelastic Scattering and QCD, DIS 98 (1998) 3, World Scientific.
20. K. Golec-Biernat and M. Wüsthoff, Phys. Rev. D59:014017, (1999).
21. K. Golec-Biernat and M. Wüsthoff, DTP-99-20.

Hard Diffraction at the TEVATRON

Kristal Mauritz

Iowa State University, Ames IA 50011, USA

(On behalf of the DØ and CDF Collaborations)

Abstract. Experimental results on hard diffraction with the CDF and DØ detectors at the Fermilab TEVATRON are reviewed and compared with theoretical expectations.

1 Introduction

Diffractive and elastic events have been studied for over 30 years and are responsible for about 40% of the $p\bar{p}$ cross-section, but still much is unknown about the nature of the process. In addition to the diffractively scattered proton or antiproton, they are characterized by the complete, or nearly complete, absence of hadronic particle activity over a large rapidity or pseudorapidity region [1,2]. This region is called a 'rapidity gap' and in elastic $p\bar{p}$ scattering, the 'rapidity gap' is complete and extends over the whole kinematically allowed rapidity range. Single diffraction is the break-up of the p or $\bar{p}$ into a typically low-mass 'diffractive system' at low momentum transfer generally assumed to be a spin-1/2 nucleon resonance with the quantum numbers of the proton. The concept of a "Pomeron" was invented by Pomeranchuk to explain this diffractive process in the context of Regge theory [3]. The pomeron, which is a color-singlet with quantum numbers of the vacuum, is the exchanged object presumed to produce a diffractive event.

Hard single diffraction (HSD) was introduced by Ingelman and Schlein[4] in 1985 as a model to probe the pomeron. The event is treated in a factorizable way, i.e. the pomeron is "emitted" by the anti-proton (or proton) with a certain probability or flux and then struck by the other proton (or antiproton) with a high p_T. 'Hard' diffraction constitutes a larger tranverse momentum process in addition to the diffractive scattering. The events studied here have a hard scattered system of at least two jets with an $E_T \geq 7\,\text{GeV}$ (usually higher) as well as a rapidity gap. The first experimental results on the subject were published by the UA8 Collaboration, which showed the existence of jets in single diffractive events and that these jets had rapidity and longitudinal momentum distributions consistent with a hard pomeron structure [5]. The field of study has expanded dramatically since then. Recent results from HERA and the TEVATRON include studies of large rapidity gap events in deep inelastic scattering [1,2,6], diffractive jet production [7–10] and diffractive W-boson production [11]. Measurements at both HERA and the TEVATRON are consistent with a predominantly gluonic pomeron, but measured rates at the TEVATRON are several times lower than predictions based on HERA data[12]. The combination of these results gives new insight into the exchanged object.

This paper will provide an introduction and brief review on the recent results at the TEVATRON of hard single diffraction and hard double pomeron exchange. Studies of hard diffraction can be used to increase our knowledge of the structure of the pomeron, especially when combined with complementary hard diffractive measurements at HERA. They can also test such basic assumptions as to the very nature of the pomeron as a particle-like object.

1.1   Hard Diffraction at the TEVATRON

At the TEVATRON, there are three types of hard diffractive processes accessible to experimentation: hard single diffraction, double diffraction or hard color singlet exchange (HCS), and hard double pomeron exchange (HDPE). Hard single diffractive jet production is a subset of single diffraction with a hard scatter in the final state. Figure 1(a) shows a two-jets final state in the process $p + \bar{p} \rightarrow j + j + X$. Because the pomeron is a color-singlet, particle production is typically suppressed between the outgoing hadron and the jets and a forward rapidity gap results in these events. Hard color singlet exchange (Fig. 1b) is a double diffractive process, but at large t, which produces a final state of a rapidity gap between jets. To produce a rapidity gap, the exchanged particle must again be a color-singlet, but the event rates are too large to be explained by electroweak boson exchange and indicate a strong interaction process [13].

Another type of diffractive event is double pomeron exchange, characterized by particle production only in the central rapidity region (Fig. 1c). Each incoming hadron 'emits' a (or interacts via an internal) pomeron (state) leading to large forward and backward rapidity gaps, and the interaction of the two pomerons results in central particle production. This class of events was first observed at the ISR in

$$pp \rightarrow pp\pi^+\pi^- \qquad (1)$$

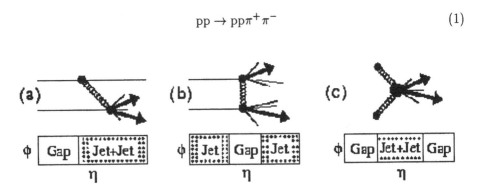

Fig. 1. Event topology of dijet production in (a) single diffraction, (b) hard color singlet, and (c) hard double pomeron exchange.

interactions [14]. At the ISR there was insufficient energy to produce hard scattering in DPE events. UA1 previously has made attempts to find "hard double pomeron exchange" using central particle production and forward rapidity gaps [15] (gaps close to the beam-pipe on either side). Although they did find a class of events consistent with double pomeron exchange, they were unable to show that these events did not arise from multiplicity fluctuations in color exchange jet events.

All three classes of events can be tagged by the rapidity gap signature. HSD and DPE can also be tagged by detecting the leading particle(s) on the gap side. Using rapidity gap tagging, the CDF and DØ collaborations have studied dijet production (two or more jets) in all three processes at $\sqrt{s} = 1800\,\mathrm{GeV}$. In addition, CDF has studied (single) diffractive W-boson, b-quark, and J/ψ production using rapidity gaps, and dijet production in HSD and DPE using a "roman pot" magnetic spectrometer to detect leading antiprotons [16]. DØ has also studied all three processes at $\sqrt{s} = 630\,\mathrm{GeV}$ and has made measurements of the dependence of HSD on jet pseudorapidity [17].

2   Hard Single Diffraction

CDF [18] and DØ [19] measure the rate of single diffraction in different processes as the fraction of events with forward rapidity gaps, called the gap fraction. Forward rapidity gaps are identified by measuring the multiplicity in the forward detectors, or those closest to the beam pipe with the greatest pseudorapidity. In both experiments, this is the forward calorimeter ($2.4 < |\eta| < 4.2$ at CDF and $3.0 < |\eta| < 5.2$ at DØ) and the forward scintillator arrays, called the beam-beam counter (BBC) at CDF ($3.2 < |\eta| < 5.9$) and the Level0 (L0) detector at DØ ($2.3 < |\eta| < 4.3$). A particle is tagged by a hit in the scintillation counters or by the deposition of energy in a calorimeter tower greater than a certain threshold energy, set to just above the noise to maximize the sensitivity to particles. At CDF, this threshold is $1500\,\mathrm{MeV}$ per calorimeter tower, and at DØ it is $150\,\mathrm{MeV}$ in an electromagnetic calorimeter tower or $500\,\mathrm{MeV}$ in a hadronic calorimeter tower.

Figures 2 and 3, respectively, show sample multiplicity distributions at DØ and CDF. The number of calorimeter towers (n_{CAL} or towers) above the energy threshold and the number of scintillator hits (n_{L0} or BBC) are measured opposite the leading two jets for the forward jet trigger (at least two jets with $E_T > 12\,\mathrm{GeV}$ and $|\eta| > 1.6$ for DØ or $E_T > 20\,\mathrm{GeV}$ and $1.8 < |\eta| < 3.5$ for CDF). The distributions show the characteristic peak at zero multiplicity in qualitative agreement with expectations for a diffractive signal component and a higher multiplicity distribution associated with non-diffractive, background events. The diffractive and non-diffractive peaks at DØ are more separated because of the lower threshold and finer calorimeter segmentation. Both experiments fit the distributions to extract the gap fraction, the number of diffractive events divided by the total number of events in the sample. CDF fits the two-dimensional multiplicity distribution to a line along the diagonal and DØ simultaneously

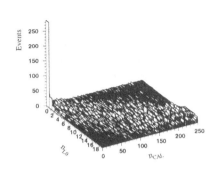

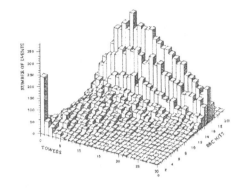

Fig. 2. DØ multiplicity distribution for 1800 GeV forward jet events.

Fig. 3. CDF multiplicity distribution for 1800 GeV forward jet events.

fits the diffractive signal to a falling exponential and the background to a four-parameter fit.

Table 1. Diffractive to total production ratios at the TEVATRON

SD hard process	$\sqrt{s}$ (GeV)	$R = \frac{\text{DIFF}}{\text{TOTAL}}$ (%)	Comments	Exp't
$W(\rightarrow e\nu)$	1800	$(1.15 \pm 0.55)\mathcal{A}^\dagger$	$E_T^e, \not{E}_T > 20$ GeV	CDF [11]
Forward Jets	1800	$(0.75 \pm 0.10)\mathcal{A}^\dagger$	$E_T^{jet} > 20$ GeV, $|\eta^{jet}| > 1.8$	CDF [8]
Forward Jets	1800	0.65 ± 0.04	$E_T^{jet} > 12$ GeV, $|\eta^{jet}| > 1.6$	DØ*
Forward Jets	630	1.19 ± 0.08	$E_T^{jet} > 12$ GeV, $|\eta^{jet}| > 1.6$	DØ*
Central Jets	1800	0.22 ± 0.05	$E_T^{jet} > 15$ GeV* *, $|\eta^{jet}| < 1.0$	DØ*
Central Jets	630	0.90 ± 0.06	$E_T^{jet} > 12$ GeV, $|\eta^{jet}| < 1.0$	DØ*
$b(\rightarrow e + X)$	1800	$(0.62 \pm 0.25)\mathcal{A}^\dagger$	$|\eta^e| < 1.1$, $9.5 < p_T^e < 20$ GeV	CDF [20]
$J/\psi(\rightarrow \mu\mu)$	1800	0.64 ± 0.12	$|\eta^\mu| < 0.6$, $p_T^\mu > 2$ GeV	CDF*

* Preliminary
$\mathcal{A}^\dagger$ Includes rapidity gap acceptance based on a hard gluon structure function
* * The diffractive jet gap fractions show no E_T dependence (shown later), so the central jet gap fractions at 1800 and 630 GeV can be directly compared.

Table 1 shows the measured gap fractions at both experiments for final state forward jet, central jet, W-boson, J/ψ, and b-quark production. There are several notable observations:

- All gap fractions for hard single diffraction are $\sim O(1\%)$.
- Forward jet gap fractions > central jet gap fractions.

- Gap fractions at $\sqrt{s} = 630\,\text{GeV}$ are at least twice as large as the gap fractions at $\sqrt{s} = 1800\,\text{GeV}$.

As can be seen by the $\mathcal{A}^\dagger$ in the gap fraction measurements in Table 1, the two experiments treat the measurement of the rate for hard diffraction slightly differently. To understand this, it is important to look at the phenomenology.

2.1 Phenomenology

The theory of diffraction relies on phenomenological models. The Ingelman-Schlein model (IS) supposed that the pomeron is a composite pseudoparticle, composed of quarks and gluons. They proposed several simple structure functions and calculated the cross-section for hard diffraction as the probability of a pomeron being 'emitted' (called the flux factor) with the QCD hard matrix element. The diffractive Monte Carlo POMPYT 2.6 [21] is based on the IS model and assumes a pomeron is emitted from the proton. The standard Donnachie-Landshoff flux factor is used at the TEVATRON, and the results can be easily rescaled for different flux factors. It allows for several possible choices of a presumed pomeron structure function, $s(x)$. The four main structure choices are a: (i) 'hard gluon', a pomeron consisting of two light gluons, $s(x) \propto x(1-x)$; (ii) 'flat gluon', $s(x) \propto$ constant; (iii) 'quark', a pomeron consisting of two light quarks, $s(x) \propto x(1-x)$; and (iv) 'soft gluon', with a gluon distribution similar to that in the proton, $s(x) \propto (1-x)^5$. The pomeron takes a small fraction of the proton's momentum called ξ, and a parton in the pomeron with a momentum fraction β to interact in the hard scattering.

The corresponding diffractive Monte Carlo multiplicity distributions with POMPYT show the same characteristic peak at low multiplicities as observed in the diffractive data with a tail extending to larger multiplicities that is dependent on the structure function choice. The expected fraction of diffractive events that are produced with a rapidity gap, called the rapidity gap acceptance, can be calculated for each structure function. The rapidity gap acceptance is treated differently between the two experiments. CDF assumes a hard gluon structure function (which has its basis in HERA data) and folds the acceptance into the data gap fraction to provide a "true" diffractive rate. DØ chooses to measure the rapidity gap fraction with no model dependence and folds the gap acceptance into the Monte Carlo predictions. For comparison, the DØ 1800 GeV forward jets gap fraction with a hard gluon gap efficiency increases the preliminary "measured" gap fraction to $(0.88 \pm 0.05)\%$.

2.2 DØ Analysis

The gap fraction can be calculated from Monte Carlo as the ratio of the number of rapidity gap events using POMPYT to the number of non-diffractive events with PYTHIA 5.7 [22] divided by the gap acceptance. The Monte Carlo gap fractions are shown in Table 2, and the trend is for the harder gluon rates to be much higher than the data, although the different data sets do not favor a particular $\sqrt{s}$-dependent scale factor as suggested by Ref. [23].

In the gap fraction ratios also shown in Table 2, the flux factor cancels for the same $\sqrt{s}$, while for different $\sqrt{s}$, any dependence on the flux factor is reduced. The Monte Carlo 630/1800 ratios are in much better agreement with the data than the numerator and denominator separately. However, the hard and flat gluon structures, which have been favored by previous measurements, are inconsistent with the forward/central jet ratio, despite the full cancellation of the flux factor. For a gluon dominated pomeron to describe the data, the addition of a significant soft gluon component, which rarely produces central jet events with rapidity gaps, would be required.

Table 2. The DØ preliminary measured and predicted gap fractions and ratios.

Sample	Data	Hard Gluon	Flat Gluon	Soft Gluon	Quark
Gap Fraction (in %)					
1800 GeV $|\eta| > 1.6$	0.65 ± 0.04	2.2 ± 0.3	2.2 ± 0.3	1.4 ± 0.2	0.79 ± 0.12
1800 GeV $|\eta| < 1.0$	0.22 ± 0.05	2.5 ± 0.4	3.5 ± 0.5	0.05 ± 0.02	0.49 ± 0.06
630 GeV $|\eta| > 1.6$	1.19 ± 0.08	3.9 ± 0.9	3.1 ± 0.8	1.9 ± 0.4	2.2 ± 0.5
630 GeV $|\eta| < 1.0$	0.90 ± 0.06	5.2 ± 0.7	6.3 ± 0.9	0.13 ± 0.11	1.6 ± 0.2
Gap Fraction Ratio					
630/1800 FWD JET	1.8 ± 0.2	1.7 ± 0.4	1.4 ± 0.3	1.4 ± 0.3	2.7 ± 0.6
630/1800 CEN JET	4.1 ± 1.0	2.1 ± 0.4	1.8 ± 0.3	2.7 ± 2.4	3.2 ± 0.5
1800 FWD/CEN JET	3.0 ± 0.7	0.88 ± 0.18	0.64 ± 0.12	$29. \pm 11.$	1.6 ± 0.3
630 FWD/CEN JET	1.3 ± 0.1	0.75 ± 0.16	0.48 ± 0.12	$15. \pm 12.$	1.4 ± 0.3

DØ also uses an inclusive jet trigger in coincidence with a rapidity gap in the LØ detector to obtain a large number of low background single diffractive candidate events with which to study the event characteristics. After reinforcing the rapidity gap in the calorimeter, the number of jets, jet width, $\delta\phi$, and E_T of the leading two jets are shown in figure 4 for 1800 GeV diffractive (solid) and non-diffractive (dashed) central jet events. The diffractive events are consistent with less overall radiation than the non-diffractive events, and the E_T distributions are similar.

The ξ distribution, or the momentum loss of the diffracted proton, can also be calculated using [24]

$$\xi \approx \Sigma \frac{E_{T_i} e^{\eta_i}}{\sqrt{s}}, \qquad (2)$$

summing over the particles with the highest η and largest E_T. The outgoing diffracted proton (or the rapidity gap) is defined to be at positive η. This formula measures ξ using only the calorimeter, emphasizing the well-measured central region close to the gap. It contains no model-dependence, and it can be defined even for non-diffractive events. After correcting for energy scale differences between Monte Carlo and data, the ξ distribution (solid) for 1800 GeV and 630 GeV forward and central jets are shown in Figure 5. The dotted and dashed curves reflect the high and low error.

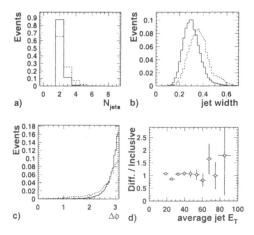

Fig. 4. The number of jets, jet width, $\delta\phi$ between jets, and the E_T distribution for the two central leading jets at 1800 GeV for the diffractive (*solid line*) and non-diffractive (*dashed line*) events (DØ preliminary)

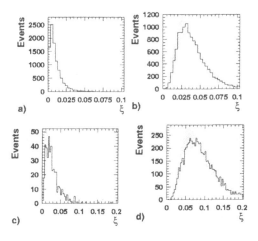

Fig. 5. The ξ distribution (*solid line*) and one sigma uncertainty for a)-b) $\sqrt{s} = 1800$ GeV and c)-d) $\sqrt{s} = 630$ GeV forward and central jets (DØ preliminary)

The forward jet ξ distribution at both 1800 and 630 GeV is peaked toward lower values of ξ although separated from zero because the pomeron requires a certain amount of momentum to create two 12 GeV jets. The central jet ξ distribution is spread over higher values, consistent with the pomeron requiring more momentum to balance the parton in the non-diffracted (anti)proton. The ξ distribution at 630 GeV is higher overall than at 1800 GeV, as expected because more momentum is needed from the pomeron at lower $\sqrt{s}$ with the same jet requirements. Although the trends of the ξ distribution are consistent with expectations, pomeron exchange is typically thought to dominate for very low values of $\xi < 0.05$. The large values of ξ observed, however, can be reproduced well with Monte Carlo.

2.3   CDF Analysis

The gluon content of the diffractive structure function is measured at CDF by comparing the hard single diffractive rates for final states with different sensitivity to quarks and gluons. This is similar to what is done by ZEUS, which compares the diffractive deep inelastic scattering (D-DIS) to dijet photoproduction rates, and by H1, through a QCD analysis of the Q^2 evolution of the diffractive structure function measured in D-DIS. Both experiments measure a high gluon content. CDF measured, with the W and dijet production rates, a gluon fraction of $f_g = 0.7 \pm 0.2$ [11], and has a new measurement including the b-quark production rate of $f_g = 0.54^{+0.16}_{-0.14}$ [20] (see Fig. 6).

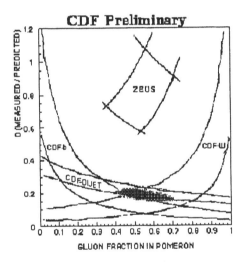

Fig. 6. The ratio, D, of measured to predicted diffractive rates as a function of the gluon content of the Pomeron. The predictions are from POMPYT using the standard Pomeron flux and a hard Pomeron structure.

In 1996 CDF installed three forward proton detectors to be able to tag the diffractively scattered proton. Data were taken at both $\sqrt{s} = 1800\,\mathrm{GeV}$ and $\sqrt{s} = 630\,\mathrm{GeV}$, requiring a good reconstructed track. The forward proton detectors have acceptance for $0.4 < \xi < 0.1$ and $|t| < 1\mathrm{GeV}^2$ at $\sqrt{s} = 1800\,\mathrm{GeV}$ and $|t| < 0.2\mathrm{GeV}^2$ at $\sqrt{s} = 630\,\mathrm{GeV}$. From these data, events are selected with two jets of $E_T \geq 7\,\mathrm{GeV}$, and the momentum fraction of the parton in the pomeron can be measured from

$$\beta = \frac{E_T^{jet1} e^{\eta_1} + E_T^{jet2} e^{\eta_2}}{\sqrt{s}\xi}. \tag{3}$$

After a correction from pile-up background and detector acceptance, the ratio of the data β-distribution in comparison to a Monte Carlo flat gluon structure is shown in Fig. 7. Although well-described at large β by a flat structure, there is an excess of events at low-β, implying a softer structure component. This is consistent with MC comparisons to data at DØ.

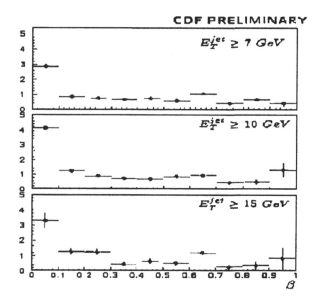

Fig. 7. The ratio of background subtracted data to Monte Carlo simulation using a flat gluon Pomeron structure and standard flux.

3   Hard Double Pomeron Exchange

Hard double pomeron exchange is another process to enable better understanding of diffraction. In the Ingelman-Schlein model, both the incoming proton and anti-proton can be said to "emit" a pomeron and the two pomerons interact to produce a massive system.

The data analysis for HDPE is analagous to hard single diffraction, except that a rapidity gap or forward proton track is required on one side and the multiplicity on the opposite side is measured. DØ triggers at 1800 GeV (630 GeV) on two 15 (12) GeV jets and a rapidity gap in the L0 detector. Figure 8 shows the opposite multiplicity distributions at 1800 GeV. The distributions at 630 GeV are similar. A clear peak at low multiplicity is observed in both distributions above a flat background in qualitative agreement with expectations for an HDPE signal. Figure 8 also shows the jet E_T distribution for events with two gaps, one gap, and an inclusive distribution. The jet E_T spectra for jets in both kinds of gap events are observed to be similar to the inclusive $p\bar{p}$ scattering. CDF finds similar preliminary results at $\sqrt{s} = 1800\,\mathrm{GeV}$ and measures the rates for hard double pomeron exchange to be $\sim 10^{-3}/\mathcal{A}_{\mathcal{HDPE}}\%$ where the gap acceptance $\mathcal{A}_{\mathcal{HDPE}} \sim 0.1$ for two jets with $E_T > 7\,\mathrm{GeV}$ and $0.04 < \xi_{\bar{p}} < 0.1$ [16]. The kinematic properties of HDPE candidates are in agreement with expectations events with very little radiation.

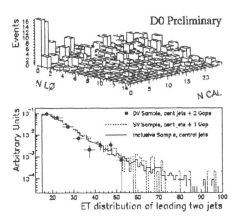

Fig. 8. The multiplicity at 1800 GeV after requiring a rapidity gap on the opposite side. The peak in the low multiplicity bins shows a clear double gap signal. The bottom figure is the corresponding jet E_T distribution for the double gap (*circles*), one gap (*dotted*), and an inclusive jet (*solid*) samples.

4   Hard Color Singlet Exchange

Both experiments have also studied events with a central rapidity gap between jets [13,25–28]. This topology could be due to very large momentum transfer pomeron exchange (a class of double diffractive events), with $|t| >> 100\,\mathrm{GeV}^2$ compared to the previously mentioned processes which typically have $|t| \simeq 0\,\mathrm{GeV}^2$. The fraction of dijet events with a central rapidity gap is typically on the order of 1%. DØ observes an increase in the gap fraction with transverse energy and pseudorapidity. Comparisions to Monte Carlo show that the data favor a soft color model [29,30], where the hard interaction is a standard QCD exchange and the rapidity gap is generated by a cancellation of the color by soft radiation. The two gluon models for the pomeron do not describe the data well. Although CDF does not observe an increase in the gap fraction with energy and pseudorapidity, the measurements are consistent with DØ.

5 Conclusion

Hard diffraction has been studied and observed at the TEVATRON in three types of processes: single diffraction, double diffraction, and double pomeron exchange at both $\sqrt{s} = 630$ and $1800\,\text{GeV}$. The analyses at DØ and CDF present consistent and complimentary results. The diffractive events are quieter in general than ordinary QCD events and the pomeron structure in single diffraction contains a significant soft component. The hard single diffractive and double diffractive gap fractions are $\sim O(1\%)$, which when compared to the HERA results $\sim O(10\%)$, represent a breakdown in the factorization of the Ingelman-Schlein model. Moreover, there does not appear to be a single scale, dependent on $\sqrt{s}$, between the $p\bar{p}$ and ep experiments.

References

1. M. Derrick et al. (ZEUS Collaboration): Phys. Lett. B **315**, 481 (1993)
2. T. Ahmed et al. (H1 Collaboration): Nucl. Phys. B **429**, 477 (1994)
3. P.D.B. Collins: An Introduction to Regge Theory and High Energy Physics, Cambridge University Press, Cambridge (1977).
4. G. Ingelman, P. Schlein: Phys. Lett. **B152**, 256 (1985)
5. A. Brandt et al (UA8 Collaboration): Phys. Lett. B **297**, 417 (1992)
6. A. Doyle: GLAS-PPE/96-01, *Workshop on HERA Physics, "Proton, Photon, and Pomeron Structure"*.
7. S. Abachi et al. (DØ Collaboration): FERMILAB-CONF-96-247-E, Proceedings of the 28th International Conference on High-energy Physics, Poland (1996)
8. F. Abe et al. (CDF Collaboration): Phys. Rev. Lett. **79**, 2636 (1997)
9. C. Adloff et al. (H1 Collaboration): Eur. Phys. J. **C6**, 421 (1999)
10. J. Breitweg et al. (ZEUS Collaboration): Eur. Phys. J. **C5**, 41 (1998)
11. F. Abe et al. (CDF Collaboration): Phys. Rev. Lett. **78**, 2698 (1997)
12. L. Alvero, J.C. Collins, J. Terron, J. Whitmore: Phys. Rev D **59**, 74022 (1999)
13. S. Abachi et al. (DØ Collaboration): Phys. Rev. Lett. **76**, 734 (1996)
14. H. deKerret et al.: Phys. Lett. B **68**, 385 (1977)
15. D. Joyce et al.: Phys. Rev. D **48** 1943 (1993)
16. K. Borras, *Hard Diffraction at CDF*, Proceedings of the Deep Inelastic Scattering Workshop, Zeuthen Germany(1999)
17. K. Mauritz, *Hard Diffraction at DØ*, Proceedings of the Deep Inelastic Scattering Workshop, Zeuthen Germany(1999)
18. F. Abe et al. (CDF Collaboration): Nucl. Instrum. and Meth. in Phys. Res. A **271**, 387 (1988)
19. S. Abachi et al. (DØ Collaboration): Nucl. Instrum. and Meth. in Phys. Res. A **338**, 185 (1994)
20. T. Affolder et al. (CDF Collaboration): submitted to Phys. Rev. Lett. (1999)
21. P. Bruni, G. Ingelman: Proceedings of the Europhysics Conference on HEP, Marseille, 595 (1993)
22. H.-U. Bengtsson, T. Sjöstrand: Comp. Phys. Comm. **46**, 43 (1987)
23. K. Goulianos: Phys. Lett. B **358**, 379 (1995)
24. J. Collins, *Light-cone Variables, Rapidity and All That,* hep-ph/9705393.
25. S. Abachi et al. (DØ Collaboration): Phys. Rev. Lett. **72**, 2332 (1994)

26. F. Abe et al. (CDF Collaboration): Phys. Rev. Lett. **74**, 855 (1995)
27. B. Abbott et al. (DØ Collaboration): Phys. Lett. **B440**, 189 (1998)
28. F. Abe et al. (CDF Collaboration): Phys. Rev. Lett. **80**, 1156 (1998)
29. G. Ingelman: *Old and New Ideas on Hard Diffraction*, Proceedings of the LISHEP Workshop, Rio de Janerio (1998)
30. O.J.B. Eboli, E.M. Gregores, F. Halzen: *(No) color in QCD: Charmonium, charm and rapidity gaps*, hep-ph/9611258 (1996)

Towards the Theory of Diffractive DIS

Wilfried Buchmüller[1]

Deutsches Elektronen-Synchrotron DESY, Hamburg

Abstract. The large rapidity gap events, observed at HERA, have changed considerably our physical picture of deep inelastic scattering during the past years. We review the present theoretical understanding of diffractive DIS with emphasis on the close relation to inclusive DIS. This includes success and limitations of the leading twist description, the connection between diffractive and inclusive parton distributions in the semiclassical approach, the colour structure of the proton and comparison with data. The progress report concludes with a list of open questions.

1 Inclusive and diffractive DIS

The intriguing phenomenon of the frequent appearance of large rapidity gaps in electron proton collisions at HERA [1] has changed our physical picture of deep inelastic scattering (DIS) to a large extent. The large rapidity gap events are very difficult to understand in the parton model where the struck quark is expected to break up the proton leading to a continuous flow of hadrons between the current jet and the proton remnant.

To develop a physical picture of diffractive DIS [2] it is convenient to view the scattering process in the proton rest frame. In this frame the virtual photon fluctuates into partonic states $q\bar{q}$, $q\bar{q}g$, ... which then scatter off the proton. From the leading twist contributions to inclusive and diffractive structure functions one obtains the parton distribution functions in a frame where the proton moves fast. This connection holds for diffractive as well as non-diffractive processes.

In the following we shall review the present status of our theoretical understanding of diffractive DIS with emphasis on the close analogy to inclusive DIS. This is appropriate since diffractive DIS is dominated by the leading twist contribution, which has been one of the most surprising aspects of the large rapidity gap events. The scattering of the partonic fluctuations of the photon off the proton will be treated in the semiclassical approach. After a comparison of theoretical predictions with data we shall conclude with a discussion of some open questions.

Inclusive DIS

Inclusive deep inelastic scattering [3] is characterized by the kinematic variables

$$Q^2 = -q^2 \,, \quad W^2 = (q + P)^2 \,, \quad x = \frac{Q^2}{Q^2 + W^2} \,, \tag{1}$$

where q and P are the momenta of the virtual photon and the proton, respectively. The cross section is determined by the hadronic tensor,

$$W_{\mu\nu}(P,q) = \frac{1}{4\pi} \sum_X \langle P|J_\nu(0)|X\rangle\langle X|J_\mu(0)|P\rangle(2\pi)^4\delta(P - P_X)$$

$$= \left(-g_{\mu\nu} + \frac{q_\mu q_\nu}{q^2}\right) F_1(x, Q^2) + \frac{1}{\nu}\left(P_\mu - \frac{\nu}{q^2}q_\mu\right)\left(P_\nu - \frac{\nu}{q^2}q_\nu\right) F_2(x, Q^2) . \quad (2)$$

Here $J_\mu(x)$ is the electromagnetic current, $\nu = q \cdot P$, and spin averaging has been implicitly assumed.

The structure functions are a sum of leading twist and of higher twist contributions which are suppressed by powers of Q^2,

$$F_i(x, Q^2) = F_i^{(LT)}(x, Q^2) + \frac{F_i^{(HT)}(x, Q^2)}{Q^2} + \cdots . \quad (3)$$

The leading twist term is dominant for Q^2 above some value Q_0^2, which is not very well known and frequently chosen to be $\mathcal{O}(1 \text{ GeV}^2)$. However, higher twist contributions are known to be important for hadronic energies $W^2 \leq 4 \text{ GeV}^2$ [4].

The structure functions $F_i^{(LT)}(x, Q^2)$ can be expressed in terms of process independent parton distribution functions,

$$F_i^{(LT)}(x, Q^2) \to f_i(x, \mu^2) = q(x, \mu^2), \ g(x, \mu^2) , \quad (4)$$

which depend on x and on the factorization scale μ^2. At small x, the quark distribution is assumed to be the same for all light flavours. The parton distribution functions $f_i(x, \mu^2)$ obey the perturbative QCD evolution equations [5],

$$\mu^2 \frac{\partial}{\partial\mu^2} f_i(x, \mu^2) = \frac{\alpha_s}{2\pi} \int_x^1 \frac{dy}{y} P_{ij}\left(\frac{x}{y}\right) f_i(y, \mu^2) , \quad (5)$$

where $P_{ij}(z)$ are the Altarelli-Parisi splitting functions. The parton distributions can be directly expressed in terms of the quark and gluon field operators. For instance, the quark distribution is given by

$$q(x, \mu^2) = \frac{1}{4\pi} \int dx_- e^{-ixP_+x_-/2} \sum_X$$
$$\langle P|\bar{q}(0, x_-, 0_\perp)U(x_-, \infty)|X\rangle\gamma_+\langle X|U(\infty, 0)q(0, 0, 0_\perp)|P\rangle , \quad (6)$$

where $U(a, b)$ is the colour matrix

$$U(a, b) = P \exp\left(-\frac{i}{2}\int_b^a dy_- A_+(0, y_-, 0_\perp)\right) . \quad (7)$$

This definition can be used as a starting point of a theoretical non-perturbative evaluation of the quark distribution.

Diffractive DIS

Diffractive DIS can be discussed in close analogy to inclusive DIS. There are two more kinematical variables which charaterize the diffractively scattered proton: the invariant momentum transfer t and the fraction ξ of lost longitudinal momentum. A complete set of variables is

$$t = (P - P')^2 , \quad \xi \equiv x_{I\!P} , \quad Q^2 = -q^2 , \quad M^2 = (q + \xi P)^2 , \quad \beta = \frac{Q^2}{Q^2 + M^2} .$$
(8)

Compared to inclusive DIS, the diffractive mass M plays the role of the total hadronic mass W, and β corresponds to x.

The hadronic tensor for diffractive DIS,

$$
\begin{aligned}
W_{\mu\nu}^D(P, P', q) &= \frac{1}{4\pi} \sum_X \langle P|J_\nu(0)|X; P'\rangle \langle X; P'|J_\mu(0)|P\rangle (2\pi)^4 \delta(P - P' - P_X) \\
&= \left(-g_{\mu\nu} + \frac{q_\mu q_\nu}{q^2}\right) F_1^{D(4)}(t, \xi, \beta, Q^2) \\
&\quad + \frac{1}{\nu}\left(P_\mu - \frac{\nu}{q^2} q_\mu\right)\left(P_\nu - \frac{\nu}{q^2} q_\nu\right) F_2^{D(4)}(t, \xi, \beta, Q^2) + \dots ,
\end{aligned}
$$
(9)

defines the diffractive structure functions $F_i^{D(4)}(t, \xi, \beta, Q^2)$. Integration over t, which is dominated by small $|t|$ for diffractive scattering, yields the extensively studied structure function

$$F_2^{D(3)}(\xi, \beta, Q^2) = \int dt\, F_2^{D(4)}(t, \xi, \beta, Q^2) .$$
(10)

Also the diffractive structure functions have contributions of leading and higher twist,

$$F_i^{D(3)}(\xi, \beta, Q^2) = F_i^{D(3,LT)}(\xi, \beta, Q^2) + \frac{F_i^{D(3,HT)}(\xi, \beta, Q^2)}{Q^2} + \dots .$$
(11)

Again it is unclear above which value of Q_0^2 the leading twist part dominates. At small x, $W^2 \simeq Q^2/x$ should be large enough, whereas the lower bound on M^2 is an open question. Our phenomenological analysis in the next section will show that the leading twist description breaks down at $M_0^2 \simeq 4$ GeV^2. This again demonstrates that in diffractive DIS M^2 plays a role analogous to W^2 in inclusive DIS.

For diffractive DIS factorization holds like for inclusive DIS [6]. The diffractive structure functions $F_i^{D(3,LT)}(\xi, \beta, Q^2)$ can be expressed in terms of 'fracture functions' [7], or 'diffractive parton distributions' [8],

$$F_i^{D(3,LT)}(\xi, \beta, Q^2) \to \frac{df_i(\xi, \beta, \mu^2)}{d\xi} = \frac{dq(\xi, \beta, \mu^2)}{d\xi} , \frac{dg(\xi, \beta, \mu^2)}{d\xi} ,$$
(12)

which depend on ξ, β and the factorization scale μ^2. The diffractive parton distribution functions $df_i(\xi, \beta, \mu^2)/d\xi$ also obey the perturbative QCD evolution equations,

$$\mu^2 \frac{\partial}{\partial \mu^2} \frac{df_i(\xi, \beta, \mu^2)}{d\xi} = \frac{\alpha_s}{2\pi} \int_\beta^1 \frac{db}{b} P_{ij}\left(\frac{\beta}{b}\right) \frac{df_i(\xi, b, \mu^2)}{d\xi}. \tag{13}$$

Note that the evolution takes now place in β and Q^2; ξ merely acts as a parameter. The physical reason for this is intuitively clear: for an arbitrary DIS event the invariant hadronic mass is W, and the quark which couples to the virtual photon can be radiated by a parton whose fraction of the proton momentum varies from 1 to $x = Q^2/(Q^2 + W^2)$. In a diffractive event, the diffractive invariant mass is M. Hence, W is replaced by M, and the quark which couples to the photon can be radiated by a parton whose fraction of the momentum ξP varies from 1 to $\beta = Q^2/(Q^2 + M^2)$. Formally, Eq. (13) follows from the fact that ultraviolet divergencies and renormalization are the same for inclusive and diffractive parton distribution functions [9]. This is apparent from a comparison of the corresponding operator definitions. The diffractive quark distribution, for instance, is given by [9],

$$\frac{dq(\xi, \beta, \mu^2)}{d\xi} = \frac{1}{64\pi^3} \int dt \int dx_- e^{-ixP_+x_-/2} \sum_X$$

$$\langle P|\bar{q}(0, x_-, 0)U(x_-, \infty)|X; P'\rangle \gamma_+ \langle X; P'|U(\infty, 0)q(0, 0, 0)|P\rangle. \tag{14}$$

Assuming 'Regge factorization' for the diffractive quark and gluon distribution functions yields the Ingelman-Schlein model of hard diffractive scattering [10] which can also be applied to deep inelastic scattering [11].

The physical interpretation of the diffractive parton distributions is analogous to the interpretation of the inclusive distributions. The function $df(\xi, b, \mu^2)/d\xi$ is a conditional probability distribution. It describes the probability density to find a parton f, carrying a fraction ξb of the proton momentum, under the condition that the proton has lost a fraction ξ of its momentum in the scattering process.

The formal definition of diffractive parton distributions tells us very little about their properties, although, comparing Eqs. (6) and (14), one may expect that diffractive DIS is a leading twist effect. However, the important physics question concerns the relation between the two types of distribution functions,

$$f_i(x, \mu^2) \longleftrightarrow \frac{df_i(\xi, \beta, \mu^2)}{d\xi} \quad ? \tag{15}$$

Both kinds of parton distributions represent non-perturbative properties of the proton and are therefore not accessible to perturbation theory. Still, one may hope that at small x, i.e. large hadronic energies W, some simple relations between inclusive and diffractive deep inelastic scattering may exist. In the following section we shall describe a picture of hadrons at small x where this is indeed the case.

2 Semiclassical approach

The phenomenon of the large rapidity gap events in DIS is very difficult to understand within the parton model. Naively, one would expect that the struck quark will always break up the proton, which should lead to a flow of hadrons between the current jet and the proton remnant without large gaps in rapidity.

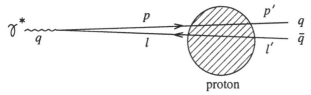

Figure 1: Diffractive or non-diffractive DIS in the proton rest frame; the proton is viewed as a superposition of colour fields with size $1/\Lambda$.

The connection between diffractive DIS and ordinary, non-diffractive DIS can be most easily understood in the proton rest frame which has frequently been used in the early days of DIS, almost 30 years ago. In this frame, DIS appears as the scattering of partonic fluctuations of the photon, $q\bar{q}$, $q\bar{q}g$ etc., off the proton. In the semiclassical approach [12] the proton is viewed as a superposition of colour fields of size $1/\Lambda$ in DIS at small x, i.e. at high γ^*p center-of-mass energies. The simplest partonic fluctuation is a quark-antiquark pair (cf. Fig. 1). Penetrating the proton, quark and antiquark change their colour. If the $q\bar{q}$ pair leaves the proton in a colour singlet configuration, it can fragment independently of the proton remnant yielding a diffractive event. A $q\bar{q}$ pair in a colour octet state will build up a flux tube together with the proton remnant, whose breakup will lead to an ordinary non-diffractive event.

The scattering amplitude for both types of events is determined by a single non-perturbative quantity, $\mathrm{tr}W_{x_\perp}(y_\perp)$. Here $x_\perp$ and $x_\perp + y_\perp$ are the transverse positions where quark and antiquark penetrate the colour field of the proton. The function

$$W_{x_\perp}(y_\perp) = U(x_\perp)U^\dagger(x_\perp + y_\perp) - 1 \ , \tag{16}$$

with

$$U(x_\perp) = P\exp\left(-\frac{i}{2}\int_{-\infty}^{\infty} dx_- A_+(0, x_-, x_\perp)\right) , \tag{17}$$

is essentially a closed Wilson loop through the corresponding section of the proton, which measures an integral of the proton colour field strength.

Diffractive DIS requires a colour singlet pair in the final state. Hence the scattering amplitude is $\propto \mathrm{tr}W_{x_\perp}(y_\perp)$ and the diffractive cross section takes the form,

$$d\sigma^D \propto \int_{x_\perp} \dots |\dots \mathrm{tr}W_{x_\perp}(y_\perp)\dots|^2 \ . \tag{18}$$

The inclusive cross section is obtained by summing over all colours, which yields

$$d\sigma^{incl} \propto \int_{x_\perp} \dots \mathrm{tr}\left(W_{x_\perp}(y_\perp) W_{x_\perp}^\dagger(y_\perp)\right) \dots$$

$$\propto \int_{x_\perp} \dots \mathrm{tr} W_{x_\perp}(y_\perp) \dots , \tag{19}$$

where the last equation follows from the unitarity of the matrix $U(x_\perp)$.

From Eqs. (18) and (19) one immediately derives the properties of Bjorken's aligned jet model [13]. For small quark-antiquark separations one has,

$$\int_{x_\perp} \mathrm{tr} W_{x_\perp}(y_\perp) \propto y_\perp^2 . \tag{20}$$

Hence, since all kinematical factors are the same for $d\sigma^D$ and $d\sigma^{incl}$, small $q\bar{q}$ pairs are suppressed in diffractive DIS. For large pairs of size $1/\Lambda$, the transverse momentum $l'_\perp$ and the longitudinal momentum fractions α and $1-\alpha$ are

$$l'_\perp \sim \Lambda , \quad \alpha \sim \frac{\Lambda^2}{Q^2} , \quad 1-\alpha \simeq 1 . \tag{21}$$

These are the asymmetric, aligned jet configurations [13] which dominate diffractive DIS.

Diffractive and inclusive parton distributions

In the semiclassical approach the evaluation of inclusive and diffractive structure functions is straightforward, in principle. One has to calculate the scattering amplitudes for the production of $q\bar{q}$, $q\bar{q}g$ ... configurations [14] in an external colour field, analogous to the production of $\mu^+\mu^-$ pairs in an external electromagnetic field [15], treat the interaction of the fast partons with the non-abelian colour field in the eikonal approximation [16], and finally integrate over all target colour fields.

The result for the leading twist part can be expressed in terms of diffractive parton distributions [17]. For the transverse structure function, for instance, one finds to leading order in the QCD coupling,

$$F_T^D(\xi,\beta,Q^2) = 2e_q^2 x \int_\beta^1 \frac{db}{b} \left\{ \left(\delta(1-z) + \frac{\alpha_s}{2\pi}\left(P_{qq}(z)\ln\frac{Q^2}{\mu^2} + \dots\right)\right) \frac{dq(b,\xi,\mu^2)}{d\xi} \right.$$

$$\left. + \frac{\alpha_s}{2\pi}\left(P_{qg}(z)\ln\frac{Q^2}{\mu^2} + \dots\right) \frac{dg(b,\xi,\mu^2)}{d\xi} \right\} , \tag{22}$$

where $z = \beta/b$, and C_F and T_F are the usual colour factors. This expression is completely analogous to the well known result for the inclusive structure function $F_T(x,Q^2)$. In the diffractive case, β plays the role of x, whereas ξ only acts as

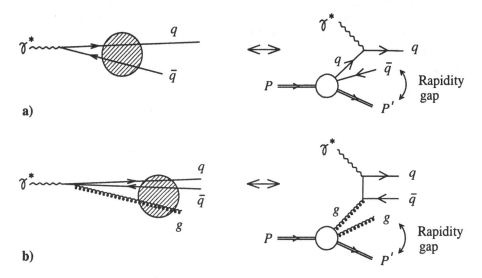

Figure 2: Diffractive DIS in the proton rest frame (left) and the Breit frame (right); asymmetric quark fluctuations correspond to diffractive quark scattering, asymmetric gluon fluctuations to diffractive boson-gluon fusion.

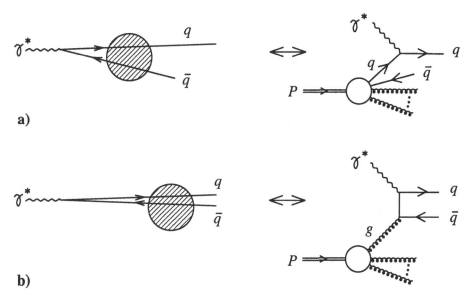

Figure 3: Inclusive DIS in the proton rest frame (left) and the Breit frame (right); asymmetric fluctuations correspond to quark scattering (a), symmetric fluctuations to boson-gluon fusion (b).

a parameter. From Eq. (22) it is obvious that, as anticipated, the diffractive parton distributions satisfy the perturbative QCD evolution equations (13).

The diffractive quark and gluon distributions have been determined in [17]. In terms of Wilson loops in coordinate space, the quark distribution can be expressed as follows,

$$
\frac{dq(\xi, b, \mu^2)}{d\xi} = \frac{2b}{\xi^2(1-b)^3} \int \frac{d^2 l'_\perp \, l'^4_\perp}{(2\pi)^6 N_c} \int_{y_\perp, y'_\perp} e^{i l'_\perp (y_\perp - y'_\perp)} \frac{y_\perp y'_\perp}{y \, y'}
$$

$$
\times \, K_1(yN) K_1(y'N) \int_{x_\perp} \mathrm{tr} W_{x_\perp}(y_\perp) \mathrm{tr} W^\dagger_{x_\perp}(y'_\perp) , \qquad (23)
$$

where N_c is the number of colours and $N^2 = l'^2_\perp \frac{b}{1-b}$.

It is very instructive to compare diffractive DIS in the proton rest frame and in the Breit frame (cf. Fig. 2). The number of partons in the final state is the same, of course, in both frames. Note, however, that the virtual parton connected to the proton changes it's direction. It appears incoming in the proton rest frame and outgoing in the Breit frame. Diffractive quark and gluon distributions correspond to asymmetric $q\bar{q}$ and $q\bar{q}g$ fluctuations with a slow antiquark and gluon, respectively.

Inclusive parton distributions can be calculated in a similar way. The inclusive quark distribution is again given by the asymmetric $q\bar{q}$ configuration (cf. Fig. 3), just with arbitrary colours in the final state. A special role is played by the inclusive gluon distribution. It is related to small symmetric $q\bar{q}$ pairs which probe the colour field of the proton directly (cf. Fig. 3). Contrary to all other parton distributions, the inclusive gluon distribution [18],

$$
xg(x, Q^2) = \frac{3\pi}{\alpha_s e_q^2} \cdot \frac{\partial F_T(x, Q^2)}{\partial \ln Q^2} \qquad (24)
$$

$$
= \frac{1}{2\pi^2 \alpha_s} \int_{x_\perp} \mathrm{tr} \left(\partial_{y_\perp} W_{x_\perp}(0) \partial_{y_\perp} W^\dagger_{x_\perp}(0) \right) = \mathcal{O}\left(\frac{1}{\alpha_s}\right) , \qquad (25)
$$

is enhanced by an inverse power of α_s in the semiclassical approach. This is the reason why diffractive DIS is suppressed. Note that the gluon distribution can be directly inferred from the cross section for a small $q\bar{q}$ pair with transverse size y [19],

$$
\sigma_{q\bar{q}}(y; x, Q^2) = \frac{\pi^2}{3} \alpha_s x g(x, Q^2) y^2 + \mathcal{O}(y^4) . \qquad (26)
$$

Integration over the target gluon fields

So far we have expressed diffractive and inclusive parton distributions in terms of Wilson loops which integrate the gluon field strength in the area between the trajectories of two fast colour charges penetrating the proton. The integration over the gluon field configurations of the target is a complicated operation depending on the full details of the non-perturbative hadronic state. However, in the special case of a very large target, a quantitative treatment becomes possible under minimal additional assumptions. The reason is that the large size of a hadronic target, realized, e.g., in an extremely heavy nucleus, introduces a new hard scale [20]. From the target rest frame point of view, this means that the typical transverse size of the partonic fluctuations of the virtual photon remains perturbative [21], thus justifying the omission of higher Fock states in the semiclassical calculation.

Within this framework, it is natural to introduce the additional assumption that the gluonic fields encountered by the partonic probe in distant regions of the target are not correlated. Thus, one arrives at the situation depicted in Fig. 4, where a colour dipole passes a large number of regions, each one of size $\sim 1/\Lambda$, with mutually uncorrelated colour fields $A_1 \ldots A_n$.

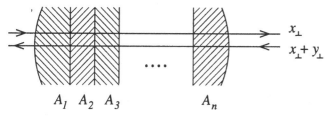

Figure 4: Colour dipole travelling through a large hadronic target.

The crucial assumption that the fields in regions $1 \ldots n$ are uncorrelated is implemented by writing the integral over all field configurations as

$$\int_A = \int_{A_1} \cdots \int_{A_n} , \tag{27}$$

i.e., as a product of independent integrals. Here the appropriate weighting provided by the target wave functional is implicit in the symbol $\int_A$. For inclusive and diffractive parton distributions we need the two colour contractions for products of Wilson loops,

$$\text{tr}\left(W_{x_\perp}(y_\perp)W_{x_\perp}^\dagger(y_\perp)\right) \longleftrightarrow \frac{1}{N_c}\text{tr}W_{x_\perp}(y_\perp)\text{tr}W_{x_\perp}^\dagger(y_\perp) . \tag{28}$$

This relation, which provides the connection between inclusive and diffractive DIS, is the essence of the semiclassical approach.

Performing the integration over the colour fields one obtains in the large N_c limit [18],

$$\int_{x_\perp}\int_A \text{tr}\left(W_{x_\perp}(y_\perp)W_{x_\perp}^\dagger(y_\perp')\right) = \Omega N_c \left(1 - e^{-ay^2} - e^{-ay'^2} + e^{-a(y_\perp - y_\perp')^2}\right) \tag{29}$$

$$\frac{1}{N_c}\int_{x_\perp}\int_A \text{tr}W_{x_\perp}(y_\perp)\text{tr}W_{x_\perp}^\dagger(y_\perp') = \Omega N_c \left(1 - e^{-ay^2}\right)\left(1 - e^{-ay'^2}\right) , \tag{30}$$

where $\Omega = \int d^2 x_\perp$ is the geometric size of the target and a plays the role of a saturation scale. Note that according to Eqs. (29) and (30) the diffractive structure function is not suppressed by a colour factor relative to the inclusive structure function, in contrast to the suggestion made in [22].

As an example, consider the inclusive quark distribution. From Eqs. (23), (28) and (29) one obtains, after changing the integration variable ξ to N^2,

$$xq(x, \mu^2) = \int_x^1 d\xi \frac{dq(\xi, b = x/\xi, \mu^2)}{d\xi} = \int_{x_\perp} \int_{l'_\perp} \frac{d\ xq(x, \mu^2)}{d^2 x_\perp d^2 l'_\perp}, \qquad (31)$$

with the unintegrated quark density

$$\frac{d\ xq(x, \mu^2)}{d^2 x_\perp d^2 l'_\perp} = \frac{N_c}{32\pi^6} \int_0^{\mu^2} dN^2 N^2 \int_{y_\perp, y'_\perp} e^{il'_\perp(y_\perp - y'_\perp)} \frac{y_\perp y'_\perp}{y\ y'}$$
$$\times K_1(yN) K_1(y'N) \left(1 - e^{-ay^2} - e^{-ay'^2} + e^{-a(y_\perp - y'_\perp)^2}\right).(32)$$

This result has recently been obtained for the quark density in a large nucleus [23] by exponentiating the amplitude for a small $q\bar{q}$ pair scattering off a single nucleon, which is described by two-gluon exchange. The effect of the varying thickness of the nucleus has also been taken into account in [23], which makes the saturation scale a dependent on the impact parameter $x_\perp$.

A Glauber type model with two-gluon exchange, similar in spirit to [24], has recently been used to study the effect of parton saturation in inclusive and diffractive DIS [25]. Although perturbative two-gluon exchange is a higher twist effect, the contribution from the soft region can be used as a model for inclusive and diffractive DIS [26,27].

Particularly close to the semiclassical approach is the light cone hamiltonian approach to diffractive processes [28], which is also based on diffractive parton densities expressed in terms of expectation values of products of Wilson lines. For a hadronic target, modelled as a colour singlet which only couples to one flavour of heavy quarks, diffractive DIS is dominated by two-gluon exchange. In the semiclassical approach, the proton is described by a superposition of colour fields. The role of classical colour fields in the case of high gluon densities has first been discussed by McLerran and Venugopalan in the case of a large nucleus [20].

Comparison with data

We now use the large hadronic target as a toy model for the proton. In case the proton can be viewed as an ensemble of regions with independently fluctuating colour fields, the model might even be realistic. We have explicitly verified that in the semiclassical approach inclusive and diffractive parton distributions satisfy the DGLAP evolution equations [5]. Hence, we can use the calculated quark and gluon distributions as non-perturbative input at some scale Q_0^2 and determine the distributions at larger Q^2 by means of the evolution equations.

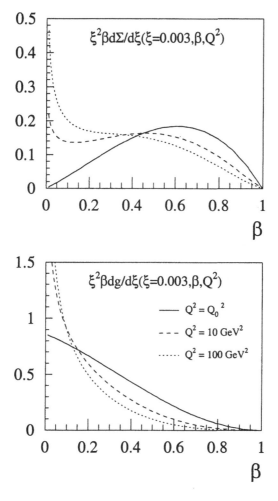

Figure 5: Diffractive quark and gluon distributions at the initial scale Q_0^2 and after Q^2 evolution. From [18].

For a given colour field, the semiclassical description of parton distribution functions always predicts an energy dependence corresponding to a classical bremsstrahlung spectrum: $q(x), g(x) \sim 1/x$. One expects that, in a more complete treatment, a non-trivial energy dependence is induced since the integration over the soft target colour fields encompasses more and more modes with increasing energy of the probe [18]. At present we are unable to calculate this non-perturbative energy dependence from first principles. Instead, we choose to parametrize it in the form of a soft, logarithmic growth of the normalization of diffractive and inclusive parton distributions with the collision energy $\sim 1/x$, consistent with the unitarity bound. This introduces one further parameter, L,

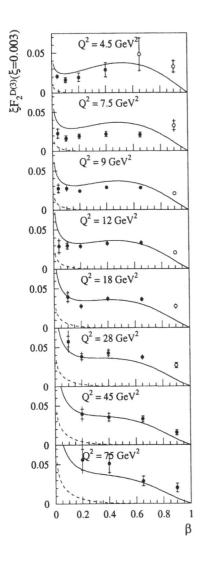
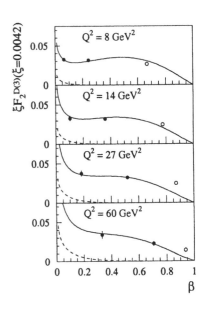

Figure 6: Dependence of the diffractive structure function $F_2^{D(3)}$ on β and Q^2, compared to data from H1 (left) [29] and ZEUS (right) [30]. Open data points correspond to $M^2 \leq 4$ GeV^2. The charm content of the structure function is indicated as a dashed line. From [18].

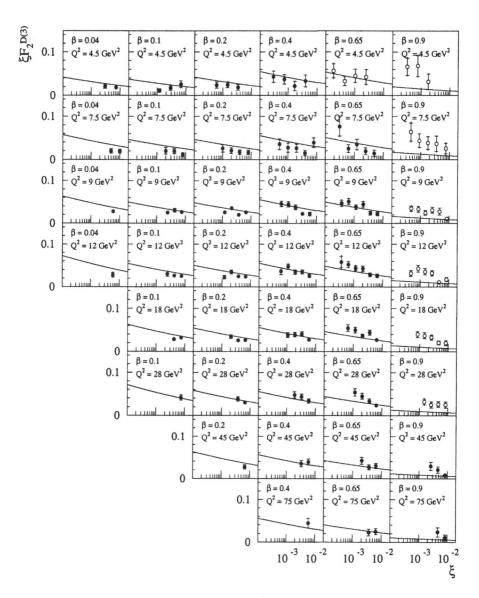

Figure 7: The diffractive structure function $F_2^{D(3)}(\xi, \beta, Q^2)$ at small ξ computed in the semiclassical approach, using the fitted parameters given in the text. H1 data taken from [29]. The open data points correspond to $M^2 \leq 4$ GeV^2 and are not included in the fit. From [18].

into the model,

$$\Omega \to \Omega \left(L - \ln x \right)^2 . \tag{33}$$

Including this energy dependence, one obtains the following compact expressions for inclusive and diffractive parton distributions [18],

$$xq(x, Q_0^2) = \frac{a\Omega N_c \left(L - \ln x \right)^2}{3\pi^3} \left(\ln \frac{Q_0^2}{a} - 0.6424 \right) , \tag{34}$$

$$xg(x, Q_0^2) = \frac{2a\Omega N_c \left(L - \ln x \right)^2}{\pi^2 \alpha_s(Q_0^2)} , \tag{35}$$

$$\frac{dq \left(\beta, \xi, Q_0^2 \right)}{d\xi} = \frac{a\Omega N_c (1 - \beta) \left(L - \ln \xi \right)^2}{2\pi^3 \xi^2} f_q(\beta) , \tag{36}$$

$$\frac{dg \left(\beta, \xi, Q_0^2 \right)}{d\xi} = \frac{a\Omega N_c^2 (1 - \beta)^2 \left(L - \ln \xi \right)^2}{2\pi^3 \beta \xi^2} f_g(\beta) . \tag{37}$$

These expressions are only applicable in the small-x region, which we define by $x \le \xi \le 0.01$. The functions $f_{q,g}(\beta)$ are parameter free predictions. The model does not specify whether, in the diffractive case, the energy-dependent logarithm should be a function of x or of ξ. However, both prescriptions differ only by terms proportional to $\ln \beta$, which can be disregarded in comparison with $\ln x$ or $\ln \xi$ in the small-x region.

The above equations summarize our input distributions, depending on a, Ω, L, and the input scale Q_0^2. At this order, the measured structure function F_2 coincides with the transverse structure function F_T. We assume all three light quark flavours to yield the same contribution, such that the singlet quark distribution is simply six times the quark distribution defined above, both in the inclusive and in the diffractive case,

$$\Sigma(x, Q^2) = 6 \, q(x, Q^2) , \qquad \frac{d \Sigma(\xi, \beta, Q^2)}{d\xi} = 6 \, \frac{dq(\xi, \beta, Q^2)}{d\xi} . \tag{38}$$

Valence quark contributions are absent in the semiclassical approach, which does not account for the exchange of flavour quantum numbers between the proton and the fast moving virtual photon state. Charm quarks are treated as massive quarks in the fixed flavour number scheme [32] (we use $\Lambda_{\mathrm{LO}, n_f=3} = 144$ MeV, $\alpha_s(M_Z) = 0.118$, $m_c = 1.5$ GeV, $m_b = 4.5$ GeV, $\mu_c = 2m_c$). A fit to the data yields for the model parameters $Q_0^2 = 1.23$ GeV^2, $L = 8.16$, $\Omega = (712 \text{ MeV})^{-2}$, $a = (74.5 \text{ MeV})^2$. The starting scale Q_0^2 is in the region where one would expect the transition between perturbative and non-perturbative dynamics to take place; the two other dimensionful parameters ΩL^2 and a are both of the order of typical hadronic scales.

The perturbative evolution of inclusive and diffractive structure functions is driven by the gluon distribution, which is considerably larger than the singlet quark distribution in both cases. The ratio of the inclusive singlet quark and gluon distributions can be read off from Eqs. (34) and (35). For the obtained fit

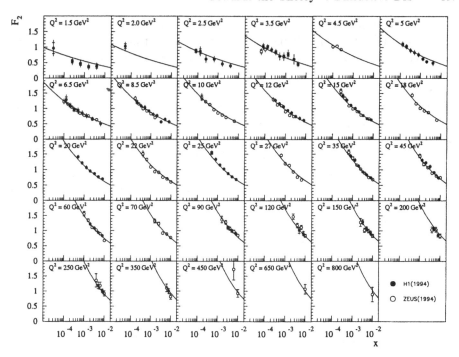

Figure 8: The inclusive structure function $F_2(x, Q^2)$ at small x computed in the semi-classical approach, using the fitted parameters given in the text. Data taken from [31]. The data with $Q^2 = 1.5$ GeV^2 are not included in the fit. From [18].

parameters, it turns out that the inclusive gluon distribution is about twice as large as the singlet quark distribution.

The relative magnitude and the β dependence of the diffractive distributions are completely independent of the model parameters. Moreover, their absolute normalization is, up to the slowly varying factor $1/\alpha_s(Q_0^2)$, closely tied to the normalization of the inclusive gluon distribution.

Figure 5 displays the diffractive distributions for fixed $\xi = 0.003$ and different values of Q^2. The β dependences of the quark and the gluon distribution at Q_0^2 are substantially different: the asymmetric quark distribution $\beta d\Sigma/d\xi$ is peaked around $\beta \approx 0.65$, thus being harder than the symmetric distribution $\beta(1 - \beta)$ suggested in [11]. The gluon distribution $\beta dg/d\xi$, on the other hand, approaches a constant for $\beta \to 0$ and falls off like $(1 - \beta)^2$ at large β. In spite of the $(1 - \beta)^2$ behaviour, gluons remain important even at large β, simply due to the large total normalization of this distribution (the β integral over $\beta dg/d\xi$ at Q_0^2 is approximately three times the β integral over $\beta d\Sigma/d\xi$). As a result, the quark distribution does not change with increasing Q^2 for $\beta \approx 0.5$ and is only slowly decreasing for larger values of β.

The dependence of the diffractive structure function on β and Q^2 is illustrated in Fig. 6, where the predictions are compared with data from the H1 and ZEUS experiments [29,30] at fixed ξ. Disregarding the large-β region, the model gives a good description of the β dependence of the diffractive structure function for all values of Q^2. It is remarkable that the qualitative features of the β and Q^2 dependence are also correctly described by the perturbative approach of [28]. This indicates that the β dependence of the diffractive structure function is to a large extent determined by the kinematics of the $q\bar{q}$ and $q\bar{q}g$ fluctuations and only partly sensitive to the details of the soft interaction with the proton. Also the ξ dependence of the diffractive structure function is rather well described for $M^2 > 4$ GeV^2, as demonstrated in Fig. 7. It has been demonstrated in [27] that higher twist contributions can account for the data at low M^2 (below 4 GeV^2). This is analogous to the breakdown of the leading twist description of the inclusive structure functions, where it occurs for similar invariant hadronic masses, namely $W^2 \lesssim 4$ GeV^2 [4].

Finally, also the data of the H1 and ZEUS experiments on the inclusive structure function $F_2(x, Q^2)$ [31] are well reproduced by the model, as demonstrated by Fig. 8.

3   Open questions

The theoretical work, stimulated by the observation of the large rapidity gap events at HERA, has led to a clear understanding of diffractive DIS as a leading twist phenomenon. Diffractive parton distribution functions can be defined completely analogous to inclusive parton distribution functions. Both kinds of distribution functions obey the perturbative QCD evolution equations. The parton distribution functions cannot be calculated perturbatively, any relation between them reflects a non-perturbative property of the proton. The leading twist description breaks down at $W^2 < 4$ GeV^2 and $M^2 < 4$ GeV^2 for inclusive and diffractive DIS, respectively.

A physical picture for diffractive DIS, and the relation to inclusive DIS, is most easily obtained in the proton rest frame where all DIS processes correspond to the scattering of partonic fluctuations of the virtual photon off the proton. In the semiclassical approach the proton is described by a superposition of colour fields. The qualitative properties of the β dependence of the diffractive structure function are well reproduced by the scattering of colour dipoles off colour fields, which are generated either by a small colour dipole or by a large nucleus. This supports our general ideas about diffractive DIS. It will be interesting to see whether a more precise measurement of the β spectrum will be quantitatively consistent with the idea of a colour field fluctuating independently in different sections of the proton.

In the semiclassical approach the proton colour field is assumed to be dominated by soft modes. Hence, diffractive and non-diffractive DIS events are kinematically very similar at small x, i.e. large hadronic energies W. This leads to an approximate relation between the inclusive and diffractive structure functions [12,33],

$$F_2^{D(3)}(\xi, \beta, Q^2) \sim \frac{1}{\ln Q^2} F_2(x = \xi, Q^2) \,, \tag{39}$$

which is in broad agreement with data [30]. For fixed momentum transfer Q^2 and diffractive mass M, both structure functions have the same dependence on the $\gamma^* p$ center-of-mass energy W. The factor $1/\ln Q^2$ reflects the suppression of small colour dipoles in diffractive scattering.

The dependence of the diffractive structure functions on ξ is not affected by the perturbative QCD evolution, contrary to the x dependence of the inclusive structure functions, and therefore a genuine non-perturbative property of the proton. Hence, Eq. (39) can only hold as long as the effect of the perturbative evolution can be approximated by a single $\ln Q^2$ factor. The ξ dependence of the diffractive structure function then plays the role of the non-perturbative input for the inclusive structure function at some low scale Q_0^2.

One expects that, due to unitarity, diffractive and inclusive structure functions satisfy a relation similar to the one between elastic and total proton-proton cross section [34],

$$\sigma_{el} = \int d^2 b \left(1 - S(b)\right)^2 , \tag{40}$$

$$\sigma_{tot} = 2 \int d^2 b \left(1 - S(b)\right) , \tag{41}$$

where $S(b)$ is the S-matrix at a given impact parameter b. Recently, it has been shown that this relation also holds for the diffractive and inclusive cross sections of a $q\bar{q}$ pair off the proton, if the diffractive cross section is defined by the colour singlet projection [35]. In the semiclassical approach, this relation can be read off from Eqs. (18) and (19) or, more explicitly, from Eqs. (29) and (30). After integration over $l'_\perp$ in Eqs. (31) and (32), which yields $y_\perp = y'_\perp$, one obtains for the cross sections of a $q\bar{q}$ pair with size y,

$$\sigma_{q\bar{q}}^D(y) \propto \frac{1}{N_c} \int_{x_\perp} \int_A \mathrm{tr} W_{x_\perp}(y_\perp) \mathrm{tr} W_{x_\perp}^\dagger(y_\perp)$$

$$= \Omega N_c \left(1 - e^{-ay^2}\right)^2 , \tag{42}$$

$$\sigma_{q\bar{q}}^{incl}(y) \propto \int_{x_\perp} \int_A \mathrm{tr}\left(W_{x_\perp}(y_\perp) W_{x_\perp}^\dagger(y_\perp)\right)$$

$$= 2 \Omega N_c \left(1 - e^{-ay^2}\right) . \tag{43}$$

Since the dependence of the saturation parameter a on the varying thickness of the target has been neglected, the integration over the impact parameter $x_\perp$ (corresponding to b in (40),(41)) could be carried out yielding the geometric size Ω as overall factor. It will be interesting to extend these considerations to more complicated partonic fluctuations.

In the coming years experiments at HERA will provide detailed information about diffractive final states, including charm and high-$p_\perp$ jets. Anticipating further support of the semiclassical approach by data, we can hope to learn a lot about the colour structure of the proton from a comparison of inclusive and diffractive DIS.

The content of this progress report is largely based on recent work with Thomas Gehrmann and Arthur Hebecker whom I thank for an enjoyable collaboration.

References

1. ZEUS collab., M. Derrick et al., Phys. Lett. B315 (1993) 481;
   H1 collab., T. Ahmed et al., Nucl. Phys. B429 (1994) 477
2. For recent reviews and references, see
   A. Hebecker, *Diffraction in deep inelastic scattering*, HD-THEP-99-12, hep-ph/9905226;
   M. Diehl, *Diffractive Interactions: Theory Summary*, DIS'99, Zeuthen, DESY 99-080, hep-ph/9906518
3. For a discussion and references, see
   R.K. Ellis, W.J. Stirling, B.R. Webber, *QCD and Collider Physics*, Cambridge University Press, 1996
4. A. Milsztajn, M. Virchaux, Phys. Lett. B274 (1992) 221;
   A.D. Martin, R.G. Roberts, W.J. Stirling, R.S. Thorne, Phys. Lett. B443 (1998) 301
5. V.N. Gribov, L.N. Lipatov, Sov. J. Nucl. Phys. 15 (1972) 438, 675;
   G. Altarelli, G. Parisi, Nucl. Phys. B126 (1977) 298;
   Yu.L. Dokshitzer, Sov. Phys. JETP 46 (1977) 641
6. J.C. Collins, Phys. Rev. D57 (1998) 3051
7. L. Trentadue, G. Veneziano, Phys. Lett. B323 (1994) 201
8. A. Berera, D.E. Soper, Phys. Rev D50 (1994) 4328
9. A. Berera, D.E. Soper, Phys. Rev D53 (1996) 6162
10. G. Ingelman, P.E. Schlein, Phys. Lett. B152 (1985) 256
11. A. Donnachie, P.V. Landshoff, Phys. Lett. B191 (1987) 309
12. W. Buchmüller, A. Hebecker, Nucl. Phys. B476 (1996) 203
13. J.D. Bjorken, AIP Conf. Proc. No.6, eds. M. Bander et al. (AIP, New York, 1972) p. 151;
    J.D. Bjorken, J.B. Kogut, Phys. Rev. D8 (1973) 1341
14. W. Buchmüller, M.F. McDermott, A. Hebecker, Nucl. Phys. B487 (1997) 283; *ibid.* B500 (1997) 621 (E)
15. J.D. Bjorken, J.B. Kogut, D. Soper, Phys. Rev. D3 (1971) 1382
16. O. Nachtmann, Ann. Phys. 209 (1991) 436
17. A. Hebecker, Nucl. Phys. B505 (1997) 349
18. W. Buchmüller, T. Gehrmann, A. Hebecker, Nucl. Phys. B537 (1999) 477
19. L. Frankfurt, G.A. Miller, M. Strikman, Phys. Lett. B304 (1993) 1
20. L. McLerran, R. Venugopalan, Phys. Rev. D49 (1994) 2233
21. A. Hebecker, H. Weigert, Phys. Lett. B432 (1998) 215
22. W. Buchmüller and A. Hebecker, Phys. Lett. B355 (1995) 573
23. A.H. Mueller, CU-TP-937, hep-ph/9904404
24. E. Gotsman, E. Levin, U. Maor, Phys. Lett. B425 (1998) 369
25. K. Golec-Biernat, M. Wüsthoff, Phys. Rev. D59 (1999) 014017; DTP/99/20, hep-ph/9903358
26. N.N. Nikolaev, B.G. Zakharov, Z. Phys. C49 (1991) 607
27. J. Bartels, J. Ellis, H. Kowalski, M. Wüsthoff, Eur. Phys. J. C7 (1999) 443

28. F. Hautmann, Z. Kunszt, D.E. Soper, Phys. Rev. Lett. 81 (1998) 3333; CERN-TH/99-154, hep-ph/9906284
29. H1 Collab., C. Adloff et al., Z. Phys. C76 (1997) 613
30. ZEUS Collab., J. Breitweg et al., Eur. Phys. J. C6 (1999)43
31. H1 Collab., S. Aid et al., Nucl. Phys. B470 (1996) 3;
    ZEUS Collab., M. Derrick et al., Z. Phys. C72 (1996) 399
32. M. Glück, E. Reya and M. Stratmann, Nucl. Phys. B422 (1994) 37
33. W. Buchmüller, Phys. Lett. B353 (1995) 335
34. A.H. Mueller, Eur. Phys. J. A1 (1998) 19
35. Yu.V. Kovchegov, L. McLerran, NUC-MN-99/2-T, hep-ph/9903246

Soft and Hard Pomerons

Peter Landshoff

DAMTP, University of Cambridge, Cambridge CB3 9EW, England

Abstract. Regge theory provides an excellent description of small-x structure-function data from $Q^2 = 0$ up to the highest available values. The large-Q^2 data should also be described by perturbative QCD: the two descriptions must agree in the region where they overlap. However, at present there is a serious lack in our understanding of how to apply perturbative QCD at small x. The usual lowest-order or next-to-lowest order expansion is not valid, at least not until Q^2 becomes much larger than is usually assumed; a resummation is necessary, but as yet we do not know how to do this resummation.

1 Introduction

Perturbative QCD has become a well-established description of hard processes. But it is not complete: it must be supplemented with other descriptions.

Regge theory is an example of another description. It was extensively developed some 40 years ago [1] and is based on our knowledge of analyticity properties of scattering amplitudes. It relates the high-energy behaviour of scattering amplitudes to exchanges of known particles. In order to describe the data, it also introduces extra terms that, at least so far, are not related to exchanges of known particles. These extra terms are called pomeron-exchange terms, after the Russian physicist Isaac Pomeranchuk. The notion of the pomeron is a very old one. We do not know whether it too describes the exchange of particles, but if it does there is general agreement that these are likely to be glueballs.

Regge theory gives an excellent description not only of soft hadronic proceses [1][2][3], but also of the behaviour of the structure function $F_2(x, Q^2)$ at small x, all the way from real-photon-induced events ($Q^2 = 0$) up to very deeply inelastic ones (large Q^2). The structure function describes an example of a semihard process. Another example of a semihard process is the quasi-elastic reaction $\gamma\, p \to J/\psi\, p$. For such semihard processes, the data reveal [4] that there exists a second pomeron, the "hard" pomeron, in addition to the original "soft" pomeron which enters into purely hadronic reactions. The soft pomeron is certainly nonperturbative in origin, though it is possible that the hard pomeron is associated with the perturbative BFKL equation [5]. However, there are many serious problems [6][7] with the BFKL equation, and so this is not sure.

Certainly, one would hope that at small x and large Q^2 Regge theory can be made to agree with perturbative QCD, in particular with DGLAP evolution [8]. However, there are problems with the DGLAP equation too at small x. The equation involves a kernel or splitting function $P(z, \alpha_S(Q^2))$. If one expands this in powers of α_S, each term is singular at $z = 0$. However [9] it is rather sure, from general considerations of the known analyticity in Q^2 of the structure function, that $P(z, \alpha_S(Q^2))$ is not singular at $z = 0$. The singularities in the terms of the expansion are a signal that the expansion is illegal near $z = 0$, and a resummation is needed to remove them. At present, we do not know exactly how to do this. It is possible that the presence or absence of the singularity is immaterial if Q^2 is large enough [10], but for most applications in the literature [11] it is likely to be a real problem.

2   Regge theory

Through the optical theorem, the structure function $F_1(x, Q^2)$ is the imaginary part of the virtual Compton amplitude $T_1(\nu, t, Q^2)$ evaluated at zero momentum transfer, $t = 0$. Here $2\nu = 2p.q = Q^2/x$. Regge theory begins [1] by considering the crossed-channel process $\gamma^* \gamma^* \to p\bar{p}$, for which $\sqrt{t}$ is the centre-of-mass energy. It first makes a partial-wave series expansion in this channel, in terms of partial-wave amplitudes $a_\ell(t, Q^2)$ and Legendre polynomials $P_\ell(\cos \theta_t)$, where θ_t is the crossed-channel scattering angle. This expansion has a definite, but limited, region of convergence in the space of the three variables $\cos \theta_t, t, Q^2$ or, equivalently, ν, t, Q^2. The partial-wave amplitude $a_\ell(t, Q^2)$ is defined initially for physical values of the angular momentum, $\ell = 0, 1, 2, \ldots$. By a well-defined procedure, its definition is extended to all values of ℓ, both real and complex, and the partial wave series is converted to an integral:

$$T_1(\nu, t, Q^2) = \frac{1}{2i} \int_C d\ell \frac{(2\ell + 1)P_\ell(-\cos \theta_t)}{\sin \pi \ell} a(\ell, t, Q^2) \tag{1}$$

The contour C is initially that of figure 1a, wrapped arround the positive real-ℓ

$$(a) \qquad\qquad (b)$$

Fig. 1. Contour C in the complex ℓ plane for the integral (1)

axis on which are located the zeros of the denominator $\sin \pi \ell$. But the properties of $a(\ell, t, Q^2)$ allow one to use Cauchy's theorem to distort it to become parallel to the imaginary-ℓ axis, as in figure 1b. It turns out that this extends the region of convergence beyond that of the original series, and the integral can be continued analytically to where we need it to calculate the structure function $F_1(x, Q^2)$:

$$t = 0 \qquad\qquad 2\nu > Q^2 \geq 0 \tag{2}$$

As we continue analytically into this region, the singularities of $a(\ell, t, Q^2)$ will move around in the complex ℓ-plane, and one of them may try to cross the contour C. We must distort the contour again so as to avoid this happening. Depending on whether the singularity is a branch point or a pole, we then have either figure 2a or 2b.

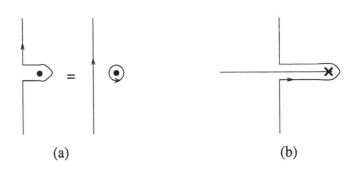

(a)   (b)

Fig. 2. Distortion of the contour C caused by (a) a pole, or (b) a branch point, of $a(\ell, t, Q^2)$ trying to cross it

From what is known about the analyticity structure, it is rather sure that the positions of the singularities of $a(\ell, t, Q^2)$ in the complex ℓ plane do not depend on Q^2, only on t. In the case of a pole

$$a(\ell, t, Q^2) \sim \frac{\beta(Q^2, t)}{\ell - \alpha(t)} \tag{3}$$

the integration around it yields a contribution

$$\frac{\pi \beta(Q^2, t) \, P_{\alpha(t)}(-\cos\theta_t)}{\sin \pi \alpha(t)} \tag{4}$$

to $T_1(\nu, t, Q^2)$. Now when $t = 0$ the analytic continuation of $\cos\theta_t$ is

$$\cos\theta_t = \frac{\nu}{iQm_p} \tag{5}$$

and when this is large the analytic continuation of the Legendre polynomial has the simple behaviour

$$\frac{\pi P_\ell(-\cos\theta_t)}{\sin \pi \ell} \sim -\frac{\Gamma(-\ell)\Gamma(\ell + \frac{1}{2})}{\sqrt{\pi}} (-2\cos\theta_t)^\ell \qquad (\text{Re } \ell > \tfrac{1}{2}) \tag{6}$$

So, with (5), we see that the integral (1) becomes just a Mellin transform, and the "Regge trajectory" $\alpha(t)$ contributes at $t = 0$

$$b_1(Q^2)\nu^{\alpha(0)} \tag{7}$$

to the large-ν behaviour of $T_1(\nu, 0, Q^2)$. Here, $b_1(Q^2)$ is a constant multiple of $\beta(Q^2, 0)$. In the case of $T_2(\nu, 0, Q^2)$ its definition includes a kinematic factor which reduces the power of ν by one unit, so since $\nu = Q^2/2x$ this gives

$$F_2(x, Q^2) \sim f(Q^2)x^{-\epsilon} \tag{8}$$

with

$$\epsilon = \alpha(0) - 1 \tag{9}$$

Regge theory gives no information about the function $f(Q^2)$, other than that it is an analytic function with singularities whose locations are known [12]. The power $(1 - \alpha(0))$ is independent of Q^2.

3 Fit to data

In the case where the singularity that crosses the contour C is a branch point instead of a pole, dragging a branch cut with it as shown in figure 2b, the simple power of x in (8) is replaced with something more complicated. One knows, from unitarity [1], that if there are poles in the complex ℓ plane, there must certainly also be branch points.

On the principle that it is usually the best strategy to try the simplest possible assumption first, Donnachie and I tested the hypothesis that at $t = 0$ the contribution from branch points is much weaker than from poles. Not everybody agrees with this strategy [13], but we applied this several years ago [3] to purely hadronic total cross-sections and found that they are all described well by a sum of just two powers ν^{ϵ_1} and ν^{ϵ_2} The two powers are

$$\begin{aligned} \epsilon_1 &= 0.08 \quad (\text{``soft pomeron'' exchange}) \\ \epsilon_2 &= -0.45 \quad (\rho, \omega, f, a \text{ exchange}) \end{aligned} \tag{10}$$

More recently [4], we tested the hypothesis that also the contribution to the structure function F_2 at small x from branch points is much weaker than from poles. We made a fit of the form

$$F_2(x, Q^2) \sim \sum_{i=0}^{2} f_i(Q^2)x^{-\epsilon_i} \tag{11}$$

We fixed the values of the powers ϵ_1 and ϵ_2 to be the same as in (10), and left ϵ_0 as a free parameter to be determined from the small-x structure-function data. Our fitting procedure had three stages. First, we arrived at a provisional value for ϵ_0 by using data only in the region

$$x < 0.07 \qquad\qquad 0 \le Q^2 < 10 \text{ GeV}^2 \tag{12}$$

which gave us

$$\epsilon_0 \approx 0.4 \tag{13}$$

Next, with this value for ϵ_0, we fitted the data to a sum (11) of three powers of x for each values of Q^2 for which there exist data, still restricting x to less than 0.07. This gave us the plots shown in figure 3 of the hard and soft pomeron coefficient functions $f_0(Q^2)$ and $f_1(Q^2)$ as functions of Q^2. The data do not constrain the (f, a)-exchange coefficient function $f_3(Q^2)$ at all well, so we retained the simple form we chose for it in the first stage of our fitting procedure.

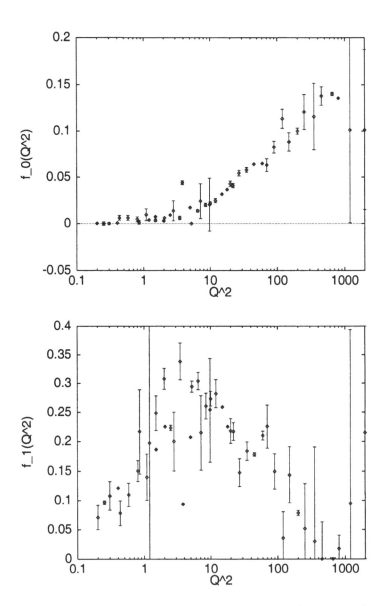

Fig. 3. The hard and soft pomeron coefficient functions $f_0(Q^2)$ and $f_1(Q^2)$ extracted from the data

We know from gauge invariance that F_2 must vanish linearly in Q^2 as $Q^2 \to 0$ at fixed ν. We see that the hard-pomeron coefficient function $f_0(Q^2)$ recovers from this rather slowly as Q^2 increases, until it begins at $Q^2 \approx 10$ GeV^2 to rise quite rapidly. On the other hand, the soft–pomeron coefficient function $f_1(Q^2)$ rises rapidly away from $Q^2 = 0$, until it peaks at between 5 and 10 GeV^2. It then falls again. This was a surprise to us: the soft-pomeron contribution to the small-x structure function apparently is higher twist.

The third part of our fitting procedure chooses functions that have the general shape of the plots in figure 3, with a number of parameters. It still retains the same form for $f_2(Q^2)$, which matters significantly only for the real-photon data at low energy — our fit includes these for $\sqrt{s} > 6$ GeV, and including the f_2 term is essential to fit them. So we use (11) with

$$f_0(Q^2) = A_0 \left(\frac{Q^2}{Q^2+Q_0^2}\right)^{1+\epsilon_0} \left(1 + \frac{Q^2}{Q_0^2}\right)^{\frac{1}{2}\epsilon_0} \tag{14}$$

$$f_1(Q^2) = A_1 \left(\frac{Q^2}{Q^2+Q_1^2}\right)^{1+\epsilon_1} \left(1 + \sqrt{\frac{Q^2}{Q_S^2}}\right)^{-1} \tag{15}$$

$$f_2(Q^2) = A_2 \left(\frac{Q^2}{Q^2+Q_2^2}\right)^{1+\epsilon_2} \tag{16}$$

The hard-pomeron power is now again a free parameter, together with the three coefficients A_i, the mass scales Q_i^2, and Q_S^2. With these 8 parameters, we obtain a χ^2 per data point of 1.0 for 595 data points. These data points have $x < 0.07$ and range from $Q^2 = 0$ to 2000 GeV^2. Sample plots are shown in figure 4.

I want to make a number of comments on these fits:

- The choice (16) for the analytic form of the coefficient functions is an economical set that describes the data well, but there are other choices that agree also with the data extracted in figure 3. Using different choices results in different values for the hard-pomeron power ϵ_0, but they are all within 10% of 0.4.
- The choice (16) makes $f_0(Q^2) \sim Q^{\epsilon_0}$ at large Q^2, which corresponds to $\beta(Q^2, 0)$ in (3) becoming constant for large Q^2. There is no general theory that explains this, though it has been predicted [14] from the BFKL equation. Parametrisations of $f_0(Q^2)$ that behave logarithmically at large Q^2 can also fit the data satisfactorily.
- The choice (16) makes $f_1(Q^2) \sim 1/Q$ at large Q^2. Fits that make it instead behave as $1/Q^2$ are also acceptable, though less good. A $1/Q$ behaviour has been predicted [15] from a combination of the BFKL equation and infrared renormalons, though this is controversial [16]. Fits in which $f_1(Q^2)$ does not go to zero at high Q^2 do not work: it really does seem that the soft-pomeron contribution to $F_2(x, Q^2)$ is higher twist. At, say, $Q^2 = 5$ GeV^2 soft-pomeron exchange dominates in $F_2(x, Q^2)$ until x is less than about 10^{-3}. The consequence of this for conventional structure-function fits [11] needs discussion.
- Without the contribution from the hard pomeron, the fit to the real-photon data agrees well [3] with the measured HERA points. However, it is seen

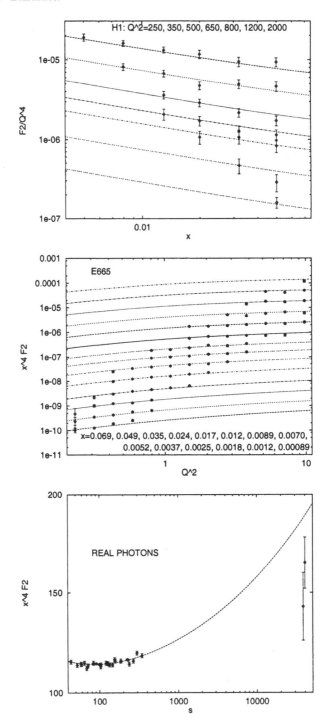

Fig. 4. The Regge fit compared with some of the data, at the largest available Q^2, small Q^2, and $Q^2 = 0$

in figure 4 that including the hard pomeron makes the fit pass significantly above these.

- Figure 5 shows a fit to the preliminary ZEUS data for the charm structure function. It shows that, to a good approximation, these data may be described well by hard-pomeron exchange alone. The fit is actually a single-parameter fit; it includes the constraint that at high Q^2 the hard pomeron is flavour blind.

- Figure 6 shows that the data for the semihard process $\gamma p \to J/\psi p$ are well described by a mixture of soft and hard pomeron exchange [4]. The figure is for $Q^2 = 0$; the HERA data indicate that, as Q^2 increases, the hard-pomeron component becomes relatively more important. One might guess that the data for $\gamma p \to \rho p$ may be described similarly, though at $Q^2 = 0$ the soft pomeron dominates.

4   Perturbative evolution

At high Q^2, parton distributions evolve with Q^2 according to a perturbative evolution equation, called the DGLAP equation. In the singlet channel, the equation is written [8] in terms of a 2-component vector u whose elements are the singlet quark distribution and the gluon distribution:

$$\dot{\mathbf{u}}(x,t) = \int_x^1 \frac{dz}{z} \mathbf{P}(z,t)\mathbf{u}(x/z,t) \tag{17}$$

On the left-hand side, u appears differentiated with respect to

$$t = \log(Q^2/\Lambda^2) \tag{18}$$

On the right-hand side $\mathbf{P}$ is the 2×2 splitting matrix.

The splitting matrix $\mathbf{P}$ may be expanded as power series in $\alpha_S(Q^2)$. It was observed long ago [17] that if one uses just the first term in this expansion,

$$\mathbf{P}(z,t) = \frac{\beta_0}{\log(Q^2/\Lambda^2)} \mathbf{p}(z) \tag{19}$$

and assumes that each element of $\mathbf{u}(x,t)$ has simple power behaviour in x as in (8), then the DGLAP equation gives a simple differential equation for the coefficient matrix $\mathbf{f}(Q^2)$ that multiplies the power. Its solution is

$$\mathbf{f}(Q^2) = \mathbf{f}(\Lambda^2) \left(\log \frac{Q^2}{\Lambda^2} \right)^{\gamma(\epsilon)} \tag{20}$$

where $\gamma(N)$ is the eigenvalue of $\beta_0 \mathbf{p}(N)$. Here $\mathbf{p}(N)$ is the Mellin transform of $\mathbf{p}(N)$:

$$\mathbf{p}(N) = \int_0^1 dz \, z^N \mathbf{p}(z) \tag{21}$$

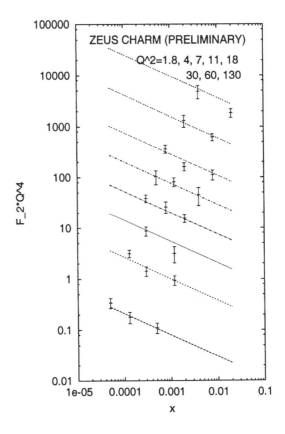

Fig. 5. Hard-pomeron fit to preliminary ZEUS data for F_2^c

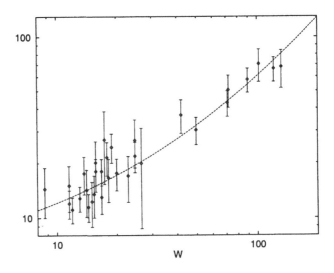

Fig. 6. Two-pomeron fit to data for $\gamma p \to J/\psi \, p$

For $\epsilon = 0.4$ the eigenvalue $\gamma(\epsilon)$ is close to 3. One may fit the data satisfactorily by requiring the hard-pomeron coefficient function $f_0(Q^2)$ to have this behaviour at large Q^2, though not as well as with the power behaviour of (16). However, in any case it is not valid to use the approximation (19) at low x, even if it is supplemented by adding in higher-order terms in the perturbative expansion. Such an expansion is not legal at small x, even though it is widely used [11].

The problem is that two of the elements p_{qG} and p_{GG} of the matrix $\mathbf{p}(z)$ have poles [8] at $z = 0$, so that their Mellin transforms have poles at $N = 0$. For example,

$$\pi \, p_{GG}(z) \sim \frac{C_A}{z} \qquad\qquad \pi \, p_{GG}(N) \sim \frac{C_A}{N} \qquad (22)$$

If one includes higher-order terms in the perturbative expansion, all the elements of the Mellin transform of the splitting matrix are singular at $N = 0$. The consequence is that, as soon as Q^2 is large enough for the DGLAP equation to be applicable, $Q^2 > Q_0^2$ say, the Mellin transforms of the parton distributions acquire rather nasty singularities at $N = 0$, something like $\exp(1/N)$. This conflicts with what we believe we know about analyticity properties in Q^2. As I explained in section 2, the Regge amplitude at high energy (or small x) is essentially just the Mellin transform: the relationship is

$$\ell = N + 1 \qquad (23)$$

So a singularity at $N = 0$ corresponds to one at $\ell = 1$. However, the analyticity properties of the amplitude whose imaginary part is $F_2(x, Q^2)$ were well studied nearly 40 years ago [1] and there was never any suggestion that it has a singularity at $\ell = 1$, let alone a nasty one. If there is no singularity when Q^2 is small, it is not possible that one suddenly appears when we continue analytically in Q^2 to $Q^2 > Q_0^2$. That is, there can be no singularity of the splitting matrix at $N = 0$ — and probably not, indeed, at any other value of N.

The appearance of the singularity at $N = 0$ arises because of the perturbation expansion. Compare, for illustration, the function

$$\phi(N, \alpha_S) = N - \sqrt{N^2 - \alpha_S} \qquad (24)$$

Its expansion in powers of α_S is

$$\phi(N, \alpha_S) = \frac{\alpha_S}{2N} + \frac{\alpha_S^2}{8N^3} + \cdots \qquad (25)$$

Each term in the expansion is singular at $N = 0$, but manifestly the complete function $\phi(N, \alpha_S)$ is not. The expansion is illegal near $N = 0$.

The timelike splitting matrix is very similar [18] in form to $\phi(N, \alpha_S)$, but in the spacelike region things are a little more complicated. If one combines the DGLAP equation with the BFKL equation, which could well be a valid thing to do when x or N is small, one finds that the element P_{GG} of the spacelike splitting matrix is given [19] in terms of the Lipatov characteristic function [5] $\chi(\omega, \alpha_S)$:

$$\chi(P_{gg}(N, t), \alpha_S) = N \qquad (26)$$

To lowest order in α_S,

$$\chi(\omega, \alpha_S) = \frac{3\alpha_s}{\pi} \left[2\psi(1) - \psi(\omega) - \psi(1 - \omega) \right] \qquad (27)$$

One may easily verify [9] that, if one inserts (27) into (26), each term in the expansion of $P_{gg}(N, t)$ in powers of α_S is singular at $N = 0$, but the complete function $P_{gg}(N, t)$ is finite there. Some authors [20] therefore advocate that one should use the solution to the full equation (26) and avoid expanding it.

This resummation is obviously sensible, but there are still serious unsolved problems. Firstly, it is highly doubtful that it is valid to use the uncorrected BFKL equation [6]: it assumes that nonperturbative effects cause no complications, and does not properly take account of energy conservation. Secondly, even if one ignores these difficulties, using the lowest-order approximation (27) to $\chi(\omega, \alpha_S)$ is not valid: the next-to-leading-order correction is huge [7]. There have recently been some interesting attempts to solve this problem [21], but more work remains to be done.

So at present, although we know that the elements of the splitting matrix cannot diverge at $N = 0$, we cannot calculate them. Any application of the DGLAP equation in which they are expanded perturbatively, and in which use is made of the fact that the terms in this expansion are large, cannot be trusted.

5   Summary

- When we use the DGLAP equation at small x it is not valid to use an unresummed perturbative expansion of the splitting matrix — at least, not until Q^2 is somewhat larger than is normally assumed. At present, we do not know how to perform the necessary resummation properly.
- Regge theory with two pomerons fits the small-x data for $F_2(x, Q^2)$ extremely well up to the highest available values of Q^2. It strongly suggests that at, say, $Q^2 = 5$ GeV^2, most of $F_2(x, Q^2)$ at small x is higher twist.
- Perturbative QCD and Regge theory are not rival theories. They complement each other and we have to learn how to make them fit together.

This research is supported in part by the EU Programme "Training and Mobility of Researchers", Networks "Hadronic Physics with High Energy Electromagnetic Probes" (contract FMRX-CT96-0008) and "Quantum Chromodynamics and the Deep Structure of Elementary Particles" (contract FMRX-CT98-0194), and by PPARC

References

1. P D B Collins, *Introduction to Regge Theory*, Cambridge University Press (1977)
2. A Donnachie and P V Landshoff, Nuclear Physics B244 (1984) 322
3. A Donnachie and P V Landshoff, Physics Letters B296 (1992) 227
4. A Donnachie and P V Landshoff, Physics Letters B437 (1998) 408
5. E A Kuraev, L N Lipatov and V Fadin, Soviet Physics JETP 45 (1977)
6. J C Collins and P V Landshoff, Physics Letters B276 (1992) 196; M F McDermott, J R Forshaw and G G Ross, Physics Letters B349 (1995) 189
7. V S Fadin and L N Lipatov, Physics Letters B429 (1998) 127; G Camici and M Ciafaloni, Physics Letters B430 (1998) 349
8. R K Ellis, W J Stirling and B R Webber, *QCD and Collider Physics* Cambridge University Press (1996)
9. J R Cudell, A Donnachie and P V Landshoff, Physics Letters B448 (1999) 281
10. R D Ball and S Forte, Physics Letters B405 (1997) 317
11. M Glück, E Reya and A Vogt, European Physical J C5 (1998) 461; A D Martin, R G Roberts, W J Stirling and R S Thorne, European Physical J C4 (1998) 463; CTEQ collaboration – H.L. La et al, hep-ph/9903282
12. R J Eden, P V Landshoff, D I Olive and J C Polkinghorne, *The Analytic S-Matrix*, Cambridge University Press (1966)
13. A Capella, A Kaidalov, C Merino, D Pertermann, and J Tran Thanh Van, European Physical J C5 (1998) 111; E Gotsman, E Levin, U Maor and E Naftali, hep-ph/9904277; K Golec-Biernat and M Wüsthoff, hep-ph/9903358
14. B Ermolaev, private communication
15. E M Levin, Nuclear Physics B453 (1995) 303
16. K D Anderson, D A Ross and M G Sotiropoulos, Nuclear Physics B515 (1998) 249
17. C Lopez and F J Yndurain, Nuclear Physics B183 (1981) 157
18. A Bassetto, M Ciafaloni and G Marchesini, Nuclear Physics B(1980)477
19. T Jaroszewicz, Physics Letters 116B (1982) 291; S Catani and F Hautmann, Nuclear Physics B427 (1994) 475
20. R K Ellis, F Hautmann and B R Webber, Physics Letters B348 (1995) 582; R S Thorne, hep-ph/9901331
21. M Ciafaloni, D Colferai and G P Salam, hep-ph/9905566; R D Ball and S Forte, hep-ph/9906222

Part VI

New Physics at HERA

Search for Physics Beyond the Standard Model at HERA

Masahiro Kuze

Institute of Particle and Nuclear Studies, KEK, Tanashi, 188-8501 Tokyo, Japan

Abstract. The latest status of searches at HERA for physics beyond Standard Model is summarized on behalf of H1 and ZEUS collaborations. Emphasis is put on production of resonant particles accessible within the HERA center-of-mass energy, such as leptoquarks, squarks in R-parity-violating supersymmetry or excited fermions. Both collaborations have accumulated a large statistics of positron-proton collisions, on which the majority of the results are based. Also preliminary results from very recent electron-proton running are presented. The contents include some updated results made available since the time of the workshop.

1 Introduction

The HERA collider at DESY is a unique facility, where electrons (or positrons) and protons interact at a high center-of-mass energy $(\sqrt{s})$, i.e. at a very short distance. It collides an electron or positron beam of 27.5 GeV with a proton beam of 820 GeV (920 GeV since 1998), resulting in $\sqrt{s} = 300$ GeV (318 GeV). Any new phenomenon in lepton-quark interactions due to physics at large energy scale could be observed at this unpreceded energy. Unlike e^+e^- or $\bar{p}p$ colliders, the initial state at HERA has non-zero lepton and baryon numbers, which makes searches at HERA most powerful in discovering particles which carry one or both of these quantum numbers.

Two collider experiments, H1 and ZEUS, have been taking data since 1992. Most of the results presented here are based on the large data sample of e^+p collisions taken between 1994–1997, corresponding to luminosities of $36.5\,\mathrm{pb}^{-1}$ and $47.7\,\mathrm{pb}^{-1}$ for H1 and ZEUS, respectively. From 1998 until the middle of 1999, HERA provided e^-p collisions with the increased proton beam energy. Preliminary results from $16\,\mathrm{pb}^{-1}$ of data taken in this period are also presented.

For most of the searches, major Standard Model (SM) background comes from neutral-current (NC) and charged-current (CC) deep inelastic scattering (DIS) processes, which is depicted in Fig. 1 (a). The NC process occurs via t-channel exchange of γ or Z boson, while the CC process turns the electron into a neutrino through W-boson exchange. The following kinematic variables are frequently used in DIS analyses:

$$Q^2 = -q^2 = -(k - k')^2 \,, \tag{1}$$
$$x = Q^2/(2q \cdot p) \,, \tag{2}$$
$$y = (q \cdot p)/(k \cdot p) \,, \tag{3}$$

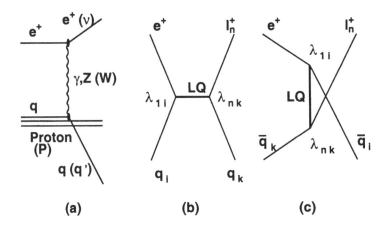

Fig. 1. Diagrams of (a) deep inelastic scattering of e^+p; (b) s-channel and (c)u-channel LQ processes involving $F = 0$ LQs

where k and k' are the four-momenta of the incoming and outgoing lepton, respectively, and p is the four-momentum of the incoming proton. Q^2 is the negative square of the momentum transfer (q), x is the Bjorken scaling variable and y is sometimes called the inelasticity parameter.

2   Leptoquarks and Related Searches

Leptoquarks (LQs) have both lepton (L) and baryon (B) numbers and couple directly to lepton-quark pairs. They appear in many models beyond SM, such as Grand Unified Theories, and naturally relate the two families of fermions: leptons and quarks. They carry color charges like quarks and have fractional electric charges. The set of LQs which preserves $SU(3) \times SU(2) \times U(1)$ symmetry has been specified by Buchmüller, Rückl and Wyler (BRW) [1]. They can be classified into scalar and vector LQs, and into $F = 0$ and $|F| = 2$ LQs, where $F = L + 3B$ denotes the fermion number.

2.1   LQ Production and Decay at HERA

Figures 1 (b) and (c) show LQ processes at HERA with s- and u-channel exchange of a LQ. The diagrams are generalized to allow for generation-changing LQs (the final-state lepton and/or quark can be of a different generation from the initial state), which will be discussed in Sect. 2.4. Here, only first-generation LQs ($n = 1$ and $i = k = 1$ for the Yukawa couplings λ in the diagrams) are discussed. According to the BRW model, the decay branching ratio of LQ to the same initial state eq is either 100% or 50%, depending on the species of LQ. In the latter case, the remaining decay is to the νq final state. At HERA, highest sensitivity to first-generation LQs can be achieved by producing the LQs via a fusion between the lepton and the valence quarks (u, d), i.e. for $F = 0$ ($|F| = 2$) LQs in e^+p (e^-p) collisions.

An individual LQ event has exactly the same topology as a NC or CC DIS event. If the mass of the LQ (m_{LQ}) is smaller than $\sqrt{s}$ and the coupling λ is not too large (order of unity or less), the s-channel resonant production dominates. In this case there appears a sharp peak at m_{LQ} in the eq or νq invariant mass distributions, or at $x_0 = m_{LQ}^2/s$ in the DIS variable x. The production cross section for this case can be simply approximated with the narrow-width approximation (NWA) :

$$\sigma(ep \to LQ\, X) = (J + 1)\frac{\pi}{4s}\lambda^2 q(x_0, m_{LQ}^2) , \qquad (4)$$

where λ is the Yukawa coupling, J is the spin of the LQ and q is the quark distribution function evaluated at the resonance $x=x_0$ with the scale $Q^2 = m_{LQ}^2$. Another characteristic of LQ process is the different y distribution as compared to DIS. The s-channel production of a scalar LQ has a flat y distribution, while it is $(1 - y)^2$ for the vector case. They are contrasted to the $1/y^2$ dependence of DIS processes, which falls more rapidly as y approaches 1. Therefore, the search strategy is to start from a selection of NC or CC DIS events and then to look for a resonance peak in the large-y region.

The signal of a LQ with mass close to or above $\sqrt{s}$ looks different from what has been discussed above, but the sensitivity on such LQs does not vanish [2]. This will be discussed in Sect. 2.3.

2.2   Mass Distribution and Limits

Figure 2 shows the mass distribution from the H1 analysis [3] of 1994–1997 e^+p data. For the NC DIS selection, the mass is calculated using the energy and angle of the scattered positron, and for the CC DIS selection using the hadronic variables with the Jacquet-Blondel method [4]. Slight excess of events is observed in the NC channel at large y around 200 GeV, which comes mainly from the data taken during 1994–1996 and was particularly noteworthy before the 1997 data were added [5]. No significant excess was seen in the 1997 data alone, and overall significance has decreased in the whole data sample. The mass distribution in the CC channel is in good agreement with DIS expectation within the uncertainty.

From the observed and expected mass spectrum, the limits at 95% confidence level (CL) on the Yukawa coupling λ can be obtained for each LQ type, as shown in Fig. 3. At an electromagnetic strength $\lambda \sim 0.3$, the limits on scalar (vector) LQs extend up to 275 GeV (284 GeV). Here only limits on $F = 0$ LQs are shown, but also results on $|F| = 2$ LQs are obtained, with weaker limits than $F = 0$.

Also ZEUS has analysed e^+p data in NC [6] and CC [7] channels and obtained preliminary limits on LQs. In the NC channel, ZEUS observes a few outstanding events at large mass and large y, coming mainly from 1994–96 data [8], but the significance has decreased after adding the 1997 data which more than doubled the total integrated luminosity. More e^+p data will be needed to clarify the origin of these high-mass, high-y events events observed by H1 and ZEUS.

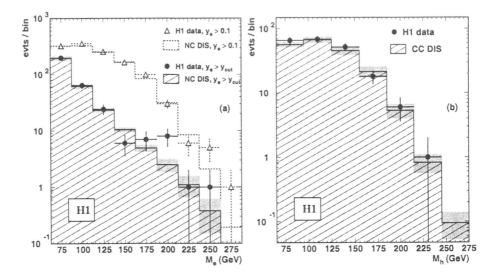

Fig. 2. Mass spectum for (a) NC-like and (b) CC-like event selection from H1 e^+p data (*points*) compared with DIS expectations (*histograms*). In (a), the distribution is shown before and after the mass-dependent lower cut in y designed to enhance the signal significance. The shaded boxes on the histograms indicate the uncertainties on the expectations

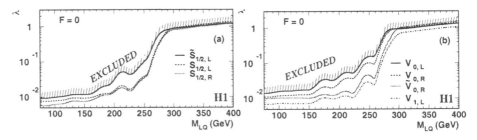

Fig. 3. Upper limits at 95% CL on Yukawa couplings λ as a function of the LQ masses for (a) scalar and (b) vector $F = 0$ leptoquarks. In the plots shown here, only NC data have been used

ZEUS has also looked at the most recent e^-p data taken during 1998–99 [9]. These data enhance very much the sensitivity to $|F| = 2$ LQs, and the increased center-of-mass energy in addition extends the sensitivity at the highest mass region ($\sim 300\,\text{GeV}$) close to the kinematical limit. Figure 4 shows the mass distribution of events from NC DIS selection. Here the mass (M_{ej}) is the invariant mass between the electron and jet, calculated directly using their energies and angles. The agreement with SM DIS expectation is very good, and the limits on Yukawa coupling for $|F| = 2$ LQs are plotted in Fig. 5. At $\lambda \sim 0.3$, the exclusion limits extend up to about $290\,\text{GeV}$ at 95% CL.

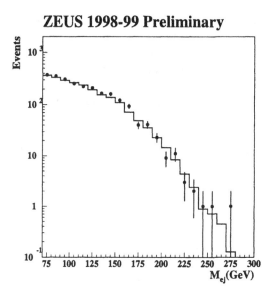

Fig. 4. Electron-jet invariant mass distribution from ZEUS e^-p data (*points*) compared to NC DIS expectation (*histogram*)

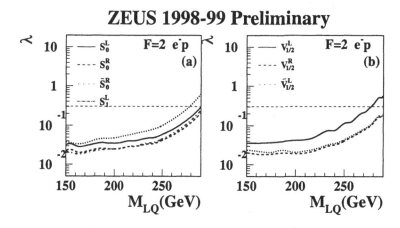

Fig. 5. Coupling limits at 95% CL as a function of the LQ masses for (a) scalar and (b) vector $|F| = 2$ leptoquarks. The horizontal lines indicate the coupling strength for an electroweak scale, $\lambda \sim 0.3$

2.3 High-Mass Leptoquark and Contact Interactions

The LQ mass range in Fig. 3 extends beyond the center-of-mass energy 300 GeV up to 400 GeV. The H1 analysis takes into account not only the s-channel contribution but also u-channel and interference terms between the LQ and SM DIS diagrams. When the LQ mass is high, these terms become important and actually the LQ is not resonantly produced but gives virtual effects on DIS-like final states [2]. In this case, the mass spectrum would not have a narrow peak but a broad mass range would be affected, and also at smaller y values. The range in mass and cut in y are thus changed accordingly. On the other hand, the ZEUS analysis uses NWA and the quoted limits stop before the coupling becomes too large.

When the mass is even higher, the effect of LQ can be described as an effective four-fermion interaction known as $eeqq$ contact interaction (CI). CI is an effective theory for physics at high mass scales in their approximation at the low-energy limit, and can represent not only heavy LQs but also a variety of models beyond SM such as new heavy vector bosons or composite models of fermions. The effect is parameterized by the ratio of the coupling and the mass scale, g/Λ, and usually the convention $g^2=4\pi$ is adopted. Depending on the chiral structure of the interaction, many CI models can be constructed and modify the $eq \rightarrow eq$ cross section differently. The common effect is the increase of cross section at high Q^2, with the interference effect at intermediate Q^2 which can be constructive or destructive depending on the model.

ZEUS has analysed 1994–97 e^+p data in terms of 30 scenarios of CI and derived lower limits on Λ. For details of the analysis, refer to [10]. The limits at 95% CL range between 1.7 TeV to 5 TeV for the scenarios considered. It is worth to note that $eeqq$ CI can be tested also at LEP2 ($e^+e^- \rightarrow q\bar{q}$) and at Tevatron ($q\bar{q} \rightarrow e^+e^-$). Generally the limits from competing colliders are comparable, and in some cases ZEUS limits are most stringent or the only existing limits.

H1 has released preliminary results on CI from e^+p data [11] and interprets the limits also in terms of LQ mass and coupling ratio, m_{LQ}/λ. The limits, valid for $m_{LQ} \gg \sqrt{s}$, range between 202 GeV and 952 GeV for various type of LQs.

2.4   Lepton-Flavor Violation

The processes in Figs. 1 (b) and (c) can allow for the case $n > 1$, where the outgoing lepton is a muon or tau lepton. This is a lepton-flavor-violating (LFV) process forbidden in SM. The event signature is very distinct and can be searched for with little background from SM processes.

The H1 analysis [3] observes no candidate event consistent with $eq \rightarrow \mu q'$ or $eq \rightarrow \tau q'$ process, with the expected background from SM being 0.12 ± 0.05 and 0.77 ± 0.30 events, respectively. The limits are interpreted in terms of a coupling for low-mass LFV LQs, and also in terms of indirect effects of high-mass LFV LQs like the CI analysis. In the latter case, the limits are given on the quantity $\lambda_{1i}\lambda_{nk}/m_{LQ}^2$. In some cases where second- or third-generation quarks are involved, and especially for the e–τ LFV case, H1 gives more stringent limits than low-energy processes such as rare decays of τ or B.

ZEUS has recently given preliminary results from a LFV search in the muon channel using 1994–97 data. No candidate is found where 0.3 events are expected from the SM background. Limits have been obtained for the low-mass LFV LQs. Under the assumption that the couplings to eq and $\mu q'$ have the same electroweak strength, 95% CL lower limit for the mass of LQs extends up to 285 GeV [12].

3   Supersymmetry with R-parity Violation

Supersymmetry (SUSY) is one of the most promising extension of the SM. Extensive searches are being performed at the high-energy colliders LEP2 and Tevatron, and HERA is not an exception. HERA's potential for SUSY discovery is maximal in the case of the R-parity-violating (RPV) extension of Minimal SUSY Standard Model (MSSM).

3.1   Phenomenology

R-parity is a multiplicative quantum number defined as $R=(-1)^{L+3B+2S}$, where S is the spin of the particle, and is assumed to be conserved in MSSM. R takes the value 1 for SM particles and -1 for their SUSY partners. The consequences of the R-parity conservation are that SUSY particles are always produced in pairs, and that the lightest SUSY particle (LSP) cannot decay.

However, the general supersymmetric and gauge-invariant superpotential contains additional terms which violate R-parity [13]

$$\lambda_{ijk} L_i L_j \bar{E}_k + \lambda'_{ijk} L_i Q_j \bar{D}_k + \lambda''_{ijk} \bar{U}_i \bar{D}_j \bar{D}_k. \tag{5}$$

Here L and Q denote the left-handed lepton and quark doublet superfields; $\bar{E}$, $\bar{D}$, and $\bar{U}$ are the right-handed singlet superfields for charged leptons, down-type quarks and up-type quarks, respectively. The indices i, j, k denote the generation. For each term and generation combination, a Yukawa coupling $\lambda(\lambda', \lambda'')$ is introduced in the model as an additional parameter.

Of particular interest for HERA is the second term with λ'_{1jk}, which makes single production of a squark ($\tilde{q}$) possible through the electron-quark fusion. This process is very much like the scalar leptoquark production discussed earlier. If the squark decays to eq with the same Yukawa coupling, the analysis is exactly the same as in the LQ search. However, there are also R-parity-conserving decays with gauge couplings $\tilde{q} \to q\chi^0$ and $\tilde{q} \to q'\chi^\pm$. These decays to neutralino/chargino compete with the LQ-like decay, and the branching ratio depends on the unknown Yukawa coupling and on the MSSM parameters.

The neutralino or chargino decays subsequently to a lighter gaugino (cascade decay), or directly through the RPV coupling. Even the LSP, usually taken to be χ_1^0, decays to SM particles via $\chi_1^0 \to e^\pm qq$ and $\chi_1^0 \to \nu(\bar{\nu})qq$. Therefore, there are variety of final states of squarks involved in the RPV SUSY phenomenology.

3.2   Experimental Results

The preliminary ZEUS analysis of 1994–97 $e^+ p$ data [14] makes a simplifying assumption that $\tilde{q} \to q\chi_1^0$ dominates the gauge decay and ignores cascade decays and charginos. The nature of χ_1^0 is assumed to be a pure photino, in which case the branching ratios to eq and $q\chi_1^0$ decays follow a simple formula [15].

The final states are classified to eq, $e^+ qqq$ and $e^- qqq$. The second topology looks like a NC event with multiple jets in the hadronic final state. The last

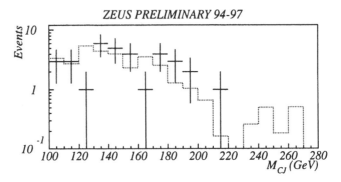

Fig. 6. Mass distribution of events selected in e^+qqq channel for data (*crosses*) and expectation (*dashed histogram*)

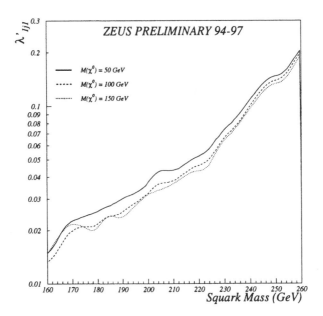

Fig. 7. The 95% CL upper limits on the RPV Yukawa couplings λ'_{1j1} as a function of the mass of the squark $\tilde{u}_j$ for different neutralino masses $M(\chi^0)$

topology has a "wrong sign" lepton for the e^+p collision and is a very clean channel with small background. The $\nu(\bar{\nu})qqq$ final states, contributing 12% of the χ_1^0 decay, is not investigated in this analysis.

Figure 6 shows the mass distribution of ZEUS events passing the cuts for e^+qqq channel. Here, 33 events are observed while 33.6 events are expected from SM, mainly from NC DIS and a small contribution from photoproduction processes. In the e^-qqq channel, no event is observed while 0.06 events is expected from the background. Since no signal for a resonance is found, limits are derived on the couplings (assuming only one λ'_{1j1} to be non-zero at a time) as a function

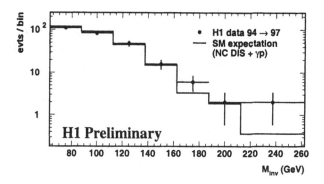

Fig. 8. Mass spectrum for $e + multijets$ final states for data (*points*) and SM expectation (*histogram*)

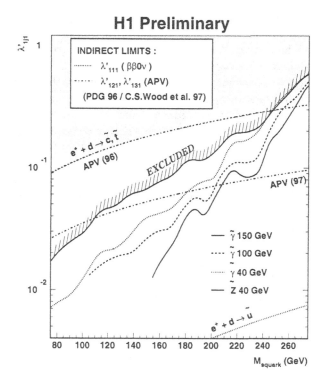

Fig. 9. Upper limits at 95% CL for the RPV couplings λ'_{1j1} as a function of the squark mass, compared with most stringent indirect limits

of the squark mass, shown in Fig. 7. Here $j = 1, 2, 3$ corresponds to $e^+d \rightarrow \tilde{u}, \tilde{c}, \tilde{t}$ production, respectively. The analysis is repeated for three different χ^0_1 masses.

The preliminary H1 analysis of 1994–97 e^+p data [16] takes neutralino mixing into account and investigates SUSY parameter space of a photino-dominated neutralino and a zino-dominated neutralino. It also calculates the branching ratio of cascade decays with χ_2^0 and χ_1^+. Figure 8 shows the mass distribution in the $e + multijets$ channel. With softer selection cuts than those used in the ZEUS analysis, 289 candidates are observed while the SM expectation is 285.7 ± 28.0 events. In the "wrong sign" channel, one candidate passes the selection while the background is expected to be 0.49 ± 0.2. No evidence for a squark is found, and the limit on the Yukawa coupling is shown in Fig. 9 for three different masses of photino-dominated neutralino and one case of zino-dominated neutralino. The coupling λ'_{111} is strongly constrained by the neutrino-less double-beta decay, but the results on λ'_{121} and λ'_{131} are competitive with atomic-parity-violation experiments.

4   Excited Fermions

The composite models of fermions regard them as being built from more fundamental particles. In such models, leptons and/or quarks can be excited to a higher-mass state and decay "radiatively" to the stable ground state (normal fermions), emitting gauge bosons such as photon, W or Z bosons. At HERA, excited states of electrons or quarks can be created through the t-channel photon or Z exchange in eq interaction, and excited neutrinos or quarks can be created through the W exchange. The excited electrons can also be produced in the elastic process $ep \rightarrow e^*p$. The search strategy is to reconstruct a photon, W or Z boson (V) in an event and look for a resonance peak in the $f - V$ invariant mass, where f is an electron, a quark or a neutrino (missing momentum).

The preliminary H1 analysis of 1994–97 e^+p data [17] searches for excited fermions in the decay channels $e^* \rightarrow e\gamma, eZ, \nu W; \nu^* \rightarrow \nu\gamma, \nu Z, eW; q^* \rightarrow q\gamma, q'W$.

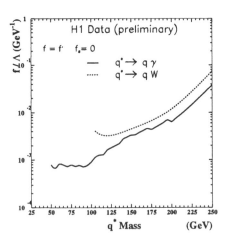

Fig. 10. Upper limit at 95% CL for the excited quark coupling divided by the compositeness scale, f/Λ, as a function of the q^* mass

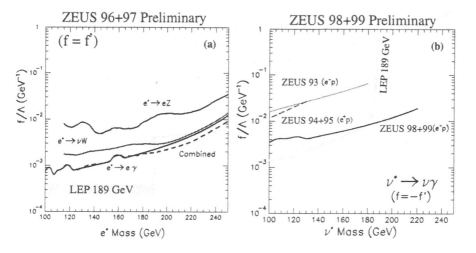

Fig. 11. Upper limits on f/Λ at 95% CL for (a) e^* and (b) ν^* production, as a function of the excited-lepton masses

The hadronic decays of W, Z and leptonic decays $W \to e\nu; Z \to e\bar{e}, \nu\bar{\nu}$ have been exploited. In all cases, the numbers of observed events are in agreement with the SM expectations and no evidence for a resonance has been found.

The derived limits are based on a specific phenomenological model of excited fermions [18] in which the cross sections depend on coupling constants f, f' and f_s for the gauge groups $SU(2), U(1)$ and $SU(3)$, respectively, and the compositeness scale Λ. The decay branching ratio of excited fermions are determined once relationships between the couplings are fixed. Figure 10 shows, as an example, the limit on f/Λ for the excited quark production under the assumption $f = f'$ and $f_s = 0$. The latter condition makes the results complementary to the searches at Tevatron, where the production of q^* is assumed to occur through the quark-gluon fusion ($f_s \neq 0$) [19].

ZEUS has made a preliminary search for e^* using the data taken during 1996–97 (37 pb^{-1}) [20]. The decays $e^* \to e\gamma, eZ, \nu W$ are exploited with the hadronic decay of W and Z. No evidence for a resonance has been found, and the limits on f/Λ using the same model [18] have been derived under the assumption $f = f'$. The limits are shown in Fig. 11 (a). A combined limit from the three decay channels is also derived. It can be seen that limits from LEP2 are more stringent below its center-of-mass energy, and HERA limits are competitive above it, where the on-shell production of e^* is not possible at LEP2 and limits come from searching for the signs of virtual e^* exchanges in the process $e^+e^- \to \gamma\gamma$ [21].

ZEUS has also performed a preliminary search for ν^* using the recent e^-p data taken in 1998–99 [20]. Using this smaller integrated luminosity (16 pb^{-1}) than e^+p data is far more beneficial in ν^* search, since the production cross section in e^+p at high ν^* mass is strongly suppressed compared to e^-p. It is due to the smaller d-quark density compared to u quark at high x and the $(1-y)^2$ suppression factor coming from the chiral nature of W exchange. The search has been done in $\nu^* \to \nu\gamma$ channel only, and two observed events are in agreement with the SM expectation of 1.8 ± 0.2. The obtained limit on f/Λ for the case $f = -f'$ (the $\nu^* \to \nu\gamma$ decay vanishes for the case $f = f'$) is shown in Fig. 11 (b).

5   Conclusion and Outlook

Extensive searches for physics beyond Standard Model are being performed by the two collaborations H1 and ZEUS at HERA. No convincing signal of new particles has been established so far, giving limits on their production which are competitive with the searches at other colliders.

HERA is planning to continue running until May 2000 with the current design and then undergoes major upgrade plans in order to increase the luminosity. There will be new final-focusing magnets close to the interaction point, which means that also the detectors will be modified during the upgrade shutdown.

The new running from 2001, with five times more luminosity than the current design value, will bring an order of $1\,\mathrm{fb}^{-1}$ of integrated luminosity. This large amount of data allows the experiments to make high-statistics analyses at large-Q^2 and high-mass regions, which will eventually unreveal the breakdown of Standard Model if the new physics is existing within the reach of HERA.

References

1. W. Buchmüller, R. Rückl, D. Wyler: Phys. Lett. B **191**, 442 (1987)
2. T. Matsushita, E. Perez, R. Rückl: hep-ph/9812481
3. H1 Collaboration, C. Adloff et al.: DESY 99-081, submitted to Eur. Phys. J.
4. F. Jacquet and A. Blondel: *Proc. of the Study for an ep Facility for Europe*, ed. by U. Amaldi, DESY 79-48, 391 (1979)
5. H1 Collaboration, C. Adloff et al.: Z. Phys. C **74**, 191 (1997)
6. ZEUS Collaboration: contributed paper #754 to ICHEP98, Vancouver
7. ZEUS Collaboration: contributed paper #546 to EPS HEP99, Tampere
8. ZEUS Collaboration, J. Breitweg et al.: Z. Phys. C **74**, 207 (1997)
9. ZEUS Collaboration: contributed paper #552 to EPS HEP99, Tampere
10. ZEUS Collaboration, J. Breitweg et al.: DESY 99-058, submitted to Eur. Phys. J.
11. H1 Collaboration: contributed paper #157f to EPS HEP99, Tampere
12. ZEUS Collaboration: contributed paper #551 to EPS HEP99, Tampere
13. V. Barger, G.F. Giudice, T. Han: Phys. Rev. D **40**, 2987 (1989)
14. ZEUS Collaboration: contributed paper #548 to EPS HEP99, Tampere
15. J. Butterworth, H. Dreiner: Nucl. Phys. B **397**, 3 (1993)
16. H1 Collaboration: contributed paper #580 to ICHEP98, Vancouver
17. H1 Collaboration: contributed paper #581 to ICHEP98, Vancouver
18. K. Hagiwara, S. Komamiya, D. Zeppenfeld: Z. Phys. C **29**, 115 (1985); U. Baur, M. Spira, P.M. Zerwas: Phys. Rev. D **42**, 815 (1990); F. Boudjema, A. Djouadi, J.L. Kneur: Z. Phys. C **57**, 425 (1993)
19. CDF Collaboration, F. Abe et al.: Phys. Rev. D **55**, 5263 (1997)
20. ZEUS Collaboration: contributed paper #555 to EPS HEP99, Tampere
21. DELPHI Collaboration: contributed paper #115 to EPS HEP99, Tampere

High P_T Leptons and W Production at HERA

Joachim Meyer

DESY, Hamburg, Germany

Abstract. The H1 and ZEUS Collaborations have observed events with high energy leptons and missing transverse momentum at the electron-proton collider HERA. While some of these events can be explained by the expected properties of W production with leptonic decay, three events with muons observed by H1 show properties atypical for Standard Model processes.

1 Introduction

In the history of high energy physics *leptons* and *missing transverse momentum* have proven to be powerful signatures in searches for new phenomena. In 1994, the H1 Collaboration has within a modest integrated luminosity of $\sim 4pb^{-1}$ observed an outstanding event [1] featuring a high P_T isolated muon, a high P_T hadronic system and a large missing P_T. Although the kinematics of the event was found to be compatible with the production of an on-shell W decaying leptonically, the event was found in a region of phase space not likely to be populated by this process.

By now the luminosity accumulated at HERA in e^+p collisions at an energy of $\sqrt{s} = 300$ GeV is an order a magnitude higher. We report on dedicated searches [2,3] performed by both HERA experiments for events of such topology, with either an isolated electron or muon. We also report on an analysis of most recent data accumulated in e^-p collisions in the years 1998-1999 at an energy of $\sqrt{s} = 318$ GeV.

2 Production of *W* bosons in electron-proton collisions

Within the Standard Model (SM), the dominant source for events with an isolated high P_T lepton and large missing P_T is the production of W bosons with subsequent leptonic decay. The lowest order diagrams for these processes ($ep \rightarrow eW^{\pm}X, W^{\pm} \rightarrow \nu\ell^{\pm}$) are shown in figure 1. From the MC simulation (using the program EPVEC [4]) one expects

- a cross section $\sim 50fb^{-1}$ per charge state and leptonic decay channel
- the cross section to increase towards low transverse momenta P_T^X of the hadronic system
- a Jacobian peak in the transverse mass of the lepton-neutrino system.

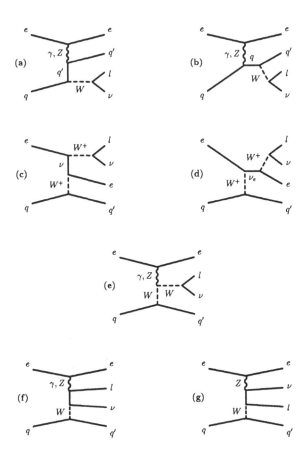

Fig. 1. Parton level diagrams for the process $eq \to eql\nu$.

In EPVEC the cross section is calculated in leading order. Higher order QCD terms might enhance W-production especially at high P_T^X. NLO calculations, which were recently made available [5] for W production processes induced by resolved photon interactions predict cross sections very close to the EPVEC numbers. However, for the high P_T^X part of the cross section, which is dominated by direct photon interactions, NLO calculations are not yet available.

W production at HERA also proceeds via the diagram in figure 1e, i.e. the Triple Gauge Boson Coupling. At HERA, photon exchange dominates over Z exchange allowing for the study of the γWW vertex. The γWW coupling is conventionally described with two parameters labelled κ and λ. In the Standard Model they take the values $\kappa = 1$ and $\lambda = 0$. Anomalous couplings may give rise to harder distributions in the transverse momentum of the W and thus in P_T^X.

3   The Data

3.1   H1 Analysis

The analysis is based on the Charged Current (CC) event selection [6] including the following criteria (for details see [2])

- calorimetric missing transverse momentum $P_T^{calo} > 25$ GeV
- a well measured track with $P_T > 10$ GeV (*High P_T track*)
- a reconstructed vertex in the interaction region
- topological and timing filters against cosmic and halo muons
- rejection of badly measured neutral current events

For the e^+p data sample, corresponding to 36.5 pb^{-1} this selection leads to 124 events.

For this data sample the isolation of the high P_T tracks with respect to jets or other tracks in the event has been investigated. The isolation is measured by the distance of the high P_T track in the $\eta - \phi$-plane (pseudorapidity-azimuthal angle-plane) to its nearest neighbour track (D_{track}) and to the nearest hadron jet (D_{jet}) . Figure 2 shows the data sample in the D_{track}-D_{jet}-plane. The bulk of the 124 events are located at small values of both distances, as expected for genuine Charged Current events where the high P_T track is associated with the hadronic shower. However, six events show a high P_T track clearly isolated in the D_{track}-D_{jet}-plane. All 6 tracks are identified as leptons, five are muons (2 μ^+, 2 μ^- and one μ of undetermined charge) and one is an e^-. The typical characteristics of these events can be seen in fig. 3, where one of the muon-events is shown as an example. Table 1 gives selected kinematic parameters of these

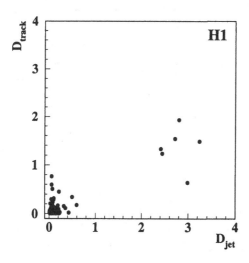

Fig. 2. Correlation between the distances D_{jet} and D_{track} (see text) to the closest hadronic jet and track, for all high P_T tracks.

$$e^+ p \rightarrow \mu^+ X$$

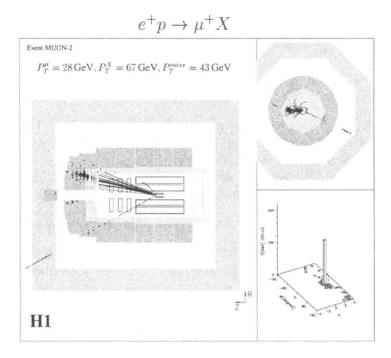

Fig. 3. Display of a H1 muon event. Shown are a longitudinal and a transverse view as well as the energy release as a function of azimuthal angle ϕ and pseudorapidity η.

six events. Although not required in the selection also the muon events show a significant imbalance in the total transverse momentum (P_T^{miss}).

The event rates from all Standard Model processes expected to contribute to the signal are summarized in table 2. W production yields for both lepton channels the highest expected rate. Whereas in the electron channel the observed rate is well compatible with expectation, in the muon channel the observed five events are to be compared with an expectation of 0.8 ± 0.2.

In order to compare the kinematic characteristics of the observed events with those expected from W - production a Monte Carlo simulation of W - production was performed with 500 times the statistics used in the data sample. Figure 4 shows for the $e-$ and $\mu-$channel the simulated events in the plane of hadronic transverse momentum P_T^X and transverse mass $M_T^{l\nu}$, together with the six events under consideration. Here $M_T^{l\nu}$ is calculated from the transverse momentum of the charged lepton and a neutrino, which is assumed to carry the observed missing transverse momentum P_T^{miss}. For W-production one would expect P_T^X to be low and the $M_T^{l\nu}$ distribution is expected to show a Jacobian peak around the W-mass. The observed e^- events shows these characteristics. Also two muon events are in a region of phase space where W- events are expected. In one of them (MUON-3) an additional positron is observed. Identifying this

[GeV]	ELECTRON	MUON-1	MUON-2	MUON-3	MUON-4	MUON-5
Q^l	neg.	pos.	pos.	neg.	neg.	unmeas.
P_T^l	$37.6^{+1.3}_{-1.3}$	$23.4^{+7.5}_{-5.5}$	$28.0^{+8.7}_{-5.4}$	$38.6^{+12.0}_{-7.4}$	$81.5^{+75.2}_{-26.4}$	> 44
P_T^X	8.0 ± 0.8	42.2 ± 3.8	67.4 ± 5.4	27.4 ± 2.7	59.3 ± 5.9	30.0 ± 3.0
P_T^{miss}	30.6 ± 1.5	$18.9^{+6.6}_{-8.3}$	$43.2^{+6.1}_{-7.7}$	$42.1^{+10.1}_{-5.9}$	$29.4^{+71.8}_{-13.9}$	> 18
$M_T^{l\nu}$	67.7 ± 2.7	$3.0^{+1.5}_{-0.9}$	$22.8^{+6.7}_{-4.2}$	$75.8^{+23.0}_{-14.0}$	94^{+157}_{-54}	> 54

Table 1. Reconstructed event kinematics of the H1 events. Given are the charge Q^l of the lepton, the transverse momenta of the lepton P_T^l, of the hadronic system P_T^X, the total missing transverse momentum P_T^{miss} and the transverse lepton-neutrino mass $M_T^{l\nu}$. In case of event MUON-5 2σ limits are quoted for the muon momentum and derived quantities.

	Electron Channel	Muon Channel
Data	$0\ e^+, 1\ e^-$	5
W production	1.65 ± 0.47	0.53 ± 0.11
Z production	0.01 ± 0.01	0.01 ± 0.01
Charged Current	0.02 ± 0.01	0.01 ± 0.01
Neutral Current	$0.51 \pm 0.10\ e^+, 0.02 \pm 0.01\ e^-$	0.09 ± 0.06
Photoproduction	< 0.02	< 0.02
Heavy Quarks	< 0.04	< 0.04
Photon-Photon	$0.09 \pm 0.03\ e^+, 0.04 \pm 0.01\ e^-$	$0.14^{+0.14}_{-0.07}$

Table 2. Observed and predicted event rates (H1). The limits given correspond to 95% confidence level. Unless stated otherwise the quoted numbers refer to the summed production of both lepton charged states.

positron with the scattered positron, allows to constrain the kinematics yielding an invariant muon-neutrino mass of $M = 82^{+19}_{-12} GeV$. The remaining three muon events (2 μ^+ and 1μ^-) are unlikely to be explained by W production or other SM processes considered.

3.2   ZEUS Analysis

The ZEUS collaboration has performed two analyses (for details see [3]) using 47.7 pb^{-1} of e^+p data.

One analysis searches for isolated tracks in events which are unbalanced in transverse momentum. The selection criteria are very similar to those applied in the H1 analysis. The main criteria are $P_T^{calo} > 20$ GeV, an isolated track with $P_T > 10$ GeV and at least one hadronic jet with $E_T > 5$ GeV. Three events

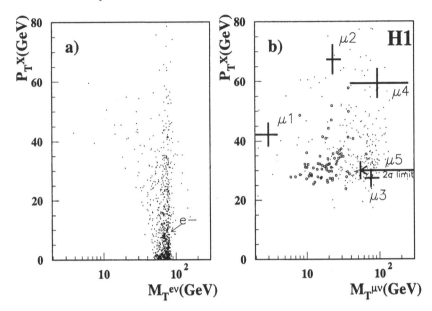

Fig. 4. Distribution of the H1-events in the P_T^X-$M_T^{l\nu}$ plane. Figure a) shows the electron channel, figure b) the muon channel. The 1-σ uncertainties on the measured kinematic parameters are indicated by the crosses. (For event $\mu 5$ the 2σ lower limit is shown for $M_T^{l\nu}$.) The dominant SM contributions (dots for W production, open circles for $\gamma\gamma$ processes in the μ channel) are shown for an accumulated luminosity which is a factor 500 higher than in the data. For the μ channel no significant contribution is expected below $P_T^X = 25$ GeV due to the event selection criteria.

pass the selection criteria, in all of them the high P_T track is identified as an e^+. The observed event yield agrees with the SM expectations of 3.5 ± 0.7 electron events and 2.0 ± 0.4 muon events, to which W production contributes 0.9 and 0.4, respectively.

In the other analysis ZEUS performed a dedicated measurement of the W production cross section. Here no hadronic activity was required but an electron or muon signature was requested in the selection. Three events were found in the electron channel (two in common with the first analysis) and zero events in the muon channel. Fig. 5 shows the display of one of the events. The imbalance in transverse momentum is clearly visible. Table 3 shows selected kinematic parameters of the 3 events. In fig. 6 the events are compared with a MC simulation of W production. They are found to be well compatible with expectation. The backgound in the event sample, mainly from CC processes, is estimated to be 1.1 ± 0.3 events. Subtracting this background from the observed 3 events yields a cross section of $\sigma(e^+p \rightarrow e^+W^\pm X) = 0.9^{+1.0}_{-0.7}(stat) \pm 0.2(syst)$ pb. This result agrees well with the SM leading order expectation [4] of 0.95 pb.

Best limits on anomalous couplings were obtained when considering only the phase space region of higher values of P_T^X and combining the electron and muon channel. For $P_T^X > 20$ GeV zero events were observed compared to a SM

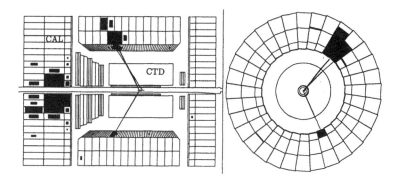

Fig. 5. Display of a ZEUS electron event. Shown are a longitudinal and a transverse view.

[GeV]	e1	e2	e3
P_T^l	22.1 ± 1.1	37.1 ± 1.6	48.8 ± 2.0
P_T^X	0.4 ± 0.1	18.7 ± 2.2	19.0 ± 2.5
P_T^{miss}	21.7 ± 1.1	32.9 ± 1.4	34.9 ± 2.4
$M_T^{l\nu}$	43.7 ± 1.6	68.2 ± 2.2	81.4 ± 3.2

Table 3. Reconstructed event kinematics of the three ZEUS events. Given are the transverse momenta of the lepton P_T^l, the hadronic system P_T^X, the total event P_T^{miss} and the transverse lepton-neutrino mass $M_T^{l\nu}$.

expectation of 1.2 events from W and 0.52 ± 0.18 events from other sources. From this an upper limit (95% C.L.) on the cross section of $\sigma(P_T^X > 20\,\mathrm{GeV}) < 0.58$ pb is derived which translates into limits of

$$-4.7 < \Delta\kappa < 1.5 \quad \text{for} \quad \lambda = 0$$
$$-3.2 < \lambda < 3.2 \quad \text{for} \quad \Delta\kappa = 0$$

on the anomalous $WW\gamma$ couplings ($\Delta\kappa = \kappa - 1, \lambda$). With the present statistics this result is not yet competitive with the limits derived at the Tevatron and LEP, but will clearly improve with larger statistics.

4   Results from e^-p HERA running

In the years 1998/1999 HERA was operated with electrons of 27.5 GeV colliding with protons of 920 GeV resulting in $\sqrt{s} = 318$ GeV. Both experiments applied the same analysis procedures as explained above.

ZEUS 1994-97

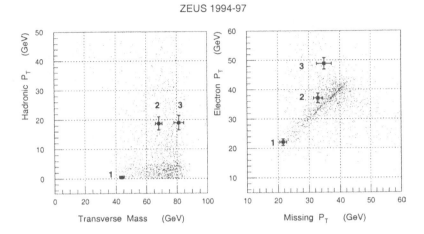

Fig. 6. Distribution of the three e^+ events observed by ZEUS in the P_T^X - $M_T^{l\nu}$ plane and in the P_T^e - P_T^{miss} plane, compared with a Monte Carlo study of W production with a luminosity of $50 fb^{-1}$.

ZEUS observed in a luminosity of $16 pb^{-1}$ two events with isolated tracks, both being identified as electrons (e^-). The result is consistent with the expectation of 1.6 ± 0.4 events from SM processes, of which 0.8 ± 0.4 and 0.8 ± 0.1 are in the electron and muon channel, respectively. Both events observed show a Neutral Current event topology where the missing transverse momentum P_T^{miss} is in one event due to a large energy leakage out of the main calorimeter and in the other event due to an energetic muon associated to the hadronic shower.

H1 analysed a data sample corresponding to $14 pb^{-1}$. No event was found fullfilling the selection criteria for the search for events with isolated tracks and P_T^{miss}. The SM expectation is 1.0 ± 0.2 events in the electron channel and 0.4 ± 0.1 events in the muon channel.

The good agreement between observation and SM expectation in both experiments implies that within the present statistical accuracy there is no evidence of an excess of events with an isolated lepton and missing P_T in $e^- p$ scattering at HERA.

5   Interpretation of the H1-Muon-Events beyond the Standard Model

Considering the low probability of the signal observed by H1 in the muon channel in $e^+ p$ data to be explained within the SM it is worthwhile investigating which processes beyond the SM could explain the observed events.

- The interpretation of a lepto-quark produced by positron-quark fusion and decaying into a muon-quark system can be discarded because the measured event kinematics is incompatible with a 2-body decay.

- In the hypothesis of a resonance decaying into a quark and an onshell W which decays leptonically, the escaping neutrino can be reconstructed using the W mass constraint. The resulting invariant mass of the system formed by the W and the measured hadronic system X is for the events MUON-2 and MUON-4 within measurement errors found to be compatible with the top mass. While the SM expectation for top production at HERA is unmeasurably small, it could be significantly enhanced in dynamical models for fermion mass generation [7].

- Another hypothesis is the production of squarks in Supersymmetric Models with R-parity violation. Stop quark production ($e^+ d \rightarrow \tilde{t}$) has been discussed by T. Kon et al. [8]. Events of the observed topology could be due to a decay $\tilde{t} \rightarrow W\tilde{b}$ with the sbottom quark decaying via $\tilde{b} \rightarrow d\nu$. This process would naturally produce events with a high P_T lepton of positive charge, a high P_T hadronic jet and significant missing P_T. The kinematics of the muon events is found to be compatible with a scenario with a stop mass of 200 GeV and a sbottom mass of 100 GeV.

  In order to investigate this hypothesis further an analysis has been performed by H1 selecting events with at least two jets and missing transverse momentum. In this class of events one would accept those stop decays in which the W decays hadronically. No events in excess of the SM expectation have been observed. From this fact a 95% upper limit on the expectation in the muon channel of $N_{observed}(W \rightarrow \mu\nu) < 0.36$ events can be derived. This result does not support the hypothesis that the three outstanding muon events could be due to this process.

6   Summary and outlook

At the HERA electron proton collider events have been observed which feature a high P_T charged lepton together with a significant imbalance in the overall event transverse momentum. Some of the observed events can be understood in terms of a production of on-shell W bosons, decaying leptonically. This constitutes the first observation of this process in lepton hadron scattering. The cross section observed agrees with Standard Model expectations.

In the H1 experiment three events have been observed in the muon channel, which are unlikely to be due to W production or other SM processes. The origin of these events will most probably only be clarified with much higher luminosities. The prospects to accumulate luminosities at least an order of magnitude higher than available now, are very good, since in 2000 the HERA machine is undergoing a major luminosity upgrade.

7   Acknowledgements

I like to thank the organizers of the Ringberg workshop for providing a stimulating atmosphere for a very interesting workshop in a beautiful surrounding.

References

1. H1 Collaboration, T. Ahmed et al. DESY Preprint 94-248 (1994)
2. H1 Collaboration, C.Adloff et al., Eur. Phys. J. **C5**,575 (1998)
3. ZEUS Collaboration, J. Breitweg et al, DESY Preprint 99-054 (1999)
4. U. Baur, J.A.M.Vermaseren, D.Zeppenfeld, Nucl. Phys. **B375**, 3 (1992)
5. M. Spira, DESY Preprint 99-060 (1999)
6. H1 Collaboration, C.Adloff et al., Z. Phys. **C67**, 565 (1995),
   Phys. Lett. **B379**(1996), 319 (1996)
7. H. Fritzsch, D. Holtmannspötter, Phys.Lett. **B457**, 186 (1999)
8. T. Kon et al. Phys. Lett. **B376**, 227 (1996),
   T. Kon et al. Mod. Phys. Lett. **A12**, 3143 (1997)

Beyond the Standard Model at HERA

Hubert Spiesberger

Institut für Physik, Johannes-Gutenberg-Universität Mainz, D-55099 Mainz, FRG

Abstract. The prospects of physics beyond the standard model in deep inelastic scattering are reviewed, emphasizing the search for contact interactions, for leptoquarks and for supersymmetry with R-parity violation. R-parity violating supersymmetry is explored as a speculative source of events with high energy muons and missing transverse momentum, but no convincing explanation for events of this type observed at H1 is found.

1 Introduction

The luminosity delivered to the experiments at HERA has now become large enough to open a new focus of physics analyses looking at processes with cross sections of the order of 1 pb and below. This is the typical value for neutral and charged current (NC and CC) cross sections at large values of Bjorken x and momentum transfer Q^2. Also measurements of rare standard model (SM) processes like the production of an additional gauge boson, are becoming possible. These low cross section processes provide a wealth of possibilities to look for deviations from the standard model predictions and constitute important backgrounds for searches for physics beyond the standard model [1].

The motivation to search for new physics at HERA has received a strong impetus by the observation of enhancements of cross sections at several places. The excess of events at large x and large Q^2 in NC and CC scattering [2] observed in the 1994–96 e^+p data has been discussed at length in the literature (see [3,4] and references therein). A similar excess was not observed in the 1997 data so that the significance in the complete 1994–97 data sample is reduced, but still there: in the mass bin $200\,\text{GeV} \pm \Delta M/2$ ($\Delta M = 25\,\text{GeV}$) and for $y > 0.4$, H1 observed 8 events, but only 2.87 ± 0.48 are expected (see Fig. 1a). In the CC channel the observed number of events is in agreement with the SM predictions within the uncertainties: H1 observed 7 events with $Q^2 > 15,000\,\text{GeV}^2$ (4.8 ± 1.4 expected) and ZEUS found 2 with $Q^2 > 35,000\,\text{GeV}^2$ (0.29 ± 0.02 expected), both in the 1994–96 data set. Notably the occurrence of five events with an isolated muon and large missing transverse momentum at H1 [5] which are seemingly not all a sign of W production presents a challenge for the understanding of the experiments.

In the following, I selected some of the alternatives to standard model physics which, if realized in nature, have a good chance to be discovered at HERA. If not, HERA is expected to significantly contribute to setting limits on their respective

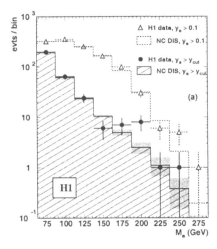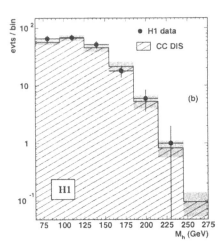

Fig. 1. Mass spectra for NC (left) and CC (right) DIS-like events for data (symbols) and SM expectation (histograms) observed at H1 in $37\,\mathrm{pb}^{-1}$ of e^+p data taken in 1994–1997 [6]

model parameters. Other related topics of interest have been discussed previously in Refs. [1,7].

2   New Physics Scenarios

Despite the great success of the standard model, various conceptual problems provide a strong motivation to look for extensions and alternatives. Two main classes of frameworks can be identified among the many new physics scenarios discussed in the literature:

- Parametrizations of more general interaction terms in the Lagrangian like contact interactions or anomalous couplings of gauge bosons are helpful in order to *quantify the agreement* of standard model predictions with experimental results. In the event that deviations are observed, they provide a framework allowing to relate different experiments and cross-check possible theoretical interpretations. Being insufficient by themselves, e.g. because they are not renormalizable, parametrizations are expected to show the directions to the correct underlying theory if deviations are observed.

- Models, sometimes even complete theories, provide specific frameworks that allow a consistent derivation of cross sections for conventional and new processes. Examples are the two-Higgs-doublet extension of the standard model, grand unified theories and, most importantly, supersymmetry with or without R-parity violation.

The following examples attained most interest when the excess of large-Q^2 events at HERA was made public [3,4].

2.1   Contact interactions

The contact interaction (CI) scenario relevant for NC processes assumes that 4-fermion processes are modified by additional terms in the interaction Lagrangian of the form

$$\mathcal{L}_{\mathrm{CI}} = \sum_{\substack{i, k = L, R \\ q = u, d, \cdots}} \eta_{ik}^{q} \frac{4\pi}{(\Lambda_{ik}^{q})^{2}} \left(\bar{e}_{i}\gamma^{\mu}e_{i}\right)\left(\bar{q}_{k}\gamma_{\mu}q_{k}\right) . \tag{1}$$

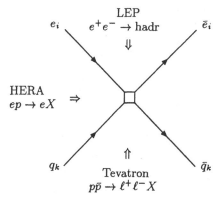

Fig. 2. Schematic view of a contact interaction term.

Similar terms with 4-quark interactions would be relevant for new physics searches at the Tevatron and 4-lepton terms would affect purely leptonic interactions.[1] In equation (1), as is usual practice, only products of vector or axial-vector currents are taken into account, since limits on scalar or tensor interactions are very stringent. Such terms are motivated in many extensions of the standard model as effective interactions after having integrated out new physics degrees of freedom like heavy gauge bosons, leptoquarks and others, with masses beyond the production threshold. The normalization with the factor 4π is reminiscent of models which predict CI terms emerging from strong interactions at a large mass scale Λ.

Equation 1 predicts modifications of cross sections for processes involving two leptons and two quarks in all channels as visualized in Fig. 2. Both enhancement or suppression are expected, depending on the helicity structure of the contact term and its sign η_{ik}^{q}. If the CI mass scale is large, the highest sensitivity is expected at experiments with highest energies, but due to the extremely high experimental precision, also atomic parity violation experiments at low energies are sensitive to parity-odd combinations of helicities [10].

Limits from single experiments at the Tevatron, HERA or LEP2 for models with one single parameter [11] are typically in the order of several TeV, and

---

[1] Contact interactions modifying CC processes can be constructed in a similar way and have been investigated in Refs. [8,9].

all present high-energy experiments have achieved limits in a very similar mass range despite their different center-of-mass energies. Consequently, with a signal at HERA one should expect visible effects at LEP2 and at the Tevatron. Moreover, global fits taking into account experimental data from these different sources give valuable additional insight. Recent global fits [12,13] have taken into account new data from HERA, LEP2, the Tevatron and CCFR. The resulting limits for single-parameter models increase from the range 1.8–10.5 TeV (derived from a single experiment) to 5.1–18.2 TeV (derived from the global analysis) [12]. In a general model where 8 independent parameters are allowed to be non-zero at the same time, the limits are of course weaker and range from 2.1 to 5.1 TeV for the various mass scales Λ_{ik}^q. A comparison of various data obtained at LEP2, the Tevatron and HERA with the prediction of a model with contact interactions as obtained in the best global fit of Ref. [12] shows that only the HERA data at highest values of Q^2 tend to support the presence of a contact term.

Assuming the presence of contact interactions with a mass scale in the range allowed by the best fit one can derive 95 % CL limits for the predicted deviations from the SM cross sections. Figure 3 shows the results for e^+p and e^-p scattering at HERA. Obviously, a possible deviation in electron scattering is much more restricted than for positron scattering; in the latter case, deviations of the cross section for $Q^2 > 15,000\,\mathrm{GeV}^2$ from the standard model of 40 % are allowed, whereas only 20 % deviations are inside the 95 % CL band for the former case. A luminosity of 100–200 pb^{-1} would suffice in e^+p scattering to observe such a deviation. On the other hand, measurements with positrons at HERA have a better chance to further improve limits on contact terms.

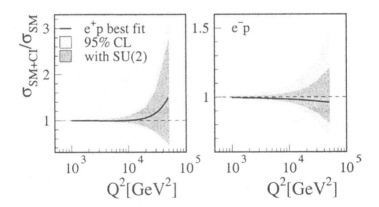

Fig. 3. The 95 % CL limit band on the ratios of e^+ and e^- cross sections for NC DIS at HERA with and without a contact term of the best fit of [14]

In the case of the observation of deviations from the standard model predictions, the combination of results obtained in different experiments and from measurements with polarized beams will be helpful to identify the helicity struc-

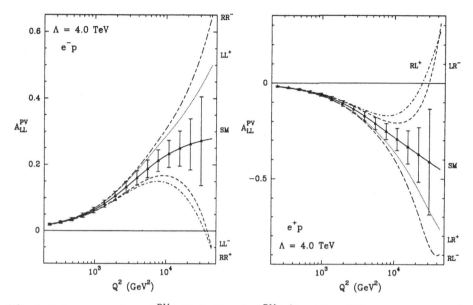

Fig. 4. Spin asymmetries $A_{LL}^{PV}(e^-)$ (left) and $A_{LL}^{PV}(e^+)$ (right). Solid lines correspond to the SM predictions; the expected errors are shown assuming a luminosity of $125\,\mathrm{pb}^{-1}$ for each configuration of beam polarizations. Non-solid lines correspond to CI scenarios with $\Lambda = 4$ TeV and helicities as indicated [15]

ture of contact interaction terms [15]. This is visualized in Fig. 4 where the parity-violating spin-spin asymmetries in $e^{\pm}p$ scattering are shown for models with various types of contact terms and a mass scale of $\Lambda = 4$ TeV. With an integrated luminosity of $125\,\mathrm{pb}^{-1}$ for each configuration of beam polarization, these models would be clearly distinguishable.

2.2 Large Extra Dimensions

In the usual contact term scenario one concentrates on interaction terms with mass dimension 6. Higher-dimension interactions are usually assumed to be less important, since they would be suppressed by a higher power of the ratio of the center-of-mass energy to the mass scale characterizing these interactions. Nevertheless it is interesting to study the effect of such terms, since models might exist where higher-dimensional interactions are the dominating deviation from the standard model framework. Recently, theories with large extra dimensions emerging in low-scale compactified string theories have been shown to constitute a viable alternative to the standard model [16]. A specific class of these theories would predict deviations from standard model cross sections through the exchange of gravitons and their Kaluza-Klein excitations [17]. The effect can be described with the help of dimension 8 NC contact terms [18], but there would also exist completely new kinds of interactions like electron-gluon contact terms.

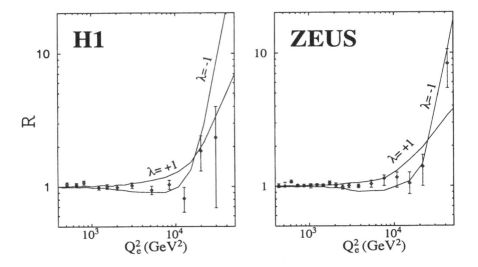

Fig. 5. Illustration of the effect of large extra dimensions on NC e^+p scattering at HERA [19]

Figure 5 shows an example [19] for the effect of graviton exchange with two choices for the relative sign of the standard model and new physics amplitudes compared to the large-Q^2 data from H1 and ZEUS. The mass scale M_s of such theories are chosen in this example to saturate the 95 % CL limits: 543 (567) GeV for H1 (ZEUS) data and $\lambda = +1$ and 436 (485) GeV for $\lambda = -1$. As discussed in [20], with an integrated luminosity of 250 pb^{-1} for each of the following channels: electron and positron scattering with left- and right-handed longitudinal polarization (i.e. 1 fb^{-1} in total), HERA could set limits slightly above 1 TeV and would thus be competitive with LEP2 (expected 1.1 TeV 95 % CL limit), but slightly worse than the Tevatron (1.3 TeV). A future e^+e^- linear collider would be sensitive to mass scales above 4 TeV and the LHC can be expected to shift the corresponding limit to 6.0 TeV [20].

2.3   Leptoquarks

Leptoquarks appear in extensions of the standard model involving unification, technicolor, compositeness, or R-parity violating supersymmetry. In addition to their couplings to the standard model gauge bosons, leptoquarks have Yukawa-type couplings to lepton-quark pairs, which allow for their resonant production in ep scattering. The generally adopted BRW-framework [21] is based on only a few assumptions concerning these Yukawa interactions, which lead to a rather restricted set of allowed states, and the branching fractions β_e for their decays to a charged lepton final state can only be 1, 0.5, or 0. States which can be produced in e^+u or e^+d scattering have $\beta_e = 1$ and for masses below 242 GeV they are excluded by Tevatron bounds [22].

Renewed theoretical work on the phenomenology of leptoquarks (see [4] and references therein) was initiated by the observation of an excess of events at large x and large Q^2 in the 1994–96 HERA e^+p data, which showed that the BRW-framework may indeed be too restrictive. The crucial and least well motivated assumption there is that leptoquarks are not allowed to have other interactions besides their gauge and Yukawa couplings. In fact, most concrete models with leptoquarks do predict additional interactions, which may lead to decay modes to other than lepton-quark final states. This would be interesting, since for example the Tevatron bounds do not exclude leptoquarks with masses above 200 GeV in scenarios with branching ratios $\beta_e \lesssim 0.7$ [3,23].

A few examples for more general scenarios have been discussed in detail in the literature. In Ref. [24] a model was proposed where two leptoquark states show mixing induced by coupling them to the standard model Higgs boson. Alternatively, interactions to new heavy fields might exist which, after integrating them out, could lead to leptoquark Yukawa couplings as an effective interaction [8], bypassing this way renormalizability as a condition, since this is assumed to be restored at higher energies. In the more systematic study of Ref. [23], LQ couplings arise from mixing standard model fermions with new heavy fermions with vector-like couplings and taking into account a coupling to the standard model Higgs. The most interesting extension of the generic leptoquark scenario

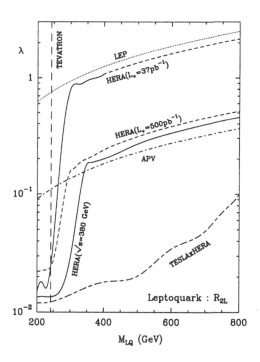

Fig. 6. Discovery limits for a scalar leptoquark at various collider experiments [27]

is, however, R_p-violating supersymmetry, which is discussed in the next subsection.

Searches at HERA [25] are essential to exclude such more general scenarios. Despite the strong dependence on the lepton-quark-LQ Yukawa coupling λ, exclusion limits from HERA experiments cover much larger mass values for small λ than those obtained from indirect searches at LEP2. Since the dependence on the branching ratio in more general scenarios can be reduced considerably by combining NC and CC data, HERA limits also supersede those from the Tevatron for small β_e. The most recent published limits from H1 [6] take into account the finite width of LQ states and the interference of their production amplitude with the standard model background, both effects which turned out to be non-negligible for very large LQ masses. Scalar leptoquarks with masses up to 275 GeV and vector states up to 284 GeV are excluded at 95 % CL for $\lambda = e$ [6]. Similar mass exclusion regions have been reported by ZEUS at recent conferences [26]. As shown in Fig. 6 [27], exclusion limits for the coupling λ at the same LQ mass values can expected to be reduced by a factor of $\sim$ 5 with 500 pb^{-1} of e^+p data. With this luminosity, limits on λ for much larger LQ masses from HERA will also come close to the limits following from atomic parity violation experiments via the corresponding induced contact interactions. To further extend the search to large LQ-masses and small Yukawa couplings, an increase of the center-of-mass energy of ep collisions (for example like at TESLA×HERA) would be essential.

2.4 R_p-violating supersymmetry

The Lagrangian of a supersymmetric version of the standard model may contain a superpotential of the form

$$
\begin{aligned}
W_{R_p} = &\ \lambda_{ijk} L_i L_j E_k^c & \not L \\
&+ \lambda'_{ijk} L_i Q_j D_k^c & \not L \quad \text{(includes LQ-like couplings)} \\
&+ \lambda''_{ijk} U_i^c D_j^c D_k^c & \not B
\end{aligned}
\tag{2}
$$

L_i and Q_i are the superfields for lepton and quark doublets and E_i^c, U_i^c, D_i^c the corresponding charge-conjugated ones for charged leptons, up and down quarks, respectively, and i, j, k are generation indices. The separate contributions in W_{R_p} violate lepton or baryon number conservation as indicated. Imposing symmetry under R-parity (defined as $R_p = (-1)^{3B+L+2S}$, $= 1$ for particles and $= -1$ for their superpartners) forbids the presence of W_{R_p}. The phenomenology of supersymmetry with R_p symmetry has been searched for at all present high energy experiments, and HERA may set interesting limits which are complementary to those obtained at the Tevatron [1].

Many low- and high-energy experiments put limits on the couplings contained in W_{R_p} [28]; however, they do not forbid interactions of the form $L_i Q_j D_k^c$ proportional to λ'_{ijk} in general, provided the λ''_{ijk} are chosen to be zero at the same time. This makes squarks appear as leptoquarks which can be produced on resonance in lepton-quark scattering. In contrast to the generic leptoquark scenarios described above, squarks do not only decay into lepton-quark final states via their R_p-violating interactions, but they can also decay into final states involving gauge bosons or gauginos. These R_p-conserving decays lead to a large number of interesting and distinct signatures (see [29] and references therein).[2] Characteristically one expects multi-lepton and multi-jet final states. Mass and coupling parameters of R_p-violating supersymmetry can be varied such that the branching ratio β_e for the decay into final states with charged leptons becomes small. In this case, the strict mass limits from the Tevatron would not exclude the existence of squarks in the mass range accessible to HERA. In fact, searches at HERA have not found a signal, and bounds on some of the λ' couplings have been derived from searches for the characteristic lepton + multijet final states, which supersede previous exclusion limits [25].

Most of the analyses done so far assume that only one of the couplings λ'_{ijk} is non-zero and only one squark state is in reach. A more general scenario with two light squark states has been considered in Ref. [31], where it was shown that $\tilde{t}_L$–$\tilde{t}_R$ mixing would lead to a broader x distribution than expected for single-resonance production. The possibility of having more than one $\lambda'_{ijk} \neq 0$ was noticed in Ref. [32] and deserves more theoretical study.

3   Events with Isolated $\mu + p_{\mathrm{T,miss}}$

R_p-violating supersymmetry has also played a role in the search for explanations of the observation made by H1[3] of five events with an isolated μ and missing transverse momentum [5] (see also [33,34]). Events of this kind can originate from W production followed by the decay $W \to \mu\nu_\mu$. Their observed number is, however, larger than expected from the standard model taking into account next-to-leading-order corrections to the dominating resolved contribution from photoproduction [35]. Moreover, their kinematic properties are atypical for W production [33]. An explanation in terms of anomalous $WW\gamma$ couplings additionally has to face limits from the Tevatron, LEP2 and ZEUS [36] and leaves the question open why a similar excess of events is not seen in $e + \not{p}_T$ events.

The observation of $\mu + \not{p}_T$ events could find an explanation in R_p-violating scenarios if it is assumed that a stop, $\tilde{t}$, is produced on-resonance at HERA. Figures 7 and 8 show examples for some of the possibilities. The process $ed \to \tilde{t} \to \mu d^k$ (Fig. 7a), which predicts μ but no large $\not{p}_T$ in the final state in gross disagreement with the experimental observation, requires two different non-zero λ' couplings [32]. The relevant product $\lambda'_{1j1}\lambda'_{2jk}$ would induce flavor changing neutral currents and is therefore limited to small values for 1st and 2nd generation quarks in the final state [40]. The analogous process with a τ replacing the μ but followed by the decay $\tau \to \mu\nu_\tau\bar{\nu}_\mu$ could also not serve as an explanation, since the decay-μ would be strongly boosted in the direction of the τ, i.e. the missing

---

[2] Monte Carlo tools needed in searches for R_p-violating supersymmetry at HERA have been improved recently [30].

[3] No event of this type was observed by ZEUS [36].

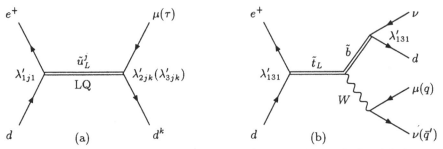

Fig. 7. Possible decays of squarks produced in e^+d scattering with R_p-violating couplings leading to isolated μ + jet final states: (a) $\tilde{u}_L^j \to \mu d_k$ through $\lambda'_{2jk} \neq 0$; (b) $\tilde{t} \to \tilde{b}W$ followed by $\tilde{b} \to \nu d$ via $\lambda'_{131} \neq 0$ and $W \to \mu^+\nu_\mu$ or $W \to 2$ jets [37].

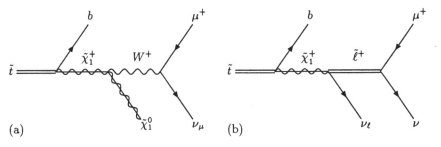

Fig. 8. Possible decay chains of the stop leading to isolated muon + jet + missing p_T: $\tilde{t} \to b\tilde{\chi}_1^+$ followed by (a) $\tilde{\chi}_1^+ \to \tilde{\chi}_1^0\mu^+\nu_\mu$ [38]; (b) $\tilde{\chi}_1^+ \to \nu_\ell\mu^+\nu$ [39].

transverse momentum would be correlated with the observed μ in contrast to the kinematic properties of the H1 events. Moreover, hadronic decays of the τ would lead to an additional outstanding experimental signature, and a search for it at H1 was negative [6].

The scenario shown in Fig. 7b [37] requires a relatively light b squark with $m_{\tilde{b}} \lesssim 120\,\text{GeV}$, and some fine-tuning in order to avoid too large effects on $\Delta\rho$ in electroweak precision measurements. It could be identified by the simultaneous presence of final states with $\not{p}_T$ and multi-jets from hadronic decays of the W. Also the cascade decay shown in Fig. 8a [38] involving R_p-violation only for the production of the $\tilde{t}$ resonance, not for its decay, seems difficult to be achievable, since it requires both a light chargino and a long-lived neutralino. This, as well as the even more speculative process shown in Fig. 8b [39], which requires R_p-violation in the $L_iL_jE_k^c$ sector ($\lambda_{ijk} \neq 0$) as well, can be checked from the event kinematics: assuming a value for the mass of the decaying $\tilde{t}$, the recoil mass distribution must cluster at a fixed value, the chargino mass.

These speculations on a possible origin of the observed events within R_p-violating supersymmetry are all linked to the presence of an excess of events in NC scattering. The basic assumption is that a squark, preferably a stop, is produced on-resonance; non-resonant stop production would be too much suppressed. Another type of explanation not relying on this assumption was proposed in Ref. [41]. Events of the observed type could emerge after the production of a single top quark followed by the decay chain $t \to bW$ and $W \to \mu\nu$. The cross section of SM top production would be much too small to explain the number of observed events, but the presence of a coupling of the type of an anomalous-magnetic moment inducing the transition $c \to t$ could enhance the cross section considerably. However, the event rate would be still too small unless a non-standard large x-behavior of the charm distribution would be present, in addition. This scenario thus requires to open two new fronts of non-standard physics.

4 Concluding remarks

There are many scenarios of new physics which are most suitably searched for with the help of deep inelastic scattering experiments. Limits on leptoquark of squark masses and their Yukawa or R-parity violating couplings obtained at HERA will stay superior to those from other experiments in many cases.

The search for new physics effects relies in most cases on trustworthy predictions from the standard model. In deep inelastic scattering this includes the necessity to know parton distribution functions as precisely as possible. It is therefore a mandatory though nontrivial task to combine the information from all available different experiments in order not to run the risk of confusing modifications of parton distributions with signs of new physics. With this in mind, the huge amount of data expected from HERA experiments in the future is guaranteed to play an indispensable role in the search for new physics — even in those cases where the most stringent limits are obtained at other experiments.

References

1. *Future Physics at HERA, Proceedings of the Workshop 1995/96* Eds. G. Ingelman, A. DeRoeck, R. Klanner (DESY Hamburg) Vol. 1, p. 237
2. H1 Collaboration, C. Adloff, et al., Z. Phys. C74 (1997) 191; ZEUS Collaboration, J. Breitweg, et al., Z. Phys. C74 1997) 207
3. G. Altarelli, *Proc. Conference on Supersymmetries in Physics (SUSY 97)*, Nucl. Phys. B Proc. Suppl. **62** (1998) 3
4. R. Rückl, H. Spiesberger, *Proc. Workshop on Physics Beyond the Standard Model: Beyond the desert*, Tegernsee, Germany (1997), p. 304, (*hep-ph/9711352*); *Ringberg Workshop on New Trends in HERA Physics, Ringberg Castle*, Tegernsee (1997), p. 113, (*hep-ph/9710327*)
5. H1 Collaboration, C. Adloff et al., Eur. Phys. J. C5 (1998) 575
6. H1 Collaboration, C. Adloff, DESY-99-081, to appear in *Eur. Phys. J.*, (*hep-ex/9907002*)
7. *3rd UK Phenomenology Workshop on HERA Physics*, Durham, England, 20-25 Sep. 1998, *J. Phys.* G25 (1999)
8. Altarelli G, G. F. Giudice, M. L. Mangano, Nucl. Phys. **B506** (1997) 29
9. F. Cornet, J. Rico (Granada U.), Phys. Lett. B412 (1997) 343; F. Cornet, R. Relaño, J. Rico, UG-FT-101/99 (*hep-ph/9908299*)

460   H. Spiesberger

10. R. Casalbuoni, S. D. Curtis, D. Dominici, R. Gatto, Phys. Lett. B460 (1999) 135; D. Dominici, DFF-347-9-99, *International Europhysics Conference on High-Energy Physics (EPS-HEP 99)*, Tampere, Finland, 1999 (*hep-ph/9909290*)

11. D. Treille, *Proc. ICHEP98* Vancouver 1998

12. A. Zarnecki, DESY-99-074, (*hep-ph/9904334*)

13. D. Zeppenfeld, K. Cheung, MADPH-98-1081, in *Proceedings of 5th International WEIN Symposium: A Conference on Physics Beyond the Standard Model (WEIN 98)*, Santa Fe, NM, 1998, (*hep-ph/9810277*)

14. A. Zarnecki, *7th International Workshop on Deep Inelastic Scattering and QCD (DIS 99)*, Zeuthen, Germany, 19-23 Apr. 1999, (*hep-ph/9905565*)

15. J. Kalinowski, H. Spiesberger, J. M. Virey, 1998, in [7], (*hep-ph/9812517*)

16. N. Arkani-Hamed, S. Dimopoulos, G. Dvali, Phys. Lett. B429 (1998) 263; Phys. Rev. D59 (1999) 86004

17. G. F. Giudice, R. Rattazzi, J. D. Wells, Nucl. Phys. B544 (1999) 3

18. T. Han, J. Lykken, R.-J. Zhang, Phys. Rev. D59 (1999) 105006

19. P. Mathews, S. Raychaudhuri, K. Sridhar, Phys. Lett. B455 (1999) 115

20. T. G. Rizzo, Phys. Rev. D59 (1999) 115010

21. W. Buchmüller, R. Rückl, D. Wyler, Phys. Lett. B191 (1987) 442; *Err.* Phys. Lett. B448 (1999) 320

22. Leptoquark Limit Combination Working Group (for the CDF and D0 Collaborations), C. Grosso-Pilcher et al., FERMILAB-PUB-98/312E, (*hep-ex/9810015*)

23. J. L. Hewett, T. G. Rizzo, Phys. Rev. D58 (1998) 55005

24. K. S. Babu, C. Kolda, J. March-Russell, Phys. Lett. B408 (1997) 261

25. M. Kuze, *these proceedings*

26. R. Galea (ZEUS Collaboration), *7th International Workshop on Deep Inelastic Scattering and QCD, DIS99*, Berlin 1999, (to appear in Nucl. Phys. B (Proc. Suppl.)); ZEUS Collaboration, *XXIX International Conference on High Energy Physics, ICHEP98*, Vancouver 1998, Abstract 754

27. J.-M. Virey, E. Tugcu, P. Taxil, preprint DO-TH-99-05, (*hep-ph/9905491*)

28. B. C. Allanach, A. Dedes, H. Dreiner, DAMTP-1999-45, (*hep-ph/9906269*)

29. E. Perez, Y. Sirois, H. Dreiner, *Future Physics at HERA, Proceedings of the Workshop 1995/96* DESY Hamburg, 1996, (*hep-ph/9703444*); H. Dreiner 1997, (*hep-ph/9707435*)

30. *Monte Carlo Generators for HERA physics, Proceedings of the Workshop 1998/99*, eds. G. Grindhammer, G. Ingelman, H. Jung, T. Doyle, http://www.desy.de/~heramc/proceedings

31. T. Kon, T. Kobayashi, Phys. Lett. B409 (1997) 265

32. A. S. Belyaev, A. V. Gladyshev, 1998, (*hep-ph/9807547*)

33. J. Meyer, *these proceedings*

34. C. Diaconu, J. Kalinowski, T. Matsushita, H. Spiesberger, D. S. Waters, in [7]

35. M. Spira, DESY-99-060, (*hep-ph/9905469*); P. Nason, R. Rückl, M. Spira, in [7]

36. ZEUS Collaboration, J. Breitweg et al., DESY-99-054 (*hep-ex/9907023*)

37. T. Kon, T. Matsushita, T. Kobayashi, Mod. Phys. Lett. A12 (1997) 3143

38. T. Kon, T. Kobayashi, S. Kitamura S, Phys. Lett. B376 (1996) 227

39. J. Kalinowski, R. Rückl, H. Spiesberger, P. Zerwas, 1997, *DESY internal note*, unpublished

40. S. Davidson , D. Bailey, B. Campbell, Z. Phys. C61 (1994) 613

41. H. Fritzsch, D. Holtmannspötter, Phys. Lett. B457 (1999) 186

Lecture Notes in Physics

For information about Vols. 1–508
please contact your bookseller or Springer-Verlag

Vol. 509: J. Wess, V. P. Akulov (Eds.), Supersymmetry and Quantum Field Theory. Proceedings, 1997. XV, 405 pages. 1998.

Vol. 510: J. Navarro, A. Polls (Eds.), Microscopic Quantum Many-Body Theories and Their Applications. Proceedings, 1997. XIII, 379 pages. 1998.

Vol. 511: S. Benkadda, G. M. Zaslavsky (Eds.), Chaos, Kinetics and Nonlinear Dynamics in Fluids and Plasmas. Proceedings, 1997. VIII, 438 pages. 1998.

Vol. 512: H. Gausterer, C. Lang (Eds.), Computing Particle Properties. Proceedings, 1997. VII, 335 pages. 1998.

Vol. 513: A. Bernstein, D. Drechsel, T. Walcher (Eds.), Chiral Dynamics: Theory and Experiment. Proceedings, 1997. IX, 394 pages. 1998.

Vol. 514: F. W. Hehl, C. Kiefer, R. J. K. Metzler, Black Holes: Theory and Observation. Proceedings, 1997. XV, 519 pages. 1998.

Vol. 515: C.-H. Bruneau (Ed.), Sixteenth International Conference on Numerical Methods in Fluid Dynamics. Proceedings. XV, 568 pages. 1998.

Vol. 516: J. Cleymans, H. B. Geyer, F. G. Scholtz (Eds.), Hadrons in Dense Matter and Hadrosynthesis. Proceedings, 1998. XII, 253 pages. 1999.

Vol. 517: Ph. Blanchard, A. Jadczyk (Eds.), Quantum Future. Proceedings, 1997. X, 244 pages. 1999.

Vol. 518: P. G. L. Leach, S. E. Bouquet, J.-L. Rouet, E. Fijalkow (Eds.), Dynamical Systems, Plasmas and Gravitation. Proceedings, 1997. XII, 397 pages. 1999.

Vol. 519: R. Kutner, A. Pękalski, K. Sznajd-Weron (Eds.), Anomalous Diffusion. From Basics to Applications. Proceedings, 1998. XVIII, 378 pages. 1999.

Vol. 520: J. A. van Paradijs, J. A. M. Bleeker (Eds.), X-Ray Spectroscopy in Astrophysics. EADN School X. Proceedings, 1997. XV, 530 pages. 1999.

Vol. 521: L. Mathelitsch, W. Plessas (Eds.), Broken Symmetries. Proceedings, 1998. VII, 299 pages. 1999.

Vol. 522: J. W. Clark, T. Lindenau, M. L. Ristig (Eds.), Scientific Applications of Neural Nets. Proceedings, 1998. XIII, 288 pages. 1999.

Vol. 523: B. Wolf, O. Stahl, A. W. Fullerton (Eds.), Variable and Non-spherical Stellar Winds in Luminous Hot Stars. Proceedings, 1998. XX, 424 pages. 1999.

Vol. 524: J. Wess, E. A. Ivanov (Eds.), Supersymmetries and Quantum Symmetries. Proceedings, 1997. XX, 442 pages. 1999.

Vol. 525: A. Ceresole, C. Kounnas, D. Lüst, S. Theisen (Eds.), Quantum Aspects of Gauge Theories, Supersymmetry and Unification. Proceedings, 1998. X, 511 pages. 1999.

Vol. 526: H.-P. Breuer, F. Petruccione (Eds.), Open Systems and Measurement in Relativistic Quantum Theory. Proceedings, 1998. VIII, 240 pages. 1999.

Vol. 527: D. Reguera, J. M. G. Vilar, J. M. Rubí (Eds.), Statistical Mechanics of Biocomplexity. Proceedings, 1998. XI, 318 pages. 1999.

Vol. 528: I. Peschel, X. Wang, M. Kaulke, K. Hallberg (Eds.), Density-Matrix Renormalization. Proceedings, 1998. XVI, 355 pages. 1999.

Vol. 529: S. Biringen, H. Örs, A. Tezel, J.H. Ferziger (Eds.), Industrial and Environmental Applications of Direct and Large-Eddy Simulation. Proceedings, 1998. XVI, 301 pages. 1999.

Vol. 530: H.-J. Röser, K. Meisenheimer (Eds.), The Radio Galaxy Messier 87. Proceedings, 1997. XIII, 342 pages. 1999.

Vol. 531: H. Benisty, J.-M. Gérard, R. Houdré, J. Rarity, C. Weisbuch (Eds.), Confined Photon Systems. Proceedings, 1998. X, 496 pages. 1999.

Vol. 532: S. C. Müller, J. Parisi, W. Zimmermann (Eds.), Transport and Structure. Their Competitive Roles in Biophysics and Chemistry. XII, 400 pages. 1999.

Vol. 533: K. Hutter, Y. Wang, H. Beer (Eds.), Advances in Cold-Region Thermal Engineering and Sciences. Proceedings, 1999. XIV, 608 pages. 1999.

Vol. 534: F. Moreno, F. González (Eds.), Light Scattering from Microstructures. Proceedings, 1998. XII, 300 pages. 2000

Vol. 535: H. Dreyssé (Ed.), Electronic Structure and Physical Properties of Solids: The Uses of the LMTO Method. Proceedings, 1998. XIV, 458 pages. 2000.

Vol. 536: T. Passot, P.-L. Sulem (Eds.), Nonlinear MHD Waves and Turbulence. Proceedings, 1998. X, 385 pages. 1999.

Vol. 537: S. Cotsakis, G. W. Gibbons (Eds.), Mathematical and Quantum Aspects of Relativity and Cosmology. Proceedings, 1998. XII, 251 pages. 1999.

Vol. 538: Ph. Blanchard, D. Giulini, E. Joos, C. Kiefer, I.-O. Stamatescu (Eds.), Decoherence: Theoretical, Experimental, and Conceptual Problems. Proceedings, 1998. XII, 345 pages. 2000.

Vol. 539: A. Borowiec, W. Cegła, B. Jancewicz, W. Karwowski (Eds.), Theoretical Physics. Fin de Siècle. Proceedings, 1998. XX, 319 pages. 2000.

Vol. 540: B. G. Schmidt (Ed.), Einstein's Field Equations and Their Physical Implications. Selected Essays. 1999. XIII, 429 pages. 2000

Vol. 541: J. Kowalski-Glikman (Ed.), Towards Quantum Gravity. Proceedings, 1999. XII, 376 pages. 2000.

Vol. 542: P. L. Christiansen, M. P. Sørensen, A. C. Scott (Eds.), Nonlinear Science at the Dawn of the 21st Century. Proceedings, 1998. XXVI, 458 pages. 2000.

Vol. 543: H. Gausterer, H. Grosse, L. Pittner (Eds.), Geometry and Quantum Physics. Proceedings, 1999. VIII, 408 pages. 2000.

Vol. 545: J. Klamut, B. W. Veal, B. M. Dabrowski, P. W. Klamut, M. Kazimierski (Eds.), New Developments in High-Temperature Superconductivity. Proceedings, 1998. VIII, 275 pages. 2000.

Vol. 546: G. Grindhammer, B. A. Kniehl, G. Kramer (Eds.), New Trends in HERA Physics 1999. Proceedings, 1999. XIV, 460 pages. 2000.

Monographs
For information about Vols. 1–20
please contact your bookseller or Springer-Verlag